A Guide to
PRACTICAL
HUMAN
RELIABILITY
ASSESSMENT

Barry Kirwan

CRC PRESS

Boca Raton   London   New York   Washington, D.C.

Library of Congress Cataloging-in-Publication Data

Catalog record is available from the Library of Congress

This book contains information obtained from authentic and highly regarded sources. Reprinted material is quoted with permission, and sources are indicated. A wide variety of references are listed. Reasonable efforts have been made to publish reliable data and information, but the authors and the publisher cannot assume responsibility for the validity of all materials or for the consequences of their use.

Neither this book nor any part may be reproduced or transmitted in any form or by any means, electronic or mechanical, including photocopying, microfilming, and recording, or by any information storage or retrieval system, without prior permission in writing from the publisher.

The consent of CRC Press LLC does not extend to copying for general distribution, for promotion, for creating new works, or for resale. Specific permission must be obtained in writing from CRC Press LLC for such copying.

Direct all inquiries to CRC Press LLC, 2000 N.W. Corporate Blvd., Boca Raton, Florida 33431.

Trademark Notice: Product or corporate names may be trademarks or registered trademarks, and are used only for identification and explanation, without intent to infringe.

Visit the CRC Press Web site at www.crcpress.com

© 1994 by B. Kirwan

No claim to original U.S. Government works
International Standard Book Number 0-7484-0052-4 HB   0-7484-011103 PB
Printed in United Kingdom by Biddles / IBT Global   2 3 4 5 6 7 8 9 0
Printed on acid-free paper

Contents

For the Seven, my family, Sheila, and operators everywhere

1

Introduction

Human error is here to stay. This perhaps obvious statement has a more profound implication for society when we consider the types of hazardous system accidents that have occurred over the past three decades, such as Three Mile Island, Chernobyl and Bhopal. Such system accidents, and others such as those reviewed in Appendix I, have all been strongly influenced by human error. Yet many similar systems are in existence, or are being built, or are being planned. Since human error is endemic to our species, there are really only two alternatives for modern society: either get rid of such systems (as advocated on risk grounds by Perrow (1984)) or try to deal with and minimise the human error problems as far as is possible.

Some of the accidents that have occurred were almost impossible to predict prior to the event, whereas others undoubtedly could have been predicted and prevented by techniques dealing with human error assessment. This book does not attempt to decide whether complex, hazardous systems should be allowed to continue, but rather attempts to define a set of useful tools for the analysis and reduction of those errors which could lead to system accidents. This book therefore deals with the subject of *Human Reliability Assessment* (HRA), as used in the study and assessment of risks involved in large, complex and hazardous systems.

The term *human error* has been pragmatically defined by Swain (1989) as follows: 'any member of a set of human actions or activities that exceeds some limit of acceptability, i.e. an out of tolerance action [or failure to act] where the limits of performance are defined by the system.' The effects of human error on system performance have been demonstrated most vividly by large-scale accidents (e.g. see Appendix I), and current accident experience suggests that the so-called high-risk industries (and some so-called low-risk ones too) are still not particularly well-protected from human error. This in turn suggests the need both for a means of properly assessing the risks attributable to human error and for ways of reducing system vulnerability to human error impact. These are the primary goals of Human Reliability

Assessment (HRA), achieved by its three principal functions of identifying *what* errors can occur (Human Error Identification), deciding *how likely* the errors are to occur (Human Error Quantification), and, if appropriate, enhancing human reliability by *reducing* this error likelihood (Human Error Reduction).

HRA can also enhance the profitability and availability of systems via human error reduction/avoidance, although the main drive for the development and application of HRA techniques has so far come from the risk assessment and reduction domain. This book therefore concentrates primarily on HRA in the risk assessment context, and is aimed at both the practitioner and the student, as well as managers who may wish to understand what HRA has to offer. It attempts to present a practical and unbiased framework approach to HRA, called the HRA process, which encompasses a range of existing HRA tools of error identification, quantification, etc. These tools are documented, along with some practitioner insights and additional useful data, to aid would-be assessors in understanding the techniques' different rationales, advantages and disadvantages – as well as how the various tools all fit together within the HRA process.

As this book is primarily practical in nature, a theoretical discussion and introduction on the nature and genesis of human error is left to other more competent texts on this subject (e.g. *Reason*, 1990), although the practical implications of such theory are embedded explicitly in appropriate sections of this book (e.g. the section on Human Error Identification). The following sections of the introduction therefore merely aim to set the terms of reference for what follows. This entails a brief discussion of the role of human error in complex systems, and how the functions of HRA have developed over the past three decades, leading to the current definition of the HRA process. Following these brief introductory sections, and two sections on the scope of the book and how to use it, the remainder of the book is concerned with practical approaches to HRA, set in the framework of the HRA process, backed up by a number of appendices containing both relevant data and real case studies.

1.1 Human errors in complex systems

Human error is extremely commonplace, with almost everyone committing at least some errors every day – whether 'small' ones such as mispronouncing a word, or larger errors or mistakes such as deciding to invest in a financial institution which later goes bankrupt. Most human errors in everyday life are recoverable, or else have a relatively small impact on our lives. However, in the work situation, and especially in complex systems, this may not be the case. A human operator in the central control room of a chemical or nuclear power plant, or the pilot of a large commercial airplane, cannot afford to make certain errors, or else accidents involving fatalities, possibly including the life of the operators themselves, may be the result.

Human error in complex and potentially hazardous systems therefore involves human action (or inaction) in unforgiving systems. For human error to have a negative system impact, there must first be an opportunity for an error, or a requirement for reliable human performance – often in response to some event. The error must then occur and fail to be corrected or compensated for by the system, and it must have negative consequences, either alone or in conjunction with another human, 'hardware', or environmental event. As a simple example, if the main (foot-operated) brakes on a car fail while the car is in motion, the operator (the driver) must bring the vehicle to a safe stop by using the handbrake or the gears (or both), or an inclined surface, or, if necessary, an object of sufficient inertia to stop the vehicle's motion. The 'demanding event' requiring reliable human performance is the brake (hardware) failure. The 'error' is in this case a failure to achieve a safe stop, and the consequences of such an error will be dependent on local population density, and the speed of the vehicle at the time of the incident, etc. The event could be compounded by other simultaneous hardware failures (e.g. the handbrake cable snapping under the sudden load), as well as by environmental circumstances (e.g. rain, ice, fog, etc.).

Two main facets of the above example are worth expanding upon. Firstly, in the event of this incident occurring, very few people would 'blame' the driver if the error occurred (assuming that the brake failure was not caused by the driver failing to have the car regularly serviced, etc.), since a sudden spontaneous brake failure is something few people are prepared or trained for. In HRA, the concept of blame is usually counter-productive and obscures the real reasons why an accident occurred. To pass a complex series of events off as a mere operator error, and then actually to replace the operator (or else just admonish the operator for his or her actions or inactions), not only crudely simplifies the causes of the event but in many cases actually allows the event (or a similar one) to recur.

Unfortunately, in large-scale accidents involving multiple fatalities, legal and natural social processes tend to encourage the desire to attribute blame to particular individual parties. Whilst there is no intention in this book of embarking on the difficult area of the ethics of culpability, it is important that the reader understand that the term 'human error', in this book as in the greater part, if not the whole, of the field of Human Reliability Assessment, has nothing to do with the concept of blame. Blame therefore does not occur within the HRA glossary. It is simply not a useful concept.

A second aspect of the above example is that the events in it are all very immediately apparent to the operator (driver). In most complex systems, however, the complexity of the system itself means that what is actually happening in the system, as a consequence both of human errors and of other failures or events, may be relatively opaque to the operator; or, at the least, this information may be delayed or reduced – or both. This is one major cost associated with the use of advanced technology, and it is one of the major factors preventing the achievement of high human reliability in complex systems.

It can be argued that human error, or rather human variability, a natural and adaptive process essential to our evolution, is so endemic to the species that high-risk systems which are vulnerable to human error should not be allowed to exist. This argument is certainly strengthened by the knowledge that many industries today are highly complex in the ways in which their potentially hazardous processes are controlled. Although this creates a very high degree of efficiency in such processes, this same level of complexity also means that estimating the ways in which systems can fail becomes difficult because there are so many interfacing and interacting components, the human operator being the most sophisticated and significant one. The argument against high-risk complex systems has been most powerfully presented by Perrow in his (1984) study of many so-called 'normal accidents', which were a natural outcome of human error in non-benign systems, and hence were very difficult to predict prior to their occurrence.

For the very reason that this book concerns itself with Human Reliability Assessment in complex, high-risk industries as mentioned above, the reader will surmise that it does not necessarily advocate the actual elimination of all such industries. Instead, it advocates a detailed assessment of risks due to human error, using the best tools available at this time – and the better ones that will develop in the future. The reason this brief discussion on errors in high-risk systems is included is to make a very simple but important point: human behaviour is intrinsically complicated and difficult to predict accurately. HRA is therefore conceptually a rather ambitious approach, particularly since it deals with the already-complex subject of human error in the additionally complex setting of large-scale systems. HRA must therefore not be used complacently, and cannot afford to be shallow in its approach to assessment. Complex systems often require correspondingly complex assessment procedures.

1.2 The roots of HRA

The study of HRA is approximately 30 years old and has always been a hybrid discipline, involving reliability engineers (HRA first arose within the field of system-reliability assessment), engineers and human-factors specialists or psychologists. HRA is inherently inter-disciplinary for two reasons. Firstly, it requires an appreciation of the nature of human error, both in terms of its underlying psychological basis and mechanisms, and in terms of the various human factors, such as training and the design of the interface, affecting performance. Secondly, it requires both some understanding of the *engineering* of the system, so that the intended and unintended human-system interactions can be explored for error potential and error impact, and an appreciation of reliability- and risk-estimation methods, so that HRA can be integrated into the 'risk picture' associated with a system. This risk picture can then act as a summary of the impacts of human error and hardware failure on system risk, and can be used to decide which aspects of risk are most important. The matter of how HRA is integrated into risk assessment is dealt with in more detail in Chapter 4 (see also Cox & Tait, 1991).

The development of HRA tools has arguably been somewhat slow and sporadic, although since the Three Mile Island incident (1979) efforts have been more sustained and better directed, largely in the nuclear-power domain. This has led to the existence of a number of practical HRA tools (see Kirwan *et al*, 1988; and Swain, 1989). There remains, however, a relative paucity of texts which attempt to bring together the more useful tools and place them in a coherent and practicable framework. This is partly because over the past three decades the focus in HRA has been purely on the quantification of human error probabilities. The human error probability (HEP) is simply defined as follows:

$$\text{HEP} = \frac{\text{Number of errors occurred}}{\text{Number of opportunities for error}}$$

Thus, when buying a cup of coffee from a vending machine, if, in 1 time in 100, *tea* is inadvertently purchased, the HEP is 0.01.

The focus on quantification has occurred because HRA is used within risk assessments, which are themselves probabilistic – i.e. involve defining the probabilities of accidents, of different consequence-severities, associated with a particular system design (e.g. a probability of 1 public fatality per 1,000,000 years). Such estimates are then compared against governmental criteria for that industry, and the risks are deemed either acceptable or not acceptable. If not, then either the risk factor must be reduced in some way or else the proposed or existing plant must be cancelled or shut down (see Cox & Tait, 1991). Risk assessment, therefore, is profoundly important for any system. And because it is firmly quantitative in nature, HRA, therefore, to fit into the probabilistic safety or risk assessment (PSA/PRA) framework, must also be quantitative, or else the impact of human error will be excluded from the 'risk picture'. This has all thus led to the need for human error probabilities above all else.

In early HRAs it appeared that what could go wrong (i.e. what errors could occur) was fairly easy to predict (e.g. operators could fail to do something with enough precision, or fail to do it at all), whereas what was difficult to predict was the human error probability. The most obvious approach to HRA in the 1960s was to copy the approach used in reliability assessment, namely to collect data on the failure rates of operators in carrying out particular tasks, much as reliability engineers collected data on, for example, how often a valve or pump failed. This led to the attempted creation of various data banks (see Topmiller *et al*, 1984, for a comprehensive review), none of which, however, remain in use today. This is largely for the now-obvious reason that whereas hardware components such as valves and pumps have very specific inputs and outputs, and very limited functions, humans do not. Humans are autonomous. They can decide what to do from a vast array of potential outputs, and can interpret inputs in many different ways according to the goals they are trying to achieve. Human performance is also influenced by a very large number of interacting factors in the work environment, and human behaviour relies on skills, knowledge and strategies stored in the memory. In short, humans are not, and never will be, the same as simple components, and should never be treated as such.

The early data-bank drive was therefore effectively a failure, and with a few exceptions it has, until recently, remained a relatively unfruitful pursuit for various reasons (see Williams, 1983; and Kirwan *et al*, 1990). This failure led to two alternative approaches: firstly, the use of semi-judgemental databases (i.e. ones partly based on what scant data were available and partly tempered by experienced practitioners' interpretations), which had little empirical justification but were held by practitioners to be reasonable; and secondly, the use of techniques of expert-judgement elicitation and aggregation (e.g. using plant operatives with 30 years' experience of their own and other's errors) to generate HEPs (as described later in Section 5.5). Both approaches therefore ultimately relied on expert judgement, their main product being the HEP.

With Three Mile Island and other more recent accidents, however, the earlier assumption that identifying what errors could occur was not a major problem was irrevocably undermined. The subject of human error identification, particularly with respect to misdiagnosis during abnormal events in systems, has become of increasing concern in high-risk industries. Therefore, in the first half of the 1980s in particular, there was a drive towards understanding human errors at a deeper, psychological level, trying to identify the causes of errors which could be measured in the work or task environment, and could therefore be used to help identify what errors should appear in the risk assessment.

This shift of emphasis from quantifying HEPs, to understanding the causes of errors, led in part to the development of techniques which could reduce error likelihood, as a function of the error's likely 'root' causes. Prior to this, the reduction of human error, if required by a risk assessment, would be more likely to have been achieved by the automation of the human's task, or by the provision of interlocks or stronger procedures, etc. It was realised, however, that such measures may not solve the human-error problem (and indeed might add new problems), and that other, more human-operator-oriented error-reduction strategies would be more effective.

These shifts away from pure quantification-oriented HRA towards an HRA approach encompassing error identification, quantification and reduction, have changed the HRA landscape in the past decade. This book is built around this type of 'new' HRA, modelling HRA as a process as delineated in Chapter 4, and defined in detail throughout the sub-sections in Chapter 5. In this

Figure 1.1 HRA process

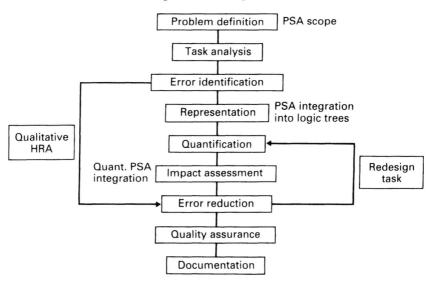

relatively new model of HRA, quantification is but one part – albeit still a critical one. The HRA process is briefly introduced below, and will be expanded upon in Chapter 5.

1.3 The HRA process

The overall HRA process is shown in Figure 1.1, and its principal components are briefly outlined below.

1.3.1 Problem definition

This refers to deciding what human involvements are to be assessed. For example, some Probabilistic Safety Assessments (PSAs) may wish to include only assessments of operators failing to deal with emergency situations, whereas others may also address the human contribution to maintenance failures in connection with critical equipment.

1.3.2 Task analysis

When the human aspect of the problem has been defined, task analysis can then define what human actions should occur in such events, as well as what equipment and other 'interfaces' the operators should use. It may also identify what training (skills and knowledge) and procedures the operators will call upon. This task-analysis phase of the study is critical in bounding the kinds of behaviour that are of interest to the study, whether they are concerned with maintenance, the monitoring of a process, carrying out control actions, diagnosing a fault situation, etc. The task analysis is used to structure the operators' tasks for further analysis, in much the same way as piping and instrument diagrams – and engineering flow diagrams – are used by engineers to define the various states of, and operations involved in, the process in question. Therefore, without some form of task analysis, the tasks under investigation will be vague, and further HRA may not be reliable.

1.3.3 Human Error Identification

Once task analysis (which defines how the task should be carried out) has been completed, Human Error Identification then considers what can go wrong. At the least, this error-identification process will consider the following types of error:

- error of omission – failing to carry out a required act.
- error of commission – failing to carry out a required act adequately: act performed without required precision, or with too much or too little force; act performed at wrong time; acts performed in wrong sequence.
- extraneous act – unrequired act performed instead of, or in addition to, required act.
- error-recovery opportunities – acts which can recover previous errors.

The Human Error Identification phase can identify many errors. Not all of these will be important for the study, as can be determined by reviewing their consequences on the system's performance. The ones which *can* contribute to a degraded system state, whether alone or in conjunction with other hardware or software failures or environmental events (or both together), must next be integrated into the risk analysis.

1.3.4 Representation

Having defined what the operators should do (via task analysis) and what can go wrong, the next step is to represent this information in a form which allows the quantitative evaluation of the human-error impact on the system to take place. It is usual for the human-error impact to be seen in the context of other potential contributions to system risk, such as hardware and software failures, and environmental events. Risk assessment typically uses fault and event trees (described later) to carry out risk assessments, and human errors and recoveries are usually embedded within such logical frameworks.

1.3.5 Human error quantification

Once the human-error potential has been represented, the next step is to quantify the likelihood of the errors involved and then determine the overall effect of human error on system safety or reliability. Human reliability quantification techniques all quantify the human error probability (HEP, as defined earlier), which is the metric of human reliability assessment.

1.3.6 Impact assessment

Once the errors have been quantified and represented in the risk-assessment logic trees, the overall system risk level can be calculated. Then, it can be determined whether or not the system has an *acceptable* level of risk. If the calculated risk is unacceptably high, then either the level of risk must be reduced or the system must be discontinued. Impact assessment involves determining not only if the risk element is acceptably low, but also which events (human, hardware, software or environmental – or any combination) contribute most to the level of risk. It is these events, or event combinations, that will then be targeted for investigation if the risk level is assessed to be too high. If human error is a significant contributor to the system risk level, and if the level of system risk is calculated to be too high, then the appropriate errors will be targeted for error reduction.

1.3.7 Error reduction analysis

Error reduction measures (ERMs) may be derived in a number of ways: according to the identified root causes of the error (from the error-identification stage); from the defined factors (called Performance Shaping Factors, usually defined in the quantification stage) that contribute to the error's HEP; or else from an assessment of the task in its system context, using ergonomics/ engineering judgement to identify how to prevent the error, or how to reduce either its likelihood or its system impact.

If error reduction is necessary to reduce risk to an acceptable level, then following such error reduction measures, the system risk level will need to be recalculated. In some cases, several iterations of impact assessment, error reduction, and re-quantification may occur until satisfactory risk levels are achieved.

1.3.8 Documentation and quality assurance

Following the impact-assessment stage or the error-reduction stage (depending on whether error reduction was appropriate), the results will be documented. Quality assurance systems will then be needed to ensure not only that any required ERMs are effectively implemented, but also that any assumptions made during the analysis remain valid throughout the lifetime of the system, or the applicable lifetime of the HRA/PSA.

The above defines briefly the HRA process, whose constituent elements will be expanded upon throughout the rest of the book – and, more systematically, in the corresponding subsections in Chapter 5. Before leaving this introductory section, however, it is necessary to introduce one of the primary research-and-development areas currently involved in HRA, one which may have long-term effects on the HRA process in the future (although those effects are currently unclear). This area concerns management and organisational influences on human performance.

1.4 Management and organisational influences

The difficulties of achieving an accurate and effective HRA have recently been compounded by the recognition of another significant influence on human performance, the management and organisational (M&O) influence. Accidents such as Chernobyl, the Herald of Free Enterprise, etc., all appeared to contain strong indications of a managerial-level influence which had an adverse effect on the vulnerability of the system to human-error-related risks. There is a good deal of effort, at present, directed at trying to derive means of protecting industries from such influences, as well as trying to estimate the quantitative impact of such influences on human reliability assessments and risk assessments themselves. These methods are discussed in Section 5.11, but the latter type of approach is not advocated at present since it is as yet relatively unproven. Instead, Section 5.11 focuses on ways of defining M&O boundaries which will enable current HRA and PSA estimates to remain reasonably accurate. This means that whilst the assessor should be aware of the potential power of M&O influences, currently that assessor can carry out a HRA according to the HRA process as it stands. It may be that in the future the HRA process will change to incorporate these influences, but for the time being M&O influences remain outside its remit.

2

How to use this book

2.1 Overall structure of the book

The overall basic structure of the book is in three main sections: introductory material; technical guidance; and supporting appendices. There is an introductory section (including this section, and one on the scope of the book), which is followed by an introduction to risk assessment and an expansion of the HRA process (Chapter 4). Chapter 5 then deals with the HRA process proper, giving useful techniques and approaches for each relevant stage. Sections 6 and 7 discuss some contemporary issues, and likely future directions of HRA, and give some conclusions on the state of the art of HRA.

The appendices have been compiled to give useful back-up information. Some of these are likely to be of use to the practitioner, while others which will be of primary interest to the more general reader – for example, Appendices I (which describes a series of accident case histories, all of which involved some degree of human error) and VI (Case Studies). Appendices II (data), III (validation evidence for quantification techniques), IV (human factors checklists) and V (descriptions of error mechanisms and associated error reduction mechanisms) will be of more use to the practitioner. The practitioner will also probably find the case-study appendix (VI) of interest, in seeing how others have approached real HRA problems.

2.2 Technical sections and appendices

The first technical section (4) describes the risk-assessment process, and discusses issues associated with carrying out a HRA within a risk-assessment framework, or outside of one. This section has the dual role both of showing how a HRA fits into PSA and, within this PSA context, of defining the human-reliability-assessment problem to be addressed. This section should, therefore, help the practitioner to determine the scope of his or her HRA, and allow the more general reader to put the HRA process into the risk-assessment perspective.

Sections 5.2–5.7 deal with the tools available for a HRA, defining in detail their rationales, mechanisms, advantages and disadvantages, and interrelationships with other parts of the HRA process. Sections 5.8 and 5.11 deal with the difficult area of ensuring that the HRA carried out remains valid during the lifespan of the system, and provides useful information for future assessments, as well as enabling the practitioner to determine when a system's M&O influences may be threatening the validity of the HRA and the safety of the plant. Section 5.10 is a central section for the practitioner since it attempts to bring all the various tools together into a single coherent framework. Section 5.9 revisits the area of problem definition to explore the different variations in the HRA process framework which can occur, depending on both the nature of the PSA approach adopted and the quantification techniques selected for the HRA. This section necessarily occurs after a discussion of HRA tools, since it is only after reviewing such techniques and seeing how they work that such potential variations involved in applying the HRA process become apparent. The practitioner would, of course, consider such more detailed aspects of problem definition at the outset of the study, i.e. as part of the problem-definition phase.

The appendices respectively detail: the human-error forms found in a series of relatively well-documented accidents; exemplary human-error data for information and for potential support in Human Error Quantification; a series of short checklists on human-factors areas, to enable the practitioner to gain a quick appreciation of the human-factors adequacy of a system; and a series of human-reliability case studies demonstrating some of the HRA tools 'in action'.

For the practitioner, therefore, all of Chapter 5 plus Appendices II–VI represent the heart of this document. The more general reader, on the other hand, will benefit from reading Chapter 4 and Sections 5.10 to 7.0, and skimming 5.2–5.7. Appendix I on accidents and the case studies (Appendix VI) will probably then be of most use in demonstrating how human error manifests itself and how HRAs look in practice.

2.3 HRA: the practitioner's framework

The objective of this book is to define all the areas involved in HRA and provide practical means of helping assessors achieve reasonably accurate and effective error predictions and reductions. The functions of HRA (e.g. Human Error Identification; Human Error Quantification, etc.) can be achieved by a range of methods. The intention of this book is not to be too prescriptive: practitioners are advised to be flexible rather than stick to only one or two techniques.

Each subsection from 5.2 to 5.7 offers a range of approaches which the practitioner can use, depending on the needs of the situation. What Section 5.10 (Practitioner's framework) attempts to do is to place these into a coherent framework based on the HRA process, with guidance for the practitioner on the selection of techniques at each HRA-process stage. As new techniques are developed and become available, the practitioner will then be able to update or tailor such guidance for her or his own use.

This book therefore presents a toolkit for HRA, plus a framework within which to use these tools (each of which naturally has its advantages and disadvantages). The tools are demonstrated via case studies in the appendices, as well as worked examples in the text – all of which come from real assessments in which the author was involved (though in some cases the facts or figures in the examples have been modified for reasons of commercial confidentiality). The author cannot state that all these techniques will be effective and accurate in all future assessments, since HRA is still a relatively young discipline and more experience is needed with many of its tools. Nevertheless, all the tools documented in this book have already been used in real assessments, and have proven effective in dealing with real human-error problems.

As a final comment on how to use this book, HRA looks best when it is being applied, because when it is being criticised from a theoretical perspective, there appear to be a great many drawbacks. In practice, HRA *does* succeed in estimating and reducing the level of risk. Probably the best way to use this book is to go through Chapter 5, stage by stage, with a real HRA scenario in mind, and work through the scenario until it has been assessed. Whilst many readers will not have the time to do this, those who do will gain a more practical understanding of the capabilities and limitations associated with a practical HRA.

3

Scope

This book is not a theoretical treatise. Instead, it focuses on HRAs. There are therefore many references to more general texts available, for example on the nature of human error (Reason, 1990; Rasmussen, Duncan, and Leplat, 1987), or risk assessment in general (Henley & Kumamoto, 1981; Green, 1983; Cox & Tait, 1991). It also aims to present practical guidance on how to use certain techniques, and their advantages and disadvantages. Some of the techniques are, it should be added, highly comprehensive methodologies in their own right (e.g. the Technique for Human Error Rate Prediction, Swain & Guttmann, 1983), and so only highly condensed versions can be discussed in this book. In such cases the aim of this book is to give the reader a sufficient appreciation of the particular technique so that the reader can decide whether to pursue it further. Similarly, with regard to some of the computerised HRA systems available, this book is limited in what it can present. The book is not, however, intended to be biased toward any particular approach. Any apparent biases will be instead largely due to the author's own – and others' – practical experiences with such techniques.

The HRA examples occurring throughout the book and in the case-studies appendix (Appendix VI) generally occurred in complex, high-risk industries (e.g. chemical, offshore, nuclear-power industries, etc.). However, the techniques outlined in Chapter 5 are applicable to a wide range of industries, including manufacturing, the service sector, defence, etc., all of which have utilised HRA. Virtually all systems are susceptible to human error and so can benefit from HRA, although they may not need so intensive an assessment as, for example, a nuclear-power plant.

Overall, this book aims to serve as a handbook or reference book for the practitioner, or as a detailed primer for the student in applied HRA. It is hoped that it may encourage the use of HRAs both in PSA and non-PSA applications, since, for all its limitations, the HRA discipline still has much to offer. And as noted at the outset, human error is here to stay.

4

Human Reliability Assessment in risk assessment

4.1 Introduction

This book is concerned with the estimation and reduction of human error in systems. This is a relevant topic in current industry because, although significant technological advances are being made in systems designs, accidents, sometimes catastrophic and large-scale, are still occurring. The systematic consideration of human error in systems designs can lead to improved safety, and indeed improved productivity in many cases. It can also be useful even for systems which are neither hazardous nor complex in nature. However, the human error 'problem' is most apparent in large-scale high-technology systems, which are both complex and 'tightly-coupled' (Perrow, 1984).

In such systems, a human error in one area can quickly propagate to and affect other areas, or interact with a hardware failure, often in ways which are difficult to predict but which ultimately place the system in a hazardous condition. Human-error identification, prediction and reduction, which together form the Human Reliability Assessment (HRA) are therefore best studied in their most challenging context, namely within the detailed qualitative and quantitative risk-assessment of large-scale systems. This risk-assessment process is generally known as a Probabilistic Safety Assessment (PSA), and such assessments, which today are carried out for all large hazardous systems, consider hardware and software failures, and environmental events, as well as human errors. This section explores the objectives and nature of the risk assessment, providing an insight into how the Human Reliability Assessment fits into the overall plant or system risk assessment process.

This section therefore describes the risk assessment process, identifying HRA inputs as this process is being discussed. It will only consider the HRA at a fairly high level in terms of its interface with the PSA, before proceeding into the more detailed aspects of the HRA techniques themselves. For more detail on the PSA itself see Green (1983) and Cox & Tait (1991). Once the

Figure 4.1 The Risk assessment process

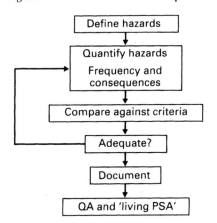

risk assessment process itself has been explored, the HRA process becomes more evident, and this process will be defined at the end of this section. The remainder of this book will then consider the various stages of the Human Reliability Assessment process plus the techniques available for achieving each stage's objectives. Such techniques can then be applied either in isolation or in unison to solve a problem or identify system vulnerabilities to human error, either within or outside the PSA process. In most cases for the foreseeable future, however, the HRA will continue to occur predominantly within the framework that is the risk assessment process.

4.2 The risk assessment process

The Probabilistic Safety Assessment (PSA) involves the evaluation of conceivable risks to worker and public safety (fatalities and injuries), and in some cases the risk of damage to the plant and the environment. The PSA identifies ways in which hazards can occur, during system operation and maintenance, which can lead to undesirable events (i.e. accidents). The PSA then calculates the probabilities and frequencies of such events, determines their consequences and compares the summarised risks estimated against industry/regulatory criteria. If the summarised risk breaches any of the criteria, then the PSA may be utilised to define how to improve the system's design so as to reduce the risk element to an acceptable level. This basic process is shown in Figure 4.1.

   The PSA is necessary for several major reasons. Firstly, as already noted, accidents continue to occur in industry which are unacceptable to the public. Additionally, system-design oversights can often lead to serious operational problems which can be damaging to the company and defeat productivity targets. These undesirable or accidental events occur because although there is good design practice and guidance, based on experience, available in industry, the complexity of many current industries means that even this quality of practice and guidance fails to take account of all possible and probable events, and combinations of events. A powerful, formal, comprehensive and systematic approach is therefore required which would be driven by a need independently to assess design and look for problems, rather than a need merely to corroborate supposed good design practice.

A second major reason is that although many identified potential system-design weaknesses will be considered improbable, such considerations will be necessarily vague until formally substantiated through calculation of their expected frequency and consequences. Only then will it be possible to see whether they really are improbable and inconsequential enough to ignore, or whether in fact a design change *will* be required. Furthermore, if design weaknesses are all looked at in isolation, then there will be no organised strategy either for prioritising the vulnerabilities or for considering alternative design-solution strategies. For example, a human error, which may warrant concern over a particular subsystem, may be eradicated by training, but if it is also likely to occur on a number of other subsystems throughout the plant, a hardware design change may in some cases be more cost-effective. The PSA will be able to give a detailed 'risk picture' of the risks involved for the entire plant in such a way that problem areas can be resolved in a more strategic and coherent fashion.

The third major reason is public acceptability. It is necessary in many countries to demonstrate, prior to building a plant or even gaining planning permission to build it, that it is (or is likely to be) acceptably safe. There is at present no viable alternative to doing this other than via the auditable PSA. Until an acceptable alternative is found, a PSA will continue to be necessary, and, given that human error is prevalent in accident causation, a HRA will also continue to be required.

The individual PSA stages are outlined below in more detail.

4.2.1 PSA hazard identification

Hazards can be identified in a number of other ways than through the use of formal investigative methods:

- By determining the various hazards associated with the intended system's materials inventory (e.g. dangerous chemicals, radioactive substances, etc.).
- By reviewing previous incident/accident experience to see what types of incidental/accidental events have occurred.
- By using the judgement of an experienced assessor.
- By reviewing hazards identified in other similar plant PSAs.

Such methods, while useful and usually employed in a PSA, are all limited, however, for a number of reasons: no two plants are ever operationally identical, even if of the same basic design; even if they are identical, the level of accident experience will be insufficient to have covered rare but significant and unacceptable accidental events; and the hazardous properties of different materials may interact in complex ways depending on the circumstances surrounding the accidental event. Therefore, it will usually be necessary to carry out a PSA for new installations.

4.2.1.1 Hazard and Operability Study

There are two major, formal fault-and-hazard-identification stages in the design life-cycle. In the conceptual and preliminary-design stages of a system's development (see Section 5.1), when the processes have been defined but only limited design engineering has taken place, a technique called HAZOP (Hazard and Operability Study, Kletz, 1974) is often utilised. This approach, which will be described in more detail in the Human Error Identification Section (5.3), basically involves a detailed consideration of plant-engineering flow-and-line diagrams by a team of designers/operators/safety personnel using a set of hazard-identification keywords. As an example, if the HAZOP team, considering a pipe carrying materials from a vessel, applied the keyword 'NONE', they would be indicating the fault of 'no flow' through the pipe. This would lead them to conclude, for example, that the pipe may be blocked, which would lead the HAZOP team to consider the installation of a high-level alarm and a high-high-level trip in the vessel, both of which would stop the filling process before overfill occurred. They would also consider whether 'no flow' could lead, for example, to cavitation or seizure of the pump at the other end of the pipe, and they would therefore also try to formulate solutions to this kind of fault (see Figure 4.2 for an illustration of these scenarios).

HAZOP studies comprehensively analyse all parts of a plant with the help of eight to ten guidewords, using a HAZOP team of about four to six personnel. HAZOP considerations are usually highly detailed and can lead to a large reduction in system faults and hazards, as well as to improvements in levels of both safety and operability (and hence to enhanced productivity). The HAZOP process, as will be argued later in Section 5.3, is a highly useful tool for considering human-error impacts in the system design, since at this early design stage improvements in human factors (or ergonomics) will be relatively inexpensive because the design is still fairly flexible. A parallel approach to HAZOP is that of Failure Modes and Effects Analysis, which is not dealt with in this book (see Henley & Kumamoto, 1981).

The next hazard-identification stage, and one both traditionally associated with the PSA, occurs during or after the detailed design stage when all the engineering design is either completed or nearing completion. By this stage, both HAZOP and the formal design and operational reviews will have eradicated the majority of hazards, and good design practices and industry

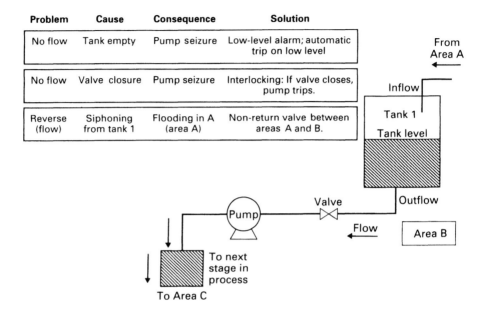

Figure 4.2 HAZOP worked example

and regulatory guidelines will have been adhered to (and such adherence documented). Two major methods of fault identification will then be applied to assess the level of risk posed by all possible identified hazards. These methods are *fault tree analysis* and *event tree analysis* (Henley & Kumamoto, 1981; Green, 1983).

4.2.1.2 Fault tree analysis

Fault-tree analysis is a top-down approach. It starts by considering an undesirable event called the 'top event'. This may have been identified by one of the informal methods mentioned earlier (e.g. inventory characteristics, experience, judgement, etc.) but still be insufficiently resolved by HAZOP – i.e. it still remains a potential hazard. The system is then investigated to define what single event, or combinations of events (human, hardware, software, environmental), could have led to the top event. An example of a fault tree is shown in Figure 4.3. The tree uses *gates*: either an 'AND' gate, which means that *all* events under the gate must occur before the event above the gate can occur, or an 'OR' gate, which means that the event above the gate will occur if any one (or more) of the events immediately below it occurs.

Fault-tree analysis requires training and experience beyond the scope of this book, but suffice to say here that it is an *explorative* approach, identifying and logically representing sometimes complex failure mechanisms, some of which will involve human errors. Although, in theory, a 'top-down' approach,

Figure 4.3 Fault Tree

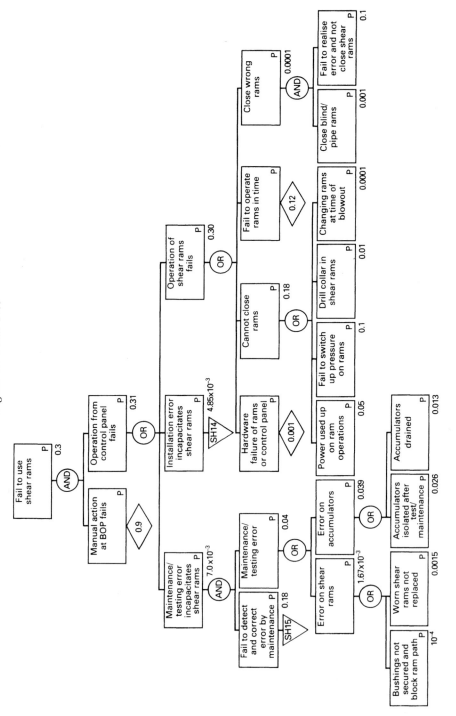

it can also become a 'bottom-up' one if, for example, an analysis of operators' tasks is used to see what errors could occur which could lead or contribute to the top event. Thus, the investigations of human errors plays a necessary and often considerable part in the construction of fault trees.

4.2.1.3 Event tree analysis

Event trees start from a basic initiating event and map out the major event sequences which can lead either to recovery to normal status or to accidental conditions. An example of an event tree is shown in Figure 4.4. The event tree proceeds from left to right in (usually) the chronological order of events, and at each node splits into two branches: the upper branch, signifying (usually – conventions vary) that the event-description in the box above that branch is TRUE; and the lower branch, signifying that it is FALSE. Event trees are used to explore accident sequences, and as such will often include the role of the operator. The operator can contribute to accident sequences in both a positive and negative sense, and such contributions as the following may be modelled in PSA accident sequences:

• Maintenance, and other, errors which would render safety systems unavailable in the event of an accident sequence (e.g. a back-up safety system in a healthy state but inadvertently left off-line, or still awaiting maintenance).
• Actions which would initiate an accident sequence.
• Actions which are required to stop the accident sequence (e.g. activation of safety systems, shutdown).
• Actions which will make the situation worse (e.g. misdiagnosis).
• Actions which can recover the system from failure and stop the accident sequence, or mitigate its consequences (e.g. evacuation, etc.).

These human-error involvements are shown in Figure 4.5 in the form of a generic accident-sequence event tree.

These two methods, fault-tree and event-tree analysis, can thus be used to identify and logically represent the failure mechanisms, together with their critical hardware failures and human errors, which can affect system risk. Moreover, they represent these event sequences and combinations in a way that will allow the calculation of the overall accidental-event frequencies or probabilities. This accident-sequence-representation stage, as will be shown, can benefit significantly from the use of human-action-description methods (called task analysis) and Human Error Identification techniques. The next stage in the PSA, having identified the hazards and accident sequences, and having represented them in a logical and quantifiable framework, is to quantify the hazards' and accidents' frequencies and consequences so that the level of risk to the system can be estimated.

Figure 4.4 Event tree

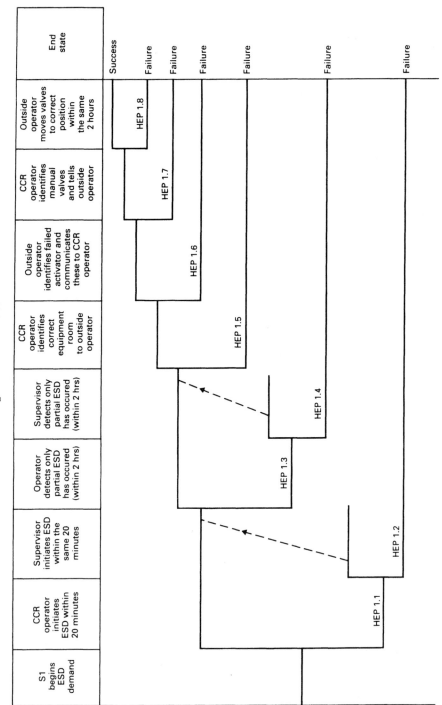

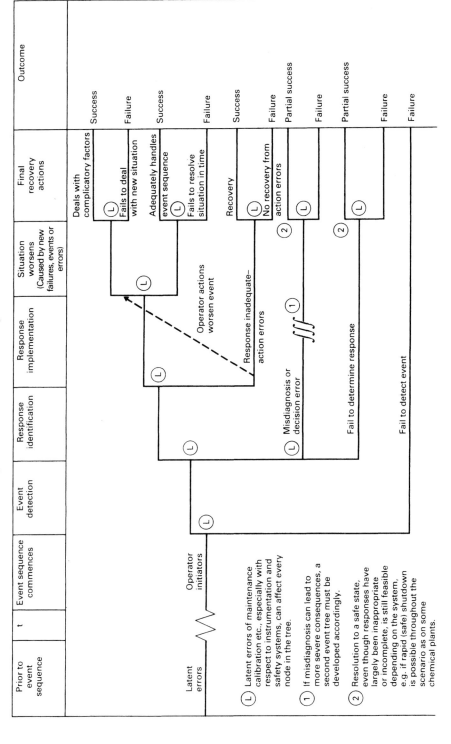

Figure 4.5 Generic accident sequence event tree

4.2.2 PSA risk estimation

The logical combinations and sequences of events which can lead to accidental outcomes, derived through fault- and event-tree analysis, must be quantitatively evaluated (using frequencies and probabilities of events, availability estimates, failure rates, etc.) to produce a single top-event frequency per year for a fault tree, and event frequencies per year for each accident-outcome category or pathway in the event tree. These frequencies and failure rates are usually either elicited from databases or calculated on the basis of physical/engineering models. In some cases expert opinion will be utilised. Human errors are calculated as probabilities via the techniques of Human Reliability Quantification described in Section 5.5.

The means of combining all these frequencies and probabilities is via reliability technology (in particular, Boolean algebra) which is beyond the scope of this book, although Section 5.4 will demonstrate the rudiments of the calculation procedures.

Once the frequencies of events have been calculated, their consequences must be assessed. This is achieved by a range of models which deal with the extent of the event (e.g. explosion and fire consequence modelling), the conditions at the time of the event (e.g. wind direction, etc., often conservatively or even pessimistically estimated) and the degree of exposure of the individual (e.g. the level of toxicity of chemicals, the effects of radiation exposure, etc.). These models are used to estimate possible consequences. The probabilities for each top event (fault tree) or accident-severity category (event tree) can then be calculated for all types of risk of concern, and these can then be added on to the total of all the other calculations for all event scenarios for the plant as a whole. This summation of the frequencies and consequences of all hazards and accidental events gives a detailed risk picture for that plant or system (see Cox & Tait, 1991).

In the PSA risk-estimation process, it is therefore the human reliability quantification techniques which are important in quantifying human error probabilities for tasks that can have an impact upon the level of risk in the system, and in quantifying factors that will affect the degree of exposure (e.g. the probability and speed of evacuation). The next stage in the PSA process is to compare the estimated risks with the risk criteria. This first requires a discussion of risk criteria.

4.2.3 PSA – accident criteria

The level of risk is broadly defined as the product of the frequency of a situation and its consequences. There are several different types of risk to be evaluated in a PSA for a plant, and these will vary from industry to industry, and company to company. The following are some examples:

- Spurious plant evacuations per year.
- The discharge of undesirable toxins or pollutants into the environment in each year.

- Injuries to the workforce in each year.
- Fatalities, per year, in the workforce.
- Fatalities, per year, in the public at large.

The above listing is in approximate order of increasing importance for industry, though variations will occur depending on the perceived severity of each of the above categories (e.g. a spurious *offshore* evacuation may be far more likely to result in direct casualties than an onshore one). See Cox & Tait (1991) for a fuller listing of potential accident and risk criteria – and for their detailed explanation.

For each of the above categories that are relevant to a particular industry there will be acceptable risk criteria as specified both by the company and by the regulatory authorities. For example, the acceptable worker fatality frequency may be stated as 1E–5 (once in 100,000 operational years). The acceptable frequency for spurious evacuations (for an onshore plant) may be once in 10 years (1E–1), while the acceptable frequency of fatalities to the public from the plant may be 1E–6, or once in 1,000,000 operational years. These criteria are designed to reflect the perceived importance of avoiding the consequences as specified in the categories.

Accident risk criteria are, not surprisingly, frequently a matter of heated debate, and it is worth briefly outlining the primary arguments in that debate, as many people opposed to the PSA are in fact opposed *not* to the PSA process itself but rather to the criteria which the PSA uses. The debate most often focuses on the fatality criteria.

Fatality criteria are implicitly based upon (or have evolved to become based upon) the concept of voluntary risk. It is argued that the public can be killed by exposing themselves, of their own free will (hence *voluntary* risk), to risks such as smoking, scuba diving, etc. Alternatively, the public may be killed by risks which are not of their own making – hence *involuntary* risks such as hurricanes and floods, crossing the road and driving, or being killed in an air crash. Statistical analysis suggests that voluntary risks cluster around a fatality frequency of 1E–5 per year. It is then argued that workers/employees in a plant are, although being paid, in effect accepting a voluntary risk associated with that plant and their employment, and hence that the 1E–5 figure can be used as the criterion. The public, however, who live near that plant, do not necessarily voluntarily accept the plant's risks, which for them makes the risk *involuntary*. Since death by involuntary risks normally has a frequency of 1E–6 to 1E–7, one of these values is usually applied to industries in calculating fatalities they cause to the public.

When criteria are being set for an industry, there are two opposing drives. The first involves making the risks as low as is reasonably practicable, or even as low as is reasonably achievable. This is a laudable goal since it militates for the continual improvement of standards of safety and operability, such that ultimately, for example, it would be safer to be in the work environment than to be at home.

However, there comes a point in any industry where the project will cease to be profitable (the second drive), because of the first drive to implement safety mechanisms and safety procedures. The resolution of this apparent conflict comes via the concept of *societal risk.* All risks and benefits (not merely those posed by the plant in question) are now included in a societal-risk equation, and in this way questions can be asked as to why a further £xm should be spent to save one life in a process plant when the equivalent sum could save 10 lives in another industry, or even 100 via improved road-safety measures, etc. The societal-risk concept, although a difficult and highly politicised potential solution to the risk-criteria debate, and one that remains somewhat elusive, is nonetheless perhaps a more logical way to allocate risk targets.

Before the problem can be solved even theoretically, however, the psychology of the public perception of risk needs to be more clearly understood than it is today. Until such a time, the debate on risk criteria and how far plant designs should be improved in safety terms will continue. During this period, the PSA will continue with whatever criteria are negotiated between industrial and governmental regulatory authorities (for more discussion see *HSE,* 1992). Having briefly, and necessarily simplistically, described risk criteria, it is worthwhile now to explore the implications of certain risk criteria on PSA and HRA objectives.

4.2.3.1 Risk criteria – impact on PSA and HRA

It appears that as far as the public perception of risk is concerned, infrequent high-consequence accidents (e.g. Chernobyl, the Herald of Free Enterprise, etc.) are far less acceptable than frequent but low-consequence accidents (e.g. mining fatalities), even when the number of fatalities, etc., is the same. As a consequence, many PSAs appear to focus disproportionately on the assessment of rare but very high-consequence events even though experience shows that the low-consequence risk may be the appropriate target area if fatality reduction is the real objective. This focus will influence the scenarios selected, during the PSA process, for a Human Reliability Assessment. Thus, for example, a plant that is being constructed may have a target worker-fatality criterion of 1E–5 for when the plant or system is operational, but several deaths may actually occur during its construction (and the construction industry, it should be added, is a particularly hazardous one). In a mathematical sense this is somewhat illogical, but it nonetheless reflects the public, and corresponding regulatory, attitudes to different types of risk. The PSA will tend to focus on the large-scale types of accident, leading to multiple fatalities, while, on the other hand, lower-consequence but higher-frequency risks (e.g. worker injury or death by, for example, falling from a temporary scaffold) may be under-assessed, or simply not assessed at all.

A related problem with respect to low-frequency, high-consequence events, one which the HRA and PSA analysts must be aware of, is the so-called 'cliff-edge' scenario. This concerns the investigation of errors or situations leading to top events – say, a single fatality with a human error probability of, for example, once in a thousand operations. The risk of such an event's occurring might be acceptably low if, for example, the event also required an independent hardware failure with a probability of 1E–3, yielding a total risk of 1E–6. However, there may be another event with a much less likely human-error sequence (e.g. 1E–5), coupled with the same hardware failure, which had far greater consequences – for example, 1000 deaths. The risk of this sequence would work out as: $1000 \times 1E{-}5 \times 1E{-}3 = 1E{-}5$, which would be unacceptable. The danger of failing to identify 'cliff edges' arises because the analysts consider the more likely failure modes and assume these to be the worst cases. Analysts must therefore be aware of this potential pitfall in PSA and HRA.

4.2.3.2 Risk criteria – limits on reliability

There is a developing concern in the area of Human Reliability Assessment about the role of accident criteria and their continually diminishing targets. Current targets in some industries are very difficult to meet in a practicable way, e.g. a fatality criterion of 1E–7 or even 1E–8. It is difficult to justify such a risk level for three main reasons. Firstly, PSA is a relatively inexact process, as is HRA, and uncertainties tend to multiply in large PSAs, so that eventually an estimate is reached, but it may have large uncertainty bounds. Whilst sensitivity analyses can be used to explore the impact of uncertainties arising from individual events on the target criteria (see Section 5.6), these rarely model the *interactions* of uncertainties, which can produce larger effects. Secondly, hardware reliability and human reliability appear to have limits, probably of the order of 1E–3 to 1E–5 failures per demand (see Section 5.4). Such failures may not be independent from other failures, and this means that the system can have a reliability 'cut-off' of the order of 1E–4 to 1E–6. Merely putting in more human supervision or more redundant/diverse hardware or software systems does not always help since dependencies will still exist.

The third problem with diminishing criteria is that it becomes necessary for the analyst to start considering highly unlikely and virtually incredible errors which would cause disaster – often via cliff-edge scenarios. These extreme and rare forms of human error can of course occur, but they verge on suicidal or sabotage actions. Such actions would not normally be considered in PSAs, but when trying to identify scenarios which could only happen once in 10 or 100,000,000 plant or system years, they begin to have credence. Such errors, needless to say, are almost impossible to predict accurately, and the analyst will have great difficulty ensuring that there is nothing a human operator cannot do in 100,000,000 years, in concert with other events, for which the plant is unprotected. This problem suggests that such criteria stretch PSAs and HRAs outside of their credible range of operation and place a virtually intractable task on the shoulders of PSA and HRA analysts. The danger is that PSAs will ignore such events, while maintaining that the ever-increasingly stringent criteria are actually being met. Such a state of affairs will only serve to push PSA predictions further from actual reality, until eventually an accident will occur with a frequency several orders of magnitude higher than that predicted by the PSA, which will simply bring the whole PSA process into disrepute.

In conclusion, risk criteria, though debatable, are currently necessary to the PSA process. The implications for the HRA are that low-frequency high-consequence events will be emphasised, and that the analyst may also have to consider so-called 'cliff-edge' scenarios. The continuing attempts to make risk criteria more and more stringent put a strain on the capabilities of both the HRA and PSA processes, and may stretch them beyond their proper field of application.

We have now considered the nature of risk criteria and the special implications for PSA and HRA. The next subsection briefly outlines the PSA risk-comparison process.

4.2.4 PSA – risk comparison

The risks estimated may all satisfy the allocated risk criteria, in which case the PSA process may move into the documentation and quality-assurance phase – or seek to reduce its risk level still further, particularly if other comparable industrial systems have further reduced their risks (this is referred to as reducing fatality risks to a level which is *reasonably practicable*, this all-important latter term usually taking on an operational meaning by reference to what is current or imminent industry practice).

If one or more risk criteria have been breached, the system design must in some way be improved. The PSA, via sensitivity analysis, will determine where the largest risk contribution is coming from (i.e. which scenario or which combination of human and hardware failures, etc.), as well as the individual contributions of risk-significant human and hardware events. The project management must then decide on what strategy to adopt which will most cost-effectively reduce the risk to an acceptable level. This strategy may involve the re-design of certain functions, the transfer of these functions to machines or humans, or the provision of extra safety barriers. The role of Human Reliability Analysis within this part of the PSA process involves defining how error probabilities may effectively be reduced, and this is achieved by error-reduction analysis, as described later in Section 5.7.

When a risk-reducing strategy has been identified, the PSA should re-examine the scenario before re-calculating the risk level, to see if the strategy has inadvertently created a new fault path. It will be particularly important to review the impact of any changes on human operations. If there are no new errors or fault paths associated with the design change, the system risk level may be re-calculated and compared against the risk criteria. This process is iterated until an acceptable design solution has been derived. This completes the PSA process; the PSA will now be documented. Quality-assurance systems must be installed to ensure that assumptions underpinning the assessment are maintained, particularly with respect to design changes. The PSA will also be periodically updated, and additionally, can be used throughout operations as a database for making operational and retrofit design decisions (e.g. if it is decided in later years to expand the system, or to modify it in some way). Such usage of the PSA is referred to as 'living PSA'.

Having defined in broad detail the PSA process, and having considered some of the more subtle implications of this process for the HRA, the HRA process, which forms the basis for the remainder of this book, and is the suggested framework for practical Human Reliability Analysis, can now be defined in more detail.

4.3 The HRA process

The HRA process is depicted in Figure 4.6, and many of the constituent elements from this diagram have already been raised in the above discussion on the PSA process. This HRA process is responsible for the structure of the remainder of the book, with major sections, in particular, on task analysis, Human Error Identification, Human Error Representation, human reliability quantification and error reduction analysis, as well as comparatively smaller sections on problem definition, impact assessment and quality assurance and documentation. All of these components of the HRA process are briefly outlined below.

4.3.1 Problem definition

The problem-definition phase is where the HRA's scope will be determined. In particular, a number of key questions will need to be answered in order to define the scope more exactly:

- Is the HRA part of a PSA or is it a stand-alone assessment?
- Will the following types of errors be considered:
  maintenance errors
  misdiagnoses
  rule violations?
- Is a quantified estimate of reliability required?
- Is the level of risk to be quantified absolutely, or will *relative* estimates be enough (the latter may suffice for a comparision of several rival designs, for example)?
- In what stage of development is the system?
- What criteria is the risk assessment trying to meet (for example, fatalities, injuries, damage to the plant)?

Figure 4.6 The HRA process

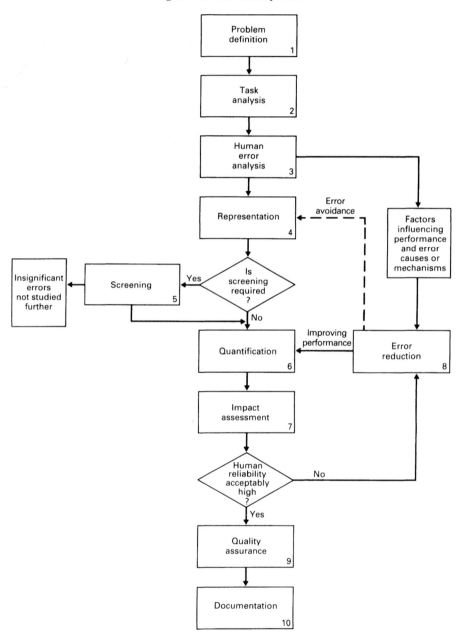

- How extensive are the resources available for the HRA?
- Is the HRA to be PSA-driven (i.e. with the PSA determining the scenarios to be analysed) or can the HRA itself determine what situations to assess?
- Has a HRA already been carried out for this system?
- How vulnerable is the system to human error?

Often the problem definition may shift with respect to the above questions as the HRA proceeds, for example due to a realisation that perhaps the system is more vulnerable to human error than was initially envisaged. Whatever the answers to the above questions are (these will all be discussed in more detail in Section 5.1), it is the last question above that will remain in the HRA analyst's mind throughout the study as the simple but critical thought, the one that is difficult to answer: 'is it safe?'

4.3.2 Task analysis

Task analysis refers to methods of formally describing and analysing human-system interactions. Task analysis defines in detail the roles of the operators within the system. A formal task analysis is therefore a critical part of the problem-definition stage, since before the analyst begins to consider what errors might occur, he or she should have a clear definition of what the operator should be doing to achieve the correct performance. The task analysis therefore forms the basic description of the human role in the system, much as a system-description document for a piece of hardware such as a pump would specify how the pump works and what it is supposed to do under a range of conditions. Just as PSA analysts need such hardware descriptions, so too does the HRA analyst need a task analysis.

In the recent past there have been a number of important developments in the field of task analysis that have proved relevant to practical HRA. These have taken the form of several new methods of describing human-system interactions, particularly in dynamic accident sequences. While there are many forms of task analysis available (Drury, 1983; Kirwan & Ainsworth, 1992), only a small number of these need be used by the HRA practitioner. All of these methods are paper-based and are fairly straightforward in terms of their rationales, procedures for application and formats. The methods most relevant to the HRA process will be defined in Section 5.2.

4.3.3 Human Error Identification

The identification of errors is perhaps the most difficult part of practical HRA, since humans have such a vast repertoire of responses, which includes those appropriate to the task at hand. Accident experience has generally shown, however, that human errors, in many situations, occur in a limited number of forms, some of which are fairly predictable. Techniques exist to help identify a range of errors, from simple errors of psychomotor control (e.g. hitting the wrong button within an array of buttons) to the far more complex errors associated with diagnosis and problem-solving during complex emergencies. These techniques are presented in Section 5.3. It should be noted that the theory underpinning such techniques is given minimal coverage in this book, which instead focuses on the tools available and how they are to be used in practice. Adequate references to more theoretical texts will, however, be given Section 5.3.

Human Error Identification (HEI) can also form the basis for human error reduction, which occurs later in the HRA process (Section 5.7). That is to say, error reduction can occur as a function of the causes and type of error predicted. Depending on the technique being utilised, the HEI tool may also identify factors affecting human performance – so-called Performance Shaping Factors (PSF) – which may then be utilised in the error-probability-quantification phase of the HRA process. Thus, HEI can be a highly influential phase of the HRA process.

The Human Error Identification process is arguably the most critical part of the HRA process after problem definition, since if an important error is not identified, then it simply will not appear in the risk assessment, which means the risk may be underestimated.

Unfortunately, Human Error Identification techniques have generally received rather less development resources than they deserve, with most effort in HRA having been expended on the quantification of human-error probabilities, for reasons explained in Section 5.9. Thus error identification remains one of the weaker areas of HRA. The techniques defined and developed in Section 5.3 have been found to be particularly useful. These have been put into a new sub-framework for Human Error Identification purposes to enable the assessor to decide which techniques to utilise, according to the error-identification goals and the scope of the HRA.

4.3.4 Human Error Representation

Once errors have been identified, a way of logically evaluating them must be found for two reasons: firstly, so that the importance of each individual error can be ascertained; and secondly, so that the combined risk probabilities of all failures and all combinations of failures (hardware, software, human and environmental) can be summed to derive the total level of risk inherent in the system. This is achieved by representing the human errors, along with other failures, in logic trees known as fault and event trees (usually run on a computer for all but small fault/event trees because of the very large number of combinations generated in such analyses). Such trees, when properly constructed, enable the use of mathematical formulae in order both to calculate all potential combinations of failures that can lead to the accidental consequences of interest, and to add together all of the individual risk probabilities of each fault combination so as to derive the total summed risk for each accident type or severity category (or both). Furthermore, when this has been done, these techniques, particularly that of fault-tree analysis, will give an indication of the degree of importance of each individual event to the total level of risk in the system. This allows the analysts to carry out detailed cost-benefit analyses of risk-reduction measures, should these be required.

There are other more sophisticated ways of carrying out a risk analysis, involving simulation techniques, etc., but these are beyond the scope of this book. Instead, the rudiments of fault- and event-tree approaches, an their calculation procedures, are defined in Section 5.4.

In addition to the basic types of logic tree in use in HRAs and PSAs, there are a number of other areas of concern to be found within what is called the *representation phase*. The first is the modelling of *dependencies* between different identified human errors. A simple example of what is meant by dependence is that of an operator who is calibrating a number of gas detectors on an offshore installation. If the operator made a random error in calibrating a single gas detector, then such an error would not be expected to recur on other detectors. If, however, the operator mishears or misreads the common setting for the detectors, then *every* gas detector on the platform may be erroneously set as a result. It is important when representing error in this scenario that, in the latter case, the errors occurring for each gas detector are not treated independently, since this would lead to an underestimation of the likelihood of many gas detectors being mis-set. Dependence is a difficult and relatively underdeveloped area in HRA, but one that can have dramatic effects on the level of risk calculated in a PSA. The few useful approaches to dependence treatment in HRA that do exist will be reviewed in Section 5.4, in the context of their use in large and small PSAs.

A third area of concern in the area of representation is that of *screening*. Screening can be used in large PSAs where a correspondingly large number of errors may have been identified, and where, as a result, significant re-sources have to be expended to quantify the probabilities for all the errors. Errors are 'screened' by assigning each error a highly pessimistic probability, in the initial 'run' of a PSA logic-tree evaluation, in order to determine whether a detailed and more accurate quantification is appropriate on an error-by-error basis. Several such screening methods will be described in Section 5.4.

4.3.5 Human Error Quantification

Human Error Quantification, or Human Reliability Quantification (HRQ), is the most developed part of the HRA process. There are a number of so-phisticated tools available, some of which have fared reasonably well in in-dependent validation studies. Furthermore, most of these tools are available in the public domain, thus making them entirely accessible to practitioners.

Section 5.5 details, examines and demonstrates a number of the more popular and effective approaches. Appendix VI, in addition, will give detailed case studies of some of these techniques – including several case studies in which error probabilities were calculated by different assessors, using different techniques – thus allowing the reader to see more clearly the similarities and differences in the various approaches. Selection criteria for use by practitioners are also summarised in Section 5.10, with Appendix III giving a review of validation studies of some of the more popular techniques.

As noted in the Introduction, ideally there would be a human-error database, but such databases have proven rather elusive. Nevertheless, in Appendix II, some representative data from a range of scenarios and industries are presented, for illustrative purposes. A 'pedigree' (e.g. real, simulator, experimental, etc.) is ascribed to each datum.

4.3.6 Impact assessment

Once the errors have been represented and quantified, the PSA risk calculations can now be carried out and the overall system risk level determined. From such calculations, the answers to two fundamental questions may be derived:

- Is the system acceptably safe?
- If not, what are the principal ways to reduce the risk to an acceptable level?

The answer to the second question may well implicate human error as a contributory or even dominant factor in undesirably high risk levels. The impact of human error on system risk levels can be determined quantitatively, and particular risk-significant human errors may then be targeted for error-reduction analysis. Impact assessment is dealt with in Section 5.6.

4.3.7 Human error reduction

Error reduction will be called for if the impact of human error on the level of risk in the system is significant, or it may be simply desirable to improve the system risk level even if the target risk criteria have been met. In either case, there are a number of methods of error reduction, and the analyst can also draw upon the considerable body of knowledge and techniques found in the field of ergonomics or human factors.

Error-reduction mechanisms can be derived for specific errors, based on information gained during the task analysis, or on the identified causes of the human error, or else on the Performance Shaping Factors utilised during the quantification phase. In addition to these HRA-based error-reduction approaches, reduction measures can be designed, either by a consideration of the engineering design of the work environment (e.g., by putting interlocks in place so that certain actions cannot occur, or by placing error checks in the software, etc.), or by placing barriers in the system to protect the operators/public from the consequences of the error/hardware-failure combinations. All of these approaches will be described in Section 5.7.

4.3.8 Documentation and quality assurance

This is the final phase of the HRA process, during which the results and methods utilised are documented. This documentation should preferably be to such a standard that the rationale is clear and the detailed methods and results can be audited by an independent agency, and then repeated by an independent assessment team. In addition, all assumptions made by the assessor(s) are documented and then made clear to the project team who will run, or who is running, the system. This is particularly important if error-reduction mechanisms are proposed: it is of quantitative benefit to the PSA/HRA process that the the mechanisms are effective as soon as they are implemented. Often, reduction measures are not as effective as envisaged, due to a number of reasons: inadequate implementation of these measures; a misinterpretation of the measures by the project-design/support group; new problems arising related to side-effects of the measure; the acclimatisation of the operators to the measure (particularly if the measure is motivational in nature). Such problems will lead to an actual net decrease in risk, but to a level less than that assumed by the PSA/HRA. Quality assurance in the HRA process is therefore primarily aimed at ensuring that the reduction measures remain effective and that the error-reduction potential is realised and maintained. This is no easy task, and indeed is often difficult, especially for external contractors who, having submitted the final draft of their PSA, will no longer have access to, or involvement with, the project. Quality-assurance issues are discussed in Section 5.8.

4.3.9 Problem definition re-visited, and M&O influences

Section 5.9 re-visits the area of problem definition, as well as different approaches to PSA. The implications of using various quantification approaches on error identification and reduction are also considered. The reason these issues are left until Section 5.9 is that to understand the above-mentioned implications, it is necessary to have first explored the 'core' of the HRA process, and to have seen the various techniques in detail. Section 5.11 considers the difficult area of Management and Organisational (M&O) influences (safety culture) on performance, and the limits this rapidly developing field places on HRAs and PSAs. Section 5.10 attempts to place the various techniques and HRA-process stages into a coherent framework – or rather,

Figure 4.7 HRA process mapped onto the PSA process

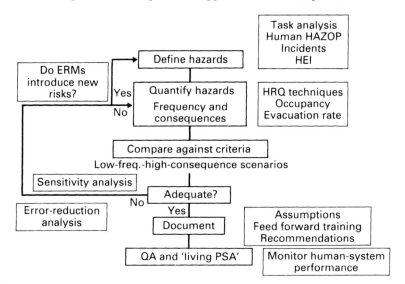

a set of frameworks, depending on the PSA 'mode' and the problem definition determining the scope of the HRA.

In summary, the PSA requires a number of significant imputs from HRA,
while the HRA process, in turn, fits neatly into the PSA cycle. Figure 4.7
attempts to show these inputs and interactions diagramatically. Since almost
all systems involve some level of human interaction with the system, the
HRA should be explicitly carried out within all PSAs so that the effect of
human error can be properly evaluated within the overall system-risk picture.
The HRA is thus essential to the PSA. Section 5 focuses on the individual
parts of the HRA process, and attempts to present workable methods for
achieving their goals.

5

The HRA process

5.1 Problem definition

Little is written on the problem-definition phase of Human Reliability Assessments. This phase basically involves setting the terms of reference for the HRA, and is usually a necessary compromise between the goals of the HRA and the constraints placed around it. Starting from the 'ideal' – or maximum – HRA approach, the analyst, when assessing a system design, may wish to analyse every possible human interaction that can occur in the system, whether it is required for the operation of the system or not (in which latter case, it usually constitutes an error). Every combination of human actions and errors, single or multiple hardware failures, and environmental events and conditions can be assessed.

Clearly, for large complex systems, such an exhaustive approach will require huge resources, and many of the interactions investigated will be found to be inconsequential, or else so improbable as to render their risk-contribution negligible. The essence of HRA problem definition is therefore to set the scope of the HRA according to the objectives of the assessors, the criteria for achievement of those goals and the existing constraints within which the HRA must work.

This section raises some of the main issues in problem definition in the HRA process. These will be returned to in Section 5.9 after the reader has seen the techniques and methodologies in more detail. In Sections 5.9 and 5.10, more practical guidance on defining and giving a scope to HRAs, which is what the problem-definition phase is all about, will then be given.

5.1.1 Problem definition issues in HRA

There are two fundamental issues that need to be addressed in any practical HRA concerning its nature and scope, and these can best be put as the following questions:

- Should the HRA be quantitative or qualitative in nature?
- How far should the scope of an HRA go? Should it focus only on the abnormal (e.g. emergency) tasks, which clearly require a significant degree of human reliability, or should it also address potentially significant maintenance contributions to the level of risk, and the kinds of error made during normal operations?

The above are the basic problem-definition questions worth reviewing at the beginning of any HRA. They can be answered according to how the following factors constrain and direct the HRA:

- The system's vulnerability to human error.
- PSA dominance in the analysis (i.e. the extent to which the HRA will be 'PSA-driven').
- PSA objectives (i.e. whether to just quantify the current risk level, or whether to try to reduce the risk level, if appropriate).
- The risk-assessment target criteria.
- The existence of prior PSA/HRA studies for the project.
- Resources available (funding, expertise, software).
- The system-life-cycle stage of the project.

Overall, the HRA analyst wants to make sure that he or she does not underestimate, nor, on the other hand, significantly *over*estimate, the impact of human error on system risk levels or performance. The analyst therefore wants the analysis to be comprehensive, penetrating and accurate, and to provide useful results, as well as (possibly) recommendations on how to improve human-system performance.

The analyst will, however, find himself or herself constrained in the pursuit of these goals, if not just by available resources then by virtue of the fact that it is impossible to predict *every* possible, potentially negative form of human impact on a complex system. This is because of the infinite variety of human responses possible, as well as the large number of potential human-system-environment failure combinations. A HRA is therefore, ultimately, an in-depth assessment of a system's risk level as a function of human performance, and should not be seen as a way of deterministically proving that a system is safe. Like the PRA, the HRA is probabilistic in nature, and is essentially indicative of future system performance. Although it may fail to predict the exact forms of incidents and accidents that actually occur, it will hopefully be able to predict both their overall and approximate form, and the approximate frequency of their occurrence.

This is an understandable state of affairs, since accidents are themselves generally stochastic (i.e. random) combinations of events that occur idiosyncratically, and in fairly complex forms, in real systems, so that their detail is usually difficult if not impossible to predict. In this light, the HRA is best seen as a tool for vulnerability protection: locating system vulnerabilities to human error or performance problems, and building in defences where they are found. The difficulty facing the HRA analyst is therefore that of setting the scope of the analysis in such a way as to offer an appropriate and reasonable level of investigation, assessment and reduction of the degree of vulnerability of the system concerned. The first step in this process of problem definition is the consideration of a system's vulnerability to human error.

5.1.2 System vulnerability to human interaction

A system's degree of vulnerability will be one of the principal determining factors in deciding the 'level' of HRA that is desirable. If the system is similar to other systems which have been in successful and relatively incident-free service for many years, and which have themselves been the subject of detailed HRAs (incorporating operational feedback (from incidents)), then unless there has been a new development in HRA or system technology, or unless a new type of risk has been identified, the desire for a full, detailed HRA will be small; and the carrying-out of any such HRA may possibly even be seen as an unjustified HRA waste of resources. If, on the other hand, the system is a new type which critically depends on human reliability for safe operation, and which involves potentially large losses (in terms of lives or profits, or both), then a full investigation of the human contribution to the level of risk will be justifiable.

The degree to which a system is vulnerable to human error is difficult to assess, but Perrow (1984) usefully defined three factors which can be considered when attempting to set a scope for the HRA: complexity, interactiveness and coupling. The complexity factor is concerned with the degree of sophistication of the physical and chemical processes underlying the system's design, and the extent to which these processes are understood by the personnel operating the plant or system. The interactiveness factor is concerned with the degree of interaction between subsystems and components. The coupling factor is the flexibility built into such interaction, such that when one process is affected negatively, there is enough flexibility in the system to allow one to use an alternative or else merely to wait until the fault in the process is repaired. A system which does not have such flexibility is 'tightly coupled'.

Clearly, a system which is complex, highly interactive and tightly coupled will be the one requiring the most intensive risk analysis. Since the human in such a system will usually have a significant role in dealing with the complexity of the system, in making links between many interactive subsystems and in compensating for the lack of flexibility in the system (or at least managing the response to losses of subsystems), then the HRA will also have to be more intensive. If such a system also has a high potential level of risk (e.g. by virtue of the potentially hazardous chemicals/processes it utilises), or is dealing with novel technology, then a full, in-depth HRA will be warranted.

Many systems or projects will not fall neatly into either of these extreme categories. As an example, there will be systems which are building on proven safe system designs, but which, for example, have novel features where a high degree of human intervention is required. The analyst must decide in such situations where to focus the most effort, and may well adopt the hybrid approach of carrying out a broad analysis of human reliability for the whole system, with more intensive analytic methods being used on the novel or potentially vulnerable system features. Little precise guidance can be given to the assessor on such decisions, although some examples are considered in the case-study appendix (VI). As for what constitutes a full and detailed HRA, and what constitutes a more generalised or broad HRA, this will become clearer as the reader proceeds through Section 5. Specific guidance on when to use these different 'levels' of HRA will be given in Section 5.10, after the various techniques of HRA (some intensive, some 'broad-brush') have been described.

Once a system's level of vulnerability has been addressed, the next main constraint to the HRA will come from PSA considerations: whether the HRA is part of a PSA; whether the HRA will be PSA-driven; whether risk-reduction measures may be required; what criteria the PSA is assessing against; and whether prior PSAs have been carried out (and are accessible). These are discussed below.

5.1.3 PSA aspects of problem definition

5.1.3.1 Non-PSA-driven HRAs

If the HRA is not being carried out as part of a Probabilistic Safety (or Risk) Analysis, e.g. if the HRA aims to look at ways of improving quality or productivity, or involves a qualitative approach to improving safety, then the need for a formal error quantification may not arise. Some HRAs have occurred which have identified errors and considered their importance (e.g. by rating their consequences and the approximate likelihood of their occurrence), and that have provided error reduction guidance, without ever using formal quantification. Therefore, if the HRA is not in a PSA framework, the applier should not automatically assume that quantification is a necessary part of HRA. For example, if a particular human-performance problem has been identified, then the qualitative tools of HRA may be used to analyse the task and perform human-error identification and error reduction. This example would be a form of targeted HRA, and would focus on a particular pre-identified task.

In many cases, quantification will nevertheless be the most effective means of prioritising errors in terms of both their impact on the system's goals of safety/productivity and their differing degrees of importance when it comes to error-reduction measures. Most HRAs are therefore quantitative in nature.

5.1.3.2 PSA-driven HRAs

PSA criteria

If the HRA is being carried out as part of a Probabilistic Safety Analysis, then there are three major aspects involved in the problem definition, which can be put as the following questions:

- What are the target criteria for the PSA?
- Are risk reduction measures required?
- Will the HRA analyst, or the PSA team, define the scope of the HRA?

The main assessment criteria for the PSA may have an influence on the problem definition of the PSA. In most cases criteria of concern will be the fatality criteria and the environmental-impact criteria (e.g. radiological releases into the biosphere, in the form of aerial or liquid discharge). The exact *extent* of these criteria can also have an influence on the problem definition, particularly in the case of the fatality criteria. If these criteria targets are very stringent (e.g.) less than 1 fatality per 1,000,000 years), then there will be more of an onus on the PSA to make sure not only that operators will react correctly to ensure safe shut-downs, etc., but also that the operators will not initiate any fault or accident sequences, or even make worse (e.g. through misdiagnoses, rule violations or errors of commission, as discussed later in Section 5.3) any such sequences that have already been started. Similarly, 'cliff-edge' scenarios may also have to be considered, which generally will require a more in-depth HRA.

Outage PSAs

The PSA may also have to consider risk levels incurred not only in normal running and in start-up/shut-down implementation, but also during plant-shut-down activities. These may be referred to as *outage PSAs*, i.e. PSAs which consider what can go wrong during scheduled or unscheduled shut-down periods. The reason for including such plant states in PSAs is that some systems have reduced safety monitoring and controls/interlocks during shut-down. If the PSA is also considering outage activities and states, then the HRA, in turn, should focus on such activities, especially since under such conditions there is usually more reliance placed on human reliability to maintain the system's integrity. Furthermore, such reliance on the human element usually goes hand-in-hand both with fewer centralised monitoring facilities and (as a result) a smaller amount of information available to the operators. On the other hand, it also entails a greater reliance on procedures and other 'administrative controls', as well as time and production schedule pressures to get the plant started up again as soon as possible.

Industrial accidents/injuries

The afore-mentioned criteria may be concerned with only particular types of fatality, e.g. nuclear-related, radiological fatalities, as opposed to 'normal' (i.e. non-radiological) industrial injuries and fatalities. If industrially related fatalities are, however, included in the criteria set for the PSA, then this will have significant impact on the scope on the HRA, in particular on the Human Error Identification analysis, because it will call for a more detailed analysis both of factors such as the workplace layout and of tasks involving manual handling or the local operation of machines with moving parts.

Risk-reduction potential
It may well be the case that the aim of the PSA is merely to quantify the level of risk, without considering reducing it (should the need become apparent from the results of the assessment). This may be because of resources constraints or because of a desire simply to quantify risk levels in the absence of any criteria at all (as, for example, with a new risk-assessment application field in which there are no established criteria). However, most PSAs today will also have the potential scope for determining how to reduce risk levels if the PSA results suggest that this is warranted.

If error-reduction measures are desirable, a greater emphasis will be put on the need for detailed qualitative analyses (e.g. detailed task analyses or error analyses, or both) as part of the HRA; and the HRA will move away from a purely quantitative, PSA-driven style.

5.1.4 Types of human interaction

Whether inside a PSA or outside of one, the problem-definition phase must decide what overall types of human interaction will be dealt with in the HRA. The most usual type involves the human response to a system demand, usually arising out of some system failure. This type of human interaction has been the focus of many PSAs, since these events are often where the system most clearly relies on human reliability to reach a safe system state. However, there are four additional human-interaction types that the HRA analyst or the PSA team (or both) should consider, and these are outlined below.

5.1.4.1 Maintenance and testing errors

Another major decision during the problem-definition phase is concerned with human involvements in maintenance and testing activities. Many PSAs assume that the error potential in these areas can be integrated into the PSA by using failure rate data based on maintained and tested systems (i.e. that the errors and their probabilities are implicitly part of the failure-rate data). However, what the failure rate data may *not* highlight are error modes which interact with other subsystems, or which could even affect an entire set of subsystems. (A calibration error, for example, could be repeated on *all* gas detectors.) This aspect of problem definition can only be fully resolved by a full, detailed task-and-error analysis of maintenance/testing activities, but this is resource-intensive and so may not be warranted by many PSAs. As with

some of the former aspects of problem definition, specific guidance cannot be given, and the personnel responsible for carrying out the PSA must make reasoned judgements of their own as to whether certain maintenance/testing activities require a detailed task/error analysis to ensure that all critical errors have been identified appropriately. One pragmatic approach is to evaluate a maintenance task on a particularly important safety subsystem and then determine whether this identifies any *additional* potential contributions to the level of risk that are not currently to be found in the subsystem failure-rate data and associated failure modes.

The degree to which maintenance and testing activities are evaluated will have an effect on the frequency of what are called *latent failures* in human-reliability analysis. Latent failures are failures which occur usually on maintainable systems, and which remain hidden until either the system is tested/maintained at a later date, or an incident occurs and the system is expected to operate. Obviously, the latter eventuality is of great concern to risk analysts, since the system could be an important emergency-back-up system. Latent failures should therefore be regarded as one of the primary human-error-related considerations in any risk assessment.

5.1.4.2 Human error initiating events

A further important consideration in risk assessment is the contribution made by human error to the initiation of accident sequences. Normally, PRAs consider accident sequences started by hardware failures or environmental events, but there may also be human errors which can actually initiate such sequences. Such errors may constitute rule violations, or be seen as gross errors of commission, but if they are credible they should be considered in the PRA, or else the overall accident-initiation frequency will be underestimated. It is therefore necessary to decide if latent errors and human error-related initiators are to be included in the scope of the HRA; or, if they are not, to justify their exclusion.

5.1.4.3 Human error exacerbators: errors of commission

Most, if not all, PRAs will also take into account human errors or tasks related to an incident-response, i.e. a case where the human operators are attempting safely to control an incident once it has begun. An important point here is whether or not the HRA will also consider whether the operator can actually make the situation worse through errors called *exacerbators, extraneous acts* or even *errors of commission*. The classic example is that of the Three Mile Island incident (1979) in which, due to a misdiagnosis, the plant involved was more damaged than it would have been if the operators had not intervened in the early stages of the accident. In effect, some human errors can place a system on a worse accident 'trajectory' than it was on prior to their intervention. In event-tree terms, this means that not only a failure to

diagnose or to act but also a *mis*diagnosis, or any act which similarly changes the course of the accident, should be modelled. This development would probably lead to an increase in the number of the final consequence categories, or even, a transition to another PSA event tree with new consequence categories. Such human errors can potentially dominate the human contribution to the level of risk, and therefore should be considered in any applied HRA. However, they are only identifiable if the HRA has involved a detailed task analysis or a detailed error-identification process (or both). The effectiveness of these two methods in the area of misdiagnosis are, however, somewhat limited (see Section 5.3). The other major type of extraneous act that can be considered is a *rule violation*, also considered in the section on Human Error Identification.

5.1.4.4 Final recovery actions and mitigating strategies

A fourth aspect of human interaction in system-risk assessments is that of final recovery actions. Often, even when there has been a bad accident, the human operator can prove remarkably resourceful in retrieving lost or failed subsystems, and in averting catastrophe – especially when in a relatively forgiving environment (unlike, for example, offshore accidents, which usually occur in a hostile and often unforgiving environment). A good example of this was possibly the Davis-Besse incident (*USNRC*, 1985), summarised in Appendix I. While HRAs typically do not take quantitative credit for such almost 'heroic' actions, because it is not conservative to do so, it can nonetheless be useful to determine how such 'rescue' actions could proceed, especially for those concerned with accident-management procedures and emergency-response training and services. Thus, even though the benefits of such potential actions may not be quantified in the HRA, they may be elucidated upon and fed forward, as part of the documentation of the HRA, for review by operations departments, etc.

The degree of operational experience can be a good indicator of the adequacy of the scope of the HRA in terms of the types of human interactions that should be addressed, since there may be incidents which have occurred which were not originally in the risk-analysis fault schedule (a list of either top events or types of accident), or which did appear but were assigned a very low probability (e.g. it is found that an event has occurred in the past 20 years which was expected to occur only once per 10,000 years). Similarly, the Critical Incident Technique (Section 5.2) can be used to check if any important near misses have occurred, which should then be entered into the fault schedule of the PSA/HRA.

5.1.5 Resources considerations

The resources available for a HRA, in terms of funds, expertise and (possibly) specialised software, will also constrain the HRA. The main constraint will be the funding available, since adequate funding can usually overcome

most other potential constraints. It is difficult to estimate the cost of an HRA, but as a guide, a HRA may be expected to cost in the region of 5–30 per cent of the total cost of the PRA, depending on the perceived importance of the human contribution to the overall level of risk in the system. The amount of available expertise in the area of HRA has generally increased, although the availability of HRA software is far more limited. This, however, is largely a function of supply and demand. One of the major constraints which interacts with that of resources is the project-life-cycle stage. The earlier the life-cycle stage, the more difficult will be the task- and human-error-identification phases, since much of the required detail concerning operator tasks and equipment will not be available. This is discussed further in the next subsection.

A primary consideration with respect to resources is the existence of any prior HRAs for the system to be assessed. Prior assessments can obviously help the HRA analyst, depending on how well-documented they are. However, it should be noted that, often, the existence of such information saves assessors less time than they anticipate, for a number of reasons, not the least of which involves the need to consider human errors with a fresh mind, even if someone has considered human-error impacts on the system in depth before.

5.1.6 System design life cycle

There are several versions of the system-design life cycle (e.g. see Kirwan & Ainsworth, 1992), but a simplified version of the stages is set out in Figure 5.1.1, which shows the key phases from the initial project concept, through its detailed design and then construction, to its operational lifetime and final decommissioning.

The system-life-cycle phase can also serve to constrain the HRA, especially if the HRA is to be carried out in the early design phase of the plant (e.g. prior to the detailed design phase). During this early phase, there will be little in the way of procedures available, or of detail on the interfaces to be utilised by the operators, or of operators who are familiar with the tasks (if it is a new system being developed), and there is also a general inability to observe operations or investigate operational/incident feedback. This will limit the ability of the analysts to assess human-reliability impacts on risk levels accurately and comprehensively: there will be more uncertainty over such an assessment than for one carried out on an existing installation with the help of many years of operational experience and many available experienced operators. In particular, the analysts will be limited in their ability to carry out detailed task and human-error analyses since there will be insufficient design detail to generate correspondingly detailed task analyses; and similarly, the PSA fault schedule – which is the list of potential accident scenarios to be addressed in the PSA – will be limited at this stage, dealing with overall failure types rather than the specific scenarios that will be developed from the detailed design phase onwards.

These limitations will have a potential effect on resources if a HRA is carried out: for example, the analysts must either make a large number of

Figure 5.1.1 System design life cycle

	Example timings for large plant
Concept Idea of process; market testing; broad concepts	yrs 1–2
Preliminary design Technical feasibility demonstrated to a higher degree	yrs 3–4
Detailed design Full details of process and instrumentation, etc.	yrs 5–7
Construction System constructed	yrs 5–8
Commissioning System tested up to full operational capabilities	yrs 8–10
Operation and maintenance System running	yrs 11–40
Decommissioning System stopped, rendered safe, dismantled and removed	–yr 42

assumptions or else spend a good deal of resources collecting and progressing formative design details which may later change anyway. However, these limitations must be traded against the potential benefits of unearthing significant design problems at a stage when it will 'still cost little' to change designs, rather than waiting until the HRA-data-collection process can proceed more easily: with the latter situation, the resultant identified design recommendations may prove prohibitively expensive because the substantive design and construction funds have already been spent.

Once the system design reaches the detailed design phase, there will be far more information on how the tasks are to be performed, and after commissioning there will be a highly detailed picture of how the plant will run in theory, soon to be supplementable with information and incident reports on how the plant actually runs in practice during its operational phase.

As an aside, as far as the author is aware, there have been few, if any, HRAs concerned with the process of commissioning a plant or system. On the other hand, there are likely to be significant risk-related human interactions implicated in the decommissioning of large-scale systems to justify that HRAs should be considered for decommissioning projects.

5.1.7 Summary on problem definition

In summary, there are a number of critical decisions facing the HRA or PSA analysts (or both) when defining the scope of the HRA. These have been explored, and what advice can be given has been outlined. In most cases the requirements of any PSA will be too specific and variable for a text such as this one to give any detailed guidance. In Section 5.9, however, once the reader has had the opportunity to view the methods of HRA in more detail, these and other more complex issues of problem definition will be revisited. The discussion in Section 5.9 culminates in a series of decision-making flow-charts, presented in Section 5.10 (Practitioner's framework), that are aimed at helping analysts to decide how to define the scope of the HRA more systematically and reliably so as to suit their goals and resources.

5.2 Task analysis

Introduction

Task analysis is a fundamental approach for the ergonomist. It is a method of describing and analysing how the operator interacts both with the system itself and with other personnel in that system. In particular, task analysis describes what an operator is required to do, in terms of actions or cognitive processes (or both), to achieve the system's goal. Task-analysis methods also usually detail the displays which prompt the operator to perform or cease an operation, and the controls with which such operations are achieved. The first primary aim of the task analysis is to create a detailed picture of human involvement using all the information necessary for an analysis of the degree of adequacy of that involvement. Thus, although all task-analysis techniques aim to describe a given task, they differ in the information that is encoded, depending on the aim of the evaluation. This section reviews task analyses generally, and then documents those task-analysis techniques which are particularly useful for the Human Reliability Assessment.

5.2.1 The task analysis process

Although, as will be shown shortly, there are many different kinds of task-analysis (TA) techniques, they all follow the same basic philosophy of evaluation, as shown in Figure 5.2.1. This process is usually fairly straightforward. For example, to determine whether or not the alarm-response performance of operators in an emergency will meet an adequate standard, the problem may be stated as whether or not the operator interface supports the successful handling of the scenario. For such an investigation, a *tabular* task-analysis approach might be selected, one which focuses on interface issues and on the degree of feedback to the operator (this technique will be described shortly). The tabular analysis could be used to define all the alarms occurring, how they are coded, etc., and (possibly) their meanings to the operators and chronological sequence of occurrence. The analyst could then determine their actual rate of occurrence and rate of required response, and compare this with ergonomics criteria for corresponding information-processing rates. The

Figure 5.2.1 The task analysis process

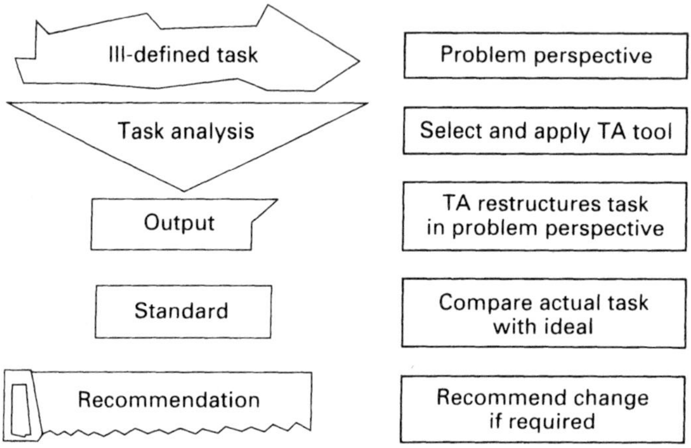

net result might be that the alarm information is coming in too fast for the operators to decode it, i.e. that an alarm 'flooding' situation is occurring. This would then require some kind of redesign (e.g. top-level alarm displays or better alarm-prioritisation systems). In this example, if it were uncertain as to whether the performance would be adequate, or if no suitable criteria could be found, then HRA quantification methods could be used to decide if the situation *was* adequate.

There are several important points to be made in connection with the simple process represented in Figure 5.2.1. Firstly, the task itself, whilst it can be observed, discussed, etc., will initially be ill-structured for investigation. A simple textual description will not deal with every conceivably important information-datum about a task that an analyst would wish to investigate. And even if it did, it would be cumbersome, difficult to use and probably unsystematic. The problem perspective in question is therefore a critical factor in the structuring of the task analysis, e.g. the time needed to perform certain actions is a critical factor when deciding on the adequacy of the operator's workload, but it is not so relevant (usually) when trying to determine how much chemistry training operators need to carry out a task. Task analysis without a problem to solve is therefore like data-collection without a purpose: it is merely an instructive exercise, and unlikely to be of much practical use to anyone. The problem perspective must therefore be fully defined before the task analysis is started.

A second major point is that in an ideal situation there would be a standard or ergonomics database in existence to judge whether the task is already adequate or whether it should be improved. In some *practical* situations as well, such decisions can usually be made, but in many others, however, the comparison will not be so clear-cut, and the ergonomist or human-reliability assessor will have to turn to judgement, experiment or, Human Reliability Quantification.

Figure 5.2.2 Task analysis application areas

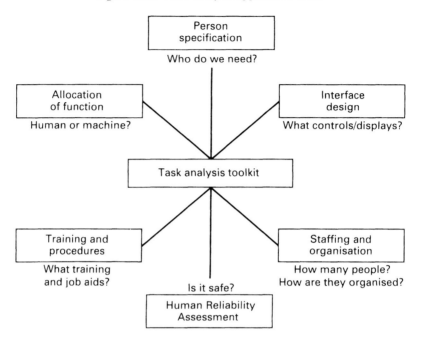

When carrying out a task analysis, the analyst must therefore keep the problem perspective in mind, and should attempt to generate first potential problems and then potential *solutions* to those problems. A task analysis is therefore not simply a passive recording of an activity but, rather, a generative process.

A third point is that a task analysis rarely occurs on its own, for its own sake, except as an academic exercise or for exploratory purposes. Instead, it is either usually part of an ergonomics or human-factors assessment (HFA), or a precursor to a Human Reliability Assessment (HRA). In order to put these two main uses of the task analysis into perspective, it is useful at this point further to define its major practical applications – i.e. the typical problems it is called upon to help solve.

5.2.2 Task analysis application areas

A task analysis is a tool which can be used to help answer a range of human-factors problems, as shown in Figure 5.2.2. 'Allocation of function' deals with a number of fundamental issues, such as whether the operation should be manual (i.e. human-operated), semi-automatic (combination of human and hardware/software-controlled actions) or fully automatic; and whether the operator is primarily engaged in operating the system or is in a largely monitoring and supervisory-control role.

'Interface design' concerns the location, arrangement, labelling, number, type and characteristics of controls and displays that comprise the operator interface – whether they are hardwired displays (as on a panel), VDU displays and controls, or other types of operating systems (for example, ones operating via automatic speech-recognition systems).

'Staffing and organisation' refers to the number of staff required and the operating structure of teams.

'Person specification' refers to the required capabilities of the operators (e.g. visual acuity for inspectors; the chemistry background of the laboratory analysts, etc.), and also possibly to the kinds of personality trait required (e.g. leadership ability for supervisors, etc.).

'Training and procedures' can obviously benefit significantly from a task analysis as this will involve formally described necessary operations, etc. There are many kinds of task analysis, therefore, that can be of direct relevance to the decision as to who must be trained, and what the required content of procedures will be.

The above five application areas are all within the scope of a human-factors assessment (HFA). For a review of 25 different task-analysis techniques that are used in the fields of HFA and HRA, see Kirwan & Ainsworth (1992).

The major principal use of task analysis of interest to us in this book is its use for HRA purposes. Any HRA investigation of a system and its component tasks will aim to determine what can go wrong during the execution of a task. Therefore, as a starting point, any HRA must first have a definition of how a task should be carried out, and this requires a task analysis. The TA-derived model of what should happen during a correct performance can then be allied to HRA techniques which will define what errors can occur at various steps in the task execution, and how likely such errors are to occur. The HRA will then determine quantitatively, against pre-defined risk criteria, whether or not the task is adequate.

The above HFA/HRA areas of investigation broadly encapsulate the major uses of a task analysis. The following subsection details the major task-analysis method used in HFAs and HRAs: that of the *hierarchical task analysis*. Subsequent subsections will then briefly review other techniques and their uses in applied HRA, as well as the timing of a task analysis in the context of the system-design life cycle.

5.2.3 Hierarchical task analysis (HTA)

The HTA is the most often-used TA technique as it is relatively straight-forward to apply but can also be very useful in addressing a large range of problems. It is also, probably, the most popular technique because it makes it easy to assimilate a large amount of information relatively quickly, whereas certain other techniques require more intensive scrutiny.

Acknowledgements

I would like to acknowledge the following people in particular (and in alphabetical order) whom I have had the pleasure of working with on various HRA and Human Factors projects, or with whom I have had discussions which have shaped, changed and inspired my thoughts on HRA: Peter Ackroyd, David Embrey, Andrew Hale, Nigel James, Mike Lihou, David Nockells, Reg Pope, Keith Rea, Julie Reed, Lynn Robinson, Helen Rycraft, Stef Scannali, Andrew Shepherd, Andy Smith, Jim Tait, Trevor Waters, David Whitfield, Barry Whittingham, Sue Whalley, Jerry Williams, Iain Wilson, and John Wilson. I would also like to thank all the operators I have worked with over the past twelve years who have been kind enough to give me an insight into their skills and knowledge.

I would like to thank all the other people I have worked with in HRA, Technica, BNFL, and Birmingham University, and all the secretarial staff over the years that have managed to decipher what could be loosely called handwriting. I should also like to thank the publishing staff at Taylor and Francis, Wendy and Richard, and the copy-editor who did an impressive job during the final drafting stages of the book, and also Christine Stapleton at EIAC who has always supplied me with information at ridiculously short notice. And my students for their patience.

Lastly I would like to thank all my friends and my family for their support over the four years it has taken to write this book, and my diving buddies at Birmingham University Diving Club, as diving has kept me sane.

Figure 5.2.3 Hierarchical task analysis – CL2 example

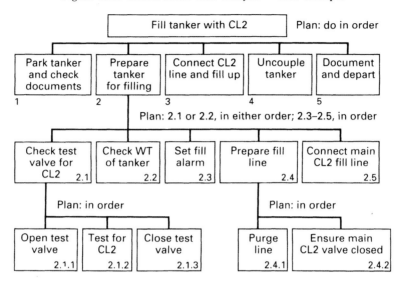

The HTA is, as its name suggests, a hierarchical approach. It describes the task from its top-level goals down to the level of individual operations. An example, illustrating all the main components of an HTA, is shown in Figure 5.2.3. The first component here is the top-level goal, which in this case is 'Fill road tanker with chlorine'. In this example, this is a goal for the operator. The goal is decomposed into five tasks (1–5 on Figure 5.2.3). These five tasks must completely define how to achieve the top goal, i.e. there must not be an additional necessary task missing at this level of the HTA; nor any *superfluous* task (the presence of which would lead to 'fuzziness' in the HTA description, which in turn would lead to problems later on – e.g. when using the HTA to define the content of training, or possible errors).

The difference between tasks and goals is that whereas a goal is a system objective which can be achieved by a varying range of tasks, a task is composed of a set pattern of operations. Sometimes, the distinction between a goal and a task is difficult to determine. The important point is that the analyst firstly uses all relevant concepts to develop properly the hierarchical, goal-driven structure of the task, and secondly uses the goal concept to explore whether there are other ways of carrying out a task.

The *operation* category refers to what is actually done in a situation, and is usually a description of the behaviour, or cognitive (i.e. mental) activity, carried out to achieve the task's objective. In the present example (Figure 5.2.3), the task located at the second level down in the HTA box numbered '2' and called 'Prepare tanker for filling' is achieved via five sub-tasks, which are in the five boxes found at the next level of the HTA. The task 'Check test valve for chlorine' has *three* operations below it. These do not really require further re-description.

It is frequently the case in applied TAs that there are at least two ways of carrying out a given task, and often the way the operators usually do it is different from how it was originally specified to be done in the procedures issued by the system designers (note that the operators may sometimes find a better method of operation which the designers themselves would have found if they had carried out a task analysis). Once again, the TA is seen here as a generative process, exploring potential work methods, rather than one which simply records what appears to be the required working method.

The HTA has three other important structural aspects. The first is the 'plan'. The plan defines the way in which the subordinate tasks are carried out. For example, at the second level in this present diagram, the five tasks 'Park tanker and check documents', 'Prepare tanker for filling', etc., are carried out according to the plan *above* them. In this case, the plan is 'Do in order' (i.e. from '1–2–3–4–5'), which means they *must* be carried out in that order. The second plan on the diagram is more flexible, allowing 2.1 and 2.2 to be done in any order to achieve the higher-level task objective of 'preparing the tanker for filling', though the sub-tasks 2.3–2.5 must also be carried out in a fixed sequence. Other plans may be more complex – for example, *algorithmic* in nature. Whatever its form, the plan dictates the constraints which must be applied before moving from one operation to the next. The construction of a plan is therefore critical to the effective use of HTA information, and it must define *when* to move from one operation or task to the next. There are various types of plan (discretionary, cyclic, cued-action, fixed sequence, etc. – see Kirwan & Ainsworth, 1992), and the analyst can use any combination of them to show how operators complete the task and operations as described in the HTA. The analyst should also consider whether the plan will remain the same under all conditions, e.g. normal and emergency conditions.

The second remaining important aspect of the HTA is known as the 'stopping rule', or the decision as to when to stop redescribing the task in terms of sub-tasks and operations. The main stopping rule is to stop redescribing when further redescription would add no further useful information to the analysis process. As an example, in the HTA in Figure 5.2.3, which is being used for HRA purposes, there is no further need to describe the task 'Open test valve', since this is a highly familiar type of operation. The operation 'Test for chlorine' could, however, be further redescribed, depending on the purpose of the study.

The actual test involves placing a piece of material soaked in a sodium-hydroxide-based solution close to the test valve and seeing if a chemical reaction occurs. If this were a commonly known operation, and the purpose of the HTA was for training needs analysis, then there would be little need for further redescription. In HRA terms it would, however, be worth further redescription since there are several errors that can occur with this simple test, as will be shown later in Appendix VI. The analyst must therefore use his or her own judgement to decide on the level of redescription required for a particular analysis; and in the HRA context, this will depend on the scope

Figure 5.2.4 Hierarchical task analysis – Shepherd (1986)

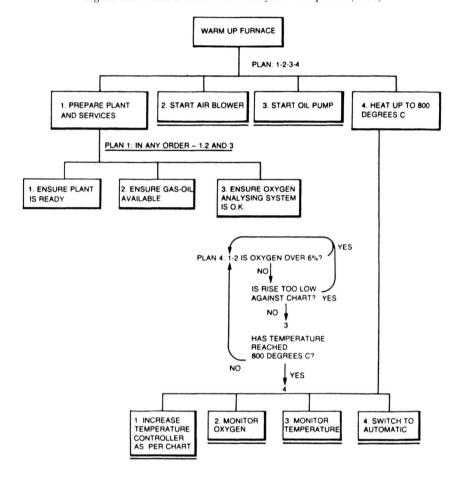

of the analysis as defined in the problem definition, and on the risk of missing potential errors in a task by failing to redescribe things down to a particular level.

Wherever the analyst does stop, he or she would then simply stop redescribing at those points. This is represented in the HTA by drawing a line under the description boxes for those tasks, as has been done in Figure 5.2.4 (adopted from Shepherd 1986). It is usual in a HTA to 'stop' at the level of *operations* rather than at the task level. Often, the decision of where to stop is an iterative process, with the analyst initially redescribing things to a very detailed degree and then 'trimming' the HTA tree, i.e. cutting off redundant detail. This approach ensures that no aspects of the problem are overlooked as a result of the analyst's assuming that redescriptions are not necessary.

The final aspect worth noting, on the HTA figure, is the *numbering*. The boxes must be numbered, and may, moreover, be numbered hierarchically (e.g. 'Increase temperature controller', in Figure 5.2.4, would be numbered '1.4.1' if 'Warm-up furnace' were numbered 'goal 1'). This may appear tedious, but it is useful when carrying out anything but a small HTA, otherwise later references to parts of the HTA will prove difficult to achieve unequivocally. HTAs usually descend to 4–5 levels, but can sometimes descend to as many as 7 levels. HTAs vary in their size, but a small HTA would probably occupy only a page, whereas a large HTA would be anything over 10 pages long. If larger than a single page, transfer arrows or triangles are used to transfer information to another page, as shown in Appendix VI for fault trees.

Below are some guidance questions, rules and instructions concerning HTA-generation:

1. What is the problem aspect of the task being analysed?
2. State the overall goal, with all its constraints. The vaguer the overall goal, the larger the corresponding HTA. For example, 'Run plant' will involve a far more extensive analysis than 'Run plant operations during normal operational campaign to produce 500 copies of product'. Set the scope of the analysis: does it need to include maintenance, abnormal tasks, system disturbances?
3. Define sub-tasks for the overall goal, and state the plan. Continue in this mode until the level of operations is reached and further redescription is not pertinent to the aims of the study.
4. For each box, consider the subordinate boxes underneath it: do these boxes (whether signifying goals/tasks/operations) completely define the superordinate one? Are there any subordinates which are superfluous, or which really belong to *another* superordinate task or goal?
5. If certain decisions need to be made again, as part of operations or tasks, it is usually worth redescribing them: if redescription is stopped and a decision is left 'implicit' within an operation, this may obscure a necessary piece of information. If an operation appears obvious to the user (as do the feedback cues that initiate it and that confirm that it has been carried out), then redescription may be stopped at this point.
6. Another 'stopping rule' is the *PxC rule*, usually appropriate for training analysis. This states that if the probability of an operation's being performed inadequately, multiplied by the cost or consequences of such an operation, is no longer of importance, then redescription can again be stopped. This is a useful criterion in theory but can be difficult to apply in practice, since neither P nor C are easy to calculate.
7. Develop plans at each level, and consider what the operator must do in those varying circumstances (normal or abnormal, etc.) which fall within the scope of the analysis. Also, consider what the operators must do if the task does not proceed according to plan; as well as what is prompting the operator to proceed to the next task/operation, or telling him that

the operation is complete. Plans must state when each of the sub-tasks is to be implemented. At such points it is also useful to note what potential 'error messages' could be given when things go wrong. All too often in system designs, when something goes wrong, the operator is supposed to somehow realise (a) that things are not going to plan, and (b) what the problem is. In most cases like this the operator will merely assume that the system *has* worked but that the feedback cue has failed, and he or she will then continue with the task.

8. There are a number of different types of plans:

   - Fixed sequences.
   - Actions to be performed on given cues.
   - Fixed action sequences 'contingent' upon cues.
   - Choices (all choice endpoints must be addressed in a comprehensive HTA).
   - Cycles – repeating a cycle of operation until a condition is reached.
   - Time-sharing two or more operations (e.g. monitoring an alarm panel while carrying out administrative tasks) – this must be explicitly stated in the plan.
   - Discretionary plans – the plan is flexible and can be carried out in accordance with the operator's preferences.
   - Mixed plans – a mixture of any of the above.

9. Reiteration throughout the TA is important, and a HTA may, for example, be developed three times before a final version is settled upon. Use other personnel to check the logic, for example, use the operators to review it: they may know of some implicit rules and plans which are difficult to detect and which therefore may have been left out of the analysis.

10. Consult operators, systems analysts, designers, assessors, etc., when developing the HTA. Do not rely on one person's 'story', and be aware of the fact that if direct observation is being used to investigate the task, then the task may not be carried out 'naturally'. This will be apparent because the operator's actions will not flow but will be 'stilted': they will be trying to remember how to execute the task as it is supposed to be done rather than how they normally do it. Shepherd (*Personal Communication*, 1990) noted some useful hints on dealing with experts when carrying out an HTA, hints which have been expanded upon below for the novice HTA analyst:

    - Inspect the work environment, but arrange for meetings away from the workplace (to avoid distractions, interruptions, etc.) and away from the experts' supervisors.
    - Explain to the experts fully the reasons why you are there, and the nature of the task analysis. Do not mislead them or attempt to obscure the real reasons for the analysis.

- One day (broken into two or more segments) is a realistic time to spend with a subject. Spend lunch with him or her if possible, as in an informal atmosphere the expert may explore doubts with you as to the 'real' nature of the study, and may also give important information which was not forthcoming in the office environment. A typical HTA day might be as follows:

8:30	Meet contact at company location
9:00	Visit work environment
10:00	Meet expert, explain role and objectives, and aim to define top two levels of HTA
12:00	Lunch
13:00	Continue work, aiming to get to the bottom of at least one area of HTA – i.e. get to bottom level of your HTA for *at least one major task*, with plans
15:00	Break
15:20	Continue broadening of HTA
16:30	Finish time with expert, and write up own comments on day's progress
19:00	Relax!

- Whenever going into the workplace, always introduce yourself to the operators (your client contact should, but may not, do this also – ask him or her to) and explain your reasons for being there.
- After initial meetings, other people will need to be brought in, as the development of a HTA is an iterative process. Warn experts of this straight away. A single day with an expert who, later on, has no time to answer queries or clarify points may end up being a wasted day.
- Let the expert see what you are doing, and preferably let them do it with you. Operators usually pick up HTAs very quickly.
- When the stating of operations becomes difficult, remember that their purpose is to attain the system's goals. Ask yourself: 'what is it that the person carrying out the task is trying to achieve?'.
- When finalising the analysis, talk through the HTA with another potential expert uninvolved in the analysis, as a verification exercise. Do not simply give it to him or her, since, in such situations, HTAs will usually just be skimmed over.

5.2.3.1 Uses of HTA

A HTA can be used in many types of human-factors assessments. Consider the following:

1. Function allocation
A HTA can describe the goals of the system without defining what equipment interfaces, etc., should be used. A degree of flexibility beneath the goal

heading will enable several different types of functional allocation to be considered. Higher-order goals can be used to prompt a consideration of alternative human and technological solutions.

2. Display design

The HTA describes what the operator needs to be able to control, and what information will prompt the operator to move to the next sub-task. Thus, content-display requirements are implicit in the HTA and can be brought out explicitly via a tabular-format task analysis (described later). One of the primary advantages of the HTA is that it aids the analyst in determining how information should be grouped together, i.e. in determining whether certain indications are all necessary for a particular sub-task.

3. Job design

The HTA can be used when considering a job design in several ways. Firstly, the content of the job is sometimes evident from the HTA tasks and operations, e.g. qualifications requirements may be ascertained from, for example, certain operations that might require a chemical background. The HTA will also be able to help delineate team relationships and responsibilities, as well as communications requirements. Job enlargement or enrichment can also be achieved, by using the HTA to expand a person's tasks and roles, organise these tasks and roles into a more coherent job or give the operator more of the responsibility for achieving the goal itself rather than just the subordinate tasks.

4. Training and procedures

A HTA can be a firm basis for defining the content of training and procedures, as well as for the documentation of the plans necessary to achieve the tasks. The HTA can also be used as a training programme for building up skills in a successive fashion. For example, personnel could be trained in bottom-level skills, or trained using a 'top-down' method, i.e. train the operators to aim for the top-level goal and the subordinate operations, etc., and then successively increase the depth of their skills/knowledge. A training matrix can also be developed based on the results of an HTA. This would have, along one axis, the types of equipment – e.g. valves, pumps – and, on the other axis, the kinds of operation involved. In each cell of such a matrix, the required training could be indicated. This would be useful, for example, for determining training requirements for a maintenance fitter. The types of equipment, types of operation, and the coincidence of the two would be derived directly from the HTA, which might show data such as: 'operator tests (operation), oxygen analyser (equipment item)'. In this way the HTA can be used to develop a detailed picture of the knowledge, skills, and abilities required to be a competent operator.

5. Error identification and quantification (HRA)

As noted earlier, the HRA, in order to identify errors and (subsequently) quantify them, requires a task analysis as a basis. The HTA can form the

Figure 5.2.5 Simple, linear format HTA

Task Analysis

Assessor: BIK
Task: CHLORINE LOADING
Safety Case: WORKSHOP
File name: WORKSHOP

3   INITIATE AND MONITOR TANKER-FILLING OPERATION
3.1   Initiate filling operation
3.1.1   Open supply-line valves
3.1.2   Ensure tanker is filling with chlorine
3.2   Monitor tanker-filling operation
3.2.1   Remain within earshot while tanker is filling
3.2.2   Check road tanker
3.2.3   Attend tanker during last 2–3 tonne filling
3.2.4   Cancel initial weight alarm and remain at controls
3.2.5   Cancel final weight alarm
3.2.6   Close supply-valve A when target weight reached

4   TERMINATE FILLING AND RELEASE TANKER
4.1   Stop filling operation
4.1.1   Close supply-valve B
4.1.2   Clear lines
4.1.3   Close tanker valve
4.2   Disconnect tanker
4.2.1   Vent and purge lines
4.2.2   Remove instrument air from valves
4.2.3   Secure blocking device on valves
4.2.4   Break tanker connections
4.3   Store hoses (not refined)
4.4.   Secure tanker
4.4.1   Check valves for leakage
4.4.2   Secure locking nuts
4.4.3   Close and secure

basic starting point for a HRA. (This will be dealt with in Section 5.3 in more detail.) However, from earlier discussions, and as will be seen with other TA approaches, many TA techniques will actually identify the error potential either directly or by the identification of mismatches between the task requirements, the human capabilities and the display/control attributes.

A simple form of HTA (in a linear format) for an elementary HRA is shown in Figure 5.2.5. This is an example of the kind of format used in the JHEDI approach that will be discussed later in the section on quantification. This is still essentially a hierarchical task description (though in the diagram there are no plans, which could be usefully added), but it is here to be found in a more readily producible format.

5.2.4 Other types of task analysis

Whilst the HTA is the most popular and arguably the most important TA technique, there are a broad range of other techniques which make useful additions to the assessor's toolkit. These techniques fit into five basic categories (see Kirwan & Ainsworth, 1992):

- **Task-data-collection approaches**: techniques which are primarily used for collecting data concerning human-system interactions, and which then feed into other techniques; for example, observational techniques may be used as a basis for collecting data for a workload assessment.
- **Task-description approaches**: these are techniques which structure the information collected into a systematic format. For example, the data collected from observation studies may be recorded in the form of flowcharts or operational-sequence diagrams. Such formats may then either serve as reference material, to enhance the understanding of the human-system involvement, or be used in a more direct way – for example, the HTA technique may be used to identify the needs and the content of training schedules.
- **Task-synthesis methods**: these approaches are aimed at 'compiling' data on human involvements to create a more dynamic model of what actually happens during the execution of a task. Such techniques (e.g. the use of a computer simulation), based on data from individual tasks, model tasks generically and can estimate such factors as the time taken to complete the task.
- **Task-behaviour-assessment techniques**: these techniques, many of which come from the engineering-risk-assessment domain, are largely concerned with system-performance evaluation, usually from a safety perspective. Primarily, they identify what would lead to a system failure, but they often also deal with hardware, software and environmental events, as well as human errors. Such techniques represent errors and other failure events together in a logical framework, thus enabling a risk assessment, and/or the identification of remedial, risk-reducing measures, to take place.
- **Task-requirements-evaluation techniques**: these techniques are utilised to assess the adequacy of the facilities available to the operator(s) for supporting the execution of the task, and also to assess both the interface (displays, controls, tools) and the documentation of procedures, instructions, etc. These techniques are usually shown in the form of checklists, or presented as special, interface-survey techniques.

A number of such task-analysis tools which are particularly useful in the HRA context are briefly summarised below. These tend to fall largely into the first three categories above. Task-behaviour-assessment techniques are dealt with in Sections 5.3 and 5.4 (e.g. HAZOP, and fault and event trees), and a task-requirements-evaluation method is available in the form of a checklist in Appendix IV.

5.2.4.1 Observation

If carrying out an HRA, it is preferable to observe the tasks in the field situation, where this is possible, or at least in a simulated environment. There will, however, be cases when this is not possible (e.g. if the tasks are designed but the equipment has not yet been built, or if no simulation exists).

Observation is thus a data-collection technique, and should be used when the task aspect of interest is relatively explicit and hence observable, e.g. a manual task or interaction with a conventional control/display (hardwired) interface. Unobtrusive observation is ideal since it allows natural behaviour to occur, but it is also very difficult to achieve.

Observation can occur with varying levels of obtrusiveness, ranging from the indirect (via closed-circuit television or video-recording), through the direct (observer present), to actual participant observation (the observer actually joins the team and does the task). The latter approach will usually the greatest degree of insight into the task, but at the same time the analyst (participant) may lose the element of objectivity in his or her analysis. At the other end of the spectrum, however, an observer sitting there with a note-book can have fairly negative effects on performance. If CCTV is used, use a system with good audio-recording facilities. Observation is best used as an adjunct to other techniques, such as interviews and documentation review (see also Drury, 1990; and Kirwan & Ainsworth, 1992).

5.2.4.2 Interviews and documentation review

Interviews and documentation review are both data-collection tools, and are probably the primary sources of HRA-task-analysis information. Documentation that could be useful in the HRA will include the following:

- Operating instructions/procedures.
- Pictures of the interface and the working environment.
- General layout diagrams.
- Engineering flow diagrams.
- Piping and instrumentation diagrams.
- Mechanical flow diagrams.
- System-specification documentation.
- System-fault schedules.
- Interlock schedules.
- Incident reports.
- Prior-risk analyses/HRA/TA documentation.

The above can be reviewed by the analyst, who will invariably develop a long list of questions on aspects of system operation, and on what the opera-tors are supposed to do and know in various situations.

The interview will then become the main information-gathering tool. The personnel interviewed will usually be experienced operational personnel, designers and managers. Often, the analyst is not in a position to choose the interviewee, since the choice will be restricted by availability. It is therefore imperative to convince the company concerned that the better the candidate made available (in terms of knowledge, experience and ability to verbalise such knowledge and experience), the quicker and more accurate the analysis will be. The best operating personnel will usually be the more experienced ones, preferably those, with 20 or more years of experience, who have seen infrequent emergency incidents and who therefore know how people have responded in at least one such emergency. Whoever is chosen, however, the reliance must not be placed on a single individual, and all information must be corroborated by other personnel. For a more general discussion on conducting interviews, see Kirwan & Ainsworth (1992) and Sinclair (1990).

5.2.4.3 Link analysis

A link analysis is a task-description technique by which the analyst records and represents (diagrammatically) the nature, frequency and importance of links in a system. Its most obvious use is in the area of workspace-layout optimisation, which is a function of the four main guiding principles of ergonomics: locating controls and displays according first to their importance, then to their frequency of use, then to their function within the system, and finally to their *sequence* of use (see Grandjean, 1988). Link analysis deals primarily with the first three of these guiding principles. An example of a link-analysis diagram format is shown in Figure 5.2.6. Its use as part of the HRA occurs principally when there are concerns that the layout of information displays and controls in, for example, a control room, might actually impede rather than support operations, and lead to a *greater* number of errors (see Kirwan & Ainsworth, 1992).

5.2.4.4 Verbal protocols

Verbal protocols are an oral commentary on what is being done by the operator while he or she carries out the task, and the verbal-protocols technique is principally a data-collection tool. Such verbalisations are particularly useful in the observation of skilled activities in which the operator is also using knowledge, and carrying out mental operations, which would otherwise be unobservable. This technique can be used to examine, for example, problem solving in a process-control task. Retrospective protocols, i.e. those made after the task is finished, are not recommended.

Protocols will not be useful if the operator cannot verbalise how he or she is carrying out the task, which may be the case where the expert is carrying out an intuitive problem-solving task, or using either visual imagery or abstract mathematical concepts. Although it is not easy to verify protocols as valid verbalisations of how the operator performs, this method *is* a useful supplement to observation, particularly where unobservable mental events are taking place (for example, where the reliability of a diagnosis made during abnormal or emergency events is being investigated). For a discussion, see Ainsworth & Whitfield (1984) and Kirwan & Ainsworth (1992).

Figure 5.2.6 Link analysis diagram representing functional systems in a control room; and table showing links made between the systems (see Ainsworth, 1985, and Kirwan and Ainsworth, 1992).

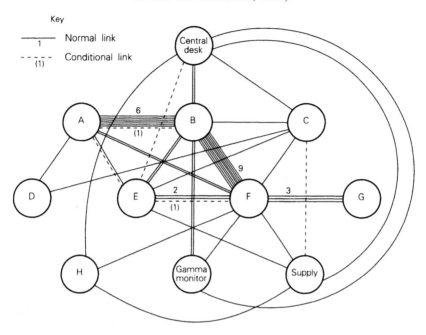

	B	C	D	E	F	G	H	Gamma Monitors	Desk	Elec. supply
A	6 (1)	-	1	1 (1)	2	-	-	-	-	-
B		1	-	2	9	-	-	2	2	-
C			1	1	1	-	-	-	1	(1)
D				-	-	-	-	-	-	-
E					2(1)	-	-	-	(1)	1
F						3	1	1	-	1
G							-	-	-	-
H								-	1	1
Gamma radiation monitors									1	-
Central desk										1

Figure 5.2.7 Decision–action diagram (Ainsworth and Whitfield, 1984)

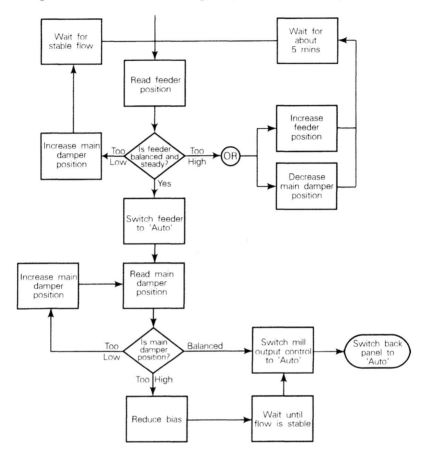

5.2.4.5 Decision action diagrams (DADs)

Decision–action diagrams are a means of describing decision-making tasks involving choices of action, and as such amount to task-description techniques. These are particularly useful for representing decisions which would otherwise involve cumbersome planning in a HTA format. They are essentially algorithmic in nature, with boxes representing states or events and diamonds representing choice points with two or more potential outcomes. An example of a decision–action diagram is shown in Figure 5.2.7 (see also Kirwan & Ainsworth, 1992).

DADs can be useful when reviewing decision-making scenarios since they define the various decisions that have to be made as well as the information and criteria required to make such decisions. In particular, when the reliability of diagnosis is being assessed, DADs can be used to consider first the points in the scenario at which the diagnosis could occur, and subsequently the points at which the operators would review their diagnosis and hence (potentially) make good the error.

5.2.4.6 Critical Incident Technique (CIT)

The CIT is a very simple data-collection approach which aims to identify, via interviews of experienced personnel, incidents or near-misses which have occurred. These will often give insights into how the task can really be done in practice, or how errors may occur. The basic approach is very simple, and involves asking a question along the following lines: 'can you think of any incident or near-miss or event which happened to you, or a colleague, which could have resulted in an accident, given other circumstances?' This question is simply being used to prompt the memory of the recipient to see if any events have occurred which could be worth including in the HRA but which may have never been formally reported because at the time they did not actually lead to a reportable accident/incident. Such information will complement that obtained from reviewing incident analyses and reports.

It will also be worth interviewing personnel involved in actual incidents, if these have occurred and are to be included in the risk analysis, since incident reports rarely capture the whole import of the real incident – which may well be embedded firmly in the operators' memories.

An analyst using the CIT approach must beware of reporting or memory biases, as well as 'hearsay' information. Ideally, the analyst will be able to corroborate CIT information with *other* sources of information. In any case, CIT information is only intended to serve as exploratory data. It may support qualitative HRA insights (e.g. concerning error identification), but it will not add any quantitative insights because of its biased nature. When investigating rule violations, however, the CIT approach may be one of the only ways of determining which rule violations are credible, since often any that have occurred (particularly ones that occur routinely) are unlikely to be reported by other means. In such situations, and in CIT generally, the respondent's identity must remain anonymous.

5.2.4.7 Tabular task analysis (TTA)

There are various forms of TTA, which amount to task-description techniques. The TTA usually follows on from a HTA, and is columnar in format. It takes each particular task-step or operation and considers specific aspects, such as: Who is doing the operation? What displays are being used? What feedback is given? What errors could occur? The column titles will vary depending on the purpose of the task analysis. The HTA-TTA combination is a very powerful one for a detailed evaluation of interfaces, since the HTA gives the analyst a firm basis for understanding the system, while the TTA, on the other hand, can be used first systematically to investigate the ergonomics aspects of the system and then to justify problems identified on the grounds of likely consequences or errors, etc. Figure 5.2.8 shows a particular TTA format example that is useful for evaluating a given interface (discussed in Kirwan & Reed, 1989). The human-error-analysis format, also shown in Figure 5.2.8, follows on from the TTA, and is used to record the derivation and justification of ergonomics improvements (see Reed, 1992).

This latter TTA/HEA tabular approach can be utilised in highly detailed HRA task analyses to identify errors arising out of problems with the characteristics of the interface. There are several features, of particular interest to the HRA practitioner when he or she is carrying out TTA or task analysis generally, which constitute important data for improving error identification in the next stage of the HRA process. These are:

- The nature of the cue telling the operator to proceed to the next task or task step.
- Feedback to the operator telling him or her that the action was correct: factors here include the speed of the feedback; the directness of the feedback (see Appendix I); the confusibility of the feedback with other signals; the diversity of the feedback signals.
- The availability of feedback on errors.
- The error-tolerance-level of the system.
- The ease of error-recovery.

The above may be posed as questions which mark the beginnings of the error-identification process:

- How does the operator know when to act (could the operator start too early or too late?)?
- How does the operator know the act was carried out correctly? Is the feedback quick enough to enable error detection? Is the feedback indirect, and thus potentially highly misleading if incorrect? Could the operator confuse signals? Are there diverse signals, or is the operator reliant on only one information source or channel, and if so, what is the reliability of that information source or channel?
- What tells the operator he or she has made an error? (Is error detection likely?)
- If an operator makes an error, is there an immediate or inevitable undesirable consequence? What are the immediate and delayed consequences of each error, taken alone and in conjunction with other human and non-human failures?
- How easily is error-recovery achieved? Who recovers, and what additional errors could happen as a result of these recovery actions?

These are some of the questions for the assessor to keep in mind when carrying out a TTA/HEA task analysis.

A third form of TTA worth describing for HRA purposes is the *tabular scenario analysis*. This is again a columnar-format approach, but it focuses

Figure 5.2.8 TTA/HEA formats

TASK STEP NO.	TASK GOAL	INFORMATION AVAILABLE TO THE OPERATOR	REQUIRED ACTION	FEEDBACK	COMMUNICATIONS	POSSIBLE ERRORS, DISTRACTIONS, TIME AVAILABLE, SKILLS/KNOWLEDGE REQUIRED

Format of the tabular task analysis

ERROR NO.	ERGONOMICS INADEQUACY	POSSIBLE ERRORS	POSSIBLE CONSEQUENCES	RECOVERY POINTS	REMEDIAL ACTIONS		
					DESIGN	TRAINING	PROCEDURE

Format of the human error analysis

instead on scenarios which are 'event-driven', and is usually only used to describe emergency scenarios. An example of a TSA is shown in Figure 5.2.9.

There are several noteworthy points about TSA. Firstly, the system status at any time is recorded as it is in reality, and can be compared with the estimated understanding of the operators at that time (as a function both of the available indications and alarms, etc., and of the operators' likely expectations). All decisions/communications/acts are recorded, and all indications noted (sometimes this is not practicable, however, because of the high density of alarms, etc.). Any secondary actions or distractions are also noted. What the TSA attempts to do is to build up a detailed, time-based picture of the scenario that shows its full complexity, as well as the flow of information to and from the operators. This enables, in particular, an insight into potential diagnosis-related errors, or at least gives a good basis for investigating them. In essence, a TSA is a paper simulation of the scenario itself, and thus a task-synthesis approach as well as a task-representation technique. It is also a particularly useful aid when investigating diagnosis made during emergencies.

Overall, TTA approaches enable the highly detailed observation of what is guiding the operator through a series of predefined events, or a series of PSA-defined events and failures. They offer both a sound task-analysis basis for a further consideration of performances and errors, and a beneficial medium for the discussion of human-error problems with other parties – whether PSA assessors, operators or designers (this is discussed further in Kirwan & Reed, 1989). Getting the information is, of course, no easy matter: TTA approaches can use up considerable resources, and usually take longer than, for example, HTAs. Once completed, however, they are often extensively utilised throughout the HRA process, and can become useful documents, moreover, for the quality-assurance phase of the project, particularly if design alterations are to be implemented.

5.2.4.8 Walk-through/talk-through (WT/TT)

A walk-through/talk-through approach utilises experts who commentate their way through hypothetical scenarios, either on-line in their workplace (i.e. *walk-through*: pointing to controls and displays which would be used), or off-line in an interview room (i.e. *talk-through*). Either the expert will run through a typical procedure or the interviewer(s) will suggest a scenario and ask the expert to commentate on how they would deal with it. The WT/TT approach can be particularly useful for examining actions in problem-solving situations, and for defining problem-solving strategies. The usual approach would be first to conduct a talk-through of a situation and then (probably) carry out a TTA/TSA to note the information displays and controls being utilised. Then a walk-through itself could be carried out, either in the actual workplace or else in a mock-up of the workplace.

There are three basic purposes to walk-through. The first is the investigation of a task which is not fully understood by the analyst – because, for

Figure 5.2.9 Example of tabular scenario analysis

T	Opr	System Status	Info Available	Operator Expectation	Procedure (written, memorised)	Decision/ Communications Act	Equipment/ location	Feedback	Secondary duties; Distractions; Penalties	Comments
0:30 mins	SS OO CRO	ESD; Cell 23 full of gas; mixture beyond explosion point at present; platform at muster status.	Gas cloud; Loud roaring noise;	Looking for leak in pipe union, flange, seal etc. Check near the gas detectors which were alarming.	1) Locate source and isolate if possible. 2) Maintain personal safety (use breathing apparatus) 3) Prevent ignition of gas	Ops search for leak in compressor module using sound and visual cues, as well as the gas detectors. Communicate to CCR.	Cell 23 gas detectors	Noise and smell of gas tactile cue if flesh exposed to gas jet path; if cold enough may see white gas plume from leak.	Maintain personal safety and avoid causing ignition. Extra delays if depressurise	Search will be more difficult in breathing apparatus. Deluge would make search safer, though less likely to succeed.
0:40	OIM CCRO SS	As at 0:30	Panel indications and from operators now outside of cell 23.	Gas leak now confirmed. Ignition possible.	Minimise chances of ignition.	CCR operator reviews vessel pressures on VDU system: gives OIM status report and confirms significant gas leak occurrence. CCR opts to consider whether to depressurise or not.			Many enquiries may block communication channels, and must be responded to by CCR operators.	Consider merits of removing compressor liquid – How long would the vessels withstand fire if ignited? Results of hazard analyses should be immediately transmitted to offshore personnel.

example, it involves the simultaneous use of several instruments (for instance, running up a back-up motor to take up drive from a larger motor which is tripping after power loss). Such tasks will be far more easily understood if the analyst can actually observe a walk-through. The second purpose is to determine how long tasks are likely to take. In such cases, the operator walks through the task in real time as far as is possible, using his or her knowledge of system-response times, etc., as a guide. The validity of these times for the real situation is something that must be judged by the analyst, and is difficult to give guidance on. However, it may be prudent to add at least 10 per cent to all timings carried out during a walk-through when estimating timings for response during a real incident. This will allow in part for factors such as unpreparedness (but *not* for reluctance, as will be seen in later sections). The third purpose of a walk-through is for the evaluation of the interface. In such an evaluation, the operator will walk through the actions fairly slowly, and the analyst will be asking questions, making sure that the operator fully understands both the significance of each indication and where the allowable operational ranges lie. In addition, the operator will be commenting on any aspects of design which are perceived to be potentially misleading, or even incorrect. He or she will also preferably be using actual (or intended) procedures, so that the veracity of the procedures can be simultaneously checked. Clearly, such an exercise is more in the realm of a *human-factors assessment* (HFA) than a HRA, but it is included here nonetheless for the sake of completeness, and also because information from such a walk-through could be highly useful to future HRA analysts, who will be able to extract it from earlier or parallel studies.

The results of a walk-through are usually documented in the TTA/TSA and/or feed into a timeline analysis.

5.2.4.9 Timeline analysis

There are two basic types of timeline analysis of interest to the HRA practitioner, namely the vertical timeline and the horizontal timeline. Both are primarily task-description tools. The horizontal timeline (Figure 5.2.10) is aimed at determining the overall time taken successfully to achieve the tasks in the scenario being assessed. Each task is entered in chronological order on the y-axis of the timeline, and a bar, representing its expected *duration*, is then placed on the graph at the appropriate starting point relative to the x-axis (indicating the passage of time). The sequence of the tasks must be determined by the HTA and the walk-through, etc. The data on system timings and accidental-event and hardware-response/failure timings must be assessed by system/PSA analysts – or via, for example, a stochastic computer model, if one is available. By carrying out this procedure for estimating timings for all scenario-relevant tasks and related events, the analyst can then calculate the 'end-time' for the scenario. It is also possible to use this approach to

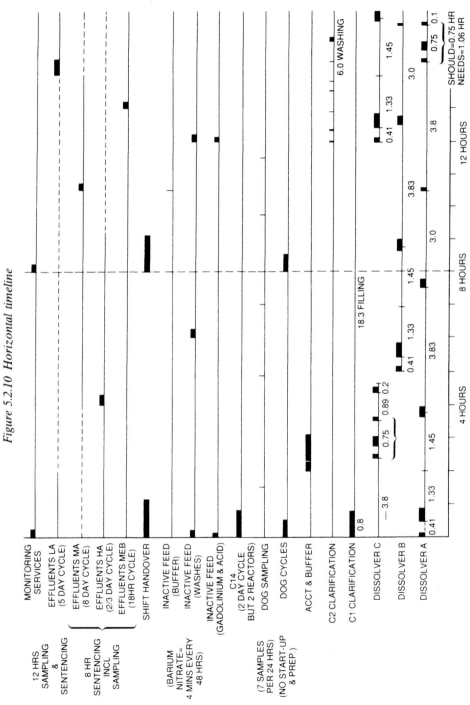

Figure 5.2.10 Horizontal timeline

Figure 5.2.11 Sequence dependency table

Task	Task dependent on	Uncertainty in Time Estimates
1. *Detect and identify problem (1)		high
2. *Ensure system shutdown (2)		moderate
3. Diesels started (3.1)		moderate
4. Back-up coolant motors working (part of 5.3)	4	moderate
5. Seal cooling temps. O.K. (part of 4.2)		moderate
6. Ensure Emergency coolant pump working (4.3)		moderate
		moderate
7. Monitor fuel + core temps (4.4)		moderate
8. *Shutdown unaffected reactor (8.5)		moderate
9. Trip biological shield device (4.6)	9	moderate
10. Close off secondary water systems (4.7)	9	moderate
11. Energise electric board for main coolant motor (4.8)	9	moderate
12. Close circuit breaker on board (5.4)	12	moderate
13. Run up main motor + switch to main motor (4.6.4 and 6.6.5)	12,13	moderate
14. Monitor cooler (7.6.6)	14	
15. Monitor reactor (10.7)		
	All dependent on 1 + 2 + 3	

Notes
1. Numbers in parentheses refer to task goals in the HTAs/TTAs in Appendix 1 + 2
* Critical task

determine a rough estimate of how busy operators are – for example, throughout a shift.

The major difficulty with the horizontal-timeline analysis has to do with determining the effects of a task whose duration does not stay not on schedule, either because of errors or because of accidental events or hardware failures. Because any one task can in reality be delayed, so, as a consequence, can any *combination* of tasks; so it is difficult to decide which delays to model. In practice, unless a simulation is being used (see later), timelines normally model timings for correct performances, though they may occasionally model additional timings that are caused by pre-identified failures, where these are moderately or highly likely to occur. In order to consider the interrelationships of timings in a *series* of tasks in a scenario, it may be useful to draw up a sequence-dependency table, as shown in Figure 5.2.11. This shows the uncertainties involved in the timings and highlights the critical tasks whose failure or delay will have significant impact on the overall timing of all the tasks (in this figure the numbers in brackets relate to an HTA).

Tasks which are largely monitoring and/or supervisory in nature may be represented as dotted lines (rather than solid bars) on the timeline, since the operator will be able to intersperse such tasks with other duties. Tasks may also run in parallel where this happens in practice, and where there is more than one operator carrying out the tasks (see Appendix VI, Case Study 7).

Vertical timelines offer another perspective on the timings in a scenario, but these focus on *who* is actually carrying out the tasks, and on interactions between all the personnel involved. The vertical timeline shown in Figure 5.2.12 is columnar in format, with the tasks and the times at which they begin or finish occupying the first three columns respectively, and the schedule of involvement (in each task) of the available personnel taking up the other columns. This timeline approach, which is particularly useful for investigating staffing and organisational aspects involved in handling a scenario, enables the analyst to assess the following:

- Whether there are enough staff.
- Whether they are appropriately located.
- Whether any staff are underutilised, or inappropriately utilised, in the scenario.
- The effects of losing one or more members of the assumed staff complement for the scenario.
- The communications requirements during the scenario.

The communications requirements can be noted on the timeline by numbering the personnel involved and then simply placing the numbers in brackets in the appropriate column at the point of time from which communication is initiated.

A timeline analysis (TLA) usually contains significant uncertainties, particularly in TLAs for emergency scenarios, but it is nevertheless a useful form of task analysis, and an essential one if the scenario is largely time-based and significantly time-constrained. In some actual applications, the TLA has demonstrated convincingly the impossibility of carrying out a set of tasks in the available time (even if carried out without error), which has meant that a quantitative HRA was unnecessary: no probability value for a failure to achieve the tasks in time was needed because the qualitative task (timeline) analysis had already in effect demonstrated this probability to be unity.

A timeline analysis can be advanced by utilising a computer-simulation package to model the timings and to run Monte-Carlo simulations of the timings, etc., to gain a range both of estimated task-completion times and of estimated uncertainty bounds. One such package, Micro-SAINT (Laughery, 1984), enables one to build up whole networks of tasks. By inputting data on the different times taken to complete each operation, the analyst can use Micro-SAINT to evaluate the time allocated for a given task, as well as the likelihood of failing to complete it in time. Micro-SAINT also has powerful sensitivity-analysis capabilities. Its one current drawback for the HRA analyst is that it requires timing estimates for each task as well as standard deviations or distribution parameters (or both) for the task times. These can of course be judgementally (and conservatively) estimated by the analyst.

Figure 5.2.12 Vertical timeline analysis

TASK NUMBER	SUB – TASK	TIME BEGIN TASK	TIME FINISH TASK	CCR BASED					LOCAL	
				SHIFT SUPERVISOR	CONTROL ROOM SUPERVISOR	DESK ENGINEER 1	DESK ENGINEER 2	OTHER	LOCAL OP. 1	LOCAL OP. 2
1.	Detect problem	0	5:00		x	x (any one else in CCR)				
2.	Ensure system is shutdown	1:00		if in CCR	x	x				
3.	Establish cooling									
3.1.1	Diesels started – check auto-start	4:00			x					
3.1.2	Start manually	4:10			x(1)			TRE (1)		
3.1.2.2	Locate buttons on panel	4:20								
3.1.3	Load power from diesels onto board	8:50	9:20					x		
3.2	Check back-up coolant motor started	10:10				x		x		
3.2.1	Check lights on panel	12:00				x				
3.2.2	Start pony motor manually	16:00				x				
3.2.3	Transfer from main coolant circulator to back-up motor	21:00				x				
3.2.4	Ensure drive taken up	21:00				x				
3.2.5	Monitor circulator speed	22:00				x				

RESPONSE TO LOSS OF COOLANT ACCIDENT

SCENARIO 3:   VERTICAL TLA

NOTES:
1. Task number refers to task as identified in the HTA/TTA
2. Items with same number in parentheses indicate communication between these individuals
3. TRE = Turbine Room Engineer
4. CCR = Central Control Room
5. Scenario starts at time zero

Micro-SAINT would probably be of benefit when carrying out a detailed, timing-analysis-related HRA, and in cases where the analyst needs to investigate both different timing arrangements and the degrees of sensitivity inherent in the tasks. Otherwise, a conventional TLA will generally suffice.

A timeline analysis can also be used to calculate workload estimates, and hence staffing requirements for jobs, and can be a useful aid for training purposes, since the timeline can explicitly show when each action must be carried out and how different tasks and operators must synchronise their actions. Workload peaks and troughs may be dealt with by a re-deployment of personel, by flexible working practices or by redesigning jobs. One problem inherent in any timeline analysis used for a workload assessment is the lack of a means of weighting tasks in terms of their cognitive 'effort', or mental workload, rather than in terms simply of the amount of time they take to complete (see Laughery & Laughery, 1987; Kirwan & Ainsworth, 1992).

5.2.5 Task analysis in the system design life cycle

Figure 5.2.13 (Kirwan, 1988) shows a framework for human-factors involvements during the various life-cycle phases of a plant, from its initial conception to the system's operation. The framework is divided into two horizontal sections: ergonomics/human-factors analysis and human-reliability analysis; and the entries in the 'cells' of this matrix show the nature of the human-factors/human-reliability objectives in each phase, as well as the types of techniques which can be used.

Functional-allocation TA techniques must be applied early, as can be seen in the diagram (see Williams, 1988, *Functional Allocation of Systems Technique*, or *FAST*). Mock-ups/simulators for design purposes are useful in the early, detailed, design phase; and the HRA begins in earnest at this stage. Following this phase, procedures, training and organisational issues must be addressed. During commissioning, final HRAs are carried out to ensure that the system is safe, and a functional TA of safety management may also be carried out. Once in the operational phase, any of these techniques may be utilised either in an evaluative function or to assist a redesign or re-organisation.

Introducing TAs late into systems designs has serious consequences. A TA can often identify the need for re-design, and the further the design has progressed, the more costly this will be. An opposing factor, however, is that the earlier in the design phase one starts, the less structured the design and operational information tends to be, and this can make the TA more difficult to carry out. When anticipating a comprehensive amount of HF-involvement in a large design project, a functional task-analysis of the TA techniques/ processes which are to be adopted will be useful, especially if it is organised into a framework such as the one shown in Figure 5.2.13. Such a TA will highlight the feedforward of information from early TAs to those later in the design's life cycle. An example of this approach is shown in Figure 5.2.14 (Kirwan, 1992).

Figure 5.2.13 Framework for a HFA/HRA in the system design life cycle (Kirwan, 1988)

Human Factors area	Concept	Detailed design	Operations support	Commissioning	Operation
Ergonomic/human factors analysis	Allocation of function • Functional analysis • Fitts List considerations • Control and other concepts • Level of automation • Other task analysis Initial staffing concepts Anthropometrics Person–machine interface philosophy • Layout Broad control and instrumentation consideration • Interlocks • Emergency shut down requirements • Environment	Detailed person-machine interface recommendations Detailed task analysis Use of CCR mock-ups. Development of advanced operator aids for design support Implementation of error reduction measures (derived from HRAs) Design for maintainability Review of design specification Software ergonomics	Development of procedures and operating instructions Development of training system Simulator usage Advanced staffing concepts • Selection and qualification • Organization structure • Sociotechnical considerations Preparedness/Emergency response plans Advanced maintainability system development	Development of QA programme Certification of operators Complete entire staffing and safety management implementation Assist with changes made during commissioning Setting up incident reporting system Revision of earlier human factors data/guidelines	Implementation of QA programme (including safety management) Refresher training (annually) Implement and evaluate retrofits/extensions as required Feedback info./data on incidents to design database
Human reliability analysis (HRA)	HAZOP Screening HRAs	HAZOP Human HAZOP Task analysis HRA Scenario anal. (inc. misdiagnosis and maintenance errors) Analysis of dependent failures Maintainability analysis Software reliability analysis Error recovery analysis	CCR audit Procedures, audits Detailed HRAs of accidental event sequences Simulator experiments • Verify HRAs • Generate human error data • Evaluate training Evacuation assessments	Management audit (requires R&D) Setting up performance monitoring systems Evaluation of changes made during commissioning Revision of human reliability techniques and data	Monitor safety progress of plant operations Feedback information and data into human reliability database Carry out operational safety assessments

Figure 5.2.14 TA programme for THORP

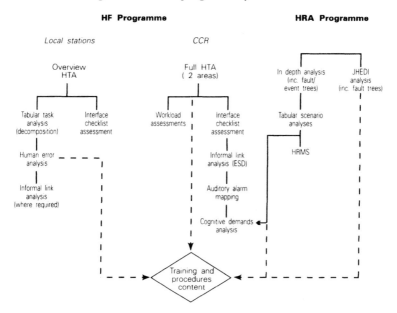

When it comes to a TA application within the context of a HRA, the later in the design cycle the HRA is taking place, the easier will be the task analysis and, as a result, the less extensive the use of resources and the more *detailed* the resultant analyses. On the other hand, error reduction may be more difficult to achieve, because of cost implications.

5.2.6 Summary of the task-analysis approach

A task analysis covers a range of tools and approaches that can be used either to assist in human-factors analyses and designs, or as the basis for a human-reliability analysis. A TA is a *generative* approach – a *process* rather than a mere passive recording of information. Many tools exist, of which HTA is the most important, which can be applied to a range of human-factors issues. When applying a task analysis, it is critical both to define the problem perspective and, if necessary, to tailor TA techniques to fit the needs of the study. The timing of TA applications can strongly determine the effectiveness of their impact. Arguably, all HFAs and HRAs start from a TA basis, whether implicitly or explicitly. In practical studies, the formal use of TA methods will yield more useful and justifiable results.

The TA techniques described in this chapter are all useful for a HRA, and have all been used in real HRAs to a greater or lesser extent. The flowchart shown in Figure 5.2.15 aims to give guidance on when to use the different techniques outlined in this section. This flowchart is not prescriptive, however,

Figure 5.2.15 Flowchart for TA usage in an applied HRA

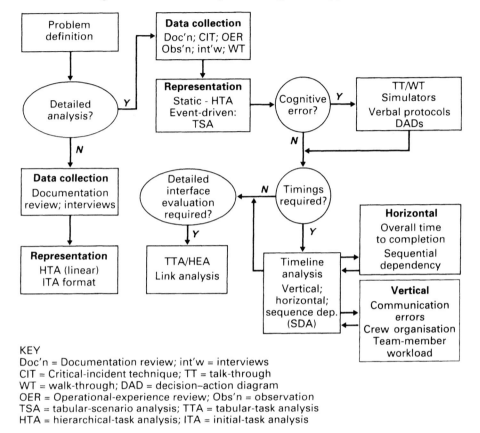

KEY
Doc'n = Documentation review; int'w = interviews
CIT = Critical-incident technique; TT = talk-through
WT = walk-through; DAD = decision–action diagram
OER = Operational-experience review; Obs'n = observation
TSA = tabular-scenario analysis; TTA = tabular-task analysis
HTA = hierarchical-task analysis; ITA = initial-task analysis

and should be supplemented by the assessor's own judgements as to the TA requirements for the investigation being carried out.

Acknowledgements

Many thanks to Dr Andrew Shepherd (Loughborough University, UK) for the use of some of his ideas on the practical aspects and uses of the HTA.

5.3 Human-error analysis

5.3.1 Introduction

Human-Error Identification deals with the question of what can go wrong in a system from the human error perspective. There are few reviews of the techniques and requirements of human-error analysis in Human Reliability

Assessments, probably for three reasons. Firstly, it is an inherently difficult subject matter, dealing as it does with the qualitative prediction of what people *might* do, instead of what they are supposed to do. Secondly, most HRA research and development has focused on the quantification of human-error probabilities for errors which are very easily identified, and so there has been little focus on systematic techniques for identifying more complex or subtle human-error contributions. Thirdly, certain PSAs may not always necessarily need the identification of human errors. The latter point, already raised earlier in the introductory section on the PSA process, will be returned to in Section 5.9 as it concerns a particular style of PSA approach.

Dealing with the first two problems, we find that these have led to only a small number of realistic techniques being developed, a paucity of published applications and hardly any validation experiments. A recent review of Human Error Identification techniques (Kirwan, 1992a; 1992b) notes the relative immaturity, and yet critical importance, of the HEI within the HRA field.

This section, which does not present anything more than a limited review of the range of HEI techniques available, attempts to provide practitioners with usable and effective tools for human-error identification/analysis (HEI/ HEA: the latter usually referring not only to the identification of human errors but also to error-recovery analysis, consequence analysis and the preliminary identification of error-reduction measures). Since any HRA carried out without an adequate Human Error Identification process will probably be inaccurate, such that the level of risk may be underestimated, it is critically important that the assessor understands the multifaceted nature of human error in complex systems and is thus aware of all the different types of human-error contributions which can ultimately make an impact upon a system's risk level. It is also necessary that the assessor be suitably well-armed with a battery of tools with which to identify such error contributions both reliably and comprehensively.

The practitioner must therefore understand the basic nature of the different types of human errors that can occur in complex working environments. This firstly requires a consideration of some of the most basic 'taxonomies' or classifications of human error, and secondly a knowledge of the currently dominant models of human error showing the different types of human-error contribution that are believed possible. The primary and secondary functions of a Human Error Identification system can then be outlined, followed by a brief review of some of the main methods that have evolved in HRAs. A new generalised HEI framework will then be described which deals with many different forms of human error at different levels of analysis (discussed later), and it will be seen that this framework and its tools draw heavily upon already-existing techniques and ideas, merely attempting to put these into a more comprehensive and coherent framework, for HRA purposes.

This section therefore contains only a *basic* discussion of different models of human error (see Reason, 1990, for a fuller theoretical review). It next contains a brief description of classification systems, as well as of the objectives of classifications in HRAs (which go beyond HEI/HEA, affecting incident investigations, data collections and Human Error Quantifications). These are useful when considering or evaluating the HEI tools presented later, which will inevitably utilise some form of classification system (see Fleishman & Quaintance, 1987, for a detailed review of classifications of human performance). The functions of a HEI are outlined, followed by a presentation of the major HEI/HEA techniques currently available in published literature. The practitioner may wish to utilise one or more of these tools for his/her analysis, and there are sufficient references for those wishing to follow up these techniques (see also Kirwan, 1992a & 1992b, and Whalley & Kirwan, 1989). The remainder of this chapter will describe a new multilevelled approach to HEI/HEA, called simply the *onion framework*. This is intended to be flexible as regards the scope and ambitions of the HRA/PSA. Worked examples of its usage are presented. The practitioner familiar with human-error theory and related techniques may wish to go straight to Section 5.3.5 where this new framework is presented.

5.3.2 The nature of human error: basic error taxonomies and human error models

5.3.2.1 Basic classifications of human errors

Note
Although it may seem strange to consider classifications *before* models, this is done because classifications usually pre-date model development, and in fact many, if not most, models will build upon or result in the development of classification systems or taxonomies. Also, since, in practice, the existing models tend to be explanatory rather than predictive, most of the actual tools used for identifying human errors resemble taxonomies more than they resemble models. Furthermore, some of the resultant approaches have overlooked some of the basic objectives and requirements of scientific classification systems, thus rendering themselves less likely to be effective. It is hoped that in the future this will change and that the HEI process will be rendered more scientific in its use of classification – and also that ultimately HEI techniques will also be model-driven. Due to the implicit use of taxonomies in almost all HEI tools, some of the basic objectives of a classification for HRA purposes are briefly defined below. This background will also be useful when actually carrying out a human-reliability assessment, and ultimately gives the assessor an insight into the limitations of the technique being utilised.

Classification in HEI
The classification of human errors is the backbone of Human Error Identification, its infrastructure and framework. Whenever a Human Error Identification procedure occurs, a classification system is also being used, whether implicitly or explicitly, model-based or intuitive in form, or in an approximate

or accurate mode. Classification is fundamental to any scientifically based approach or discipline, and the area of human reliability is no exception. But in practice, this area is rarely dealt with in human-reliability books or courses, and is even rarely to be found in relevant journals on human factors. This is because human-error modelling and classification is so fundamentally difficult, dealing as it does with the mechanics of human behaviour in virtually all its possible industrial manifestations. This area is simply not well understood as far as explicit scientific modelling is concerned.

A classification itself is simply the partitioning of a group of phenomena (e.g. human behaviours) into categories or types (e.g. error types) that are based upon a set of descriptors (e.g. omission, commission, etc.) which can be applied relatively reliably (e.g. by different assessors) to the group of phenomena in question. The principal advantage of a classification is that without it the intellectual process grinds to a halt, since what one person is talking or writing about is opaque to any other person. It is difficult to describe, for example, the colour of a rose without using a classification system (involving colour, hue, saturation, warmth, etc.); indeed, the word 'rose' is itself part of a rather large classification system or taxonomy.

Similarly, in human reliability, if one wishes to use a human-error datum to quantify a human-error probability, it is necessary to know if the datum matches the identified error, and thus whether it is appropriate to use this particular datum or select another one instead. Such a judgement can only strictly be made on the basis of a classification system. This point may seem laboured, but it is a recurring theme throughout this and other sections, and is fundamental to any robust and scientific human-reliability assessment.

The principal advantages of a classification system, or taxonomy, of human error are as follows:

- It offers a *structure* to the assessor, such that categories can be used either predictively, to identify errors, or retrospectively, to classify events.
- If robust, it offers a means of rendering Human Error Identification both reliable and repeatable, to the extent that two independent assessors are able to identify the same errors.
- It can add meaning to the assessment by defining the causes and reasons underlying the errors, and thus making the assessment more lucid.
- It offers a means by which human-error data can be collected and/or generated, matched with identified human errors and then quantified predictively.
- In theory, a good classification system will at least enhance (and sometimes even guarantee) the comprehensiveness of the error-identification process, such that no critical errors are omitted from the assessment.

Classification is thus a worthwhile objective since it offers so many advantages, and can make the process of error-identification less of a 'black box' affair. However, there *are* two major pitfalls:

- If the classification system is not complete, important errors may be omitted from the analysis.
- If the classification of errors is incorrect or superficial, critical differences in error typology may be overlooked, leading later to an inappropriate and incorrect quantification of human reliability.

Classification systems of human error, we can now see, are used in three main ways:

- For the analysis of incidents: to identify what happened and then prevent its recurrence; as well as to derive *data* from such incidents.
- For Human Error Identification purposes: to identify errors which may have an impact upon a system's goals, as well as the level of risk inherent in it.
- Human Error Quantification: to match existing data to identified human errors in order to quantify the likelihood of the latters' occurrence.

The identification of human errors is the main concern in this section, but if the assessor is attempting to classify an incident, or to use real human-error data to quantify a scenario, then the classification system used will be a critical consideration. The purpose of a system of classification in a Human Error Identification procedure is as follows:

> to aid the assessor in identifying reliably (i.e. repeatably) the total set of human errors and actions which could affect the system.

The significant parts of this definition are the words 'reliably' and 'total'. If the classification system is unreliable or non-comprehensive, then the system assessment may be incorrect: for example, the level of risk may be underestimated.

The Human Error Identification stage is arguably the most critical stage in the Human Reliability Assessment process: if errors are omitted at this stage, they will not be quantified or reduced but will remain as hidden, latent errors in the assessment itself. The bulk of effort in this area has focused on trying to develop comprehensive systems, with only a secondary effort going into improving their reliability when in use – which is perhaps understandable since a comprehensive but unreliable system may ultimately perform better than a reliable but non-comprehensive system, particularly if more than one assessor is utilised.

Having briefly outlined the importance and rationale of a classification, plus its objectives, advantages and limitations, it is now appropriate to consider some of the basic systems themselves.

Basic HEI classification systems
Heuristic classification systems, or taxonomies, are of major interest to the assessor since they have largely been developed through the experience and

expertise of those assessors carrying out human-reliability and systems analyses. They do not offer the advantages of the more theoretically sound model-based approaches (later in this section), but they are often of more practical use, nonetheless, and they in fact lend themselves more easily to the context of technological systems such as nuclear-power plants or offshore installations. The most famous example is the simple but prevalent taxonomy shown below (Swain & Guttmann, 1983):

- Error of omission – acts omitted (not carried out).
- Error of commission – acts carried out inadequately; or in the wrong sequence; or too early or too late. This category also includes errors of quality, where an action is carried out to too great or too small an extent or degree; or in the wrong direction; etc.
- Extraneous act – wrong (unrequired) act performed.

This simple approach encompasses virtually all kinds of error that are likely to occur, although in a fairly gross and relatively non-discriminating fashion. It can be advantageously used with another 'system oriented' taxonomy (Spurgin *et al*, 1987) which involves the following categories:

- Maintenance-testing errors affecting (safety) system availability (so-called *latent errors*).
- Operator errors initiating the event/incident.
- Recovery actions by which operators can terminate the event/incident.
- Errors which can prolong or aggravate the situation (e.g. via a misdiagnosis).
- Actions by which operators can restore initially unavailable pieces of equipment and systems.

These two approaches can be used by an assessor (with a good understanding of the system being investigated) to identify those errors which could have an impact upon the system goals/tolerance levels under investigation. The way in which these approaches work is by first considering the task (or the event sequence) being undertaken, and then considering, at each stage, what the operator could do wrong: for example, the operator could omit to turn a valve, or could turn it in the wrong direction. These two approaches involve classifications of errors, and as such must first be applied to a task description and then used to 'operate' on the action verbs within the description so as to identify credible error modes.

An alternative approach is to have a taxonomy of behaviours, which would then be used in the task-analysis process itself. Such a taxonomy is shown in Figure 5.3.1 (Berliner *et al*, 1964). If this hierarchical taxonomy is used in the task-description phase, the analyst can merely consider which of the action verbs in the task description could be misinterpreted or ignored by the agent (e.g. the operator 'calculates' incorrectly, or mis-'identifies'), and these then become the human errors of interest in the assessment.

Figure 5.3.1 Berliner's taxonomy

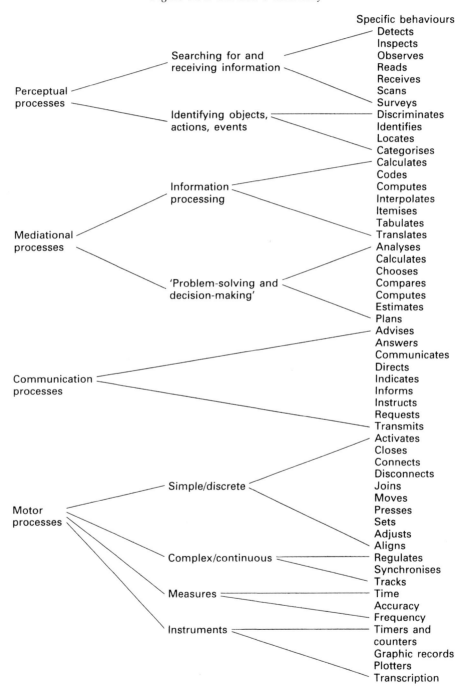

Figure 5.3.2 Information processing model (adapted from Wickens, 1992)

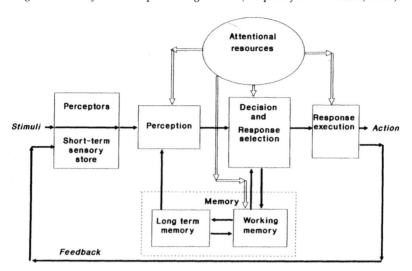

The Berliner (*et al*) approach is in part a model-based approach as it is loosely based on an early version of the information-processing model (see Wickens, 1992). However, it is basically a taxonomy of human behaviours that outlines what an operator is required to do in an industrial (process-control) context.

Many of the techniques that will be shown later are essentially taxonomic in nature, i.e. appearing simply to be lists of error types that the assessor must apply with a large degree of his or her own judgement. Many of these approaches are, however, also related to models of human behaviour, such as the information-processing model depicted in Figure 5.3.2, which is fairly self-explanatory (see also Wickens, 1992). However, one model which has been highly influential in shaping the format of several HEI techniques in the past decade is that developed by Rasmussen and his co-workers (1981). This model, and other important contributions by workers such as Reason and Dorner, are reviewed in the next subsection (for a more detailed review, see Reason, 1990; and Rasmussen, Duncan & Leplat, 1987).

5.3.2.2 Human error models

BACKGROUND

A human error is often simply described as a 'failure to carry out an act as required'. This error can arise from a number of sources, for example:

- A lack of precision, due to manual or skill variability: for example, pressing the wrong button inadvertently.
- A pattern-recognition failure: for example failing to realise that the state of the system has changed.

- A memory failure: for example, forgetting, misremembering.
- A cognitive failure: for example, a misinterpretation or failure of judgement (due to biases, etc.).
- An attentional failure: for example, place-losing errors caused by an interruption, or a lack of vigilance.

Human errors are not intrinsically different from any other form of human behaviour, rather they are actions which are misplaced. Reason (1990) usefully distinguishes between *slips* and *lapses* (e.g. pressing the wrong push-button, or forgetting a step in a long procedure), which are 'unintentional' or 'inadvertent' errors, *mistakes,* which are errors of intention (e.g. a misdiagnosis), and *violations* (e.g. taking an 'illegal' short-cut during a procedure). The latter contains a risk-taking element.

The point of interest here is that all errors are a natural by-product of human behaviour; either they are actions which normally work, but occasionally fail, or else they are *strategies* which often work but may also fail: even rule violations are a class of behaviour which may be successful many times before leading to failure. Thus, a human error is no different from a normal human action. Therefore, ideally, a model of human error should also be a model of action; or rather, a true model of human action should *also* be able to identify any human errors that are likely to occur. In practice, such a model has proven rather elusive (see Watson & Oakes, 1988). Probably the most influential model in the field of HRA since the early 1980s has been the skill-, rule- and knowledge-based behaviour model of Rasmussen and his co-workers, described next.

THE RASMUSSEN *ET AL* (1981) SKILL-, RULE- AND KNOWLEDGE-BASED MODEL
It has now become standard practice for most writers on human reliability to mention, if not depict graphically, the decision-making model developed largely by Rasmussen (see Rasmussen *et al*, 1981). This model is empirically based on the analysis of licensees' reports, made by licensees, containing information on human errors committed in plants. The model, which is essentially qualitative in nature, outlines the hierarchical 'step-ladder' processes which Rasmussen believes are associated with actions and decision-making.

Rasmussen's model considers that behaviour may be classified into three levels: skill-, rule- and knowledge-based behaviour. In the skill-based regime, the model assumes that a performance is controlled by stored patterns of behaviour, and that the operator reacts to stimuli with little conscious effort or consideration: essentially, therefore, the operator is in an 'automatic' mode of operation. Rule-based behaviour is regarded as involving a performance in familiar settings, using stored or readily available rules. This level of behaviour requires both an initial appreciation of the necessity of consciously applying such rules (rather than just giving an automatic response) and the subsequent actual conscious selection and application of those rules, which may invoke skill-based behaviour. Knowledge-based behaviour, finally, is

Figure 5.3.3 Levels of performance: skill-, rule-, and knowledge-based behaviour (after Rasmussen et al, *1981)*

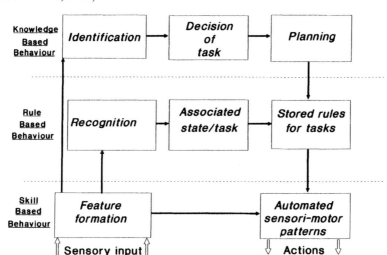

presumed to involve event-specific behaviour that is based on a knowledge of the system at the point in time when a demand is placed on the operator. This level of behaviour usually requires 'higher-level' cognitive processes – such as problem-solving, goal selection, and planning – that are based on a functional understanding of what is happening in the system.

Rasmussen's model is shown schematically in Figures 5.3.3 and 5.3.4. Figure 5.3.3 shows the different levels of performance, with the top area indicating knowledge-based behaviour, the middle rule-based, and the bottom skill-based. It should be noted that this is fundamentally a hierarchical model, since any plan developed using knowledge-based behaviour will call both upon rule-based procedures (e.g. coming from memory) and upon skill-based operation (to execute the procedure).

Figure 5.3.4 shows the overall structure of the model. The model identifies what error occurred (external mode of malfunction), how it occurred (internal mode of malfunction) and *why* it occurred (a Performance Shaping Factor, etc.).

This model was derived from analyses of licensee event-reports, and so is empirically based. Although it was designed as a retrospective technique, it can also be applied predictively, in combination with a task analysis/description. By a consideration of the type of task (and hence the probable level of behaviour), certain error modes can be specified. Thus, for example, an operator in a well-known and proceduralised situation might omit a procedural step via the psychological mechanism of 'slip of memory'.

This model has a considerable explanatory power and, as has been mentioned, can be used predictively, though the reliability of such a use is not

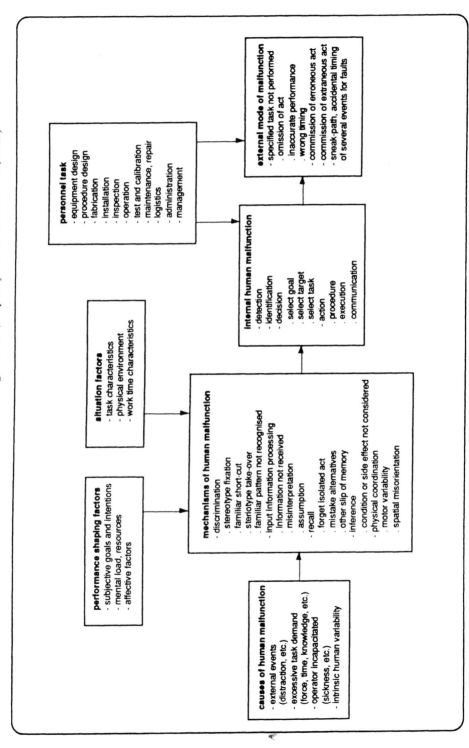

Figure 5.3.4 Framework for skills-rules-knowledge model (adapted from Rasmussen et al, 1981)

assured. The skill-rule-knowledge distinction, nevertheless, remains prevalent in human-reliability studies, and Rasmussen's model is the first of its kind successfully to blend psychology and engineering together into a single credible framework.

This model also gives a useful framework for classifying errors. Indeed, it investigates, in considerable depth, the causes of human error (i.e. the internal model of malfunction), not just the overt or externally observable manifestation of error. As such, it is much better equipped for judging whether or not a given datum can be extrapolated to an identified human error.

A second classification system of this genre, developed by Reason (Reason & Embrey, 1985; Reason, 1987; 1990), is the generic error-modelling system (GEMS), which also focuses on the underlying mechanisms giving rise to the error.

GENERIC ERROR MODELLING SYSTEM (REASON, 1987; 1990)

The GEMS model concerns cognitive error modes, and, though partly model-based, is largely taxonomic in nature. The classification is based on the traditional information-processing-related error-forms, such as *resource limitations* (e.g. overload, loss of information), but adds two more particularly useful error tendencies. The first is the inappropriate use of *schemas*. These are internal representations of external events and phenomena which we utilise to handle the large influx and output of information which occurs in modern society. Schemas are like subroutines, and can often be used 'automatically', i.e. with little conscious effort. A schematic operation is resource-effective, but it can be prone to error if, for example, an abnormal situation occurs which is *similar* to a usual one. In this case, if the operator fails to see the difference and believes this is the usual situation, then a frequently-used, but here inappropriate, schema will be made to come into operation.

An important aspect of Reason's model involves the natural drive towards economy of cognitive effort, such that if a schema exists which roughly matches the task demands, then this will be used in preference to deriving a new solution, since taking the latter path would require significant mental effort. This aspect of minimising cognitive effort is also partly implicit in the SRK model – which is hierarchical in nature, only making use of 'higher' mental functions (e.g. KBB) when necessary.

A second useful aspect of the GEMS system is its consideration of heuristics used by operators when dealing with difficult situations (e.g. emergencies). Operator strategies or heuristics, or 'rules of thumb' may be inappropriately applied – e.g. that of oversimplification when dealing with a complex situation.

The GEMS system is considered in more detail in the Human Error Identification section. It is a useful model particularly when considering cognitive-error situations such as are dealt with during a diagnosis of events.

ORGANISATIONAL AND SOCIOTECHNICAL ERRORS

Dorner (1987) adds a useful additional set of error types related to problem-solving, emergency and team situations; for example:

- An increase in the tendency to try and 'escape' from having to deal with a difficult situation.
- A decreasing willingness to take decisions.
- Formulating an increasingly global hypothesis.
- Thinking in causal series rather than in causal nets.
- Experiencing difficulties in dealing with exponential developments, etc.

Some of the error types identified involve difficulties in dealing with complexity, but others are what are known as 'socio-technical' errors. These are difficult to use predictively, but can occur in reality. At present, probably the best way to deal with sociotechnical errors on the *qualitative* level is by developing a good safety-management structure in the organisation, training to avoid such biases or behaviour patterns. The best way to deal with them on the quantitative front, on the other hand, is by incorporating a certain degree of pessimism into any (quantitative) assessment of diagnostic performance.

A brand of errors which has yet to be properly classified is the set of errors which affect an *organisation*, and which do so at a high level: examples here include the problems which were encountered in the Challenger Space Shuttle disaster, Bhopal, and on the Herald of Free Enterprise. These 'management' errors, which can have a severe effect on safety levels, are the subject of ongoing research (see Section 5.11).

This subsection (5.3.2) has described the usefulness, and indeed necessary predominance, of taxonomic approaches in HEI processes, has outlined two of the more important models in human-error theory, and has noted the existence of socio-technical errors – though these are currently beyond the scope of existing HRA techniques. The next subsection deals with what the HEI techniques need to be able to do, i.e. their principal functions.

5.3.3 Functions of HEI techniques

During the risk-assessment of a new or existing plant, it is critical that the full impact of human error on the system's safety level is properly evaluated. Probabilistic Safety Assessments (PSAs) and Probabilistic Risk Assessments (PSAs) utilise fault- and event-tree approaches to calculate risk levels, and typically these approaches include descriptions of human errors and their associated probabilities (HEPs).

However, whilst there are formal and accepted approaches to the quantification of human error, the actual identification of human errors that can have an impact upon safety levels, as has already been noted, has received less attention in the PSA world. This raises the serious question, when carrying out an HRA, of whether all significant human errors have been identified, and hence of whether the PRA calculation is correct or could in fact be underestimating the level of risk that is present.

In the field of Human Reliability Assessment (HRA), a good number of quantification methods have been developed, criticised by peer reviews and validated via comparative experiments (Swain, 1989; Kirwan, Rea & Embrey, 1988; Waters, 1989; Kirwan, 1988). In contrast, the Human Error Identification process, which must occur prior to quantification, is often unstructured, involving no empirically or theoretically based model of the process of human error. Typically, a rudimentary task analysis is undertaken which specifies what the operator has to do in a situation. The analyst then simply considers what could go wrong in terms of errors of omission and commission. This approach, which may not be applied systematically, pays no attention to why an error occurs, nor indeed necessarily to the way in which it may manifest itself in an actual incident. Since this simple approach leads only to a crude appreciation of the role of human errors in scenarios, it is by no means certain that this modelling of the impact of human errors is either consistent, comprehensive or accurate.

Whilst the main goal of a PRA may be traditionally considered to be the assessment, and not the reduction, of risk, at least one recent paper has, however, concentrated on this latter aspect, specifying six goals for what it calls an 'advanced human reliability assessment' (Vuorio and Vuaria, 1987):

- An identification of possible weak points affecting the level of risk inherent in the plant.
- An assessment of the need for, and the feasibility of, improvements.
- An identification of previously unrecognised ways of preventing or mitigating accidents.
- An identification of factors affecting human reliability, together with a consideration of their direction and relative degree of influence.
- A quantitative assessment of operator- and maintenance-action reliability.
- Development of data for comparative studies between different methods and theoretical models, as a means to estimating human-error probabilities.

The first four of these goals can most easily be satisfied by carrying out an in-depth Human Error Identification approach. Essentially, if the level of risk is undesirably high (as must arguably be found in at least some PRAs – e.g. for novel system designs), then, in a proportion of such cases, human error would also probably be highlighted as the main problem area to be solved. The logical step is to reduce the likelihood of human error, but to do this without actually understanding what causes the error, and how it manifests itself, may lead to an ineffective solution, and one that is erroneously credited with reducing the HEP. Clearly, the process of reducing human error requires a proper understanding of the human error itself, rather than just a simple, gross description such as 'Operator presses wrong button'. There are, for example, several reasons why such an error may occur: for example, inadequate training, inappropriate stereotypes adhered to by operator, inadequate labelling, or workspace design/layout. And each may have a different ergonomics solution.

A further important aspect of human-error modelling, often ignored, is the correct recording of assumptions made during risk assessments, both for reference purposes and for possible use later on during the operational phase of a plant (Kirwan & James, 1989). The 'auditability' of many studies is inadequate in this respect for the reason that operator errors are specified only in their 'external error mode', or as they appear in the fault or event tree ('Operator fails to respond to alarm', etc). What will be of interest to operations personnel during the life of the plant will be the analyst's assumptions as to why this event could occur (e.g. operator absent, stress levels high) and how its likelihood of occurring could be reduced. If an accident actually happens and its external form was predicted by a PRA, is it in fact for the same reasons that the analyst predicted? The answer to this question will often be unknown, this being perhaps one reason why human-reliability techniques and data banks are rarely advanced by actual historical events in industry. The lessons which could be learned cannot be learned because the format of the predictions is far too crude and does not reflect the very specific feedback gained from real incidents and accidents.

There are, therefore, at least three reasons why more rigorous, systematic, detailed and theoretically/empirically valid Human Error Identification-based techniques are needed:

- To ensure a reliable and comprehensive error-identification system, and thus the accuracy of PSA/PRA modelling.
- To provide an accurate understanding of potential human errors, in case error-reduction measures are required.
- To ensure that the assessment provides an effective database for use during the whole lifetime of the plant.

The above are three important functions of a HEI technique, placed in order of priority for HRA purposes. As will be seen in the next two subsections, different techniques will vary in the degree to which they attempt to fulfil these functions. At a more specific level, however, there are three other functions that a technique can fulfil. These functions ultimately determine the theoretical depth of the tool, and can influence the accuracy of the quantification of identified errors, as well as determining the usefulness of the technique for error-reduction purposes:

- An external-error-mode-identification function: *all* human errors can be categorised into external error modes, which comprise the level of description which is used in a Probabilistic Safety Assessment. For example, 'Error of omission', or 'Not done' if the operator fails to respond to an alarm.
- An error-mechanism-identification function:
  These functions serve to define, in psychologically and/or ergonomically meaningful terms, how the error actually occurred. For example, the above 'failure to respond' may be due either to a 'misinterpretation' of the signal or to 'reduced capabilities'.

- A function that identifies Performance Shaping Factors (PSF). These can obviously be usefully considered during the Human Error Identification phase, although frequently they are not identified until the quantification phase. Examples of PSFs include time pressures, adequate training or adequate procedures.'

All HEI techniques must, at the very least, identify the external error mode, as this is a necessary step for the integration of the human-error element into a PSA. If the PSFs are also identified, this can have an impact upon those quantification techniques which utilise PSFs as part of their quantification process. If the PSFs and error mechanisms are both identified, it is then possible to consider using real data to quantify a human-error probability, as long as it is known what mechanisms and what PSFs are underpinning the data (assuming such data is available).

The distinction between the external error mode and the error's actual cause or mechanism is also particularly important if an error-reduction is to be achieved. If the error mechanisms are known, effective error-reduction mechanisms will be more readily specified. It is likely that future Human Error Identification techniques will follow this basic pattern, making the process of identifying errors' both more psychologically/ergonomically meaningful and more helpful in reducing the potential for accidents.

The next section reviews some of the major Human Error Identification techniques currently available to see both how well they achieve their objectives and how much use they are to the human-reliability practitioner. A number of techniques are reviewed, ranging from simple pragmatic approaches to more complex flowchart-based systems. These may be utilised by the assessor, followed up by the interested reader or used together with, or instead of, the 'onion-framework' approach to be described in Section 5.3.5.

5.3.4 Review of Human Error Identification methods

The following subsections describe some of the more well-known approaches in HEIs/HEAs. For a fuller and more evaluative review, see Kirwan (1992a; 1992b).

5.3.4.1 *Technique for human error rate prediction (THERP)*

The simplest approach is to consider the following possible external error modes (Swain & Guttmann, 1983) at each step of the procedure defined in the task analysis:

- Error of omission – acts omitted (not carried out).
- Error of commission – act carried out inadequately; in the wrong sequence; too early or too late; to either too small or too great an extent (or degree), or in the wrong direction (errors of quality).
- Extraneous error – wrong (unrequired) act performed.

The identification of error-recovery paths is largely carried out using the judgement of the analyst.

The task analysis should also highlight those points in the sequence at which the discovery of an error will be possible, for example via indications (especially the occurrence of alarms) and checks, or via interventions by other personnel. This approach is rudimentary but nevertheless can identify a high proportion of the potential human errors which can occur, as long as the assessor has a good knowledge of the task, and a good description in his or her hand of the operator–system interactions.

The above approach is best used with a systematic task-behaviour taxonomy, such as that of Berliner *et al*'s (1964) shown earlier in Figure 5.3.1. In the THERP manual (Swain & Guttmann, 1983), there are many sections which expand on the types of problem which can occur in a range of behavioural situations. These sections essentially imbue the reader with a fair variety of human-factors perceptions on good and poor task and interface designs. In addition, the THERP methodology considers a range of Performance Shaping Factors (PSFs: see THERP technique in quantification section) which can give rise to human error. Lastly, the human-error data tables themselves (found in Chapter 20 of the THERP manual) are effectively an expanded human-error taxonomy, i.e. the result of applying the simple error classifications listed above to a set of behaviours (defined by Swain's version of the taxonomy system of Berliner *et al*) in the context of nuclear-power-plant scenarios.

Therefore, in total, the THERP-HEI methodology comprises the following:

- An error taxonomy (omission, commission, etc.).
- A behavioural taxonomy.
- A basic engineering perspective on human factors.
- A consideration of Performance Shaping Factors.
- Human-error tables.

If a practitioner actually takes the trouble to read all the way through NUREG/CR-1278 (the THERP handbook), he or she will find that the HEI approach is a powerful one, and is not just the simple error taxonomy which it sometimes appears to be.

The disadvantages of this methodology are its lack of rigorous structure – which means that there can be a considerable variation in how different assessors model scenarios (see Brune *et al*, 1983) – and its failure to consider underlying psychological mechanisms – i.e. the reasons why errors occur. Its principal advantage is that it is straightforward and simple to use, and has the potential to model all the errors that can affect a system.

5.3.4.2 Human error HAZOP (hazard and operability study)

Another method, one which has arisen more from the engineering approach than the psychological domain, is the *human-error HAZOP* approach. This is rooted in the hazard-and-operability-analysis (Kletz, 1974) tradition, and

can range from a simple biasing of the HAZOP procedure towards the identification of human errors to an actual alteration of the format of the HAZOP
procedure itself.

The hazard and operability study (HAZOP) is a well-established technique
in process-design audits and risk assessments in engineering. It is typically
applied in the early stages of the design of a system or subsystem, although
it can equally be applied to existing systems, particularly when the design is
being modified, or if 'retrofits' are being introduced. It also translates the
experience both of the HAZOP chairman and of selected system-design and
operational personnel into a powerful analysis of a new design.

The HAZOP is primarily applied to process and instrument diagrams. Small
sections of these diagrams are picked out, and a set of guidewords are applied
to each section to detect design flaws. For example, guidewords such as 'No
flow' or 'Reverse flow' can be applied to a piece of piping feeding into a
pump, after which 'what if?' questions can be asked. The HAZOP team
might be considering the question: 'what if a reverse flow occurred through
the valve?'. This might lead to a recommendation to install a non-return
valve to prevent reverse flow if it is considered that, were this not done, there
would be adverse consequences.

There is an obvious parallel between this approach and the one that combines a task description with an error taxonomy. One important advantage of
extending a HAZOP to a more systematic identification of human errors is
that the technique is applied at a relatively early design stage, when human-
factors considerations can be considered in the most cost-effective way.

Whalley (1988) usefully related the traditional HAZOP guidewords to the
human-error types found in the Potential Human Error and Cause Analysis
(PHECA) system. The human-error modes derivable from HAZOP are as
shown in Table 5.3.1:

Table 5.3.1 PHECA error types related to HAZOP key words (Whalley, 1988)

HAZOP guideword	PHECA error type
No	Not done
Less	Less than
More	More than
As well as	As well as
Other than	Other than
–	Repeated
–	Sooner than
Reverse	Later than
–	Misordered
Part of	Part of

As part of a project in Norway (Comer *et al*, 1986), the author of this book
was involved in creating an expanded set of HAZOP guidewords to aid the
identification of human errors during an offshore-drilling exercise. Sixteen

new types of error *not* covered by the traditional HAZOP were identified by a process of studying several other taxonomies of error and applying a rudimentary knowledge of drilling operations. Some of these additional guidewords are shown below, with brief contextual examples from the offshore-drilling world:

- Calculation error: e.g. calculates wrong mud volume.
- Installation error: e.g. installing BOP (blowout-preventer) rams upside down; or installing empty hydraulic accumulators.
- Scheduling error: e.g. a poor scheduling for maintenance leading to delays in repairs.
- Interlock bypassed: no current example.
- Reading error: e.g. 'Measurement-while-drilling' read-out is incorrectly read.
- Diagnostic error: e.g. a developing blowout situation is not recognised.

It was also considered that information utilised by the operator in the form of communications and indications should be given special attention in the HAZOP process, since these can critically affect both human performance and strategies adopted in drilling practices. An example of an offshore-drilling HAZOP record sheet is shown in Figure 5.3.5 (from Comer *et al, op. cit.*).

Although HAZOP participants frequently identify human errors without additional guidewords, it is nonetheless felt that additions such as those above could promote a greater awareness not only of human error but also of the importance of considering, and taking prompt action against possible human errors at an early design stage. This extension of the traditional HAZOP technique may only be useful for specific, largely human-operated situations. The technique itself therefore requires further development of its range of application. The HAZOP technique is returned to later in Section 5.3.5, as part of the onion framework.

Analysts' judgements are a useful aid the identification of errors. Many risk-analysis practitioners build up a knowledge of how to identify errors in safety studies, as a result either of their operational experience or their involvement in many different safety analyses. Whilst this is perhaps an 'art' rather than a science, and as such is less accessible to the novice than would be a more formal method, its value must not be overlooked. Due to the specificity of every new safety analysis carried out, the human-reliability practitioner 'standing on the outside' may often be more disadvantaged in terms of error-identifications than the hardware-reliability analyst who knows the system's details intimately. It is therefore worthwhile to adopt a hybrid, team approach to the identification of error, as well as to make use of formal, systematic methods. The adage of 'two heads is better than one' is especially true in the area of Human Error Identification.

The primary advantages of HAZOP are therefore not only that it occurs at an early design stage but also that it identifies errors in a system-knowledge-rich environment (thanks to the hybrid nature of the HAZOP team itself).

Figure 5.3.5 Driller's HAZOP: extract (Comer et al, 1986)

Study ref. no.	Deviation	Indication	Causes	Consequences	Action notes questions recommendations	Follow up comments
	5A/5D. Lift and set aside kelly. Ensure hook is free to rotate.					
5.1	No movement.		Kelly valve not closed.	Mud on the drill-floor.	A new (mud-saver) valve was discussed. (A pre-set pressure valve which prevents flow for pressure below 200 psi, i.e. a mud column equal to the length of the kelly assembly.)	
			Forgot to open hook to ensure free rotation.	With stabiliser in the hole, the drill string will rotate when pulled out, which may cause damage to the hook.		
5.2	Reverse movement.		Lifting kelly too high.	Damaged hoses, piping, etc. Injuries to operators.	Q.7 Should sensor alarms be provided on guide dolly?	
			Failure of lifting equipment; e.g. drill line breaks/ snaps, or operator error.	Falling objects causing damaged equipment and injuries to operators.	Q.8 How do we ensure that kelly is not lifted too high? R.2 Consider safe location of driller's cabin in designing new rigs.	

This means that errors can be detected early and then rectified at a minimal cost to the design project. The disadvantages are that HAZOP is resources-intensive, and is only as good as the team and chairman. In addition, it is not always effective at resolving human-error problems if a human-factors-orientated person is not present. This is because HAZOPs often solve error problems by suggesting procedural and training solutions which would reduce error but would not *remove* it as a design change could do.

5.3.4.3 SRK (skill-, risk- and knowledge-based behaviour) approaches

The work by Rasmussen *et al* (1981) produced a powerful taxonomy of error classifications, and paved the way for later techniques described below. The logical extension of the SRK model to derive a predictive tool has, however, not apparently arisen – not, at least, in the shape of an effective tool. One attempt (Pedersen, 1985), called simply 'work analysis method', appeared promising but was resource-intensive. Approaches based on the SRK model need to determine not only what could go wrong, called the 'external mode of malfunction' (e.g. error of commission), but what the 'internal human malfunction' would be (e.g. detection failure) not to mention the 'underlying mechanism of human malfunction' (e.g. 'stereotype fixation'). Whilst this can be done for incidents which have already happened, it is much more difficult to predict possible future incidents, given the large number of possible combinations of these three variables. Thus, a great effort is needed to derive a 'tractable' human-error analysis. Nevertheless, the SRK model *could* be developed to be predictive, and in fact several of the models considered below do stem from such an approach.

As far as the practitioner is concerned, probably the most useful aspect of the SRK model for predictive HEI purpose is the flowchart in Figure 5.3.6, which can be used to derive the psychological mechanism of human malfunction. If this flowchart is used predictively, the assessor must then define the external error modes which will actually appear in the PSA. For example, a stereotype fixation may be described in the PSA as: 'Operator fails to detect [deviant condition]'. Even if the assessor never utilises the SRK system in a practical way, the SRK paradigm is sufficiently influential for any serious practitioner in the field of human reliability to be reasonably familiar with it.

5.3.4.4 Systematic human error reduction and prediction approach

Another method for human-error analysis is embedded within the systematic human-error reduction and prediction approach (SHERPA, Embrey, 1986). This human-error-analysis method consists of a computerised question–answer routine which identifies likely errors for each step in the task analysis. The error modes identified are based both on the SRK model and on the generic error-modelling system (Reason, 1987). Table 5.3.2 shows the psychological error-mechanisms underlying the SHERPA system. An example

Figure 5.3.6 SRK error mechanism flowchart (adapted from Rasmussen et al, 1981)

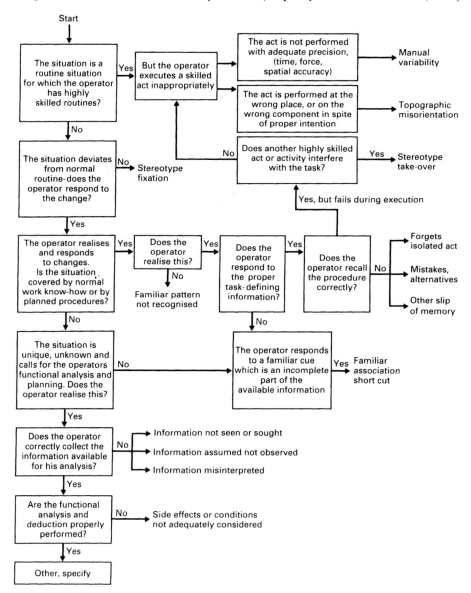

Table 5.3.2 Classification of psychological error mechanisms underlying SHERPA system

1. Failure to consider special circumstances: a task is similar to other tasks, but special circumstances prevail which are ignored, and the task is carried out inappropriately.

2. Short cut invoked: a wrong intention is formed based on familiar cues which activate a short cut or inappropriate rule.

3. Stereotype takeover: owing to a strong habit, actions are diverted along some familiar but unintended pathway.

4. Need for information not prompted: a failure on the part of external or internal cues to prompt a need to search for information.

5. Misinterpretation: the response is based on wrong apprehension of information, such as a misreading of the text (or an instrument) or a misunderstanding of a verbal message.

6. Assumption: a response is inappropriately based on information supplied by the operator (by recall, guesses, etc.) which does not correspond with information available from the outside.

7. Forget isolated act: the operator forgets to perform an isolated item, act or function, i.e. an act or function which is not prompted by the functional context, or which does not have an immediate effect upon the task sequence; alternatively, it may be an item which is not an integrated part of a memorised structure.

8. Mistake among alternatives: a wrong intention causes the wrong object to be selected and acted on. Or, the object presents alternative modes of operation, and the wrong one is chosen.

9. Place-losing error: the current position in the action sequence is misidentified as occurring *later on*.

10. Other slip of memory: (as can be identified by the analyst).

11. Motor variability: a lack of manual precision, too great or too small a force applied or inappropriate timing (including deviations from 'good craftmanship').

12. Topographic or spatial orientation inadequate: in spite of the operator's correct intention, and correct recall of identification marks, tagging, etc., he unwittingly carries out a task/act in the wrong place or on the wrong object. This occurs as a result of following an immediate sense of locality where this is not applicable (or has not been 'updated', perhaps due to surviving imprints of old habits, etc.).

of the tabular output from such an analysis is shown in Figure 5.3.7 (Kirwan & Rea, 1986). This 'human-error-analysis table' has similarities to certain approaches adopted by engineers to identify the failure modes of hardware components. One particularly useful aspect of this approach is its determination of whether errors can be recovered immediately, at a later stage in the task, or not at all. This information is useful if error-reduction measures are required later in the analysis. This particular tabular approach also attempts to link error-reduction measures to the underlying causes of the human error,

Figure 5.3.7 *HEA table (SHERPA) (Kirwan & Rea, 1987)*

TASK 51: TERMINATE SUPPLY AND ISOLATE TANKER (SEQUENCE OF REMOTELY - OPERATED VALVE OPERATIONS)

TASK STEP	ERROR TYPE	RECOVERY STEP	PSYCHOLOGICAL MECHANISM	CAUSES, CONSEQUENCES AND COMMENTS	RECOMMENDATIONS		
					PROCEDURES	TRAINING	EQUIPMENT
51.1	ACTION TOO LATE	NO RECOVERY	PLACE LOSING ERROR	OVERFILL OF TANKER RESULTING IN DANGEROUS CIRCUMSTANCE	OPERATOR ESTIMATES TIME/RECORDS AMOUNT LOADED	EXPLAIN CONSEQUENCES OF OVERFILLING	FIT ALARM-TIMING/VOLUME/TANKER LEVEL
51.2.1	ACTION OMITTED	5 2 4	SLIP OF MEMORY	FEEDBACK WHEN ATTEMPTING TO CLOSE CLOSED VALVE OTHERWISE ALARM WHEN LIQUID VENTED TO VENT LINE			MIMIC OF VALVE CONFIGURATION
51.2.2	ACTION TOO EARLY	5 2 2	PLACE LOSING ERROR	ALARM WHEN LIQUID DRAINS TO VENT LINES	SPECIFY TIME FOR ACTIONS	OPERATOR TO COUNT TO DETERMINE TIME	
51.2.2	ACTION OMITTED	5 2 2	SLIP OF MEMORY	AS ABOVE AND POSSIBLE OVER-PRESSURE OF TANKER (SEE STEP 5 1 2 3)			MIMIC OF VALVE CONFIGURATION
51.2.3	ACTION TOO EARLY	NO RECOVERY	PLACE LOSING ERROR	IF VALVE CLOSED BEFORE TANKER SUPPLY VALVE OVERPRESSURE OF TANKER WILL OCCUR		STRESS IMPORTANCE OF SEQUENCE AND EXPLAIN CONSEQUENCES	INTERLOCK ON TANKER VENT VALVE
51.2.4	ACTION OMITTED	5 2 6	SLIP OF MEMORY	AUTOMATIC CLOSURE ON LOSS OF INSTRUMENT AIR			MIMIC OF VALVE CONFIGURATION
51.2.4	ACTION OMITTED	5 2 2	SLIP OF MEMORY	AUDIO FEEDBACK WHEN VENT LINE OPENED		EXPLAIN MEANING OF AUDIO FEEDBACK	MIMIC OF VALVE CONFIGURATION
51.3	ACTION OMITTED	NO RECOVERY	SLIP OF MEMORY	LATENT ERROR	ADD CHECK ON FINAL VALVE POSITIONS BEFORE PROCEEDING TO NEXT STEP		MIMIC OF VALVE CONFIGURATION

Table 5.3.3 SRK external error modes

Action omitted
Action too early
Action too late
Action too much
Action too little
Action too long
Action too short
Action in wrong direction
Right action on wrong object
Wrong action on right object
Misalignment error
Information not obtained/transmitted
Wrong information obtained/transmitted
Check omitted
Check on wrong object
Wrong check
Check mistimed

on the grounds that treating the 'root causes' of the errors will probably be the most effective way of reducing the frequency of those errors.

The method comprises a set of flowcharts which have been computerised to lead the assessor through a task analysis to identify the principal external error modes and psychological error-mechanisms. In practice, the assessor needs to be familiar with the task context and the meanings of the psychological error-mechanisms.

The original computerised SHERPA system has been superseded by a later version called *SCHEMA* about which little has been published by its developer Embrey, and SHERPA itself is also being further developed by its funders, UKAEA. Therefore, at present, it is difficult to evaluate its current status. However, one principal advantage which it does enjoy is that of giving the assessor a means of carrying out a finer-detailed analysis which has both a higher degree of resolution of external error modes (see Pedersen, 1985, and Table 5.3.3), and an expanded set of psychological error-mechanisms. Since it utilises task analysis and a HEA tabular format, it can represent a powerful, in-depth analytical tool. Its disadvantage is that despite its title, it is not clear that it is systematic since it may not always be reliably used by two different assessors. It also used to be fairly 'jargon-based', and initially assessors found it difficult to use. It is possible, however, that these problems, which meant that in practice it was really more of an expert's tool, may have since been overcome.

5.3.4.5 Generic error modelling system (GEMS)

Developed by Reason (Reason & Embrey, 1985; Reason, 1987; 1990), this is a model which assists the analyst in understanding the process by which an operator might move from the skill-based and automatic level of task-implementation to the rule-based level, and, if necessary, to a higher, knowledge-based diagnosis.

GEMS (Reason, 1987), as mentioned before, classifies errors into two categories: *slips* and *lapses* at the one level, and *mistakes* at the other level. It further postulates that 'slip'-type errors occur most frequently at the skill-based level, whereas mistake-type errors usually occur at the higher, rule- and knowledge-based levels. In this context, a slip or lapse can be considered unintentional, whereas a mistake (or rule violation) could in broad terms be thought of as an actual error of judgement or intention.

The technique gives guidance on the types of 'error-shaping factors' which are likely to apply to the above two categories of error, such as mind-set, overconfidence, an incomplete mental model, etc. In total, these amount to factors for all three categories of action. However, it must be noted that it is very much left up to the analyst's insight and imagination first to ascribe particular error-shaping factors to any individual task step and then propose measures by which these negative factors may be overcome.

The generic error-modelling system approach is a necessary 'partner' tool to use with SHERPA (although it may now be superseded by CADA; see Gall, 1989), since SHERPA was only designed for skill- and rule-based errors (once again, this may have been resolved by Embrey's SCHEMA system) while GEMS was concerned with cognitive errors, i.e. the knowledge-based behavioural domain in Rasmussen's terms. As has already been noted, GEMS is basically a classification system (lately, Reason has suggested that it can be computerised as a model, although it is not clear how this would actually be done). GEMS is therefore still treated as a taxonomy, and the error types are shown in Table 5.3.4.

The GEMS system is useful in that it defines a set of cognitive-error modes, including biases in judgement, which may affect human performance in error-sensitive scenarios such as a case where there is a misdiagnosis during abnormal events or incident sequences. As with a good number of techniques in this area, however, the guidance on how to choose these underlying errors is quite limited and relies, ultimately, on the assessor's own judgement. Nevertheless, the GEMS system arguably offers a more useful set of error modes for cognitive-error analysis than the SRK model, and so is accordingly more useful to the practitioner. Anyone considering using the GEMS classification should, as a minimum requirement, consult Reason (1990) for a detailed description of what each error-form refers to and what factors influence these errors' occurrence.

5.3.4.6 Murphy Diagrams

Developed by Pew *et al* (1981) as part of a study commissioned by the Electric Power Research Institute in the United States, Murphy Diagrams are named after the proposer of the well-known axiom 'If anything can go wrong, it will'.

Table 5.3.4 The generic error modelling system (GEMS) (Reason, 1987)

Performance level	Error-shaping factors
Skill-based I	1. Recency and frequency of previous use 2. Environmental control signals 3. Shared schema properties 4. Concurrent plans
Rule-based II	1. Mind-set ('It's always worked before') 2. Availability ('The first to come is preferred') 3. Matching bias ('like relates to like') 4. Oversimplification (e.g. 'halo effect') 5. Overconfidence ('I'm sure I'm right')
Knowledge-based III	1. Selectivity (bounded rationality) 2. Working-memory overload (bounded rationality) 3. Out of sight, out of mind (bounded rationality) 4. Thematic 'vagabonding' and 'encysting' 5. Memory prompting/reasoning by analogy 6. Matching bias revisited 7. Incomplete/incorrect mental model

Essentially they are pictorial representations, very much following the pattern of fault trees, which show the error modes and underlying causes associated with cognitive decision-making tasks. Each stage of the decision-making process is represented separately, such stages being:

- Activation/detection of system-state signal
- Observation and data collection
- Identification of system state
- Interpretation of situation
- Definition of objectives
- Evaluation of alternative strategies
- Procedure selection
- Procedure execution

In each diagram the 'top event' is the failure to carry out the task stage correctly (e.g. a failure to detect a signal), whilst the immediately following intermediate events are the higher-level error causes (e.g. a signal obscured). These are followed by lower-level causes which consist of either 'hardware deficiencies' (e.g. in the control-room layout) or, in some cases, psychological mechanisms (e.g. distraction). Within the Murphy Diagrams, the terminology employed for these three levels of event are Outcome (Top), Proximal Sources (Intermediate) and Distal Sources (Bottom) (see example in Figure 5.3.8).

As a technique for representing decision-making errors, the Murphy Diagram is easily followed. Recently, it has been resurrected to an extent in the form of the Critical Action and Decision Approach (CADA: Gall, 1990), in

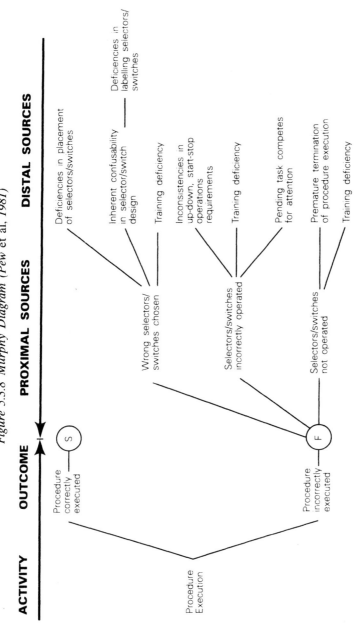

Figure 5.3.8 Murphy Diagram (Pew et al, 1981)

ACTIVITY   OUTCOME   PROXIMAL SOURCES   DISTAL SOURCES

S   Procedure correctly executed

F   Procedure incorrectly executed

Procedure Execution

Wrong selectors/switches chosen

Deficiencies in placement of selectors/switches

Inherent confusability in selector/switch design

Deficiencies in labelling selectors/switches

Training deficiency

Selectors/switches incorrectly operated

Inconsistencies in up-down, start-stop operations requirements

Training deficiency

Selectors/switches not operated

Pending task competes for attention

Premature termination of procedure execution

Training deficiency

an attempt to update the model and use it predictively for HEI (in the 9 years since it was first developed, it had started to show some deficiencies in its scope and comprehensiveness). The Murphy-Diagram approach is yet another offshoot of the SRK model.

5.3.4.7 Human reliability management system (HRMS)

The HRMS is not currently available in the public domain, but is included here to show how some of the more sophisticated models are developing in the field of HRA.

The HRMS (Kirwan, 1990; Kirwan & James, 1989) is a fully-computerised HRA system which contains a HEI module which is used by the assessor on a previously prepared and computerised task analysis. Both the task-analysis module and HEI modules are defined below. Figure 5.3.9 shows the basic interrelationship of all the HRMS modules.

TASK ANALYSIS MODULE

The task-analysis module has the following facilities:

- a simplified hierarchical and sequential task description, input by the assessor, based on an off-line hierarchical task analysis (see Figure 5.3.10). This is the minimum task analysis for a given scenario, and it specifies what the operator has to do in terms of goals and operations in a fixed or flexible sequence. This task description feeds through into the identification and quantification modules, and work in those modules cannot be carried out unless a task-analysis file has already been prepared. Hence, quantification without a task description (which amounts to a misuse of the system) cannot take place within the HRMS.
- a more detailed tabular-format task-analysis description. Within the program are guidelines on how to develop a tabular-format task description which will identify interactions between personnel and equipment, communications, etc. Guidelines are given on when to use such a task analysis – for example on more sensitive scenarios, and especially those involving critical time-dependence factors (e.g. responses to alarms). See Figure 5.3.11 for an example of a tabular task analysis.

An example of the use of such task analysis is given to the user. The facility actually to carry out such a task analysis on the computer has not been judged to be cost-effective at this stage, and as a result the user develops the tabular task analysis *off-line* instead. The same sequence of task steps in the tabular task description is then entered into the sequential-format task description, as described above.

HUMAN ERROR IDENTIFICATION (HEI) MODULE

The Human Error Identification module has the following components within it:

Figure 5.3.9 Human reliability management system framework (Kirwan, 1990; Kirwan & James, 1989)

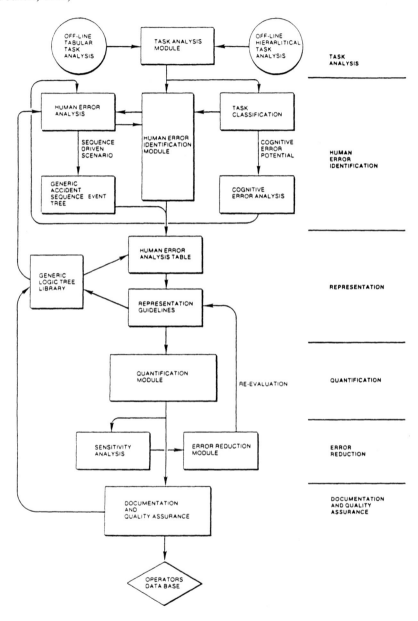

Figure 5.3.10 HRMS linear HTA format

Justification of human error data interpretation

FILE NAME:
ASSESSOR:
DATE:
SAFETY CASE:

Scenario

The enrichment monitoring and blending sequence reaches the point where a daily sample of UN is required. The CCR operator requests that a local operator go to the plant room and take a sample from the correct tank. The local operator does this and despatches the liquor to the laboratory.

In the laboratory, the analyst performs the test, records the result and transmits it back to the CCR. Here the CCR operator inputs the result into the DCS. If this result differs markedly from the on-line measurement, then the DCU draws this to the operator's attention. The operator makes a decision on sentencing the liquor, and must obtain a Management Authorisation before proceeding. It is assumed that this supervisor's check should reveal any mistake by the CCR operator while inputting the result, as well as any difference between this result and a corresponding on-line measurement which has not been noticed by the CCR operator.

Task analysis

q1. INITIATE SAMPLING
1.1 CCR op recognises need for daily UN sample
1.2 CCR op communicates with local op
1.3 Local op goes to plant room and takes sample
1.4 Local op despatches sample for test
2. ANALYSIS
2.1 Analyst carries out test
2.2 Analyst records results
2.3 Analyst transmits results
3. INPUT OF RESULT TO DCU
3.1 CCR op enters lab result into DCU
3.2 DCU warns op of discrepancy between measurements
3.3 CCR op obtains Management Authorisation
3.4 CCR op sentences liquor

- A task-classification module. This module helps the user to identify whether or not there is a knowledge-based component to the task. If, and only if, there is, the cognitive-error sub-module is then selected before the normal-error module. The task-classification module primarily uses Rasmussen- and GEMS-type classifications to formulate a set of questions. The answers to these questions, achieved via a simple algorithm, determines whether a knowledge-based or cognitive error (e.g. such as a misdiagnosis) could occur. The classification module errs on the conservative side by identifying all knowledge-based and rule-based diagnostic tasks as having a cognitive-error potential. The essence of this module is presented in flowchart form later in this section.
- A cognitive-error-analysis sub-module. This sub-module is aimed at detecting any possible cognitive errors within the whole scenario. A set of 10

Figure 5.3.11 HRMS tabular task analysis format

Task goal	Time	Staff	System status	Information available	Decision/action/ communication	Equipment/ location	Feedback	Distractions/ other duties	Comments and operator expectations
Normal operations		CRO, SS	Normal operating envelope			CCR		Busy on SC/HHC activities	Single CRO. SS not busy. Assume beginning of leach sequence (worst case timescale).
Respond to first alarm	T + 10s	CRO	Both DOG fans fail for a cause other than C&I failure	*Alarm*: (yellow) on low DOG fan shaft speed	CRO targets 'fetch alarm'. Goes to level 4 mimic	VDU, DOG console	Flashing yellow alarm message	Previous tasks; Other alarms begin to occur	
Respond to further alarms	T + 20s	CRO	Further alarms occur due to: • Low DP across fans • Low flow in DOG fan leg • Low DP across dissolver • Low pressure fall across columns	*Alarms*: Yellow Red Red Red	CRO targets next alarm. Goes to level 4 mimic. CRO goes to group alarm listing. Looks at printer	VDU, printer	Flashing, audible messages	Other alarms	
CRO identifies fan failure & calls Supervisor (accept alarms)	T + 60s	CRO/ SS	Alarms on: • Low flow DOG duct at Stack Monitoring Room • Hardwired in DOG duct manifold	*Alarms*: Red Hardwired	CRO monitors level 3 mimic CRO acknowledge alarms, calls Super. SS detects alarm from own console	DOG VDU DOG VDU and BUP SS VDU console	Audible and visual Audible and visual Audible and visual	SS involved other tasks	

Figure 5.3.12 Human error analysis HRMS module – example of EEM questions and PEMs

Human error analysis

⇒ 1   DETECT AND ACKNOWLEDGE VDU ALARMS
   1.1 Detect first alarm (fan shaft low speed)
   1.2 Acknowledge alarm from level-4 VDU mimic
   1.3 Respond to next alarm (n = 4)
   1.4 View group-alarm mimic
   1.5 View alarm listing
  2   RESPOND TO HARDWIRED ALARM
   2.1 Detect hardwired alarm
   2.2 Acknowledge alarm
   2.3 Contact Shift Supervisor (SS)
  3   INFORM SS OF SCENARIO HISTORY
   3.1 Control Room Operator (CRO) explains prior operations
   3.2 CRO suggests fan failure as scenario
   3.3 SS checks alarm listing and level-3-and-4 mimics
  4   SS/CRO DETERMINE RESPONSE

Considering each task step above. . . .
Could the operator fail to detect plant-state indications within the required timescale?
Potential error: CRO fails to respond to all signals
  Press <F1> for instructions and help

Task Step: 1   DETECT AND ACKNOWLEDGE VDU ALARMS

  Could this be caused by . . .
  1. Lapse in operator's attention.
→ 2. Operator's missing indications because too many are present.
  3. Absence of important signals due to maintenance/calibration error or hardware/
     software failure.
  4. Operator's misperceiving indications as the usual signals.
  5. Failure to discriminate adequately and detect that a signal is different.
  6. Concentration on another task causing a failure to perceive signal(s).

Could the operator fail to detect plant-state indications within the required timescale?
Potential error: CRO fails to detect first alarm
  Press <F1> for instructions and help

questions, with slight overlaps/repetitions, is put to the user, one at a time; these look for potential misdiagnoses made as a result of various psychological mechanisms.

- A human-error-analysis sub-module. This sub-module aims to identify all potential errors associated with a task (excluding cognitive errors). As for the previous sub-module, the questions are asked individually, but this time the user must attach the error, via a menu selection, to one of the sequential task steps. Questions are asked which denote particular human-failure modes. These are based on a number of error taxonomies. See Figure 5.3.12 for an example from this module.

Also provided is a screening-analysis table, based on the task description and the errors identified during this module. This table encapsulates the information gained so far, and allows the user to identify error-recovery points in the scenario, as well as dependencies between identified errors. This table (see Figure 5.3.13) will also allow the user to note which error descriptions are redundant, i.e. which such descriptions have been identified more than once due to the heterarchical structure of the HEI module. The HEI table or the listing of human errors, or both, can be printed out as hard copy. The HRMS HEI module is based on a combination of the SRK-, SHERPA- and GEMS-type models, but also incorporates additional psychological error-mechanisms derived from the assessor's experience.

The HEI process in the HRMS relies on a high degree of assessor expertise and judgement in deciding which errors are likely and which of the possible psychological error-mechanisms are principally responsible. Ideally, the HRMS would consider PSFs or the type of task involved (or both factors) in helping the assessor to decide these parameters, but the developer believes this is not yet feasible given the current state of knowledge. Hence, rather than risk ruling out errors based on *ad hoc* theory, the system allows a large proportion of them to be identified. Thus, a trade-off is implicit in this system in favour of comprehensiveness rather than resources-effectiveness.

However, it has a powerful set of error modes and psychological error-mechanisms, the latter being allocated to particular stages in SRK functioning, which can lead in turn to an exhaustive HEI procedure for a scenario. One main advantage is that after using the system, the assessor has a high degree of confidence that all errors have been identified. One disadvantage, however, is that although jargon is avoided, the reliance on the assessor's own judgement is high, and more errors than are required for quantification purposes may initially be identified. This latter aspect is resolved, however, by the representation module in the HRMS, which makes use of a screening method (see the case study in Appendix VI).

The HRMS has only recently been developed, and there is not enough practical evidence so far to determine its level of effectiveness. However, in internal studies it *has* been shown to be comprehensive, and hence to offer a (potentially) rigorous framework for identifying errors at all SRK levels.

5.3.4.8 Summary on existing methods

The above has reviewed some of the state-of-the-art techniques involved in the HEI process. For a review, including two comparative reviews (one of which was a validation), see Kirwan (1992a & 1992b), and Whalley & Kirwan (1989).

Clearly there is scope for new methods, or for methods which attempt to deal with different types of human-error contribution to risk levels, and there is certainly a tremendous scope for more validation and testing of these methods. What the next section does is present a framework of techniques (some old and some new) by which the assessor can address most human-error contributions to risk levels, with the exception of sociotechnical and

Figure 5.3.13 Summary table of screening in HRMS

STEP	No. 1	DETECT & ACKNOWLEDGE VDU ALARMS
ERROR	No. 01090	CRO fails to respond to all signals
ERROR MECHANISM		Cognitive/stimulus overload
RECOVERY		SS/HARDWIRED ALARM/OTHER ALARMS
DEPENDENCY/EXCLUSIVITY		
SCREENING		
COMMENTS		REC HIGHLY LIKELY TO BE QUANTIFIED AS CRO CONFUSION DUE TO HIGH DENSITY ALARMS.

STEP	No. 2	RESPOND TO HARDWIRED ALARM
ERROR	No. 07090	CRO fails to detect hardwired alarm amongst others
ERROR MECHANISM		Discrimination failure
RECOVERY		
DEPENDENCY/EXCLUSIVITY		
SCREENING		SUB
COMMENTS		HARDWIRED ALARM VERY SALIENT. 'COPIED' TO SS CONSOLE. SUB 01090: ALARM QUANTIFICATION MODULE QUANTIFIES TOTAL RESPONSE TO ALARM SET.

STEP	No. 3.2	CRO suggests fan failure as scenario
ERROR	No. 13010	CRO/SS fail to identify scenario in time
ERROR MECHANISM		Cognitive/stimulus overload
RECOVERY		SS, EMERGENCY PROCEDURES
DEPENDENCY/EXCLUSIVITY		
SCREENING		
COMMENTS		THIS ERROR IS USED TO REFER TO THE FAILURE TO IDENTIFY THE OVERALL GOAL OF PREVENTING BOILING BY ACHIEVING RECIRCULATION VIA CHANGING VALVE STATUS.

organisational errors. This framework is called simply the *onion framework*, for reasons that will become apparent below.

5.3.5 The onion framework

5.3.5.1 Background to a multi-levelled approach to HEA

There are currently no totally comprehensive HEA tools – i.e. tools which will deal with skill- and rule-based errors, as well as cognitive errors (misdiagnoses), rule violations, maintenance errors and management failings leading to potential disaster on the commercial market. In short, there are no techniques which address the full potential impact of human error on one system or another. Instead, there are only tools which deal with particular types or subsets of potential human errors. What the human-error analyst has at most, therefore, is a set of limited HEA tools in his or her tool-box. The achievement of a comprehensive error analysis at present is therefore likely also to be relatively dependent on the assessor's own individual judgement. This is not a particularly desirable state of affairs, but it is, unfortunately, the current state of the art. For this reason, HEA technology is a primary target for research.

Notwithstanding these limitations, the HEA approach can be highly effective in identifying a system's vulnerabilities to human error, and, in some cases, in identifying how to curtail them. The problem for the assessor is knowing what tools to use, when to use them and how to integrate them into a coherent approach. To facilitate these aims, this chapter has cast a range of tools in a multi-levelled framework; this is called the *onion* framework because at each successive 'level' of assessment, the assessor is, in a sense, conducting a deeper analysis of errors, their causes and the subtlety/complexity of their potential impact: much like peeling the layers of an onion. This framework is depicted in Figure 5.3.14. Each layer is outlined below and then described in full detail in a series of subsections in this chapter.

LEVEL '0'

The outer level, level 0, is the lowest level of the HEA since in effect it does not actually identify errors at all but considers simply task successes and task failures. It is, however, necessary to consider this most basic level first since this is a primary interface between the HEA and the PSA. Initially, a PSA analyst may simply want to know whether a particular task is likely to be achieved or not, and not be concerned with why the task may not be achieved. For example, in a loss-of-coolant scenario in a nuclear-power plant (NPP), the PSA may require the quantification of the task 'Achieve re-circulation'. Such a task may involve anything from a few single steps such as activating a pair of push-buttons, to a complex and extensive series of value manipulations. At this stage, however, the PSA analyst may simply want a single

Figure 5.3.14 Onion framework

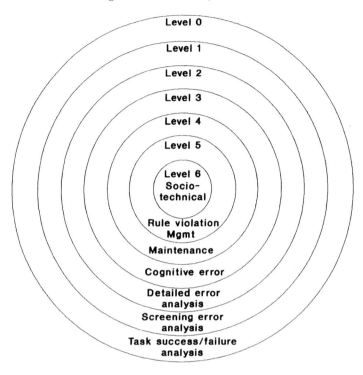

task-success/failure probability to feed into the PSA event tree. If the task can be thus quantified 'holistically', then the HEP can be put into the PSA and its sensitivity determined.

If the PSA risk calculations in such an example *are* in fact sensitive to this probability, then the PSA analyst will be likely to demand a more detailed analysis of the task, one breaking it down into errors and quantifying each of these, both to gain a more robust quantification of the task-failure likelihood and to determine the factors affecting performance for later error-reduction purposes.

Level '0' can thus be considered a 'screening' level (to be dealt with more fully in Section 5.4), or else a level fulfilling the needs of a purely quantitative PSA (i.e. not requiring any risk reduction within its scope). There are, however, two potential limitations inherent in this approach. Firstly, the model assumes quantification can take place at a holistic level. While a number of quantification techniques purport to be able to do this, their actual ability in this respect remains relatively unverified, however; this will be discussed in later sections. Secondly, this approach assumes that there are no significant errors overlooked which could either invalidate the resulting quantification (by lying outside the database supporting the quantification technique) or lead to different and more severe consequences than those modelled in the PSA.

These limitations render the level-0 HEA a somewhat uncertain approach. Indeed, when the level-0 approach is adopted for reasons of resources (e.g. in a very large PSA for a complex installation) it is usually accompanied by selective, detailed analyses of key tasks (e.g. required responses from the operator following plant trips). Some practitioners, however, will simply feel uneasy about the level-0 HEA, and will always proceed to level 1, or even deeper, in setting a starting point for their HRA.

CRITICAL TASK IDENTIFICATION

Implicit in level-0 analysis, and in some other levels, is the selection of key tasks to be incorporated into the PSA. In a NPP, for example, it would be very resource-intensive to analyse every task the operators had to do, during normal, abnormal, shutdown, start-up and maintenance operations, to see what errors could occur and what consequences could arise. But in a system for loading chemicals into a tanker, for example, such a comprehensive and 'bottom-up' analysis *could* be done. With a NPP PSA, therefore, certain key tasks will be targeted for inclusion in the analysis, e.g. tasks during which a *trip shutdown* (i.e. a very rapid or 'crash' shutdown of a system) would be initiated, tasks related to the functioning of back-up and emergency safety systems and finally required post-trip actions. The identification of such tasks sets the scope of the remainder of the analysis, and so is a highly critical stage. This area of 'critical task identification' is, however, even less well researched than the HEA process itself. PSAs will typically utilise the following (and other) approaches:

- A review of similar/previous PSAs.
- A review of operational experience.
- The carrying-out of formal hazard analyses – including a HAZOP, a fault-tree analysis, an event-tree analysis, a FMEA, etc – to model all the system's hazards.
- The carrying-out of a functional analysis of the system, and of its defences.

Most of these approaches are described in this section, and they will demonstrate a number of key human involvements. One limitation inherent in HRAs and PSAs (which currently just has to be tolerated) is that in a large PSA, not all tasks can be considered, such approaches may overlook a potentially significant human error – This limitation is a justification for conservatism in failure-likelihood estimates for HRAs and PSAs. It is also a justification for detailed but selective HEAs and error-reduction strategies, in the latter case so that some of the error-reduction mechanisms (ERMs) implemented may be generic enough to mitigate against the effects of any unforeseen errors.

Although these limitations in PSAs and HEAs may at this point seem dismaying, it is unreasonable to expect that a HEA system could identify every single potential human error that could one day occur in the conceivable lifetime of a complex system. The extraordinary adaptability and variability of human behaviour argues against this very strongly. However, what a HEA *can* do is give an indication both of the extent of human error likely and of the level of vulnerability of the system to human error. If the system is a detailed one, it can also outline ways of lowering that level of vulnerability.

A human HAZOP, as outlined earlier, can also be applied at this stage, as well as at subsequent stages.

LEVEL '1'

Level 1 of the onion model is still not a HEA at a 'detailed' level (which occurs from level 2 onwards) but it does mark the *first step* in the HEA process, and may be used as a more refined screening approach which attempts to gain more of an insight into the task, its PSFs and any potentially significant errors. Either such errors can then be quantified holistically (i.e. the task-failure probability is still quantified, but the experts, or the assessor, using the quantification techniques know what errors and weaknesses are inherent in the task), or else each error can be quantified and then the task-failure likelihood can be assessed as the aggregate of these error probabilities. Once again, if the error probabilities are found to be important in the PSA, then a more detailed HEA should ensue. In a large PSA, this level may well be the starting point for most practitioners.

LEVEL '2'

Level 2 is the first detailed level of analysis, and this is the level at which several of the available techniques exist. This level aims only at the skill- and rule-based levels of behaviour, leaving cognitive errors to the level-3 analysis. Many HEAs have taken place at level 2, and usually a large number of potential errors are yielded which, following representation, are quantified, and fed into detailed fault and event trees. It is also at this level that the causes of errors (PSFs) and their underlying mechanisms are assessed and recorded, thus enabling an error-reduction analysis to take place either at this stage or later in the analysis (following quantification). Error-recovery for all levels is dealt with in the subsection on level-2 assessment.

LEVEL '3'

Level 3 considers cognitive errors, or undesirable impacts on the system, occurring as a result of mistakes and misconceptions on the part of the operator or the operating team. These are the errors such as occurred in the Three Mile Island incident and other incidents, and they are recognisably pernicious due both to their ability to degrade safety systems and to their relative resistance to recovery. Cognitive errors are a relatively difficult field, and the tools available are relatively wide-ranging – from taxonomies or classifications of the cognitive error-forms to more formalised means of comparing the characteristics of potentially confusable symptoms in a scenario. Other more sophisticated approaches are under development (Woods *et al*, 1990).

LEVEL '4'

Level 4 considers the impact of maintenance on a system's risk level. This raises an old HRA question concerning the degree to which the component-failure-rate and maintenance data implicitly include an element of human error, on the grounds that component-failure-rate data will have come partly from human-maintained items. The problem involved with assuming that such data *do* include a human-error factor is that the component being assessed in the current study may now fall under a different maintenance regime, and this fact could change its failure rate significantly. Furthermore, the component in the current study could also be embedded within a system in a *different* way from the ones which were the subject of a data collection, and as a result human errors on the current component could involve different system aspects from those which an existing data source assumes belongs to them.

Maintenance errors are rarely assessed in PSAs, yet 40–80 per cent of incidents contain a maintenance-caused factor. Level 4 is an attempt, therefore, to bring some of the key maintenance contributions under HRA scrutiny.

LEVEL '5'

Level 5 deals with two even more difficult areas, namely those of rule violation and management failures. It may at first sight seem odd to put them at the same level, but currently, at least, this is appropriate since the occurrence of a rule violation is indicative of a safety-culture problem, and hence of a management error or oversight. In practice, the management 'errors' are left out of level 5 at this stage in a HRA's development since this is currently an area of intense but still relatively formative research. This area is discussed further in Chapter 5.11. Rule violations are considered, in this section, however, together with some brief guidance on how they may be assessed (this is an area requiring further research).

LEVEL '6'

Level 6 is concerned with 'sociotechnical errors'. These are errors which will usually be highly idiosyncratic, and are a combination of fairly personal factors in a relatively vulnerable organisational system. They are of particular concern, for example, in scenarios where a single individual has the potential to kill a large number of persons – for example, with transport systems or missile defence systems. Tools at this level are virtually non-existent, but some basic

Table 5.3.5 The onion framework: question to help determine the required level of analysis

---

Is the PSA and the number of tasks undertaken by the operator very large?

If so, consider a screening approach, at level 0 or level 1, backed up by a detailed task and error analysis of sensitive tasks (levels 2–5).

Is the PSA concerned with a novel system which has a high hazard potential and a high degree of human involvement?

If yes, carry out levels 2–4; also, implement a safe culture (see Chapter 5.11).

Does a cognitive-error potential exist (see flowchart at level 3)?

If yes, carry out a level-3 analysis.

Is there any evidence (from operational experience) of rule violations?

If yes, carry out a level-5 analysis.

Is reliable maintenance-failure data available for this plant or an identical one?

If no, carry out a level-4 analysis for a subset of tasks to determine the appropriateness of the maintenance data currently available.

Do any personnel carry out tasks during which many people's lives rest in their hands alone?

If yes, carry out a level-6 analysis.

---

guidelines *can* be given for the identification and reduction of errors in these scenarios.

The onion framework is hopefully a temporary stand-in in the field of HEA until a more 'global' model of human action and error is developed. Nevertheless, the tools described below in the following subsections *can* help the assessor to identify and define human impacts on a system's risk level, and ultimately reduce them. No strict guidance can be given on which levels to use and where to use them, but some of the key determinants for the practitioner can be summed up by the questions shown in Table 5.3.5.

Finally, before presenting the techniques themselves, the author cannot guarantee these techniques will work, alone or in unison, and in almost all cases, wishes there were more theoretically robust and empirically validated techniques available. All that can be said is that in a number of applications, these tools *have* proven useful for HEAs, and have helped significantly to reduce the level of vulnerability of systems to human error.

5.3.5.2 *Level 0: task success/failure identification*

As noted above, explicit error-identification does not take place at this stage. Nevertheless, a basic level of task analysis (called 'Initial Task Analysis, ITA) must be undertaken so that subsequent quantification will benefit from some kind of understanding of the task. This basic or minimum task description will need to encompass the following, and can be put into a tabular format, or retained simply as descriptive information in prose format – if more than a handful of scenarios are being considered, the former is suggested:

- The scenario's initial cause.
- The goal of the task.
- The approximate extent of the task, in terms of the number and type of tasks comprising the overall goal or task being assessed (e.g. a CCR vs a local operation; a simple vs a complex task; skill-, rule- or knowledge-based task components).
- The time available to perform the task.
- The personnel available to perform the task.
- Any adverse conditions present (e.g. the fact that the task has to be done at night-time, the weather, environmental hazards, etc.).
- The availability of equipment for performing the task, including displays that prompt the operator to act, appropriate controls and the presence or absence of feedback as to whether or not the action has occurred and how effective it was.
- The availability of written procedures (i.e. written operating instructions).
- The degree of training relevant to the task.
- The frequency and severity of the event.
- The basic sequence of events (in a simple linear HTA format)

If the above, basic information requirements are not met, the analyst will find it difficult to quantify the degree of human-error contribution accurately. This basic information, which will give the analyst an overall 'feel' for the feasibility of the task, is really gathered to ensure that there is someone available to perform the task who has enough time, equipment and knowledge, written procedures to achieve it. Without such basic information, the quantification of human reliability becomes a fairly optimistic or even bogus activity. An example of a task described at this level is shown in Table 5.3.6. Further information may be required depending on which quantification approach is utilised.

As noted in the HEA's introductory section, the most important matter as regards a level-0 HEA is that of the identification of critical tasks – i.e. deciding which tasks to address. This is usually determined during the PSA through the means already noted earlier. Some additional guidelines are as below:

- Consider all phases of human interaction:

    Maintenance, testing and calibration
    Normal running
    Start-up and shutdown
    Abnormal and emergency tasks.

- For each of the above, consider the key safety roles played by the operator(s), in particular where: an action on the part of the operator(s) is needed to

Table 5.3.6 Level 0 – task description

Initial cause of scenario: the rupture of a pressurised half-inch pipe containing HF (hydrogen fluoride, which has toxic and corrosive properties).

Task goal: the CCR operator has to evacuate the plant and make the plant safe.

Main tasks: the CCR operator has to initiate the evacuation of the area and stop the flow of HF. This requires detecting the event (via HF-detection alarms, loss-of-pressure alarms and CCTV); activating the emergency-evacuation push-button; activating the emergency-shutdown (ESD) button; manipulating three valves remotely via the VDU control system; and finally sealing off ventilation to and from the area (via the VDU system).

The evacuation and ESD responses are required within 1 minute of the alarm's occurrence, and isolation is required within 5 minutes. The CCR is permanently staffed. The event is assumed to occur at night, with a 30 per cent chance of occupancy (one operator only) in the affected area. Operators are trained in the required response, and emergency procedures are available. The frequency of the event is estimated (from the PSA) at once per 20 years.

---

initiate or back up an engineered safety function; a control procedure is performed by the operator(s) to maintain a safety parameter (e.g. in connection with temperature or pressure) within certain limits (e.g. during shutdown or certain maintenance activities); the barrier to an accident is largely a *human* one, perhaps involving a series of human 'administrative' controls (rules or procedures rather than physical barriers); for safety purposes, critical maintenance (e.g. the calibration of gas detectors offshore) is required.
• Carry out operational-experience reviews to consider what types of failures/errors/near misses have occurred in the past. This may be supplemented by a use of the Critical Incident Technique (see Kirwan & Ainsworth, 1992).
• Review emergency procedures for essential post-trip actions.
• Run a table-top discussion involving PSA and systems analysts, HR analysts and system personnel (if the system exists), together with designers and operations personnel (if possible), to consider what human interactions should be targeted for analysis.

These methods can be used as a supplement to the more traditional ones mentioned earlier as a means to furthering (or ensuring) the comprehensiveness of the critical-task-identification process that is carried out. Ultimately, this is an area of HRA that requires further development in collaboration with PSA analysts.

5.3.5.3 Level 1 HEA

The first major difference between level 0 and level 1 is that a basic HTA is required for level 1. An example of this form of HTA, called simply a *linear HTA*, is shown on Figure 5.3.15. This figure is shown in two formats, the

Figure 5.3.15 HTA and corresponding Linear HTA

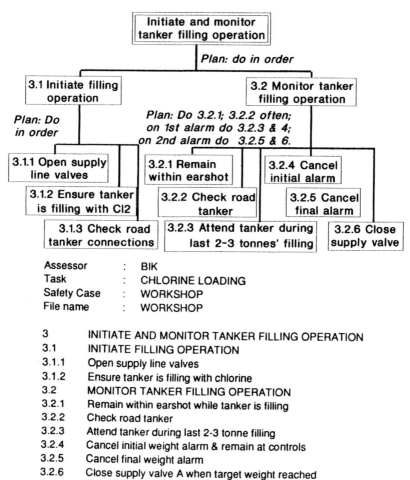

Assessor	: BIK
Task	: CHLORINE LOADING
Safety Case	: WORKSHOP
File name	: WORKSHOP

3	INITIATE AND MONITOR TANKER FILLING OPERATION
3.1	INITIATE FILLING OPERATION
3.1.1	Open supply line valves
3.1.2	Ensure tanker is filling with chlorine
3.2	MONITOR TANKER FILLING OPERATION
3.2.1	Remain within earshot while tanker is filling
3.2.2	Check road tanker
3.2.3	Attend tanker during last 2-3 tonne filling
3.2.4	Cancel initial weight alarm & remain at controls
3.2.5	Cancel final weight alarm
3.2.6	Close supply valve A when target weight reached

diagrammatic HTA format (5.3.15a) which the analyst may actually use when developing the HTA, and the condensed linear format (5.3.15b), which may be used, subsequently, for a HEA. The HTA, whichever format is used, serves to encode both the goal and all the tasks involved (together with their individual operations and plans). In addition, more information-gathering occurs at this stage, including the observation of tasks (where possible), interviews with operators concerning error potentials and PSFs (see later) and more specific references to procedures and instructions for determining task sequences.

The primary aim of this level of the HEA process is to define the task and identify major error-forms within the task. This can be achieved by using a range of tools:

- Interviews/CITs/OERs.
- Generic accident-sequence event-tree/system-impact models.
- Cognitive-error-potential flowchart.
- A consideration of basic EEMs.
- An error checklist.
- An error-consequence-recovery table.

These are described below. The major difference between this level and level 2 (described in the next section) is that level 1 is less formalised than level 2, and is both more discretionary and more judgemental. It is intended for a relatively quick assessment, prior to screening/quantification, whereas level 2 is intended to be exhaustive and systematic.

INTERVIEWS/CRITICAL INCIDENT TECHNIQUE (CIT)/OPERATIONAL EXPERIENCE REVIEW (OER)

When trying to determine both significant errors and the overall feasibility of the task, the databases recorded formally, and in operators' memories, are an important source of information. Interviews may be structured or unstructured, but are usually best if structured at first, until a rapport is built up between the analyst and the subject. It must be stressed that the interviews are assessing not individuals but the system. The nature and objectives of the PSA, and the role of the HRA within the PSA, must both be explained. It is then best to start with a particular representative scenario, preferably one previously judged to be feasible. Some typical generic questions are listed in Table 5.3.7. It is helpful to select subjects who are experienced, cooperative and good at verbalising. It is also essential that the interviewer (analyst) understand a good deal about the system prior to interviewing anyone. This can be ensured in a number of ways (see the documentation review in the level-2 discussion, next section).

At least two operators, preferably from differing operating teams, should be interviewed. The interview can take between 20 minutes and 1 hour (if longer is needed, take a break). For more guidance on interviewing, see Sinclair (1990) and Kirwan & Ainsworth (1992).

The Critical Incident Technique (CIT) can be highly useful in two ways. Firstly, it may unearth near-miss incidents which have been overlooked during the PSA identification of critical task/hazard sequences. Secondly, it gives the analyst an insight into real incidents, indicating not what should have happened but what did happen, and how people reportedly reacted in a stressful situation. A CIT can also give a good insight into the real extent to which PSFs are at work in a system. However, it must always be borne in mind that a CIT produces evidence of an anecdotal nature, which often cannot be accurately corroborated. Nevertheless, it is a quick technique to apply, and can be highly beneficial particularly if, for example, one of the scenarios in the PSA has actually happened at the plant.

Table 5.3.7 Sample interviewing questions for level-1 HEA

1. Considering the scenarios in question, and given the required time-scale, how feasible would you consider the task?

2. Who would carry out the task, and where would they be based? How would they communicate, and do they rely on oral communication?

3. How do the operators know what to do? What cues prompt them to act? What cues prompt them to stop? What tells them if they've gone wrong?

4. What are the symptoms prior to an action's taking place? Are these similar to any other events?

5. Are you trained for this work? How often? As a team?

6. Which procedure would be used? (Ask to see it, then or later.)

7. Are there any adverse conditions, occurring either coincidentally with the event or in a causal relationship to it (e.g. a loss of power occurring at the same time as, and possibly as a result of, a storm), which could affect the level of performance significantly?

8. How stressful do you think the scenario would be for the operating team? Have you been in any events like this one, or in any other emergencies/abnormalities? Would you anticipate this being more or less stressful?

9. What do you believe would be the most credible way in which this task could fail?

10. Can you think of any errors or unintended actions which could delay the task's completion or jeopardise it entirely? Do they rely on any crucial maintainable systems?

The operational-experience review (OER) involves an analysis of accident/incident records together with an interviewing of site-incident-analysis personnel. Some systems will have useful HEA-type information encoded in incident reports, while others will not. Information recorded following incidents is more likely to be accurate if the incident-investigation team can respond very rapidly when an incident occurs and if the incident-investigation system is non-punitive towards those found to have committed a (system) error.

These three approaches are relatively rudimentary, yet for level 1 and more so for level 2, they constitute important groundwork for identifying potentially significant errors. The type of information that may be made available through such methods is illustrated by the hypothetical extract in Table 5.3.8.

These three methods are useful in assessing an existing installation with some operational background. If the HRA is for a new design, they will be largely inappropriate, but in such a case a level-2 assessment should still be adopted.

THE SYSTEM RISK PERSPECTIVE

The system-risk perspective involves the consideration of significant error opportunities from the PSA perspective, and is shown diagrammatically via

Table 5.3.8 Example of information derived from CIT/OER/interviews

The analyst may have found that the PSA scenario anticipated has actually occurred in connection with the installation being assessed, and that the evacuation was initially delayed because the operator believed it to be a false alarm and did not wish unnecessarily to 'crash-shutdown' the plant (see the example being developed in this chapter). However, since that event, the reliability of the detection system has been improved, and operators now have a clear safety mandate to carry out an evacuation, given any two confirmatory alarm signals or a single group alarm in the CCR. Even so, in terms of isolating the ventilation to and from the area, it became clear from interviews that the operator could mistakenly seal off the wrong area since the ventilation mimics on the VDU were on separate 'pages' from the alarm and HF-process information, and the four HF areas looked fairly identical on the VDU.

Table 5.3.9 System-risk perspective: human-failure modes

Are there any human actions/errors that can directly cause an accidental event, independent of hardware failures?

What possible errors could occur in conjunction with hardware, software or environmental events/conditions which could lead to an accident?

Are there any combinations of independent human errors which could lead to an accidental event?

What errors could occur during abnormal (maintenance or emergency) events which could (alone or in conjunction with other non-human events) lead to accidental events?

Are there any maintenance errors or actions which could place the system in a more vulnerable state?

the generic accident-sequence event tree (GASET) in Figure 4.5 (Kirwan & James, 1989), as well as via the table of questions in Table 5.3.9. These approaches simply prompt the assessor to search for particular generic human involvements which can progress the sequence of events in an accidental-event chain. The GASET can be used in connection with the task, whereas the system-risk-perspective human-failure modes will be used in conjunction with the task analysis. An illustrative example of the use of these aids is given in Table 5.3.10(a).

COGNITIVE ERROR POTENTIAL (CEP) FLOWCHART

Figure 5.3.16 shows this flowchart, which is returned to in the discussion of level-3 HEAs. The flowchart is used when considering the whole task, though if any part of the task has a cognitive-error potential then the analyst should consider a level-3 analysis, either for that task step or (more likely) for the whole task.

If any cognitive-error potential is identified during a level-1 or level-2 analysis, this does not necessarily imply that a level-3 assessment must also

Table 5.3.10(a) Example of application of GASET/SRP human-failure modes

Within the continuing development of the hypothetical scenario, these aids could prompt the assessor to consider the implications of isolating the wrong ventilation area. Consequences would include possible problems for occupants carrying out maintenance in an unaffected area, but more seriously the transmission of HF gas into the ventilation system itself, and then out into the atmosphere and the main warehouse. The assessor may also consider initiators such as maintenance on pipes adjacent to the HF pipework, either whilst the pipe contains HF (thereby promptly initiating the rupture itself, and probably also leading to an instant casualty) or whilst the HF pipe is correctly isolated (possibly leading to a latent error which will cause a leak when the piping is later de-isolated and used).

Table 5.3.10(b) Application of cognitive-error-potential flowchart

Although the event in this chapter's main example is unusual, it is well-rehearsed, and covered by procedures, etc. No 'knowledge-based reasoning' is required. The potentially significant error of isolating the wrong area has already been identified, and whilst this is a misconception leading to additional consequences, it has arisen out of a skill-based or rule-based slip, rather than a knowledge-based mistake. This example indicates the importance of utilising diverse tools to look for a significant error potential. No cognitive-error potential is assessed at this level.

Table 5.3.10(c) Error modes identified from Table 5.3.11

In the current example, the most important additional major EEMs selected by the analyst, in addition to those already identified, include a failure to initiate the evacuation and the ESD (errors of omission) and a failure to manipulate values in time ('Too late'; or error of omission).

Table 5.3.10(d) Error modes additionally identified via the checklist (Table 5.3.12)

In the HF-release scenario, the assessor may consider crew-functioning errors or communication errors, which could both lead to a delay in initial evacuation, but then reject these upon determining that both operators are required to cover each other so as to ensure rapid evacuation; this is achieved primarily by the operator's being at the chemical reactor console at the time. However, the assessor may decide that a systematic miscalibration of all HF-detectors should be investigated further, as this would be a significant error delaying the detection of the leak.

now be considered and quantified in the HRA as a potentially significant error. It will be up to the analyst to consider whether, for example, a misdiagnosis is a possibility, leading to a different or worse consequence category than that already modelled in the PSA. An example of the application of this flowchart is given in Table 5.3.10(b).

CONSIDERATION OF BASIC EEMS

These are defined in Table 5.3.11, adopted from the technique for human-error-rate prediction (THERP, Swain & Guttmann, 1983). They are considered

Figure 5.3.16 Cognitive-error-potential flowchart (proceeds top-down)

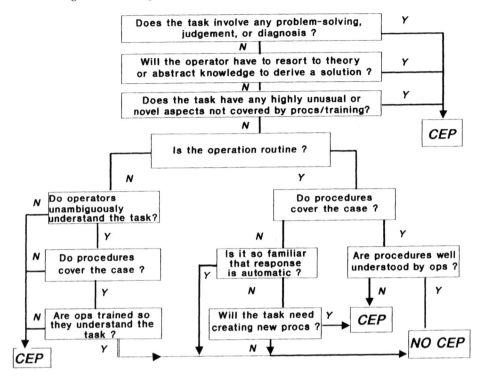

in the context of the task analysis, and the analyst elects those considered to be of major importance. Examples of error modes that could be considered by the analyst are listed in Table 5.3.10(c). For error-recovery-mode analyses, see the level-2 assessment in the next subsection.

ERROR CHECKLIST

The next step is to consider a more comprehensive checklist that looks at errors in all types of behaviour, including rule violations. This kind of checklist is shown in Table 5.3.12. Whilst this encompasses the previous errors, it goes further and adds more context to prompt the assessor in his error-identification activities. This stage acts as a final check to a level-1 HEA. Examples of additional failures identified by this approach are given in Table 5.3.10(d).

The above analyses, despite appearances, would not actually be expected to take very long to carry out, once the initial task-analysis phase has been completed. The final stage in a level-1 analysis is to document the potentially significant errors in a tabular format as shown in Figure 5.3.17, recording also the consequences and likely recovery points of these errors.

Once the table is completed, the analyst can then proceed to a representation in the form of fault/event trees and a quantification, or he or she may

Table 5.3.11 Basic error modes (adapted from Taylor-Adams & Kirwan, 1994)

Errors of Omission

- Omits entire task
- Omits step in task

Errors of Commission

Time Errors
- Action too early
- Action too late
- Latent error prevents execution
- Action too long
  - Accidental timing with other event/circumstance
- Action too short
  - Accidental timing with other event/circumstance

Qualitative Errors
- Act incorrectly performed
  - Too much of action
  - Too little of action
  - Action repeated

Selection Errors
- Right action on wrong object
- Wrong action on right object
- Wrong action on wrong object
- Information not obtained/transmitted
  - Communication Error
- Wrong information obtained/transmitted
  - Communication Error
- Substitution/intrusion error

Sequence Errors
- Incorrect sequence
- Action in wrong direction
- Misalignment/orientation error

Extraneous Acts

- Rule violations

carry out more detailed HEA for particular tasks or task steps. It should be noted that the analyst may still decide to represent and quantify human errors holistically in the PSA, rather than quantify the likelihood of each individual, identified error mode. If this route is taken by the analyst, following a level-1 analysis, the analyst will still enjoy a better understanding of the human-error contributions to risk levels than he or she would have via a level-0 analysis, and can begin, indeed, to make error-reduction suggestions, if required, following the PSA calculations. However, if error-reduction measures are required, a level-2 analysis is a more appropriate approach for what are obviously risk-sensitive error-contributions.

A level-2 analysis may be undesirable on the grounds that it is too resource-intensive, or because the system under assessment is relatively benign (i.e.

Table 5.3.12 Level-1 HEI checklist

---

1. Could a misdiagnosis or misinterpretation occur, possibly due to: a misperception of signals, or expectations; the complexity of the situation itself; a failure to notice special circumstances?

2. Could a short-cut or even a rule violation occur, due to production pressure or a failure to be aware of the hazard?

3. Could a previous (latent) maintenance or calibration error lead to difficulties or errors in responding to the scenario?

4. Could a check fail, be omitted or occur on the wrong system, or simply be the wrong kind of check?

5. Could an information-communication error occur?

6. Could a crew-functioning error occur, such that, for example, it is not clear who should be carrying out the task?

7. Could a required act be omitted because it occurred at the end of the task, because of a slip of memory/distraction or because the operator failed to respond (e.g. due to his or her workload)?

8. Could the task be carried out in the wrong sequence, or a sub-task be repeated, or the operator lose his or her place in the procedures?

9. Could an action or check be carried out either too late or too early?

10. Could an action be carried out with insufficient precision or force, or just inadequately?

11. Could a *non*-required action be carried out by operating on the wrong object or in the wrong location, or by carrying out an incorrect but more familiar task, or by selecting or acting on the wrong information?

---

non-hazardous), with little involvement on the part of the operator, and these are all valid reasons why level 1 may have been selected instead. This depends on the judgement of the assessors (PSA and HRA), and on any regulatory guidance. Once again, if level 0 or level 1 are selected as the analysts' major HEA approach, it is advised that some detailed analysis (e.g. at level 2 or above as appropriate) be carried out for those tasks where the potential human-error impact appears to be the highest.

5.3.5.4 Level 2 HEA

TASK ANALYSIS

The level-2 human-error analysis involves the detailed analysis of tasks and actions involved in an operation. It requires a full task analysis, using the following tools as required:

- Hierarchical task analysis (HTA) – to define the overall goal-task-plan structure of the task, particularly if no emergency response is involved. If an emergency response *is* involved, a HTA can be used to define the top-level goal-task structures.

Figure 5.3.17 Level-1 human-error-analysis table

Error	Consequences	Recovery
Failure to initiate evacuation.	Increased likelihood of casualties.	Back-up operator in CCR will initiate; local initiation will eventually take place.
Failure to initiate ESD.	Continued supply of HF.	Back-up operator in CCR will initiate; plant manager will initiate when evacuation alarm occurs.
Miscalibration of HF detectors.	HF not detected, or detection delayed: rise in casualties, and release of HF.	Next check/test/calibration of monitors; supervisory inspection.
Hole created in pipe during welding.	Instant or delayed HF-release; with the former, immediate casualty likely; increased frequency of leak initiation.	Rapid detection if instant; if latent, detection as per main scenario above.
Valves not operated in time.	There are back-up actions in case ESD not effective.	ESD itself should be effective, as it has a high level of design-reliability. Back-up manual valves exist.
Area not isolated in time.	Leaking of HF into ventilation system and out into atmosphere.	Back-up operators/local rescue team will arrive at affected area in BA within 20 minutes.
Wrong area isolated.	Continued leakage. Possible problems for personnel in unaffected but now isolated area.	Back-up operator and Plant Manager will check that operations have occurred correctly; personnel in unaffected area have access to BA locally, and can effect manual exit. Evacuation alarm sounds in all areas.

- Tabular task analysis (TTA) – to consider interface-related errors – e.g. errors associated with not knowing when to act, or whether an act has been performed correctly – which derive from the kinds of cue and the level of feedback available to the operators via the interface.
- Tabular scenario analysis (TSA) – to define the sequence of tasks in an abnormal/emergency-event-driven situation.
- Timeline analysis (TLA) – to determine the overall feasibility of the task with respect to time constraints (as usually found in connection with emergency-response tasks), as well as both the overall sequence of events and task-personnel interrelationships, by plotting the relevant information

onto a time axis. Problems associated with workloads can be identified and addressed, albeit in a rather crude fashion.

- Sequence dependency – by way of an extension of the TLA, the sensitivity of the overall-task-completion probability to individual task failures or delays can be reviewed or, via simulation, be statistically modelled.
- Walk-through/talk-throughs – if task-timing estimates need to be derived, or if the interface needs to be investigated in its relationship to training and procedures, a walk-through can be utilised. A talk-through is particularly useful for exploring emergency scenarios.
- Link analysis (LA) – to be used if a consideration of detailed movements in a confined location is required; e.g. it can be used to determine if interference to the task will occur, or if signals will be missed due to the poor location of instruments, etc., or to consider the possible accidental activation of critical controls.
- Decision–action diagrams (DADs) – these are used to model decision-making sequences, and are important for all but simple/straightforward decisions.

All of the above, which have been defined in the earlier section on task analysis, provide rough guidelines on when/where they should be used. This will also, however, depend on the judgement of the analyst, as well as on the kinds of scenario and assessment involved. As a general rule, an HTA plus one form of tabular analysis (TTA or TSA) should be considered a minimum for the purposes of a level-2 analysis, with DADs utilised as appropriate. TLAs are also useful, though not always easy to generate, particularly if the TLA is modelling scenarios within specified time constraints. The most difficult part of a TLA will be gaining timing estimates for individual tasks (which can be done via walk-throughs/talk-throughs; direct observation, if feasible; or failing these, estimates both from the operators/engineers and, as a last resort, from the analyst). It will also be necessary for the analyst to 'temper' these task timings according to his or her estimation of the following: how smoothly the tasks will run and be coordinated; whether there will be interference from secondary tasks (especially non-required communications); any delays in decision-making caused either by reluctance or simply by operators double-checking, etc., before deciding to act. (As an example of reluctance, it may in theory take only a single push-button operation to shut down a plant, but in reality, operators will first want to make sure that such a shutdown, with all its associated cost penalties, is *absolutely* necessary before pushing that button.) It must also be recognised that even a walk-through will tend to only give minimum estimates for the time required, particularly for emergency-response tasks which may, in reality, occur under adverse conditions, and which will still tend to be unexpected even if well-rehearsed. The analyst must therefore make the estimates conservative.

The task analyst at level 2 will also make use of methods specified for level 1 – in particular, documentation analyses and, more specifically, operating instructions and emergency operating procedures (EOPs). The analyst must take care, however, to see whether the procedures are utilised by the operators or not, and to examine the extent of the operators' actual use of/reliance on them. This will be particularly important with respect to the generation of decision–action diagrams. There may in fact be flowcharts of decision-making processes in the EOPs themselves, but the analyst must verify that the operators would both utilise and be constrained by them before he or she actually adopts them as the proper definition of how decisions are made.

ERROR ANALYSIS

The major advances in level 2 over level 1 have had to do with both a comprehensiveness in error-identification and a new depth in the modelling of the causes and mechanisms of errors which has enhanced the ability to carry out error-reduction measures where required. The practitioner can utilise any of the techniques detailed earlier (e.g. SHERPA, etc. in 5.3.4; also, see section 5.3.6 and 5.10). There are two additional approaches documented in this section; the first is called simply, *human error identification in systems tool* (HEIST); the other is a form of the HAZOP (hazard and operability study: Kletz, 1974) called *human HAZOP*. These approaches are prototypical in nature, and in fact HEIST has not been used in a real assessment, and is intended here more as a theoretical approach showing comprehensiveness and the facilitation of error reduction analysis. To be more useful, HEIST would need to be computerised. However, it is included here in full, to show the potential extent of a comprehensive and functional error reduction-based human error analysis approach. The reader not interested in these developments is advised to turn to error recovery analysis.

The major differences between these two approaches are that the HEIST, which is intended to be a structured and well-defined (and hence systematic) approach, could be carried out by a single assessor (though all error-analysis techniques are best performed by more than one analyst, or two analysts working independently – see Kirwan, 1992b), and without the aid of any task-domain experts, whereas the human HAZOP, on the other hand, is more open-ended, allowing for more creative thought processes in the identification of errors; this latter is also a group-judgement exercise, requiring relevant task-domain experts. A HAZOP can be systematic, but in part its strength lies in its explorative open-endedness. As will be noted later, the human HAZOP can be used both for rule-violation assessment and for a consideration of human-error impacts incurred through maintenance procedure, and as a result it has a broader scope than HEIST. The two approaches are parallel to the approaches of FMEAs and HAZOPs carried out in general reliability assessments. Each will be described in turn below.

HUMAN ERROR IDENTIFICATION IN SYSTEMS TOOL

This approach has strong affinities with a number of techniques, namely SRK, SHERPA, CADA, HRMS and Murphy Diagrams (see Kirwan, 1992a

& 1992b). It aims to identify the EEM, either through 'free-form' identification or via a set of question–answer routines (the expert may prefer the former) which link the EEMs to particular PSFs. All EEMs are then linked to psychological error-mechanisms (defined in detail in Appendix V). Finally, the recovery potential is noted in a tabular error-analysis format, as are the consequences and error-reduction mechanisms (the latter may be expanded later if a more detailed ERMA is carried out).

The 'free-form' EEM analysis would basically use the expanded list of EEMs already documented for a level-1 analysis. The alternative approach, however, is to utilise the question–answer approach shown in Table 5.3.13. This table contains the following:

- A code for each error-identifying question.
- The error-identifier prompt.
- The external error mode (EEM).
- The identified system cause or psychological error-mechanism.
- Brief error-reduction guidelines.

The tables each refer to one of the eight principal stages of human information-processing outlined in the SRK model, as utilised in Murphy Diagrams, CADA and HRMS. The error-identifier prompts are a set of PSF-based questions; each is coded (via the second or third letter in the code) as indicating one of six 'global' PSFs:

- Time (T)
- Interface (I)
- Training/experience/familiarity (E)
- Procedures (P)
- Task organisation (O)
- Task complexity (C)

The analyst must consider each table for all tasks involved, noting potential errors (the analyst must decide between EEMs and PEMs) in the human-error-analysis table, along with consequences and selected ERMs (these are more fully explained in Appendix V), as has been shown in the example in Table 5.3.14.

HUMAN HAZOP

HAZOPs and human HAZOPs have already been discussed above in Section 5.3.4. It remains only to reiterate the usefulness of this technique if suitable expertise is available in the form of participants and the chairperson. The human HAZOP has been used effectively in a small number of applications in which it proved highly penetrating in identifying significant errors which participants often believed would not have been otherwise identified. This is probably due to the fact that other techniques tend to be forced to focus on a specified task analysis, errors thus identified are supplemented by the PSA

Table 5.3.13 HEIST table

1. ACTIVATION/DETECTION

Code	Error-identifier prompt	External error mode	System cause/psychological error-mechanism	Error-reduction guidelines
AT1	Does the signal occur at the appropriate time? Could it be delayed?	Action omitted; performed too early or too late	Signal timing deficiency; failure of prospective memory	Alter system configuration to present signal appropriately; generate hard copy to aid prospective memory; repeat signal until action has occurred
AI1	Could the signal source fail?	Action omitted or performed too late	Signal failure	Use diverse/redundant signal sources; use a higher-reliability signal system; give training and ensure procedures incorporate investigation checks on 'no signal'
AI2	Can the signal be perceived as unreliable?	Action omitted	Signal ignored	Use diverse signal sources; ensure higher signal reliability; retrain if signal is more reliable than it is perceived to be
AI3	Is the signal a strong one, and is it in a prominent location? Could the signal be confused with another?	Action omitted; or performed too late; or wrong act performed	Signal-detection failure	Prioritise signals; place signals in primary (and unobscured) location; use diverse signals; use multiple-signal coding; give training in signal priorities; make procedures cross-reference the relevant signals; increase signal intensity

AI4	Does the signal rely on oral communication?	Action omitted or performed too late	Communication failure; lapse of memory	Provide physical back-up/ substitute signal; build required communications requirements into procedures
AE1	Is the signal very rare?	Action omitted or performed too late	Signal ignored (false alarm); stereotype fixation	Give training for low-frequency events; ensure diversity of signals; prioritise signals into a hierarchy of several levels
AE2	Does the operator understand the significance of the signal?	Action omitted or performed too late	Inadequate mental model	Training and procedures should be amended to ensure significance is understood
AP1	Are procedures clear about action following the signal or the previous step, or when to start the task?	Action omitted or performed either too early or too late	Incorrect mental model	Procedures must be rendered accurate, or at least made more precise; give training if judgement is required on when to act
AO1	Does activation rely on prospective memory (i.e. remembering to do something at a future time, with no specific cue or signal at that later time)?	Action omitted or performed either too late or too early	Prospective memory failure	Proceduralise task, noting calling conditions, timings of actions, etc . . .; utilise an interlock system preventing task occurring at undesirable times; provide a later cue; emphasise this aspect during training

Table 5.3.13 (Continued)

Code	Error-identifier prompt	External error mode	System cause/psychological error-mechanism	Error-reduction guidelines
AO2	Will the operator have other duties to perform concurrently? Are there likely to be distractions? Could the operator become incapacitated?	Action omitted or performed too late	Lapse of memory; memory failure; signal-detection failure	Training should prioritise signal importances; improve task organisation for crew; use memory aids; use a recurring signal; consider automation; utilise flexible crewing
AO3	Will the operator have a very high or low workload?	Action omitted or performed either too late or too early	Lapse of memory; other memory failure; signal-detection failure	Improve task and crew organisation; use a recurring signal; consider automation; utilise flexible crewing; enhance signal salience
AO4	Will it be clear who must respond?	Action omitted or performed too late	Crew-coordination failure	Emphasise task responsibility in training and task allocation amongst crew; utilise team training
AC1	Is the signal highly complex?	Action omitted, or wrong act performed, or act performed either too late or too early	Cognitive overload; inadequate mental model	Simplify signal; automate system response; give adequate training in the nature of the signal; provide on-line, automated, diagnostic support; develop procedures which allow rapid analysis of the signal (e.g. use of flowcharts)

AC2	Is the signal in conflict with the current diagnostic 'mindset'?	Action omitted or wrong act performed	Confirmation bias; signal ignored	Procedures should emphasise disconfirming as well as confirmatory signals; utilise a shift technical advisor in the shift-structure; carry out problem-solving training and team training; utilise diverse signals; implement automation
AC3	Could the signal be seen as part of a different signal set? Or is, in fact, the signal part of a series of signals which the operator needs to respond to?	Action performed too early or wrong act performed	Familiar-association short-cut/stereotype take-over	Training and procedures could involve display of signals embedded within mimics or other representations showing their true contexts or range of possible contexts; use fault-symptom matrix aids etc.

Table 5.3.13 (Continued)

2. OBSERVATION/DATA COLLECTION

Code	Error-identifier prompt	External error mode	System cause/psychological error-mechanism	Error-reduction guidelines
OT1	Could the information or check occur at the wrong time?	Failure to act; or action performed too late or too early; or wrong act performed	Inadequate mental model/inexperience/crew coordination failure	Procedure and training should specify the priority and timing of checks; present key information centrally, utilise trend displays and predictor displays if possible; implement team training
OI1	Could important information be missing due to instrument failure?	Action omitted or performed either too late or too early; or wrong act performed	Signal failure	Use diverse signal sources; maintain back-up power supplies for signals; have periodic manual checks; procedures should specify action to be taken in event of signal failure; engineer automatic protection/action; use a higher-reliability system
OI2	Could information sources be erroneous?	Action omitted or performed either too late or too early; or wrong act performed	Erroneous signal	Use diverse signal sources; procedures should specify cross-checking; design system-self-integrity monitoring; use higher-reliability signals
OI3	Could the operator select a wrong but similar information source?	Action omitted or performed either too late or too early; or wrong act performed	Mistakes alternatives; spatial misorientation; topographic misorientation	Ensure unique coding of displays, cross-referenced in procedures; enhance discriminability via coding; improve training

		Communication failure	Use diverse signals from hardwired or softwired displays; ensure back-up human corroboration; design communication protocols
OI4	Is an information source assessed only via oral communication?	Action omitted or performed either too late or too early; or wrong act performed	Use diverse signals from hardwired or softwired displays; ensure back-up human corroboration; design communication protocols
OI5	Are any information sources ambiguous?	Misinterpretation; mistakes alternatives	Use task-based displays; design symptom-based diagnostic aids; utilise diverse information sources; ensure clarity of information displayed; utilise alarm conditioning
		Action omitted or performed either too late or too early; or wrong act performed	
OI6	Is an information source difficult or time-consuming to access?	Information assumed	Centralise key data; enhance data access; provide training on importance of verification of signals; enhance procedures
		Action omitted or performed too late; or wrong act performed	
OI7	Is there an abundance of information in the scenario, some of which is irrelevant, or a large part of which is redundant?	Information overload	Prioritise information displays (especially alarms); utilise overview mimics (VDU or hard wired); put training and procedural emphasis on data-collection priorities and data management
		Action omitted or performed too late	

Table 5.3.13 (Continued)

Code	Error-identifier prompt	External error mode	System cause/psychological error-mechanism	Error-reduction guidelines
OE1	Could the operator focus on key indication(s) related to a potential event while ignoring other information sources?	Action omitted or performed too late; or wrong act performed	Confirmation bias; tunnel vision	Training in diagnostic skills; enhance procedural structuring of diagnosis, emphasising checks on disconfirming evidence; implement a staff-technical-advisor role; present overview mimics of key parameters showing whether system integrity is improving or worsening or adequate
OE2	Could the operator interrogate too many information sources for too long, so that progress towards stating identification or action is not achieved?	Action omitted or performed too late	Thematic vagabonding; risk-recognition failure; inadequate mental model	Training in fault diagnosis; team training; put procedural emphasis on required data-collection time frames; implement high-level indicators (alarms) of system-integrity deterioration
OE3	Could the operator fail to realise the need to check a particular source? Is there an adequate cue prompting the operator?	Action omitted or performed either too late or too early; or wrong act performed	Need for information not prompted; prospective memory failure	Procedural guidance on checks required; training; use of memory aids; use of attention-gaining devices (flash; alarms; central displays and messages)

OE4	Could the operator terminate the data collection/observation early?	Action omitted or performed either too early or too late; or wrong act performed	Overconfidence; inadequate mental model; incorrect mental model; familiar-association short-cut	Training in diagnostic procedures and verification; procedural specification of required checks, etc.; implement a shift-technical-advisor role
OE5	Could the operator fail to recognise that special circumstances apply?	Action omitted or performed either too late or too early; or wrong act performed	Fail to consider special circumstances; slip of memory; inadequate mental model	Ensure training for, as well as procedural noting of, special circumstances; STA; give local warnings in the interface displays/controls
OP1	Could the operator fail to follow the procedures entirely?	Action omitted or wrong act performed	Rule violation; risk-recognition failure; production–safety conflict; safety-culture deficiency	Training in use of procedures; operator involvement in development and verification of procedures
OP2	Could the operator forget one or more items in the procedures?	Action omitted or performed either too early or too late; or wrong act performed	Forget isolated act; slip of memory; place-losing error	Ensure an ergonomic procedure design; utilise tick-off sheets, place keeping aids, etc.; team training to emphasise checking by other team member(s)

Table 5.3.13 (Continued)

Code	Error-identifier prompt	External error mode	System cause/psychological error-mechanism	Error-reduction guidelines
OO1 (AO2)	Will the operator have other duties to perform concurrently? Are there likely to be distractions? Could the operator become incapacitated?	Action omitted or performed too late	Lapse of memory; memory failure; signal-detection failure	Training should prioritise signal importances; develop better task organisation for crew; use memory aids; use a recurring signal; consider automation; use flexible crewing
OO2 (AO3)	Will the operator have a very high or low workload?	Action omitted or performed either too late or too early	Lapse of memory; other memory failure; signal-detection failure	Better task and crew organisation; utilise a recurring signal; consider automation; use flexible crewing; enhance signal salience
OO3 (AO4)	Will it be clear who must respond?	Action omitted or performed too late	Crew-coordination failure	Improve training and task allocation amongst crew; team training
OO4	Could information collected fail to be transmitted effectively across shift-hand-over boundaries?	Failure to act; or wrong action performed; or action performed either too late or too early; or an error of quality (too little or too much)	Crew-coordination failure	Develop robust shift-hand-over procedures; training; team training across shift boundaries; develop robust and auditable data-recording systems (logs)

| OC1 | Does the scenario involve multiple events, thus causing a high level of complexity or a high workload? | Failure to act; or wrong action performed; or action performed too early or too late | Cognitive overload | Emergency-response training; design crash-shutdown facilities; use flexible crewing strategies; implement shift-technical-advisor role; develop emergency operating procedures able to deal with multiple transients; engineer automatic information recording (trends, logs, print-outs); generate decision/diagnostic support facilities |

Table 5.3.13 (Continued)

3. IDENTIFICATION OF SYSTEM STATE

Code	Error-identifier prompt	External error mode	System cause/psychological error-mechanism	Error-reduction guidelines
IDT1	Could identification of system state occur too early or too late?	Omission of action; or action performed either too early or too late	Overconfidence; thematic vagabonding; inadequate or incorrect metal model	Training (including team training) and procedures should specify time frames; implement shift-technical-advisor role; install system-integrity-degradation alarms; implement other interface cues
IDI1	Could the operator forget some of the early symptoms, or be unaware of prior conditions, and hence fail correctly to identify the system state?	Omission of action; or action performed either too early or too late; or wrong act performed	Slip of memory; cognitive overload; prospective memory failure; inadequate mental model	General hard-copy recording (e.g. print-outs); design system-status 'freeze' indicators; engineer plant logs and automatic trend monitoring; procedures; training; use memory aids
IDE1	Could the operator focus on one key indication related to a potential event whilst ignoring other information sources?	Action omitted or performed too late; or wrong act performed	Confirmation bias; tunnel vision	Training in diagnostic skills; procedural structuring of diagnosis, emphasising checks on disconfirming evidence; implement shift-technical-advisor role; provide overview mimics of key parameters showing whether system integrity is improving or worsening or adequate

IDE2	Could the operator fail to recognise that special circumstances apply?	Action omitted; or wrong act performed; or action performed either too late or too early	Fail to consider special circumstances; slip of memory; inadequate mental model	Training for, and procedural noting of, special circumstances; implement shift-technical-advisor role; local warnings in the interface displays/controls
IDE3	Could the operator misidentify the state, favouring a similar state which is more memorable to the operator or perceived as more likely/frequent, or which is only based on a subset of the available information?	Omission of action; or wrong action performed	Inadequate mental model; stereotype takeover; similarity matching; frequency gambling; overconfidence; risk-recognition failure;	Training in fault diagnosis; procedures based on symptoms, but also noting disconfirming signals; generate decision/diagnostic support facilities; use fault-symptom matrices; implement shift-technical-advisor role; give very high-level information on, and alarms for, the system's critical functions or safety parameters (so-called critical-function-monitoring systems, or CFMSs)
IDE4	Could the operator fail to integrate the information available in time to identify the state?	Omission of action; or action performed too late	Cognition overload; bounded rationality; freezing; memory failure; familiar pattern not recognised	Training in the use of fault-diagnosis aids; emergency (stress) training; provide symptom-based procedures; utilise simulator training, provide decision/diagnostic support facilities; implement shift-technical-advisor role

Table 5.3.13 (Continued)

Code	Error-identifier prompt	External error mode	System cause/psychological error-mechanism	Error-reduction guidelines
IDE5	Could the scenario present the operator with a novel event not covered by the procedures?	Omission of action; or action performed either too late or too early; or wrong action performed	Bounded rationality	Provide function-based procedures; give problem-solving training for novel faults; team training; provide crash-shutdown facilities and/or ultimate procedures; provide automatic protection
IDP1	Could the operator apply the wrong identification procedure or set of rules?	Omission of action; or wrong action performed, or action performed either too late or too early	Misinterpretation; slip of memory	Use symptom-based rather than event-based procedures; implement shift-technical-advisor role; CFMS; training in state-identification procedures
IDP2 (OP1)	Could the operator fail to follow the procedures entirely?	Action omitted or wrong action performed	Rule violation; risk-recognition failure; production–safety conflict; safety-culture deficiency	Training in use of procedures; operator involvement in development and verification of procedures
IDO1 (AO2)	Will the operator have other duties to perform concurrently? Are there likely to be distractions? Could the operator become incapacitated?	Action omitted or performed too late	Lapse of memory; memory failure; signal detection failure	Training to prioritise signal importances; better task organisation for crew; memory aids; recurring signal; consideration of automation; flexible crewing

IDO2 (AO3)	Will the operator have a very high or low workload?	Action omitted or performed either too late or too early	Lapse of memory; other memory failure; signal-detection failure	Better task and crew organisation; recurring signal; consideration of automation; flexible crewing; signal salience
IDO3 (AO4)	Will it be clear who must respond?	Action omitted or performed too late	Crew-coordination failure	Training and task allocation amongst crew; team training
IDO4 (OO4)	Could information collected fail to be transmitted effectively across shift-hand-over boundaries?	Failure to act; or wrong action performed; or action performed either too late or too early; or an error quality (either too little or too much)	Crew-coordination failure	Robust shift-hand-over procedures; training; team training across shift boundaries; robust data-recording systems
IDC1 (OCI)	Does the scenario involve multiple events, thus causing a high level of complexity or a high workload?	Failure to act; or wrong action performed; or action performed too late or too early	Cognitive overload	Emergency-response training: provide crash-shutdown facilities; flexible crewing strategies; implement shift-technical-advisor role; develop emergency operating procedures able to deal with multiple transients; utilise automatic information recording (trends, logs, print-outs); provide a decision or diagnostic support system (DSS)

Table 5.3.13 (Continued)

4. INTERPRETATION

Code	Error-identifier prompt	External error mode	System cause/psychological error-mechanism	Error-reduction guidelines
INT1	Could the operator fail to interpret the situation in time?	Omission of action; or action performed too late	Bounded rationality; cognitive overload; thematic vagabonding; inadequate mental model; insufficient time	Automatic protection training-and-procedures support; DSS; STA; symptom-based procedures
INT2	Could the operator prematurely interpret the situation?	Omission of action; or action performed; too early	Overconfidence; inadequate mental model; risk-recognition failure; oversimplification	Training in fault diagnosis; procedures; STA; DSS; CFMS
INI1 (IDI1)	Could the operator forget some of the early symptoms, or be unaware of prior conditions, and hence fail to correctly interpret the situation?	Omission of action; or action performed either too early or too late; or wrong action performed	Slip of memory; cognitive overload; prospective memory failure; inadequate mental model	Hard-copy recordings (e.g. print-outs); system status 'freeze' indicators; logs and trends; procedures; training; memory aids; STA
INI2 (OI1)	Could important information be missing due to instrument failure?	Action omitted; or action performed either too late or too early; or wrong action performed	Signal failure	Use diverse signal sources; maintain back-up power supplies for signals; have periodic manual checks; procedures should specify action in event of signal failure; engineer automatic protection/action; use a higher-reliability system

ID	Question	Failure mode	Error mechanism	Recommendations
INI3 (OI2)	Could information sources be erroneous?	Action omitted or performed either too late or too early; or wrong action performed	Erroneous signal	Use diverse signal sources; procedures should specify cross-checking; design system-self-integrity monitoring; use higher-reliability signals
INI4 (OI7)	Is there an abundance of information in the scenario, some of which is irrelevant, or a large part of which is redundant?	Action omitted or performed too late	Information overload	Prioritise information displays (especially alarms); utilise overview mimics (VDU or hardwired); training and procedural emphasis on data-collection priorities and data management
INE1 (OE1)	Could the operator focus on key indication(s) related to a potential event whilst ignoring other information sources?	Action omitted or performed too late; or wrong action performed	Confirmation bias; tunnel vision	Training in diagnostic skills; enhance procedural structuring of diagnosis, emphasising checks on disconfirming evidence; implement shift-technical-advisor role; present overview mimics of key parameters showing whether system integrity is improving or worsening or adequate

Table 5.3.13 (Continued)

Code	Error-identifier prompt	External error mode	System cause/psychological error-mechanism	Error-reduction guidelines
INE2 (OE5)	Could the operator fail to recognise that special circumstances apply?	Action omitted, or wrong action performed; or action performed either too early or too late	Fail to consider special circumstances; inadequate mental model	Training for, and procedural noting of, special circumstances; STA; local warnings in the interface displays/controls
INE3 (IDE3)	Could the operator misidentify the state, favouring a similar state which is more memorable to the operator or perceived as more likely/frequent, or which is only based on a subset of the available information?	Omission of action, or wrong action performed	Inadequate mental model; stereotype takeover; similarity matching; frequency gambling; overconfidence; risk-recognition failure	Training in fault diagnosis; procedures based on symptoms, also noting disconfirming signals; generate decision/diagnostic support facilities; use fault-symptom matrices; implement shift-technical-advisor role; provide very high-level information on, and alarms for, the system's critical functions or safety parameters (so called critical-function-monitoring systems, or CFMSs)
INE4 (IDE4)	Could the operator fail to integrate the information available in time to identify the state?	Omission of action; or action performed too late	Cognitive overload; bounded rationality; freezing; memory failure; familiar pattern not recognised	Training in the use of fault-diagnosis aids; emergency (stress) training; provide symptom-based procedures; utilise simulator training; give decision/diagnostic support facilities; implement shift-technical-advisor role

INE5 (IDE5)	Could the scenario present the operator with a novel event not covered by the procedures?	Omission of action; or action performed either too late or too early; or wrong action performed	Bounded rationality	Provide function-based procedures; give problem-solving training for novel faults; team training; provide crash-shutdown facilities or ultimate procedures (or both); provide automatic protection
INE6	Could the operator fail to consider all the implications of the scenario?	Error of quality (partial diagnosis); omission of action; wrong action	Overconfidence; bounded rationality; inadequate mental model; side-effects not considered; memory failure	Procedures should deal comprehensively with the scenario; training; use checksheets and flowcharts; team training; STA; enhance allocation and coordination; provide problem-solving training
INP1 (IDP1)	Could the operator apply the wrong identification procedure or set of rules?	Wrong action performed; or omission of action; or action performed either too late or too early	Misinterpretation; slip of memory	Use symptom-based rather than event-based procedures; implement shift-technical-advisor role; CFMS; training in state-identification procedures
INP2 (OP1)	Could the operator fail to follow the procedures entirely?	Action omitted, or wrong action performed	Rule violation; risk-recognition failure; production–safety conflict; safety-culture deficiency	Training in use of procedures; operator involvement in development and verification of procedures

Table 5.3.13 (Continued)

Code	Error-identifier prompt	External error mode	System cause/psychological error-mechanism	Error-reduction guidelines
IN01 (AO1)	Does activation rely on prospective memory (remembering to do something at a future time)?	Action omitted, or action performed either too late or too early	Prospective memory failure	Proceduralise task, noting calling conditions, timings of actions, etc . . .; utilise an interlock system preventing task occurring at undesirable times; provide a later cue; emphasise this aspect during training
IN02 (AO2)	Will the operator have other duties to perform concurrently? Are there likely to be distractions? Could the operator become incapacitated?	Action omitted or performed too late	Lapse of memory; memory failure; signal-detection failure	Training should prioritise signal importances; improve task-organisation for crew; use memory aids; use a recurring signal; consider automation; utilise flexible crewing
IN03 (AO3)	Will the operator have a very high or low workload?	Action omitted or performed either too late or too early	Lapse of memory; other memory failure; signal-detection failure.	Improve task and crew organisation; use a recurring signal; consider automation; utilise flexible crewing; enhance signal salience
IN04 (AO4)	Will it be clear who must respond?	Action omitted or performed too late	Crew-coordination failure	Emphasise task responsibility in training and task allocation amongst crew; utilise team training

INC1 (OC1)	Does the scenario involve multiple events, thus causing a high level of complexity or a high workload?	Failure to act; or wrong action performed; or action performed either too late or too early	Cognitive overload	Emergency-response training; design crash-shutdown facilities; use flexible crewing strategies; implement shift-technical-advisor role; develop emergency operating procedures able to deal with multiple transients; engineer automatic information recording (trends, logs, print-outs); generate decision/diagnostic support facilities
INC2 (AC2)	Is the signal in conflict with the current diagnostic 'mindset'?	Action omitted, or wrong act performed	Confirmation bias; signal ignored	Procedures should emphasise disconfirming as well as confirmatory signals; utilise a shift technical advisor in the shift-structure; carry out problem-solving training and team training; utilise diverse signals; implement automation

Table 5.3.13 (Continued)

5. EVALUATION

Code	Error-identifier prompt	External error mode	System cause/psychological error-mechanism	Error-reduction guidelines
EVT1	Could the operator/team fail to evaluate the situation in time?	Omission of action; or action performed too late	Cognitive overload; insufficient time available; inadequate mental model; multiple events; insufficient time available	Automatic protection; training; emergency operating procedures; decision support systems; STA
EVT2 (INT2)	Could the operator prematurely interpret the situation?	Omission or action, or action performed too early	Overconfidence inadequate mental model; risk-recognition failure; oversimplification	Training in fault diagnosis; procedures; STA; DSS; CFMS
EVI1 (IDI1)	Could the operator forget some of the early symptoms or be unaware of prior conditions and hence fail correctly to identify the system state?	Omission of action; or action performed either too early or too late; or wrong act performed	Slip of memory; cognitive overload; prospective memory failure; inadequate mental model	Hard-copy recording (e.g. print-outs); system-status 'freeze' indicators; logs and trends; procedures; training; memory aids; STA
EVI2	Are information sources on criteria relevant to the evaluation process available and accurate?	Omission of action; or error of quality (inadequate evaluation)	Insufficient information; false signals	Diverse and redundant information channels which are prioritised; determination of what information they need; EOPs and other memory aids
EVI3	Are there too many information sources, so that flooding of information into the central evaluation location occurs?	Omission or action; or error of quality (inadequate evaluation)	Cognitive overload	Prioritisation of information sources; adequate control of communications; function-based EOPs and displays; training

EVI4	Does evaluatory information require integration and mental interpolation for a decision on the current state and target goal?	Wrong evaluation; omission of action	Cognitive overload	Prior processing of information; special displays and EOPs for evaluation process; training
EVO1	Could the team/supervisor/operator omit key parameters in the evaluation process (i.e. fail to check them)?	Error of quality (inadequate evaluation); wrong action	Failure to consider side effects; inadequate mental model; bounded rationality	Procedural evaluation aids; team training; function-based displays and procedures
EVO2	Could the team/supervisor/operator check but ignore key parameters during the evaluation process?	Error of quality (inadequate evaluation); wrong action	Risk-recognition failure; reluctant rationality; risk-taking	Training and supervision; function-based EOPs and displays; stress training; remote incident-monitoring
EVO3	Could the operating team fail to allocate responsibility correctly, for purposes of making evaluations and subsequent decision, in the 'command chain'	Omission; premature action (too early); wrong action	Responsibility-allocation failure; crew-coordination failure; crew member(s) injured/incapacitated/killed by event	Flexible responsibility structure (command chain) in event of key-member incapacitation; team training; clarity of responsibility; stress training
EVO4	Could the team fail to consider the full complexity of the situation with respect to decision criteria, leading to a simplistic or unidimensional solution?	Error of quality (inadequate or partial solution); wrong action	Oversimplification; cognitive overload; bounded rationality	Accident-management training involving multiple criteria; stress management; problem-solving training; accident-management procedures and EOPs; function-based procedures

Table 5.3.13 (Continued)

Code	Error-identifier prompt	External error mode	System cause/psychological error-mechanism	Error-reduction guidelines
EVC1	Could the team be placed in a novel and complex situation outside of existing procedures (e.g. a BDBA)?	Error of omission; wrong action	Bounded rationality	Problem-solving training; stress training; team training; function-based displays; and EOPs/accident-management procedures
EVE1	Could the team be predisposed to misevaluate the situation due to industry over-or underconcern about particular consequences?	Error of quality; wrong action	Inadequate mental model; risk-recognition failure; availability bias; out-of-sight bias	Training to ensure operator concerns are coherent with assessed risk concerns; EOPs note criteria/parameters of concern in accident sequences; team and stress training
EVO5	Could the team be overloaded with tasks, so that evaluation does not occur in time, or is made when necessary but only after insufficient time for a proper evaluation?	Error of omission; or error of quality (inadequate evaluation)	Cognitive overload; crew coordination failure: integration failure	Allocation of responsibility; team training and stress training; instrumentation prioritisation and function-based displays; time-monitoring; accident-management procedures; remote incident-monitoring
EVE2	Could the team, having all data present, make an incorrect evaluation of the situation?	Wrong action	Imperfect rationality	Decision support; EOPs; training in problem-solving; remote incident-monitoring; team training

6. GOAL SELECTION/TASK DEFINITION

Code	Error-identifier prompt	External error mode	System cause/psychological error-mechanism	Error-reduction guidelines
GTT1 (EVT1)	Could the operator/team fail to evaluate the situation in time?	Omission of action; or action performed too late	Cognitive overload; insufficient time available; inadequate mental model; multiple events; insufficient time available	Automatic protection; training; emergency operating procedures; decision support systems; STA
GTT2 (EVT2)	Could the operator prematurely interpret the situation?	Omission of action; or action performed too early	Overconfidence; inadequate mental model; risk-recognition failure; oversimplification	Training in fault diagnosis; procedures; STA; DSS; CFMS
GTI1 (EVI3)	Are there too many information sources, so that flood of information into the central evaluation location occurs?	Omission of action; or error of quality (inadequate evaluation)	Cognitive overload	Prioritisation of information sources; adequate control of communications; function-based EOPs and displays; training
GTI2 (EVI4)	Does evaluatory information require integration and mental interpolation before a decision on the current state and target goal can be made?	Wrong evaluation; omission of action	Cognitive overload	Prior processing of information; special displays and EOPs for evaluation process; training
GTE1 (INE6)	Could the operator fail to consider all the implications of the scenario?	Error of quality (partial diagnosis); omission of action; or wrong action performed	Overconfidence; bounded rationality; inadequate mental model; side-effects not considered; memory failure	Procedures; training; checksheets and flowcharts; crew training, STA; crew allocation and coordination

Table 5.3.13 (Continued)

Code	Error-identifier prompt	External error mode	System cause/psychological error-mechanism	Error-reduction guidelines
GTE2 (EVE2)	Could the team, having all data present, make an incorrect evaluation of the sections?	Wrong action	Imperfect rationality	Decision support; EOPs; training in problem-solving; remote incident-monitoring; team training
GTE3 (EVE1)	Could the team be predisposed to misevaluate the situation due to industry over or underconcern over particular consequences?	Error of quality; wrong action	Inadequate mental model; risk-recognition failure; availability bias; out-of-sight bias	Training to ensure operator concerns are coherent with assessed risk concerns; EOPs note criteria/parameters of concern in accident sequences; team and stress training
GTE4 (IDE3)	Could the operator misidentify the state, favouring a similar state which is more memorable to the operator or is perceived as more likely/frequent, or which is only based on a subset of the available information?	Omission of action or wrong action	Inadequate mental model; stereotype takeover; similarity matching; frequency gambling; overconfidence; risk-recognition failure.	Training in fault diagnosis; procedures based on symptoms and also noting disconfirming signals; DSS; fault-symptom matrices; STA; CFMS.
GTO1 (EVO3)	Could the operating team fail to allocate correctly responsibility for making the evaluation and subsequent decision in the 'command chain'	Omission of action; or premature action (too early); or wrong action	Responsibility-allocation failure; crew-coordination failure; crew member(s) injured/incapacitated/killed by event	Flexible responsibility structure (command chain) in event of key-member incapacitation; team training; clarity of responsibility; stress training

GTO2 (EVO4)	Could the team fail to consider the full complexity of the situation with respect to decision criteria, leading to a simplistic or unidimensional solution?	Error of quality (inadequate or partial solution); wrong action	Oversimplification; cognitive overload; bounded rationality	Accident-management training involving multiple criteria; stress management; problem-solving training; accident-management procedures and EOPs; function-based procedures
GTC1 (EVC1)	Could the team be placed in a novel and complex situation outside of existing procedures (e.g. a BDBA)?	Error of omission; wrong action	Bounded rationality	Problem-solving training; stress training; team training; function-based displays and EOPs/accident-management procedures
GTP1 (IDP1)	Could the operator apply the wrong identification procedure or set of rules?	Wrong action; or omission of action; or action performed either too late or too early	Misinterpretation; slip of memory	Symptom-based rather than event-based procedures; STA; CFMS; training in state identification
GTO3 (EVO5)	Could the team be overloaded with tasks so that evaluation does not occur in time, or is made when necessary but only after insufficient time for a proper evaluation to be made?	Error of omission; error of quality (inadequate evaluation)	Cognitive overload; crew coordination failure; integration failure	Allocation of responsibility; team training and stress training; instrumentation prioritisation and function-based displays; time-monitoring; accident-management procedures; remote incident-monitoring
GTO4 (EVO1)	Could the team/supervisor/operator omit key parameters in the evaluation process (i.e. fail to check them)?	Error of quality (inadequate evaluation); wrong action	Fail to consider side-effects; inadequate mental model; bounded rationality	Procedural-evaluation aids; training; team; coordination; function-based displays and procedures

Table 5.3.13 (Continued)

7. PROCEDURE SELECTION

Code	Error-identifier prompt	External error mode	System cause/psychological error-mechanism	Error-reduction guidelines
PST1	Could the operator fail to select the procedure in time?	Omission of action	Insufficient time available; thematic vagabonding; cognitive overload; inadequate mental model	EOPs; automatic protection; training
PST2	Could the operator select the procedure prematurely?	Action performed too early; or wrong action performed	Stereotype takeover; familiar-association short-cut; similarity-making/ frequency gambling	Training; function-based EOPs; ergonomic procedures; STA
PSC1	Could the operator select the wrong procedure?	Wrong action	Intrusion; inadequate recall; familiar-association short-cut; stereotype takeover; similarity-making or frequency gambling	Function-based procedures; training; good procedure; STA; function-based displays
PSP1	Could the operator select a procedure involving a rule violation?	Wrong action	Risk-recognition failure; risk-taking; side-effects not considered	Clear warnings in procedures; training; crew organisation and supervision; EOPs and function-based procedures
PS01	Could the operators be too busy to select the procedure?	Omission of action	Attention failure; cognitive overload; excessive demands; crew-coordination failure	Team and stress training; flexible crewing; interface and procedural prompts during events; STA; remote incident-monitoring

PSC2 (EVC1)	Could the team be placed in a novel and complex situation outside of existing procedures (e.g. a BDBA)?	Error of omission; or wrong action	Bounded rationality	Problem-solving training; stress training; team training; function-based displays and EOPs/accident-management procedures
PSC3 (EVO4)	Could the team fail to consider the full complexity of the situation with respect to decision criteria, leading to a simplistic or unidimensional solution?	Error of quality (inadequate or partial solution); or wrong action	Oversimplification; cognitive overload; bounded rationality	Accident-management training involving multiple criteria; stress management; problem-solving training; accident-management procedures and EOPs; function-based procedures
PSE1 (IVE6)	Could the operator fail to consider all the implications of the scenario?	Error of quality (partial diagnosis); or omission of action; or wrong action	Overconfidence; bounded rationality; inadequate mental model; side-effects not considered; memory failure	Procedures; training; checksheets and flowcharts; crew training; STA; crew allocation and coordination

Table 5.3.13 (Continued)

8. PROCEDURE EXECUTION

Code	Error-identifier prompt	External error mode	System cause/psychological error-mechanism	Error-reduction guidelines
PET1	Could the operator fail to carry out the act in time?	Omission of action	Insufficient time available; inadequate time perception; crew-coordination failure; manual variability; topographic misorientation	Training; team training and crew-coordination trials; EOPs; ergonomic design of equipment
PET2	Could the operator carry out the task too early?	Action performed too early	Inadequate time perception; crew-coordination failure	Training; perception (time) cues; time-related displays; supervision
PEP1	Could the operator carry out the task inadequately?	Error of quality; or wrong action; or omission of action	Manual variability; random prompting; random fluctuation; misprompting; misperception; memory failure	Training; ergonomic design of equipment; ergonomic procedures; accurate and timely feedback; error-recovery potential; supervision
PEP2	Could the operator lose his/her place during procedure execution, or forget an item?	Omission of action; or error of quality	Memory failure; interruption; vigilance failure; forget isolated act; misprompting; cue absent	Ergonomic procedures with built-in checks; error-recovery potential (error-tolerant system design); good system feedback; supervision and checking

PEP3	Could the operator act on the wrong equipment?	Wrong action	Topographic misorientation; manual variability; misprompting; mistakes alternatives; intrusion; random fluctuation; similarity-matching or frequency gambling	Ergonomic design of controls and displays; ergonomic procedures; training clear labelling, etc.
PEP4	Could the operator carry out the wrong task on the right item of equipment?	Wrong action	Similarity-matching or frequency gambling; stereotype takeover; misperception; intrusion; recognition failure	Training; supervision and checking; ergonomic procedures with checking facilities; error-recovery potential; prompt system feedback
PEP5	Could the operator carry out the wrong task on the wrong item of equipment?	Wrong action	Random fluctuation	Ergonomic design of equipment; training and procedures; checking; system-error feedback; error-recovery potential
PEP6	Could the operator fail to carry out the task?	Omission of action	Memory failure; excessive demands; place-losing; forget isolated act; habit intrusion; signal-discrimination failure; assumption not checked; operator incapacitated; cue absent	Memory aids in procedures; perceptual cues; checking and supervision; feedback; function-based displays; automatic procedure tracking; training; flexible crewing

Table 5.3.14 Worked sample example (in condensed format: no consequences or PEMs) of level-2 analysis using HEIST

Error	Recovery	Error reduction
Signal perceived as unreliable (A12) – delays response	Checks CCTV	Ensure operators know how to provide ESD for two diverse alarms; automate ESD (very-short-response timeframe)
Operator selects wrong information-source (CCTV) (O13) – delays response	Communication from local area, if occupied; other alarms	Place CCTVs in same functional grouping as alarms for that area, i.e. dedicate a CCTV to each area
Operator fails to consider implications – does not isolate area ventilation (INE6)	Other CCR ops; procedures; continuing HF-flow/pressure alarms	Flowchart procedure and team training
Wrong valve operated upon – area not sealed off via valves, and HF-escape continues (PEP3)	VDU indications; continuing alarms	Ergonomic design of valves; good labels, readable through breathing apparatus, required

team's deliberations over what hardware failures could be caused by human error. Alternatively, in a HAZOP, the HRA analyst and the PSA analyst are reviewing the system together, and the resulting output involves a hybrid and potentially more powerful analysis, one in which one analyst does not have to guess or interpret what would seem important to the other. Furthermore, since someone with operational experience will be present, aspects such as rule violations can also be considered, since this person will have an insight into what types of violations are credible – and have actually happened.

The human HAZOP approach is currently being researched in a small number of European projects, and will probably emerge in the near future in a refined form specifically for HEI/HEA purposes. In the meantime, analysts wishing to carry out human HAZOPs can do so within the traditional HAZOP format, or they may wish to adapt it in the way discussed earlier, or else into other 'assessor-customised' formats. A specific human-HAZOP format for a level-4 analysis, called an error-of-commission analysis (EOCA – for the identification of latent errors, etc.) is presented in the Section 5.3.5.6.

ERROR RECOVERY ANALYSIS

Many human errors are recoverable, and it is important to model such recovery opportunities appropriately so as not to overestimate the human contribution to risk. Therefore, error-recovery modes, with their associated failure probabilities, should be entered into the PSA representation where appropriate (the actual means of representing recovery in fault and event trees is dealt with in Section 5.4).

There are several types of recovery, as follows:

- Internal recovery: the operator, having committed an error or failed to carry out an act, realises this immediately, or later, and corrects the situation.
- External recovery: the operator, having committed an error or having failed to do something that is required, is prompted by a signal from the environment (e.g. an alarm; an error message; some other non-usual system-event).
- Independent human recovery: another operator monitors the first operator, detects the error and either corrects it or brings it to the attention of the first operator, who then corrects it.
- System recovery: the system itself recovers from the human error. This implies a degree of error tolerance, or of error detection and automatic recovery.

These types of recovery can all be modelled in HRAs, though the first type is usually not modelled unless there is a significant reason for so doing – e.g. the operator gets some form of prompt which aids recovery (in which case it becomes a Category-II recovery mode). Instead of an internal recovery being modelled, it is usually assumed that the error probability attached to the error of interest incorporates implicitly, or via a consideration of PSFs (such as the amount of time available), some chance of recovery. Therefore, the internal recovery is not usually modelled but instead taken account of quantitatively in the HRA.

The other three types of recovery can be represented as events with attached probabilities in the PSA logic trees. However, an external recovery is generally only modelled if it is believed to have a significant recovery-promoting effect; e.g., a high-level alarm (audible and visual) would be seen as significantly promoting recovery chances, whereas a footnote in a procedure would not. Similarly, the frequency of independent human recovery-enhancing behaviour should not be overestimated: people often do not check each other's performance in any depth, and as a result it cannot be assumed that simply because two people are present they will check each other effectively. Therefore, judgement must be exercised by the analyst. (This will be dealt with in more detail in the section on representation, as it requires a consideration of what are known as *dependent failures*.)

The identification of recovery modes is not a well-elucidated subject, and it relies heavily on the assessor's own judgement. The reader is referred to the THERP technique (Swain & Guttmann, 1983), which is one of the few methodologies which models error-recovery in detail and gives guidance on such modelling. The best time to identify error-recovery modes, however, is as each error is being identified, or shortly after the set of error modes has been identified for a particular scenario. Once an error has been identified, the consequences of the error can be considered, as well as whether or not there are any in-built recovery paths (i.e. system prompts or tolerance to errors) or whether or not the task is designed so as to promote human recovery. Such information is best recorded in the form of a human-error-analysis table (as shown earlier). If such a tabular-analysis approach is adopted for a level-2 analysis, the error-recovery analysis will proceed systematically for all errors identified. Alternatively, when conducting a HAZOP, it is similarly relatively easy to amend the HAZOP recording sheet format in such a way as to make a record of error-recovery opportunities (though, in a formal HAZOP, it must be remembered that such error-recovery opportunities may themselves have high failure probabilities, e.g., of 0.1–0.3).

When error-recovery modes have been identified, their credibility should be investigated by the assessor – for example, via the following questions:

• Is there enough time to correct the error once it has been detected?
• Does the prompt itself have a high failure rate, or is it potentially ambiguous, or even liable to a false alarm?
• Is the prompt salient, or could it be overlooked or masked by other signals?
• How is the effectiveness of independent checking by other personnel actually ensured?
• Is it in the operator's interests to ignore a prompt (e.g. because a later, higher-level prompt will occur when action is required, or because the operator knows the system will deal with the error)?

The answers to these questions may suggest that an error recovery may be less likely than would be hoped for. Nevertheless, this does not mean that such recovery should not be modelled but rather that its own probability of failure may be high. It is useful to note the potential pitfalls of error-recovery modes at the HEA stage.

5.3.5.5 Level 3 analysis

This level is concerned with the potential cognitive errors, i.e. the chance that misdiagnosis, misconception, diagnostic failure or incorrect judgement will occur. It assumes that earlier on in a level-0, -1, or -2 analysis, a cognitive-error potential was identified, either because it was defined as an area of concern in the problem-definition phase or because its potential became apparent either during the various task and error analyses or via the cognitive-error-potential flowchart shown earlier. This section therefore deals with methods used to identify some of the basic cognitive errors.

When decision-making or a diagnosis is required, there are generally only a few error-modes possible at the typical fault- or event-tree basic-event level (discussed in more detail in Section 5.4), namely the following:

(i) A failure to diagnose (diagnostic failure) – this leads to inaction, and is a cognitive 'error of omission'.

(ii) Partial or incomplete diagnosis – this means that the solution implemented will be inadequate in some way; e.g. it may leave a residual amount of risk, and thus lead to the occurrence of a hazard. The diagnosis may, however, be insufficient to the extent that the full consequences still occur (i.e. injury, contamination, damage to plant/equipment); or it may simply be an inefficient strategy (e.g. one which leads to a longer post-event plant shutdown procedure).

(iii) Misdiagnosis – this is when the operators have made an incorrect diagnosis of the situation, so that the strategy they select will be inappropriate. In this case, a misdiagnosis may lead to a failure to tackle the problem at all (i.e. the operators believe the situation is healthy or will rectify itself of its own accord). The implementation of an ineffective strategy can lead to scenario consequences that are *worse* than those that would have occurred if the operators had done nothing (e.g. as arguably happened in the Three Mile Island incident in 1979: see Appendix I).

(iv) Thwarted diagnosis – the initial diagnosis is correct but events occur (hardware failures, environmental events or human errors such as slips) which transfer the operators' initial successful diagnostic plan or trajectory out of one particular well-defined accident-event sequence and into a new accident-event-sequence pathway, perhaps one which will be more difficult to recover from, and which will require a re-diagnosis of the scenario.

(v) Premature diagnosis – the operators react too quickly and make a diagnosis before they actually have enough information to do so correctly. This type of diagnosis can usually be subsumed by (i)–(iv) above.

Only the first of the above points is relatively easy to identify in PSA work. The other four, particularly the thwarted diagnosis, are far more problematical, for two major reasons. The first is that (ii)–(v) can cause new consequential outcomes, or even transfers from one logic tree in the PSA to another potential failure path, possibly not currently in existence within the PSA. This may cause problems for the integration of the HRA into the PSA.

The second major problem is the identification of such errors. The process of diagnosis or of problem-solving in complex work environments is itself a complex one: operators use a variety of problem-solving and search strategies when looking for a solution, or when trying to diagnose an event, and they call upon a huge database of knowledge. This knowledge database encompasses not only engineering, physics and chemical knowledge (based on system designs) but also their own historical databases of events and their likelihoods. The diagnostic or problem-solving process also requires a knowledge of what particular patterns of indications mean in different combinations, and under various conditions. Diagnosis also requires knowledge of the reliability both of particular indicators and of the predicted (via the PSA) likelihoods of particular accident events and sequences. Added to this is the fact that a diagnosis of events in real accident conditions (e.g. the Davis-Besse incident – see

Appendix I) often involves redundant information (i.e. secondary and some-times irrelevant indications), incomplete information (e.g. as a result of a temporary loss of signals) and incorrect or spurious signals – all of which happen in a particular time sequence where the chronology of occurrence of different indicators can itself have varying degrees of diagnostic importance.

Finally, a diagnosis of events also often occurs in emergency situations, and is often associated with safety such that the plant or system is under impending threat of a hazard. In other cases, such a hazard (and associated fatalities) may have *already* occurred, and here the operators' job is to manage the accident and reduce further casualties as far as possible, rather than to prevent it. In such a situation, the operators will be under significant stress, aware of the critical importance of their actions as regards their own and others' safety.

Given the above description of the complexity and multidimensional na-ture of diagnoses made during accident sequences, the difficulties facing the analyst become clearer. An analyst is never likely to have a full understanding of the operators' various knowledge databases, and even if such information *were* available, the amount of resources required to determine all possible combinations of indications and subsequent diagnoses, in real time and under stressful conditions, will in most cases prove unobtainable. The one possible exception to this is research, currently in progress, that involves the develop-ment of both a large simulation of an NPP (including subsequent indications of accident sequences in real time) and a simulation of an expert operator diagnosing the event(s) as they occur (Woods *et al*, 1990). Such cognitive-reliability research is most laudable. (At the time of writing, it is not yet available to most HRA practitioners involved in industrial assessments.)

The HRA assessor, therefore, has a difficult task in the form of cognitive-reliability analysis (CRA), and the dangers of misdiagnoses, and associated error forms, are significant. Given that cognitive errors will largely arise from mis-interpretations, or else mental models of the plant's response to various events that are inappropriate, insufficient or inaccurate, it makes sense for the HRA analyst to go straight to the source of such knowledge: the operators themselves.

The two main approaches utilised in this section, both of which can be implemented by an assessor and do not require computer models, are as follows (see Kirwan, 1992a & b for a review of two alternative computer models under development);

- Fault-symptom-matrix-approach (FSMA) analysis.
- Confusion-matrix analysis (CMA) (see Potash *et al*, 1981).

These two methods can be utilised together, the FSMA making the CMA easier to carry out. They are both described below.

FAULT SYMPTOM MATRIX ANALYSIS (FSMA)

The FSMA is a simple but useful approach, one in which a series of prespecified faults are examined to see what indications, signals or symptoms would appear for each fault, and whether such signals are local or central (i.e. central to where the diagnosis/decision-making is actually occurring). Additionally (in the case of the existing plant), the reliability of indications can be noted (i.e. whether they may be unreliable, or subject to false alarm, etc.), as well as their perceived priority *before* a diagnosis has been made. And also, the reliance on oral communication can be noted, as this can obviously lead to an error of misconception via a communication failure. A sample FSM, based on a real NPP system, is shown in Figure 5.3.18.

The FSM shows the indications that would occur within 15 minutes, which is the critical period of time in this particular scenario (the critical time-frame is determined by the expected diagnostic response time for a particular installation or system). If desirable, other time-frames could also be considered (e.g. 5 minutes, 1 hour, 3 hours, etc.) so as to make allowances for a premature diagnosis, or the availability of stronger diagnostic clues later in the scenario – particularly those which would make the operators reconsider their diagnosis if it were incorrect. Alternatively, the FSM could note the time after the onset of each indication, as is done in a TSA. This modelling of time in an FSMA becomes important when the chronology of indications is critical to the determination of what is happening. This must be judged at the outset by the HRA assessor, system analyst and operators, and then re-evaluated throughout the CRA (particularly during the CMA, as described below).

In the example of an FSM, it can be seen that some scenarios are clearly discriminable, whereas others are fairly confusable, being distinguishable only by one or two non-priority alarms, or even only by oral communication. It should also be noted that in the first column, 'environmental signals' are noted. These are often not officially recognised or recorded, and yet at the same time they often act as clear and obvious indicators to the operators concerned (e.g. a large seismic event will be felt by those in the area). Lastly, an FSM can also be used to indicate an absence of signals. Sometimes, events are characterised more by *missing* information than by alarms which are present. For example, if some alarms are apparently indicating a gas release of a certain type, then it should be possible to smell the gas locally. If there is no such smell, however, then a gas release may be less likely as the cause of these indications and another diagnosis must be derived.

What the FSM gives the analyst is a set of detailed scenario 'pictures', as well as an overview of the discriminability of major, threatening accident-sequences for the plant in question. This information can then be utilised in the next phase of the CRA, the confusion-matrix analysis.

Some of the limitations of the FSM approach are as follows:

- It may be resource-intensive to verify the FSM's accuracy by contrasting it with alarm-design schedules. Often, therefore, the process of verification may involve carrying out the same exercise with an independent operator or team of operators.
- It is far easier to do an FSMA with an existing plant and experienced

Figure 5.3.18 Fault-symptom matrix: hypothetical example*

	Environmental signals	Reactor-trip indication	Control core rods	Power drop	Temperature	Emergency pumps start	Emergency diesels start
Loss of grid	Switch to emergency lighting	Y	Y	Y	Drops	Y	Y
Spurious trip	None	Y	Y	Y	Drops	N	Y
Seismic event	Earth shudders	Y	Y	Y	Drops	Y	Y
Loss of all diesels	Probably fire local to diesel house	Manual trip	Manual trip	N	—	N	N
Steam break	Steam in turbine hall	Y (Low steam pressure)	Y	Y	Drops, then rises	Y	N

* A typical nuclear-power-plant FSM might have, for example, 50 indications, central and local, and 20–30 scenarios

operators, otherwise data on timings and signal reliabilities will be uncertain, or simply non-existent.
• The FSM is a static reproduction of a dynamic scenario during which different operators or teams may develop different interpretations of the signals as they occur. No matter how clearly discriminant scenarios look on paper (i.e. via the FSM), they may not be perceived so by the operators in 'the heat of the moment'.

The FSM will tend to focus on clear-cut scenarios, i.e. the textbook version of scenarios. In reality, each scenario can occur with a wide range of signals in an often-unspecified dimension, or even set of dimensions. A clear example is a seismic event, which can vary in its characteristics: in some cases, for example, operators may not even be sure if an earthquake has in fact happened at all. Similarly, a NPP loss-of-coolant accident (LOCA) is usually defined (quantitatively, in terms of breach size) as a small break or a large break, but many intermediate values are possible, and such variations will have subtle but perhaps important effects on the indications given to the operators. The problem is that whereas a small-break LOCA may be clearly discriminatable from a large break and many other non-LOCA scenarios, an intermediate-break LOCA may *not*. Of necessity, the FSMA must focus on a limited set of faults which will endeavour to set the 'bounds' for all other 'intermediate' faults which can occur. The success of this will depend on the knowledge and experience of the PSA team, the HRA assessor(s) and the operators.

Despite these limitations, the FSM is undoubtedly a useful tool for CRA, and can be used to demonstrate to other members of the PSA team, designers, or operators themselves, in a relatively objective fashion, areas where a misdiagnosis is unlikely and areas where scenarios may be confusable. To decide how confusable scenarios are is the subject of the next subsection on the CMA.

CONFUSION MATRIX ANALYSIS (CMA)
Like the FSMA, the CMA is a fairly straightforward concept, and one that is also relatively easy to apply. It is aimed specifically at two of the diagnostic error-forms, namely misdiagnoses and premature diagnoses. The following discussion of the CMA assumes that a FSMA has already been carried out.

Once all potentially confusable scenarios have been derived, these are put into a confusion matrix, such as the one in Figure 5.3.21. Each event or scenario is then simply placed on the x and y axes of the matrix, and the expert panel then decides how confusable each scenario is with every other scenario. This panel should consist of at least two operators who have an extensive operational experience of the system under investigation, or a similar one. It is useful to have at least one expert who is at the operator level – i.e. used to physically controlling the plant – and one who is at the supervisory level – i.e. supervising the operators. Above this latter level, experience may

Figure 5.3.19 Confusion matrix: hypothetical example

Scenarios	S1	S2	S3	S4	S5	S6
S1	X	LN	LN	LN	LN	LN
S2	LN	X	HI M/N	LN	LN	LN
S3	LN	HI M/S	X	LN	MS	LN
S4	LN	LN	LN	X	MS	LN
S5	LN	LN	MN	MS	X	LN
S6	LN	LN	LN	LN	LN	X

Each cell entry denotes   L   = low confusability
a row event (actual)   M   = moderate confusability
confused with a column   H   = high confusability
event   HI = high initial confusion (1st 15 minutes)
   N   = negligible consequences if confused
   S   = significant consequences if confused

be too far removed to be useful, while below the former level, experience may be insufficient. Having these two level together in one panel is productive, since the operator knows how the indicators react, while the supervisor, in turn, knows how the operators react, as well as the broad relationships between the indicators and the scenarios (this assumes the supervisor is also an *ex-operator*). The HRA assessor should also have reviewed all operational experience to see if any of the faults, or their intermediates, have actually occurred on plant.

An example of a CM is shown in Figure 5.3.19, with scenarios placed along each axis. In the diagonal cell an 'X' is placed. In the case of Cell 'N (row N), M (column M)', for example, the event 'row N' has happened but the event 'column M' is diagnosed. The letters in the respective cells have the following meaning:

L   =   Low: negligible confusability.
M   =   Moderate: the potential for confusion exists.
H   =   High: scenarios have a strong potential for confusion.
X   =   No comparison required: row scenario = column scenario.
HI =   Initially high: the potential for a premature diagnosis exists since symptoms are initially confusable, but as the scenario develops, contradictory symptoms will occur (which may or may not be interpreted correctly by the operators).
S   =   Significant impact in the accident sequence if a misdiagnosis or premature diagnosis occurs.

N = Negligible impact on the scenario if a misdiagnosis or premature diagnosis occurs.

The expert judges consider each scenario-pairing (i.e. each cell) in turn and rate the potential confusability as 'L', 'M' or 'H', as defined above. If the judges rate the potential confusability as 'L', no more is done and they move onto the next cell. If it is rated 'M' or 'H', however, the judges (and preferably a PSA analyst taking part in the CMA scenario) must then decide what would be the likely impact of the misdiagnosis, in the event of its occurrence. In some cases, the misdiagnosis or premature diagnosis will have a negligible impact on the handling of the accidental-event sequence because the resultant operator actions would have been the same whether or not the diagnostic error had occurred, and so would be rated 'N', as defined above. If, on the other hand, the impact *would* be significant, leading to a decline in the level of safety, then this misdiagnosis or premature diagnosis would have to be considered more formally in the HRA, and would be designated with an 'S' in the matrix.

It should be noted that the matrix-cell designations are not necessarily symmetrical; that is to say, confusability may be high in one 'direction' but moderate in the other, while similarly, the consequences may be significant where A is confused for B but negligible where B is confused for A. In practice, often one diagonal half of the matrix is used for the main part of the CMA, while the other diagonal half is used only to record asymmetries of this nature.

What the CMA enables the HRA assessor to do is focus on those confusable scenarios that have significant consequences and then feed these forward to the representation and quantification phase of the HRA. The resultant CMA, as with all information derived from the judgements of experts, should be verified by at least one independent, expert judge (operator).

It is sometimes best in practice if the faults in the FSMA are, by way of an initial review, put into 'clusters', each of which involves similar sets of symptoms. Otherwise, the CMA can become so large that it places a burden on the operators/experts when they are trying to decide if events are confusable or not, since they will have to consider so many different pairs of events (most of which would be non-confusable). These clusters would be conservatively chosen, and for each cluster a 'link' event from another cluster is inserted into the 'target' cluster to see if it is truly non-confusable. If this link event is found, during the CMA, to be confusable with another, then this violates the clustering process and all the events from the link event's original cluster must now be considered in the 'target' cluster (see case study No. 8 in Appendix VI). If, on the other hand, the link is found to be *non*-confusable by the CMA, then the clustering process is upheld. Clearly, the link event must be chosen conservatively – i.e. the event, from the non-confusable cluster, which is most confusable with the target cluster, would be chosen.

It is useful to gain an appreciation of the operators' perceived likelihoods of all the various events involved. Although the anticipated likelihoods can be derived from the PSA itself (based on the PSA's prediction methods or historical information), for the purposes of HRA quantifications, and to decide if operators would really find the scenarios confusable, it is worth knowing what the operators' own personal expected frequencies are for the various scenarios. If the scenarios are confusable but one is deemed likely and the other incredible, this is valuable information for determining the likely direction of the misdiagnosis.

The operators can therefore be asked to categorise the different events/ scenarios into a set of categories, such as the following:

E1 – Expected relatively frequently (e.g. up to 10 times during the operating lifetime (e.g. 30 years) of the plant.
E2 – Expected infrequently, e.g. 2–4 times.
E3 – Expected once in the lifetime of a plant.
U1 – Not expected to occur at all, but could conceivably happen.
U2 – Never actually expected to occur. It is perceived as incredible, or so low in frequency that it will be regarded as of a negligible likelihood by the operators.

As an aside, the PSA team who predict the actual likelihoods of events may also find such categorisation by the operators interesting.

The advantages of the CMA are its simplicity and the utility of its output, turning what originally seemed an insurmountable problem into a focused set of confusable scenarios, or demonstrating that a misdiagnosis is unlikely. Its disadvantages, on the other hand, are its heavy reliance on operational expertise – which, ultimately, is difficult to validate – and its focus only on misdiagnoses and premature diagnoses – it largely ignores other diagnostic-error forms, and, as with the FSMA, only deals with well-defined scenarios. The following discusses those residual CRA approaches which may be used to attempt to deal with the full impact of a cognitive error.

RESIDUAL CRA APPROACHES
At the outset of this section, five types of diagnostic error were noted:

• Diagnostic failure.
• Partial/incomplete diagnosis.
• Misdiagnosis.
• Thwarted diagnosis.
• Premature diagnosis.

The first category is something that can be considered for each scenario being addressed and then taken through the quantification phase. If, at the outset of the CRA, it has already been confirmed that there is a cognitive-error potential in the scenario, then 'Failure to diagnose in time' is the usual starting point for any HRA. A diagnostic failure will usually be fed forward into the quantification phase unless it can be demonstrated that such a failure is highly unlikely. Such a demonstration would usually require a detailed task analysis (e.g. as found at level 2) showing that there are prompt, unambiguous and compelling signals that give the operator plenty of time to react. Even if this were the case, many HRAs would nevertheless go ahead and quantify the diagnostic failure, albeit giving it an appropriately *low* probability of occurrence.

One of the best ways to investigate the potential both for diagnostic failures and for partial diagnoses is via an FSMA and a tabular scenario analysis, the latter providing a clearer representation of the dynamics of the situation, showing both the likely workload on the operators and the rate, sequence and chronology of the alarm's occurrence. As well as using a TSA, decision–action-diagram flowcharts can be highly useful in pinpointing when critical decisions will need to occur, and what the anticipated decision outcomes are likely to be. Sometimes, these may be gleaned from emergency procedures, particularly if these are in a flowchart format.

The identification of a thwarted diagnosis is also best carried out via a tabular scenario-analysis, since by means of such an analysis action errors (slips) during critical actions on the part of the operator can be reviewed to see what effects they have on the accidental-event sequences. This type of analysis, and the preceding ones in this section, are all best carried out by a small team comprising an HRA assessor, operational staff and a PSA analyst.

SUMMARY ON CRA

The CRA is still relatively under-developed, and the beginning of this section was intended to show the inherent difficulties involved in this method in relation to cognitive diagnostic performance. The thrust of any CRA at this stage in the development of HRA is therefore not to be comprehensive but instead to identify the most critical cognitive error-forms. FSMAs, CMAs, TSAs, DADs and checklists, and, possibly, the error-of-commission analysis technique (EOCA) described in the next section, are all useful and practicable tools for doing just this. Furthermore, as will be shown later, these tools, and others, can help to defend against cognitive-error vulnerabilities by generating error-reduction guidelines, some of which will be more generic, and will play a role in fostering effective decision-making via decision-support systems.

5.3.5.6 Level 4 and 5 analysis

This section is concerned with three types of human-error contributions to the level of risk:

- Errors of commission (i.e. non-required acts: level 4 and 5).
- Latent failures (e.g. miscalibration of equipment during a test or maintenance: level 4).
- Rule violations (routine or extreme: level 5).

These are all treated in a single sub-section as, although they cut across the boundary between levels 4 and 5 in the onion framework, the techniques available address these issues at the same time. It is therefore easier to present the techniques only once, in this sub-section.

Each of these three error types can influence the level of risk in a system, but they are all difficult to assess in HRAs and PSAs. Many PSAs to date have largely ignored their possible contribution to risk levels, assuming that such contributions are implicitly included within the quantitative calculation of the overall level of risk to the system, even if the individual failure modes are not identified. Such an assumption is both questionable and unproven, and ignores, moreover, the option of trying to identify and reduce such contributions via an error-reduction analysis (ERA). However, techniques for dealing with these three human error types are relatively immature, and themselves relatively unproven and exploratory in nature. Nevertheless, this section will describe one such approach, similar to a HAZOP in format, which can be used to consider all three of these error contributions, though it is largely aimed at the first and is thus referred to as an error-of-commission analysis (EOCA).

It should be noted that latent failures can be investigated by carrying out a level-1 or -2 analysis of maintenance and testing operations. Particular maintenance operations can be analysed – e.g. those concerned either with the availability of back-up, engineered safety systems or with the calibration of event-detection equipment – rather than *all* maintenance tasks, which, in most cases, would prove impossible given restricted resources.

There are two types of rule contravention defined by Reason (1990), namely *routine* and *extreme* rule-violations. The routine violations are those which occur habitually, tend to go 'unpunished' by the task environment and are usually a consequence of the operator's taking the path of least effort to achieve a system goal. In some cases, such rule violations are known by the management but ignored as long as the system functions adequately (in the production sense). Extreme violations, on the other hand, are more difficult to predict, and involve known risk-taking or the infringement of important safety rules. These are much rarer, but they may occur, for example, when there is an intense safety–production conflict.

Rule violations are very difficult to predict, and the analyst is probably best advised to predict these on the basis of historical information on rule contraventions which have already occurred. This will give the analyst an idea of the types, frequency and severity (i.e. routine vs extreme) of different violations that have actually happened. The analyst may then discuss potential rule violations with operational staff so as to gain their reactions to the credibility of such violations. Rule-violation assessments, for plants at the design stage, for which there is *no* relevant historical database would be difficult indeed to carry out.

Before going on to discuss the technique of the EOCA itself, the area of 'errors of commission' itself needs to be explored.

ERRORS OF COMMISSION

One of the most difficult areas in human-error analysis is that of identifying those human actions which can lead to unsafe consequences. Such actions, technically speaking, should be called extraneous actions, but in the HRA field they have more recently come to be known as errors of commission (EOCs). A simple example would be one where an operator is required to push a button but pushes another, wrong one instead, possibly one whose function is totally unrelated, or even contradictory, to that of the correct button. A more complex example would be where a safety system is initiated automatically when an incident occurs but then turned off (via a manual override) by the operators concerned due to a misconception about the state of the system – as, indeed, happened at Three Mile Island. An even more complex set of errors of commission would be the events that took place prior to the explosion at Chernobyl (1986), where the actions responsible were not only unrequired but reflected a marked ignorance of reactor safety practices.

The problem with such errors of commission is twofold. Firstly, they are inherently difficult to predict, and are usually the result of misconceptions, rooted in either knowledge inadequacies or misleading or confusable indications, all of which cry out for detailed analyses that would identify what could go wrong. And secondly, such errors can have a dramatic impact, often more so than the more typical errors and failures that are modelled in PSAs and HRAs.

The HRA analyst therefore faces a dilemma. A detailed analysis is required if errors of commission are to be identified, but an EOC can occur at *any* stage in a task, and may or may not be related to the task's objectives – which makes an EOC analysis (EOCA) a potentially daunting task itself. And yet on the other hand, if an EOCA is not carried out in some form, key human-error factors may remain overlooked.

This dilemma has been problem for some time in HRAs, and practitioners have often simply used their own judgement and 'detective work' to root out EOCs that can have a significant impaction safety. Recently, more systematic approaches have been developed (e.g. the sneak-path analysis, Gertman *et al*, 1991) which tend towards detailed analyses of both actions and interfaces, and which consider 'what if?' – type questions in relation to controls, etc. Such an analysis can be carried out as part of a large task and human-error analysis of control systems, where each control is considered in all its modes of operation, and where 'maloperations' are postulated for each control state or configuration. This has in fact already been carried out with a measure of success for a number of control systems (e.g. see Kirwan & Reed, 1989). However, such an assessment is resource-intensive, and may be expected to become more so almost exponentially as the system complexity increases and the possible combinations of control/display states multiply. Additionally, such a fine-grained analysis may tend to miss the larger, wider-scope errors of commission that occur as a result of mistaken intentions at the *goal* rather than at the task level.

The HRA analyst, fortunately, has a powerful ally in the PSA itself, which from the very outset builds an increasingly sophisticated model of the safety integrity of a system (and of how it *stays* safe), with its various designed features and its redundant back-up and protection systems. Such a model of safety integrity can be investigated to see how a human error could lead to the failure of more than on safety 'barrier', for it is these kinds of EOC that are of principal interest.

The other principal ally for identifying EOCs is the plant operator (provided, of course, a plant actually exists). Although the HRA analyst can identify (via his or her knowledge of EEMs and PEMs) what could in theory happen, and although the PSA analyst can then rapidly determine the significance of any potential error, often only the *operator* can say whether or not such an event is credible. The HRA analyst will find it difficult to decide alone what EOCs are credible, or which would have a negligible likelihood. The risk, therefore, of doing an EOCA *without* the added involvement of a plant operator is that some identified EOCs may be treated with scepticism by sundry operational staff, the result that the PSA may end up omitting EOCs which are both credible and important in risk terms.

Given the foregoing discussion, the use of a HAZOP-type approach now looks to be a practicable proposition, and this is in fact the major approach recommended here. The following subsections describe the basis of the approach and the type of results that can be gained from it, with comments on its utility in practice.

ERROR OF COMMISSION ANALYSIS (EOCA)

The objective of an EOCA is to identify critical, unrequired actions on the part of the operator, caused via slips, lapses, mistakes or violations, which would lead to a significant impact on the level of risk, a level that is currently *under*modelled in the PSA/HRA. An error-of-commission analysis (EOCA) is difficult to achieve since, by definition, it is trying to identify what *might* happen, over and above what *should* happen – the latter being defined by the HTA and by other TAs. To achieve such an analysis, it is necessary both to have a detailed knowledge/understanding of the operational system and to be able to exploit such knowledge from an *error* perspective. An EOCA, therefore, requires, at the very least, an EOC-analysis framework, as well as operational expertise or a detailed task analysis (or even both). It also (preferably) requires hazard-analysis experience that would help the analyst to determine, 'on-line', the approximate significance of different EOCs: many, if not most, EOCs will in fact be of little or no PSA significance.

The EOCA has only recently been seen as a viable HRA option, with exploratory approaches such as INTENT only appearing in prototype form (Gertman *et al*, 1991), and with many PSAs still leaving this potential error source relatively unaddressed.

What is clear is that EOCs often take the form of relatively straightforward slips, e.g. a confusion between two values (or two controls) or a simple failure to carry out actions in the right sequence, leading to a new outcome. Such errors can be classified into three basic PSA types:

- Slips/omissions/violations leading to significant error-opportunities/impacts.
- Misdiagnoses (mistakes).
- Latent maintenance errors (i.e. those unaccounted for by failure-data for the PSA).

Such error impacts can be addressed, for identification purposes, from two different major perspectives:

- The PSA perspective: requiring a review of 'sensitive' equipment (e.g. redundant/diverse systems, control systems or back-up systems).
- The HRA perspective: calling for a review of critical tasks and functions carried out by the operator.

At this stage in the development of HRA technology, it is sensible to adopt the pragmatic approach of utilising *both* perspectives during any EOCA. To achieve this in a resource-effective way, as well as in a way which is open-ended and which allows for creative insight, rather than relying on unproven and relatively *ad hoc* taxonomies of error, a HAZOP-style approach is advocated as below.

This approach involves both a review of the system's components or tasks (using a small group of experts) and a consideration of those failure modes and PSFs which could yield a significant error of commission. The elements required for such an analysis is as follows:

- Procedures
- A task analysis (including a walk-through/talk-through and a knowledge, via a tabular task analysis, of equipment design).
- A PSA fault schedule, plus associated information on the PSA failure modes and their causes (e.g. hardware failures, critical actions, alarms, etc.).
- Operational expertise.

The 'keywords' must be based in HRA/HFA theory, and must fall into two basic categories:

- Error types (including maintenance contributions).
- PSF considerations.

The former is based primarily on the JHEDI system (Kirwan, 1990), involving a set of 11 error descriptions which are relatively comprehensive. The latter, on the other hand, is based on an examination of PSFs that are relevant to EOCs.

The format of the HAZOP and its output are as follows. The HAZOP team is chaired by the HRA analyst, who also records the output onto pre-printed record sheets. The other two members of the HAZOP team are an operator with significant (e.g. a minimum of 20 years') operational experience and a PSA analyst who is familiar both with HAZOPs and with fault and event trees as utilised in PSAs. The tabular output contains the following information:

- The task/action required (and whether it is pre-trip, trip or post-trip).
- The item of equipment involved.
- The error of commission (EOC).
- The cause of the EOC.
- The recoverability of the EOC.
- Consequences.
- Error-reduction measures.

The 'Consequences' above should also state whether the EOC in question significantly increases the probability of an already identified fault path (P) or whether it instead gives rise to a *new* event sequence (E).

As part of the EOCA, it is useful to consider again the major stages of an accident sequence via a generic accident-sequence event tree (Kirwan & James, 1989). The main stages in this tree yield the following types of errors:

- Latent Errors (pre-event errors).
- Initiators.
- Detection and decision-making failures (as well as diagnostic errors).
- Slips/lapses that worsen the scenario.

This GASET-based approach can be useful in structuring the EOCA for each scenario, taking the basic error types above in order.

The EOCA method, therefore, would probably utilise the PSA's main scenarios as the starting point in order to decide, for each scenario, whether any EOCs could occur, and whether these were latent errors, scenario initiators, etc. The results of such an analysis would then have to be fed into the PSA as appropriate: perhaps new event sequences would have to be considered, or merely new human events within already-identified sequences; or maybe the frequency rating of the event which *initiates* the PSA's main-scenario event would have to be increased if a human-error initiator is found which is credible and unaccounted for in the current frequency rating for the scenario. An example of the use of an EOCA is given in Table 5.3.15.

5.3.5.7 Levels 5 and 6 analysis

Level 5 concerns management failures, and level 6 the so-called sociotechnical error forms. Essentially, there are few, if any, approaches to the identification and prediction of such errors because management failures, and the sociotechnical errors themselves, can take so many different forms, and are

Table 5.3.15 Example of EOCA table

Date:	14/7/92	HAZOP Team	BIK/DB/SS	Page 6 of 6
Bounding fault		Boiler-tube leak		
Task/action required		Identify leaking boiler		
Item of equipment involved		Feedwater		
Pre-trip Trip/SD Post-trip		 Y		
Error of commission		Shut off feed to all boilers by inadvertently closing the wrong feed-system valve		
Guideword		Misinterpretation – cognitive tunnel-vision, perseverance		
PSF		Lack of clear feedback; workload involved in having simultaneously to shut down the reactor		
Causes (other)		Operators trying to identify leaking boiler. If they cannot decide via instrumentation, they may therefore try to consider one at a time with feed-off		
Consequences		Boilers boil dry. Loss of heat removal.		
Affects probability?		Yes		
New event/sequence FTA/ETA?		CMF – of FW system		
Recoverability		Would have to be within 2 hours.		
Error-reduction measure(s)		Data-log moisture indicators in CCR to indicate first-up. Would record time at which first boiler failed.		
Comments		Recovery potential high		
Actions		Investigate qualitatively for ERMs – recommend control measures preventing inadvertent valve closure, e.g. valve-locking mechanism, permit-to-work.		

dependent on the safety-management system in place in a plant. No tools are therefore presented for these two types of human-error risk-contribution. However, management-audit tools and approaches which do exist, and which are being developed for PSAs, are discussed in Section 5.11. Error-reduction devices for combatting sociotechnical errors are also discussed, albeit in a limited fashion, in Section 5.7. These two areas are, not surprisingly, primary targets for research.

5.3.6 Summary on human error analysis

This section has presented a large amount of information. Taxonomies of error and error models have been discussed, the functional requirements of HEI tools have been defined and a selection of current available tools for HEIs has been presented. The onion framework, a multi-levelled approach to human-error analysis, has been described in depth, together with some worked examples. A summary table of the onion method, linking aspects of problem definitions, task analyses and Human Error Identifications, is presented in Table 5.3.16. This table can help the analyst to decide what tools are appropriate to what types of analysis. Some examples of HEA approaches used in real applications are given in the case studies (Appendix VI).

In addition, there is Table 5.3.17, which summarises the relative strengths and weaknesses of 12 different HEA methods (Kirwan, 1992b) according to a number of criteria. The first is *comprehensiveness*, i.e. the degree to which the technique deals with skill-, rule-, knowledge- and rule-violation-type errors. The second is concerned with the degree to which the technique is structured, as opposed to being reliant on the assessor's own judgement. *Usefulness* refers to the ability to generate error-reduction measures (ERMs); and *resourcefulness* and *documentability* are self-explanatory. *Acceptability* refers to the degree to which the technique has already been used, and whether or not it is available on the open market at the time of this book's going to press. (For more background information, see Kirwan, 1992a and 1992b.) Also, more guidance on selection of techniques for human error analysis is given in 5.10, after quantification has been discussed, along with revisiting problem definition issues, which can influence selection of HEI tools.

Once the errors (*and* recoveries, if appropriate) have been identified, the analyst, in almost all cases, will want to consider their consequences (which will probably be recorded in some form of HEA tabular format, as found in a level-2 analysis), and thus determine if and how they should be inserted into the PSA. At this point, various issues in the representation of human errors in PSAs must be considered. These are discussed in the next section.

5.4 Representation issues in HRA

Representation refers to ways of integrating the identified human contributions to risk (both positive and negative) with other relevant contributions

Table 5.3.16 Human-error-analysis modes and interactions

	PSA input to HRA	Relevant task analysis	Human-error-analysis tools	PSA evaluation of tasks/errors	HEA output format	HEA functional outputs
Level 0 Task success/ failure	Level 0 is PSA-driven: FTA; ETA; HAZOP-define tasks requiring quantification	Initial task analysis (ITA)	HAZOP Human HAZOP EOCA CRA	Significant errors identified in screening analysis using conservative HEPs (levels 1 or 2)	Task-success and task-failure definitions; ITAs Significant errors	Task-success or task-failure; Significant errors Critical tasks
Level 1 Error screening	PSA/HRA-driven: FTA; ETA; HAZOP. Overview of PSA information; plant visits; design information; Procedures	OER Interviews CIT ITA	CEP (level 3) (CRA) GASET; Basic EEMs; HAZOP; Human HAZOP/EOCA	Significant or sensitive errors requiring ERA (level 2)	HEA table; HAZOP table(s); EOCA table	Errors; Consequences; Significant error-potential
Level 2 Detailed error analysis	HRA-driven: FTA; ETA; HAZOP. Detailed PSA information available; access to plant/design	HTA; TLA; TTA; TSA; DAD; WT/TT; LA; OER; CIT; Interviews	CEP (level 3) (CRA) Human HAZOP EOCA Techniques in Table 5.3.17 HEIST; Appendix V	Significant and sensitive errors go to ERA, then re-quantified	Detailed HEA table; ERMs; EOCA/HAZOP table	Error-reduction required; errors and consequences
Level 3 Cognitive error	HRA-driven: PSA 'fault schedule'; plant-design information on the interface	HTA; TSA; DADs; TTA; OER; CIT; Interviews; WT/TT	FSMA CMA Appendix V	Diagnostic errors may lead to new events/event trees in the PSA	Diagnostic errors; CM; FSM	Diagnostic and decision-making error-potential

Table 5.3.16 (Continued)

	PSA input to HRA	Relevant task analysis	Human-error-analysis tools	PSA evaluation of tasks/errors	HEA output format	HEA functional outputs
Level 4 Maintenance error analysis	PSA/HRA-driven: PSA safety-modelling; PSA information and maintenance reliability databases	OER; CIT; Interviews; ITA; DTA (level 2)	EOCA HAZOP Human HAZOP	May lead to extra maintenance errors in logic trees or to revision of maintenance reliability estimates.	Maintenance-related errors in EOCA/HAZOP table	Significant error-contributions from maintenance
Level 5 RVA and Management-error analysis	HRA-driven: significant assumptions about rules, compliance, administration controls and unusual plant modes	CIT; OER; interviews	EOCA Human HAZOP Safety culture Analysis tools (see 5.11)	Sensitivity of PSA to RVs etc. may be determinable	RVs and SC/SM vulnerabilities in tabular format	RV and SC/SM vulnerability
Level 6 Sociotechnical errors	HRA-driven: PSA model of safety integrity	–	–	–	–	–
Level 7 Sabotage analysis	HRA driven: PSA model of safety integrity	–	–	–	–	–

* RVA = Rule Violation Analysis
SC/SM = Safety culture/safety management

Table 5.3.17 Comparative evaluation, based on qualitative criteria, of HEI techniques (Kirwan, 1992b)

Criteria	THERP	HAZOP	SRK	GEMS	PHECA	Murphy	SHERPA	CADA	HRMS	IMAS	CM	CES
Comprehensiveness S,R,K,Rv	S,R	S,R,Rv	S,R,K	S,R,K Rv	S,R,K Rv	S,R,K	S,R	S,R,K	S,R,K Rv	R,K	R,K	R,K
Consistency 'Structuredness'	Low/moderate*	Low	Moderate	Low	High	Moderate	High	Moderate	Moderate	Low	Low	Moderate**
Theoretical validity Model-based EEMs/PEMs/PSF	Low EEM	Low EEM	High All	High All	High All	Moderate EEM/PSF	High All	High All	High All	Moderate All†	Low EEM	High All†
Usefulness ERMs facility	Low	High	Moderate	Moderate	High	High	High	High	High	Moderate	Low	Moderate
Resources Resources usage Experts required Experts tool	Moderate No No	High Yes No	High No Yes	High No Yes	High No Yes	High Yes Yes	Moderate No No	High Yes Yes	High No Yes	High Yes Yes	Moderate Yes No	High No Yes
Documentability	Moderate	High	High	Low	High	High	High	High	High	Moderate	Moderate	High
Acceptability Usage Availability	High Yes	Moderate‡ Yes	Low Yes	Low Yes	Moderate No	Low Yes	Moderate Yes	Moderate Yes	Moderate No	Low Low	Moderate Yes	Low Low

Notes

* THERP's rating on this criterion depends on whether the assessors' usual approach is utilised or whether the formal THERP-HEI approach is utilised (the latter deserving a rating of 'Moderate')

** CES consistency depends on how the assessor decides to affect the simulated operator's available information etc. to produce errors

† These techniques implicitly identify the PSF and PEMs, which admittedly comprise a rather limited set in comparison with those sets produced by other methods

‡ HAZOP's usage-rating as a HEI tool is only moderate, but its general usage as a PRA-related tool is nonetheless extensive

Rv = Rule Violation

(hardware, software and environmental) in a logical and quantifiable format. Representation allows the overall level of risk to the system to be accurately calculated, and enables the risk assessors and managers to see what are the relative hardware or human (etc.) contributions (described in the impact-assessment section, 5.6). Thus, without adequate representation, it is difficult, and in many cases impossible, to gain a coherent and balanced picture of how the various failures *interact* in such a way as to cause accident scenarios un- less the scenario is a very simple one. In short, the 'risk picture' is dependent on the *adequacy* of the representation. If the representation is inaccurate, then the appreciation of the system's risk level afforded by the PSA will at best be fuzzy, or simplistic, and at worst incorrect.

Representation covers a number of interrelated issues, and is itself a de- veloping field, in connection with both HRAs and PSAs. The first is the format of the representation itself. Usually one or both of two formats are utilised in a PSA: the fault tree and the event tree, as already introduced in Section 4. The second issue is that of the dependence between human errors, also already alluded to in earlier sections. There are two types of dependence in HRA, and whilst the subject of dependence is a theoretically difficult one to unravel, in practice there *are* some simple and practicable approaches for the assessor.

The third issue is that of screening, which is only usually of importance for large PSAs and HRAs where limited resources may make it necessary to restrict the process of quantification to only those HEPs that contribute above a certain level to the overall risk level inherent in the system. The fourth rep- resentation issue is that of test and maintenance errors, and the degree to which the PSA's hardware-failure data already includes a human-error con- tribution. Finally, the difficult area of the level of decomposition is briefly addressed. This refers to the decision as to when to stop breaking down human errors into yet more detailed causes.

The assessor should be aware of all these issues, and of the approaches that in some cases solve them, and in other cases work around them to achieve a 'requisite' representation, i.e. one which though not as theoretically sound as we would wish for, nevertheless does the job satisfactorily.

5.4.1 Fault tree and event tree representation

Fault trees and event trees are the major formats for representation used in PSAs currently. Whilst other methods can be used (e.g. simulation methods), these are beyond the scope of this book.

5.4.1.1 Fault trees

A fault tree is a standard way of representing a set of human errors and their effects on the goals of the system. A fault tree is a logical structure which defines what events (human errors; hardware/software faults; environmental

Figure 5.4.1 Brake-failure fault tree

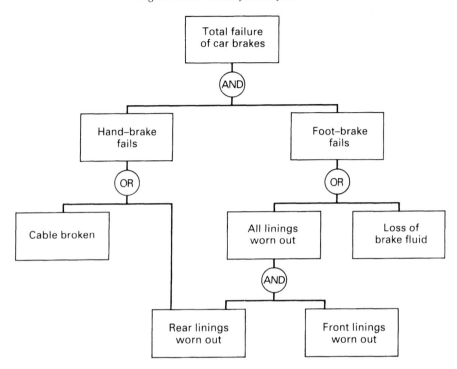

events) must occur in order for an undesirable event (e.g. an accident) to occur. The undesirable event or outcome, usually placed at the top of the 'tree', and hence called the 'top event', may for example be 'Failure to launch a lifeboat successfully at first attempt', or 'Failure to achieve recirculation of primary coolant', etc. The tree is constructed primarily by using two types of 'gate', by means of which events at one level can proceed to the next level up, until finally they reach the top event.

Figure 5.4.1 shows a simple fault tree (Bellamy, 1986; personal communication). The top event is the car-brake failure, and this is broken down logically into the constituent failures that need to occur for the top event to occur. An 'AND' gate means that all events *under* the gate must occur for the event *above* the gate to occur. Therefore, for a total car-brake failure to occur, *both* brake systems must fail within the same time-frame. An 'OR' gate means that the event *above* the gate will occur if any *one* of the events immediately below the gate occurs. Therefore, in the example in Figure 5.4.1, any one of the events beneath the foot-brake-failure box can *alone* cause the foot-brake to fail; and similarly for the hand-brake. There are other types of gates (e.g. see Henley & Kumamoto, 1981), but these gates are the most important and useful ones. Figure 5.4.2 shows a set of symbols commonly used in fault trees. Fault trees can run to many 'pages' through the use of transfer arrows/triangles to show the links from one page to another (as for a HTA).

Figure 5.4.2 Symbology for fault trees

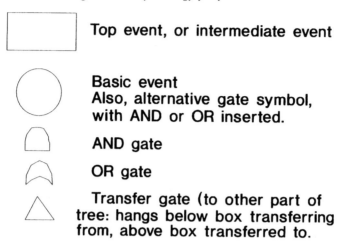

Top event, or intermediate event

Basic event
Also, alternative gate symbol,
with AND or OR inserted.

AND gate

OR gate

Transfer gate (to other part of
tree: hangs below box transferring
from, above box transferred to.

A detailed treatment of the mathematical evaluation of fault trees is beyond this book (see Henley & Kumamoto, 1981; Green, 1983; Cox & Tait, 1990), but the basics can be outlined. Firstly, fault trees make use of frequencies and probabilities. Typically, a fault tree will have as its output a frequency for a top event (e.g. the number of offshore blowouts, caused by drilling incidents, per 1000 years), but there may only be one frequency in the entire fault tree (in this example, the number of wells drilled per year), all other 'events' in the tree being probabilities. This may not be the case in many fault trees, but for simplicity of explanation it is assumed hereafter that the events being quantified and mathematically evaluated are probabilities, with the exception of a frequency that is one level down from the top event. The reason for this assumption will become apparent shortly.

The basic procedure in fault-tree evaluations by hand (most companies now utilise sophisticated computer codes to evaluate fault trees) is that events' probabilities under an OR gate are *added* together, and the events' probabilities under an AND gate are *multiplied* together, to arrive at the superordinate event's (i.e. the one above the gate) probability of occurrence. This simple procedure may require a correction for the OR gate if the HEPs are relatively large (≥ 0.1), for reasons explained in Figure 5.4.3. But in many, if not most, cases the overall top-event frequency for a small fault tree can be calculated fairly simply by adopting the simple procedures outlined above.

One point that needs to be mentioned is that a probability multiplied by another probability is still a probability at the end of it, while a probability multiplied by a *frequency* is now a *frequency* instead. However, a frequency multiplied by another frequency is a dimensionally meaningless or uninterpretable quantity; therefore, one frequency cannot appear together with another frequency under an AND gate.

Figure 5.4.3 Fault-tree mathematics

Bicycle brake failure

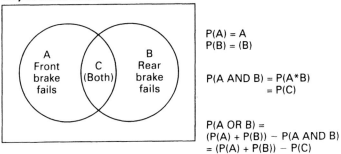

If A and B are *mutually exclusive* events (i.e. they cannot both exist at the same time), then:

P(A OR B) = P(A) + P(B) (Since P(C) = 0)

In most cases 'C' (the 'cross product') will be far smaller than P(A) and P(B) and, for approximate calculations, may be ignored where P(A) and P(B) are each less than 0.1. If the cross-product C is not subtracted, it will be counted twice when 'A' and 'B' are added unless C = 0 or is very small.

The one remaining facet of fault trees which needs to be dealt with in this section is that of *cut-sets*. A cut-set (or *minimal* cut-set) contains a combination of the *minimum* number of events required for the top event to be reached. A list of cut-sets for Figure 5.4.4 is shown in Table 5.4.1. These are the minimal cut-sets for the figure, and they have been prioritised by the fault-tree-evaluation program. In addition, the program has generated a measure of the importance of each event: that is to say, a measure has been derived of the relative contribution of each event, summed for all calculated cut-sets, to the overall level of risk inherent in the system. Obviously, this type of information is extremely useful when considering risk-reduction measures, or when considering where sensitivity analyses should be carried out (see Section 5.6). Cut-sets will also be returned to when dependence and screening methods are discussed later in this section.

A fault tree can be used to represent a simple or complex pattern of system-failure paths, and may comprise human errors alone, or a mixture of human, hardware and environmental events, depending upon the scenario. Once structured, the events (including human-error probabilities) must be quantified to determine the overall top-event frequency (or probability), together with the relative contributions of each error to this undesirable event.

EVENT TREES

An event tree is (usually) a binary logic tree which proceeds from an initiating event to the logical set of outcomes or consequences that can happen to

Figure 5.4.4 Fault-tree example

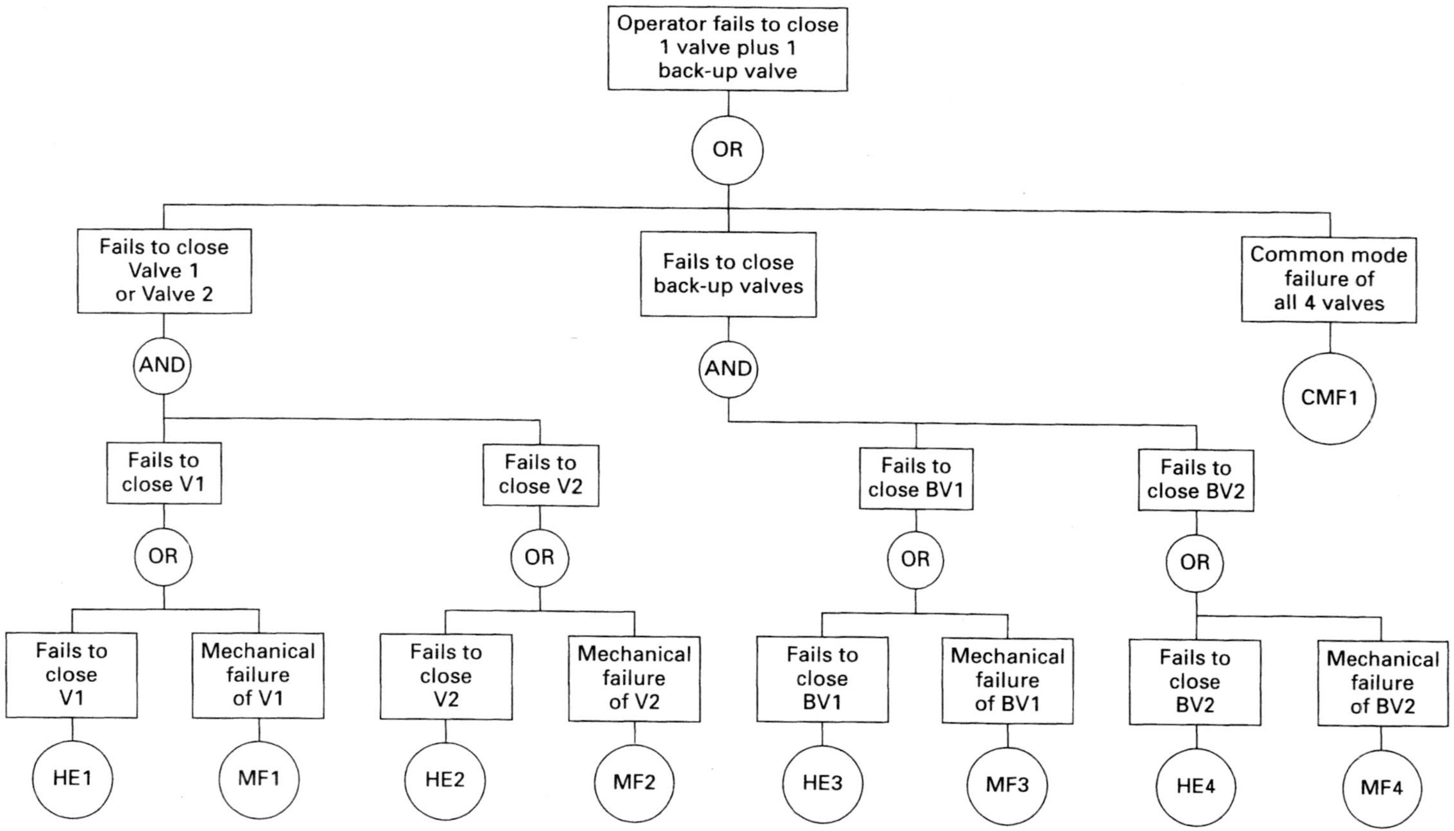

Table 5.4.1 Cut-sets analysis for fault-tree example

Probabilities for fault tree	Importance measure ($P(x)$ / P (top event))
CMF1 = 0.001	0.8052
HE1 = 0.01	0.0805
HE2 = 0.01	0.0805
HE3 = 0.01	0.0805
HE4 = 0.01	0.0805
MF1 = 0.001	0.00805
MF2 = 0.001	0.00805
MF3 = 0.001	0.00805
MF4 = 0.001	0.00805

Top-event probability = 1.242 E-03 [0.001242]

Minimal cut-sets

CMF1	= 0.001
HE1 × HE2	= 0.0001
HE3 × HE4	= 0.0001
HE1 × MF2	= 0.00001
HE3 × MF4	= 0.00001
HE2 × MF3	= 0.00001
HE4 × MF1	= 0.00001
MF1 × MF2	= 0.000001
MF3 × MF4	= 0.000001
Sum	= 0.001242

the system depending on which events occur subsequent to the initiating event. An example of an event tree is shown in Figure 5.4.5. In this representation, in which the event sequence proceeds from left to right, an upper 'branch' indicates that the event in the box at the top of the page has occurred, while a lower branch indicates that it has *not* occurred. The calculation procedure is simpler than that for a fault tree (although the two formats are in fact mathematically interchangeable) since each probability associated with a branch is simply multiplied by the next branch probability, etc., so that an *outcome probability* is derived for each success or failure path. These outcome probabilities will always sum to unity since at each 'node' in the tree, the two probabilities of the successful and unsuccessful event-occurrences respectively will themselves be complementary and will come to unity when added together. A useful check to the veracity of the tree's probability calculations is therefore simply to ensure that the outcome probabilities do in fact add up to unity.

Event trees are particularly useful for considering sequences of events involving many different potential outcomes (i.e. rather than the single top event of the fault tree). Event trees and fault trees can also be used together effectively, e.g. with the fault tree calculating the probability of the major

Figure 5.4.5 Event-tree example

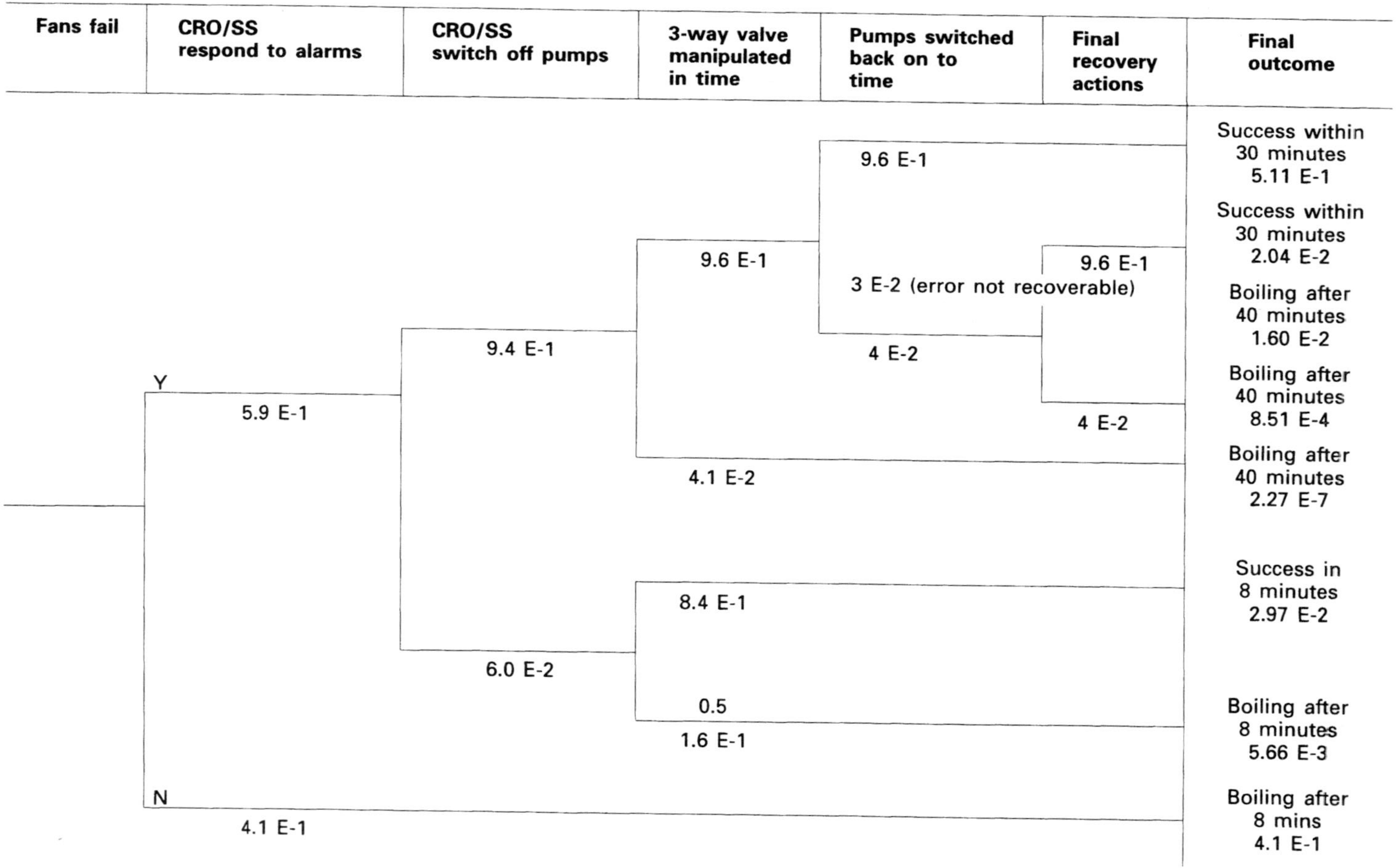

Fans fail	CRO/SS respond to alarms	CRO/SS switch off pumps	3-way valve manipulated in time	Pumps switched back on to time	Final recovery actions	Final outcome
	Y 5.9 E-1	9.4 E-1	9.6 E-1	9.6 E-1		Success within 30 minutes 5.11 E-1
				3 E-2 (error not recoverable)	9.6 E-1	Success within 30 minutes 2.04 E-2
						Boiling after 40 minutes 1.60 E-2
				4 E-2		Boiling after 40 minutes 8.51 E-4
			4.1 E-2		4 E-2	Boiling after 40 minutes 2.27 E-7
		6.0 E-2	8.4 E-1			Success in 8 minutes 2.97 E-2
			0.5			Boiling after 8 minutes 5.66 E-3
N 4.1 E-1			1.6 E-1			Boiling after 8 mins 4.1 E-1

Figure 5.4.6 Mathematical equivalence of fault tree and event trees (Whittingham, 1990)

Fault tree

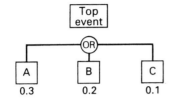

Fault-tree mathematics

$$P(X) = P(A) + P(B) + P(C)$$
$$- P(A)P(B) - P(A)P(C) - P(B)P(C)$$
$$+ P(A)P(B)P(C)$$
$$= 0.3 + 0.2 + 0.1 - 0.06 - 0.03 - 0.02 + 0.006$$
$$= 0.496$$

Event tree

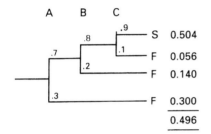

Event-tree mathematics

$$P(X) = P(A) + P(1-A)P(B) + P(1-A)(1-B)P(C)$$
$$= 0.3 + 0.140 + 0.056 = 0.496$$

event's occurring (e.g. offshore blowout) and the event tree calculating the various outcomes once the event has occurred (e.g. the probabilities of a successful evacuation, depending on the rapidity of the reaction to the blowout, weather conditions, evacuation methods, etc.). Figure 5.4.6 shows the mathematical equivalence of the two trees by means of a simple example.

A HRA-specific type of 'tree' is called an *operator-action event tree* (OAET). The OAET is in essence an event tree which puts more emphasis on human operations, required or unrequired, following an initiating event, and which may also include the contribution, prior to the initiating event, of latent failures to the event sequence.

Event trees are especially useful when dynamic situations are being considered, and in general their use is preferable to the use of fault trees in scenarios where human performance is dependent upon previous actions/ events in the scenario sequence (as will be discussed shortly in the next subsection). Since fault trees do not depict the passage of time, they are ideal for faults which involve a relatively random and unfortunate co-occurrence of events, but they become more opaque, on the other hand, when trying to represent sequences of events where the exact sequence is what determines the outcome. Thus, whilst the trees may be mathematically equivalent, in terms of meaningfulness they are not equivalent.

A recent variant on the OAET involves representing the emergency actions of operating personnel in the form of an event tree (Spurgin *et al*, 1987)

Figure 5.4.7 HRA event tree (adapted from Swain & Guttmann, 1983)

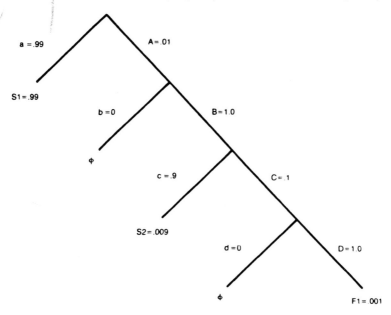

A = FAILURE TO SET UP TEST EQUIPMENT PROPERLY
B = FAILURE TO DETECT MISCALIBRATION FOR FIRST SETPOINT
C = FAILURE TO DETECT MISCALIBRATION FOR SECOND SETPOINT
D = FAILURE TO DETECT MISCALIBRATION FOR THIRD SETPOINT
φ = NULL PATH

which uses three different branches representing, in turn, a successful operation, a failure to respond and a mistake or misdiagnosis. This type of OAET
is capable of representing important potential misdiagnoses that may have
been identified in the human-error-analysis stage. If significant misdiagnoses
are identified, the main event tree may branch off to another *sub* event tree
which considers the alternative consequences that may occur as a result of
the operator's mistake. A fourth type of event tree is the *HRA* event tree, as
utilised in the THERP method (Section 5.5). This *type* looks dissimilar, but
as with all the other event-tree variants, it follows both similar construction
rules and identical mathematical evaluation methods (see Figure 5.4.7).

FAULT VS EVENT TREES
The above are formal methods used for representing moderately complex
patterns and sequences of failures. Whilst the examples given here are simple,

in practice such trees can become quite complex, with many 'nodes' and events, and, in the case of event trees, with a large number of possible final outcomes. It is something of an art to develop such trees in such a way that the human errors are adequately and accurately represented, on the one hand, without letting the trees become too complex and unwieldly on the other hand.

The following recommendations are relevant for event-tree vs fault-tree assessments:

- If the scenario is time-dependent (i.e. sequential in nature, or 'event-driven'), use an event tree. If the sequence of errors in relation to other (e.g. hardware) events does not matter, use a fault tree.
- Select the method which gives most *meaning* to the scenario.
- If the PSA is heavily biased towards fault or event trees, then the use of the PSA's favoured format will help to ease the integration of the HRA into the PSA, and to facilitate a sensitivity analysis, etc.
- If there are *many* possible outcomes from an event sequence, an event-tree analysis may prove problematical: and in these such cases, although a fault-tree analysis may be less meaningful, it can be made mathematically equivalent, and may prove a more tractable form of analysis.

In summary, the major criteria to take into account when choosing a representational format are time/sequence-dependence, meaningfulness, ease of integration into the PRA and tractability of analysis.

A more detailed version of fault/event-tree-selection considerations is given in Table 5.4.2. (Whittingham, 1990).

5.4.2 Human error dependence

There are two types of dependence, both dealt with separately below.

5.4.2.1 Type 1 dependence

The first type of dependence, called a *type-1* or *direct* dependence, is concerned with the direct dependence between two actions – e.g. the response to the first alarm and the response to the second alarm – such that it is intuitively obvious that if one and the same operator is involved in both actions, then the errors associated with each of these events are unlikely to be independent. Dependence at this level can be dealt with explicitly in two ways, firstly by using a formal dependence model, and secondly by the use of *conditional probabilities*.

Conditional probabilities have to do with the adjustment of the human-error probability in those situations where human intervention is possible at two or more places in the fault sequence. For example, the results of sampling a tank's contents may be checked by both the operator *and* the supervisor

Table 5.4.2 Fault trees vs event trees (Whittingham, 1990)

No	Attribute	Fault tree	Event tree/operator-action tree	HRA event tree
1	Sequential events	Unable to represent sequences explicitly	Represents sequences of events, but will not represent parallel sequences	Represents sequences of events, including recovery from human error
2	Order of events	Unable to represent order	Represents order of events	Must represent order of events, for meaningfulness
3	Information needs	System success-or-failure outcomes from status of components/human performance; probabilities; failure rates	As for fault tree, together with sequential order of events, for meaningfulness	As for event tree, together with knowledge of recovery paths for determination of logical event sequence
4	Dependence	Implicit only, in point values of probability	Implicit only, in point values of probability	Explicitly, in terms of values of probability defined at point of occurrence
5	Recovery	Can be implicitly shown as a non-basic conditional event 'AND' human-error basic event	Shown explicitly as a conditional event probability, with effect of recovery success clearly depicted	Shown as an explicit conditional event, with recovery success-or-failure routes clearly identified
6	Frequency data	Can be used at any point in the tree; care must be taken to observe rules of logic	Can only be used as a single initiating-event frequency at front of tree, or by converting *low* frequency to probability	Can only be used as a single initiating-event frequency at the top of the tree

7	**Mathematical validity**	Can be solved mathematically, without knowledge of sequence or order of events, using probability formula	Single fault tree can be represented by many event trees of different sequential orders – all give same mathematical result	Single fault tree can be represented by many event trees, one of which is the nearest representation of the true system – all give same mathematical result
8	**Data availability**	Failure rate/probability data usually available with knowledge of system-failure modes	As for fault tree, but knowledge of event order is required to construct a meaningful tree	As for event tree, but knowledge of recovery paths and dependencies for construction of a meaningful tree
9	**Complexity**	Involves complex logic which is difficult to understand with large trees, particularly when sequences are represented implicitly	Represents complex sequences more explicitly, but results in many alternative outcomes, the consequences of which are not always clear from the tree	Represents complex sequences of human activity explicitly, with recovery and dependency clearly shown, but does not easily include component failures
10	**Scrutability**	Scrutability is poor with large trees	Scrutability is better if the tree does not have too many complex outcomes	Scrutability is excellent as regards representing human-activity sequences
11	**Face validity**	Well proven, acceptable technique in many industries, and includes limited Human Error Representation	Because it represents sequential events in order, it is an easily understood and acceptable method for including a fair proportion of human-error events	As a method for representing sequences comprising *all* human error, it is not well known or understood; subject to criticism as a quantification technique rather than as a model

Table 5.4.2 (Continued)

No	Attribute	Fault tree	Event tree/operator-action tree	HRA event tree
12	Auditability	Difficult to audit unless cut-sets are clearly presented and described	Easier to audit than a fault tree when representing simple sequence, but more difficult when wrongly used in place of a fault tree, due to multitude of outcomes	Easy to audit human-activity sequences and failure modes and paths
13	Meaningfulness	Less meaningful than event trees when attempting to represent sequential events, but more meaningful for instantaneous events	Meaningful when representing ordered event sequences with limited number of outcomes	Meaningful when representing sequences of human activity
14	Usefulness	Only provides useful information if the cut-sets can be analysed easily (presentation, importances etc.)	Provides useful information about the contribution made by each event to the top event, but only for simple event trees	Provides useful information about the relative importance of each human error-mode to the failure of the complete sequence
15	Resources	Dependent on complexity of system, particularly in time-related aspects, but generally less resources than *equivalent* event tree	More difficult to construct than a fault tree because right sequential order of events needed if the tree is to be meaningful; resources are greater	More difficult to construct than event tree because of need to represent often complex recovery and dependence factors. Limited to important sequences. High resources

before the contents of the tank are discharged from the plant – e.g. into the sea. An appropriate human-error probability for the detailed and systematic checking by a competent person may be, for example, 1E–2. This would be applied to the first check by the operator, while a conditional probability of 1E-1 would be applied to the supervisor's check, conditional on the failure of the first check. This is because, in practice, one person checking another's work may not scrutinise it as if it were his or her own but instead implicitly assume that the first operator has done his or her job, and that the main responsibility for the adequacy of the task rests with the person carrying out the task, not with the checker.

This is a simple enough example, but in practice dependence-modelling using conditional probabilities faces two major problems. The first is that some data-based quantification techniques (e.g. HEART), as will be seen later in Section 5.5, do not contain conditional probabilities. Similarly, quantification techniques based on the judgements of experts must rely on the experts' being able to incorporate the dependence effects into their quantitative judgements. This leads to the second problem, which is that it is sometimes not clear whether dependence is there or not, or how one error could actually affect another; i.e. dependence between errors is known to exist through experience, but it is not well understood, and a subjective judgement concerning something that is not well understood is hardly likely to be reliable, nor necessarily conservative. For these reasons, there is clearly a need for a formal method. Only one such method exists to date, and this is described below.

This type of dependence can be dealt with through the use of the dependence model embodied in the technique for human-error-rate prediction (THERP, Swain & Guttmann, 1983).

The THERP system has five dependence levels, from zero to complete dependence, the intermediate valves being *low*, *medium* and *high*. (See Table 5.4.3.) Guidelines are given on how to decide which level should be carried out, and a simple mathematical model is applied for altering the HEPs. This model is dealt with in full, in NUREG/CR-1278. Here is an example from that source:

Moderate dependence (MD) = P(failure on task N/failure on task N – 1)
$$= (1 + 6N)/7, \text{ where } N = P(\text{failure on task N})$$

If, for example, N(independent) = 0.1, then $P(N/MD) = \left(\dfrac{1 + 6 \times 0.1}{7} \right) = 0.23$

If there is zero dependence, then N = 0.1; if there is complete dependence, N = 1. For low and high dependence, the equations are (respectively):

$$LD = \frac{1 + 19N}{20} \qquad HD = \frac{1 + N}{2}$$

Table 5.4.3 THERP dependence model

Level of dependence	Failure equations
ZD	$\Pr[F_{"N"} \mid F_{"N-1"} \mid ZD] = N$
LD	$\Pr[F_{"N"} \mid F_{"N-1"} \mid LD] = \dfrac{1 + 19N}{20}$
MD	$\Pr[F_{"N"} \mid F_{"N-1"} \mid MD] = \dfrac{1 + 6N}{7}$
HD	$\Pr[F_{"N"} \mid F_{"N-1"} \mid HD] = \dfrac{1 + N}{2}$
CD	$\Pr[F_{"N"} \mid F_{"N-1"} \mid CD] = 1.0$

This model is a useful one despite the fact that the actual values for 'low', 'moderate' and 'high' are relatively arbitrary based both on judgements made by its authors and on an assumed logarithmic relationship. What is meant by low, medium and high dependences should ideally be more fully defined, either generically or in the context of specific industries.

Therefore, whilst there *are* debatable aspects of this model (e.g. whether the model is conservative enough in its assumptions concerning the log-linear relationship between dependence effects, i.e. the increases in HEPs, and the five dependence categories), the basic approach does offer a means of assessing dependence in a reasonably practicable way. In practice, it appears to be the only model currently in wide use in HRAs. Appendix VI contains a case study showing the application of dependence to a series of HEPs. It should, however, be pointed out that a number of quantification approaches do not model dependence.

5.4.2.2 Type 2 dependence

The second type of dependence considers dependence at a 'higher level' than that of 'sub-task error dependences'. The problem brought on by this second type of dependence is very similar to that found with hardware. With this fact in mind, the following will discuss first of all the approaches developed in hardware PSAs for dealing with common-cause failures and then the potential for a similar approach to human-dependence problems that makes use of 'human-performance limiting values'.

COMMON MODE FAILURES IN HARDWARE ASSESSMENTS

A common-mode failure (CMF), or common-cause failure (CCF), is best described by way of an example. Let us suppose that three water-pumps are available for use in the event of a fire. Only one pump is required to supply enough water for all the sprinklers in a building, which means that the other two are effectively 'redundant' (i.e. are the identical supporting) back-up systems, to the used when the first does not start on demand. If the probability of one of these pump's failing to start on demand were 0.01, then a first, simple estimate of the unreliability of the whole system would stand at 1E–6 (i.e. $0.01 \times 0.01 \times 0.01 = 0.000001$). However, on further examination, it is realised that all pumps are supplied, by way of a 'starting signal', by one and the same power source, which itself has a failure probability, at the outset of a fire, of 0.0001, and a failure on the part of this power source to operate would prevent *any* of the pumps from starting automatically. To overcome this 'common cause' of failure of the pumps, it is decided to use three different designs of pump with different power-line-supply configurations in *separate* locations in the building, so as to make the common failure of all three a highly unlikely event. Unfortunately, at present, all three 'diverse' pumps are located in the *same* room in the basement, which in the case of an oil fire in the basement area would mean that all three pumps would trip (for their own protection) on the detection of smoke entering their air-intake system. As this example is attempting to show, the drive for ever-increasing levels of reliability for systems is impeded by common-cause effects (in this example, power supply failure; oil fire).

For over three decades in the field of hardware reliability, as a result of initial exploration into how to make missile systems highly reliable (mission-controlling systems on board the missile had unacceptably high failure rates), the idea of redundancy was tried out – i.e. the use of multiple identical back-up systems. This meant that if one system failed, another would be able to take over control of the system. This approach, however, was vulnerable to any (undetected) common design fault, which would render all redundant and main systems more likely to fail in the same way, and often at the same time – or else within a damagingly similar time-frame. This meant that although the use of redundant components improved the level of reliability, the gain in reliability was not as much as predicted or needed.

The idea of *diversity* – that is, of using back-up systems which had the same function as the main system but which were of a different design and sometimes in a different location, or again of using a different power system, etc. – was then tried. However, although this again improved the level of reliability, the presence of common modes of failure, or common causes, meant that, as with the redundancy idea, the improvement in reliability fell short of expectations. Even today this is a problem since, until they have actually been experienced, common failure paths are notoriously difficult to predict (see SRD, 1987). In practice, it appears difficult to make a system highly reliable to the extent that it will have an unreliability of, e.g., 1E–6 or even 1E–7, which although attainable in theory, experience suggests that it is difficult, in practice, to achieve greater than 1E–4, due to common-mode effects.

This has led to the practice of using 'cut-offs', or limiting values, in PRAs/PSAs to ensure that reliability predictions do not become unduly optimistic. (These values are used for particular cut-sets according to which *multiple* failures must occur before the intermediate failure or the top event can occur). Another method used is the 'beta factor' method, but this is outside the scope of this book. (See Martin & Wright, 1987.)

Of course, one of the most significant potential common modes of failure in a system is the *human operator*, as demonstrated in accidents such as Three Mile Island and Chernobyl, where protective systems were disengaged by the operators due to a misconception of what was happening in the system. A PSA must therefore endeavour to take account of such potential human-error contributions to the level of risk, either by including such contributions in its own limiting values or beta factors, or by making the HRA system employ its *own* human-performance limiting values. Alternatively, again, the PSA could incorporate both of these methods (while taking care to avoid double-counting).

HUMAN PERFORMANCE LIMITING VALUES

Human-performance limiting values (HPLVs: Kirwan, Martin, Rycraft & Smith, 1990) can be seen as representing a quantitative statement of the analyst's uncertainty as to whether all significant human-error events have been adequately modelled in the fault tree. Due to the increasing sophistication of large systems, rendering them highly complex and interactive in nature, this cautiousness is entirely warranted. HPLVs *should*, therefore, be used in addition to other CCF limiting values in the fault tree, so that all forms of dependent failure are modelled and accounted for.

The quantification of human reliability is not as advanced as that for system/hardware reliability. What is also evident is that some human-reliability analyses may fall into the same kind of trap that hardware analyses used to fall into by treating basic human-error events as *independent*. Accidents such as those at Seveso, TMI, Bhopal and Chernobyl have demonstrated the folly of such assumptions. Having said that, the field of Human Reliability Assessment is currently still concerned mainly with emergency responses and total crew-response-reliability values of the order of 1E–2 to 1E–3; and the modelling of dependences between individual errors at this level is still sometimes carried out. The arguments for a limiting value on human performance only become strong when the highly *reliable* end of the human-reliability spectrum is being considered – for example, highly important but infrequent and low-time-stress tasks, such as those performed during a shutdown, in which scenario most of the safeguards and interlocks are inoperative, their leaving the safety factor ultimately in the hands of the human operators. It is in these tasks that reliabilities at the value of 1E–5 may be desirable, but such reliabilities may not be justifiable.

THE MEANING OF A HUMAN PERFORMANCE LIMITING VALUE

It is useful to elucidate on what a human-performance limiting value actually represents. The first and most obvious definition of a limiting value is literally, 'the greatest degree of reliability that can be expected from a task'. Human beings are themselves subject to random variabilities in performance, and 'learning' itself can be thought of as the process of reducing the range of such variabilities so that the performance gradually falls within acceptable tolerance limits. However, some degree of variability will always still occur, and eventually even the most proficient or skilled person will make an error – i.e. a response which falls outside the tolerable range (unless, of course, the tolerance limits are excessively large). Errors which are random in this sense, or which are due to more systematic causes, can also be induced by external circumstances, events or disturbances that affect the normal operating environment, or again by internal events (such as fatigue, illness or psychological 'ups and downs'). Systematic errors may be due in particular to a mismatch between the operator's mental model of what is happening and the actual events that are occurring in the system. Other equally pernicious errors may be due, for example, to stress, caused by time or production pressures, which leads to a failure properly to check a chemical-sampling report on the part of all of the 'independent' personnel required to do so. These types of error are largely predictable, and occur as performance failures. Clearly, a human-performance limiting value *is* applicable to these types of error, and data summarised later in this report suggests, for example, that a limiting value for a single operator performing a single task probably falls around the 1E–3-to-1E–4 region.

The second meaning of the limiting value relates to the uncertainty involved in the modelling of the degree of complexity of a real system. Bhopal and Chernobyl are examples of failures in which various largely incredible errors (or so they seemed prior to the accident) occurred as a result of management oversights or, in the case of Chernobyl, rule violations. These two accidents are often, however, seen as unrepresentative of those occurring within a 'traditional plant'. Accidents such as the Challenger Space Shuttle disaster, an incident in 1987 at a Swedish reactor (Reason, 1988: this latter near miss, which occurred in a 'well-run' plant, apparently resembled some of the events in Chernobyl) and the Davis-Besse incident (USNRC, 1987) in the USA are perhaps closer to the types of incident/accident which could occur within a 'traditional plant' but which are still 'undermodelled' in PRAs.

Although such complex accidents are difficult to predict, it is clear from nuclear-power-plant experiences, and from risk assessments, that such accidents can and do occur, and hence that PSAs which do not model such possibilities may be underestimating the level of risk involved. Thus, even when no failure mode can be conceived, during the specification of a high-reliability scenario, which could lead to an error, this does not mean that a serious error could not actually occur. A limiting value in this context, therefore, would be equal to a quantified value for the level of the uncertainty involved in the modelling of the scenario.

Thus, performance limitations and the modelling of uncertainty are the two principal factors that go towards determining a human performance limiting

Table 5.4.4 Human-performance limiting values (Kirwan et al, 1990)

Basic scenario	Limiting value**
Single operator carrying out task(s)	1E–4
Operators on plant carrying out tasks	1E–4 to 1E–5*
Control-room-based team (operator plus supervisor plus shift manager)	1E–5

* Use 1E–4 unless 1E–5 can be justified as a result of exceptionally good procedures, checks, etc.
** Note: These should not generally be applied instead of detailed assessments identifying particular errors. In many cases, such assessments will derive human reliabilities of, say, 1.0 to 1E–3 for different scenarios, and hence the limiting values will be redundant. These limiting values are therefore primarily for the purpose of preventing the assessment safety cases from becoming too optimistic. They are not to be used as a sole means of quantifying scenarios.

For existing plants commissioned prior to 1980, the conditions for human error may be less than optimum. Hence, HPLVs applied on PRAs for such plants are, accordingly, more conservative:

1. 1E–3 is used where only one operator is involved.
2. 1E–4 is used where a *human system* is involved – for example, an operator, a supervisor or two shifts.
3. 1E–5 is used where a human system exhibits demonstrable relations of independence between personnel; for example, cognitive independence, or different responsibilities, or both.

value. They become necessary primarily when one is considering high-human-reliability situations.

Table 5.4.4 summarises the application of limiting values in fault trees (Kirwan *et al*, 1990), while Figure 5.4.8 shows examples of HPLV applications in both a fault tree and an event tree.

DISCUSSION OF HPLVS

The HPLVs derived will prevent the occurrence of over-optimistic assessments, via fault-tree logic, of human-error impacts. The HPLVs will not replace proper HEP estimations by means of HRA techniques, but they will add an additional element of conservatism into the PRA. In many cases, the HPLV will in fact be redundant; for example, if the HRA estimates suggest a human-error contribution of 1E–2 in a given scenario, then a HPLV of 1E–4 becomes inconsequential. However, in certain types of scenarios, often those involving long time-scales and maintenance operations, HPLVs can sometimes actually *dominate* the fault tree.

The HPLV is a first step in dealing with a human CCF/dependence, and taken alongside existing methods, such as the THERP, which deal with dependence at the sub-task level, it represents a real step forward towards more accurate and realistic PSA. The actual HPLV values suggested in this report

Figure 5.4.8 HPLV fault- and event-tree examples (latter includes limiting value 'event': Whiltingham, 1990)

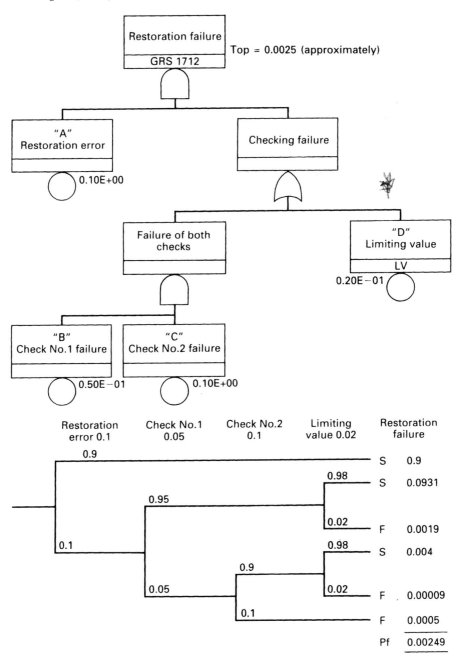

$$0.1 \times 0.05 \times 0.9 \times 0.02 \approx 0.00009$$

have not been derived rigorously but are held to be reasonable values. Their validation is unlikely through data collection, and given that they represent the modelling of uncertainty, it is difficult to see that expert-judgement validation would help. Nevertheless, what is important is that they are utilised in a PSA as an interim measure, to counter an effect which detracts from the validity of the PSA process.

5.4.3 Screening

A screening analysis identifies where the major effort in the quantification analysis should be directed. There may be, for example, particular tasks which are *theoretically* related to the system goals being investigated but which, *in fact*, make little contribution to the level of risk when they fail to be carried out, for example due to the presence of diverse and reliable back-up systems which adequately compensate for human errors, or to the trivial nature of the tasks themselves with respect to the overall level of risk involved. It is more useful, therefore, to spend little effort on such tasks and instead focus on those in which the *human*-reliability factor is the critical one. The identification of those errors which can be effectively ignored by the rest of the study is the purpose of a screening analysis.

The systematic human-action-reliability procedure (SHARP: Spurgin *et al*, 1987) defines three methods of screening the logically structured human errors. The first method 'screens out' those human errors which can only affect the system goals if they occur in conjunction with an extremely unlikely hardware failure or environmental event (i.e. if they occur in the same cut-set as an extremely unlikely event or events). The decision as to what is unlikely enough to justify screening will depend on the PSA criteria. As an example, if a human error occurs in a cut-set that also includes a large earthquake (unlikely in the UK) together with an independent failure of a high-reliability system, plus its independent back-up, then the screening process may well rule out a further consideration of this error if, for example, the probability of this cut-set were 1E–7 before the human-error probability is included, and the target criterion were 1E–4.

The second method involves allocating each human error a probability of 1.0 and then examining the effects of the various errors on the level of risk inherent in the system – i.e. the risk is calculated using fault- and event-tree methods. Those errors that have a negligible effect, even with a probability of unity, are not considered further.

The third method assigns broad human-error probabilities to human errors according to a simple error-categorisation framework (such as the one shown in Table 5.4.5). This method works in the same way as the previous method, but is a 'finer-grained' analytic method, and for this reason is known as *fine screening*.

With most screening methods (particularly the third method above) there are two main dangers. The first is that of inadvertently ruling out of the study

Table 5.4.5 Generic human-error probabilities

Category	Failure probability
Simple, frequently performed task, minimal stress	1E−3
More complex task, less time available, some care necessary	1E−2
Complex, unfamiliar task, with little feedback and some distractions	1E−1
Highly complex task, considerable stress, little time to perform it	3E−1
Extreme stress, rarely performed task	1E−0

errors and interactions that are of some importance. Fine screening, in particular, may underestimate the error probability. In practice, however, an assessor is usually able to judge the conservatism of fine-screening values for those individual human errors that are being represented.

The second danger with screening, one which is more difficult to detect and deal with, is that of *under*modelled dependence. For example, a single HEP may be given a fine-screening value of 0.1, but if it is, in fact, highly dependent on another error elsewhere in the tree, then its value according to THERP would be closer to 0.5. Similarly, errors that occur subsequently to, and are dependent on, an error that has already been removed via screening will no longer have that dependence modelled, and as a result, the level of risk will be underestimated. This problem is less likely to occur in event trees than in fault trees, because in the former the dependence and chronology of the scenario sequence are clearly defined, whereas in fault trees, two errors may be dependent but appear in different parts of the tree, so that the relation of dependence may be less obvious.

One of the only ways (albeit a rather laborious one) to address this problem is first to represent the errors and then derive and investigate the *cut-sets*. The analyst can see what errors appear together in each cut-set, and as a result dependence relationships can be identified prior to screening. Furthermore, if the PSA or HRA analysts are adopting a cut-off limit for cut-sets (e.g. if the target criterion for risk is 1E−4, the analysts may have decided to ignore cut-sets with a probability of less than 1E−7), the cut-sets must be reviewed for dependence before being excluded. This is because there may be, for example, a cut-set of magnitude 5E−8 which has perhaps three human errors in it, two of which are highly dependent on the first; such a cut-set's probability, in reality, taking account of screening, could be much higher (e.g. 1E−5), and therefore should not be screened out from the analysis. As a rough guide, when a cut-off limit for probabilities is being applied, cut-sets with two human errors should be examined for dependence two orders of magnitude below the cut-off value, while cut-sets with three or more human errors in them should be examined four orders of magnitude below the cut-off limit.

This will ensure, in most cases, that dependences are *not* underestimated in the PSA. A quantitative example of such an analysis is given in Appendix VI.

These dangers of screening must be balanced against the need to reduce the complexity of the analysis, and the amount of resources required to a manageable level. A 'manageable level' for screening purposes means PSAs with, for example, 50–100 HEPs requiring quantification. As a general rule, when applying any screening technique at any level in the study, if in any doubt, leave the human error in the fault or event tree.

5.4.4 Test and maintenance (T&M) errors

There is little doubt that errors in test and maintenance activities contribute significantly to the level of risk found in complex systems. This is because such errors can leave back-up, protective systems disabled without the system controllers' (the operators') being aware of it (often) until an incident occurs. However, many PRAs do *not* model such errors to any significant extent. The primary justification for not identifying and representing such latent errors is that failure-rate data for hardware components implicitly (and in some cases explicitly) *already* take account of the human-error contribution – a proportion of the failure rate has always to do with human error.

Test and maintenance errors can cause a number of principal impacts on a system's reliability:

* An increase in the failure rate of a component.
* An insertion of an incorrect component (which will cause failure).
* Miscalibration.
* A failure to align the system back to its operational configuration.

Whilst failure-rate data for maintained components may well take account of the first of these factors, it may *not* take account of the other forms, for the following reasons: the second factor above may result in a simple failure, or a change in the system's functioning, which is not modelled in either the hardware data or the PSA; the third factor above may lead to a common-mode failure (e.g. a failure on the part of all gas detectors to detect unsafe levels of gas) which again may not be accounted for by the PSA modelling, and lead to a system state that again may not be modelled in a PSA.

It is not sufficient, therefore to assume that T&M errors are taken into account, in existing safety cases, through hardware failure data, and it is at least necessary, as a result, to scrutinise potential T&M contributions to the risk level and consider to what extent the hardware data and PSA modelling will account for such contributions. If this is not certain, then T&M error contributions should be modelled and then represented in the logic trees.

Representation of T&M errors

T&M errors which are detected during subsequent start-up procedures do not usually require modelling in the PSA (unless the level of risk involved in a *shutdown state* is being modelled – this is a so-called *outage PSA* since such errors *will* be recovered. Latent errors, however, will remain in a system either until they are recovered by subsequent checking or testing or until there is a demand for the system to start operating. The degree of un-availability of the back-up system must then accounted for in the PSA – i.e. an estimate of how long it will be before the latent error is recovered.

A safety-system T&M is usually subject to fairly stringent quality-assurance and control procedures, and these, in turn, provide to a number of opportun-ities for detecting latent errors. Such procedures include:

- System testing (monthly).
- Routine maintenance (half-yearly or yearly).
- Walk-round plant-inspections (daily or weekly).
- Control-room checks (each shift).
- System realignment (during shutdown, e.g. annually).

The likelihood of detecting a latent error will depend both on its ease of detection (of course) and on the depth of detail of the above checks.

The unavailability of a system can be calculated according to the following relationship (Smith, 1990):

$$U = Fi \times Di/T$$

where Fi = the error-initiation frequency of the type-i error
  Di = the duration of the type-i error in the system prior to its recovery
and T = the total time for which the system needs to be available.

Di is given by the following relationship:

$$Di = Pij \times dij$$

where Pij = the probability that an error of type i will be recovered by recovery-type j
and dij = the duration of error-type i before it is detected and recov-ered by recovery-type j.

Due to the recursive nature of the checking paths, Pij * dij will have to be recalculated for each successive opportunity for error recovery. It will also be necessary to incorporate a model of dependence, and this has the (usual) result that the probability of error-recovery becomes less likely with each successive recovery opportunity.

The treatment of T&M errors and recoveries is thereafter no different from the treatment of other errors, although there are some quantification techniques that have been specifically developed to deal with maintenance and testing failures (e.g. MAPPS; see Section 5.5).

Figure 5.4.9 Levels of decomposition

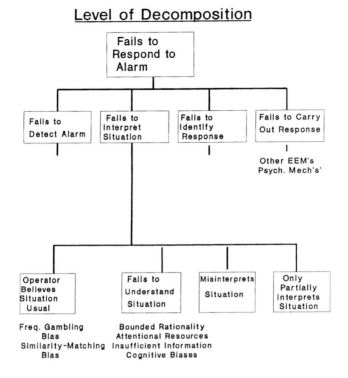

Level of Decomposition

5.4.5 Level of decomposition in representation

The problem of the level of decomposition is best illustrated with an example such as the one shown in Figure 5.4.9. At the simplest level of decomposition, called Level A, the human-error contribution to a particular scenario is defined simply as 'Operator fails to respond to alarm', assuming that it is a single alarm that must be responded to. Most techniques of quantification would have little trouble in quantifying the probability of such an external error mode as long as there is sufficient data to satisfy the technique's own information needs, e.g. information on PSFs such as the time taken to respond. This error descriptor could be broken down further – for example, to Level B in the figure – and a number of sub-errors such as 'Fails to interpret situation' could be specified. At this level of decomposition, it becomes apparent that there are several reasons why the simple failure to respond could occur. At the next level (C), the meaningfulness increases, for example via descriptive psychological error-mechanisms such as 'Fails to understand situation due to bounded rationality'. Some techniques – and many PSAs – would not, however, decompose the situation below Level A. This is because for many years, in applied PSA, the primary objective was to quantify the human contribution to risk levels, with little emphasis on being able to reduce errors where the level of contribution proved undesirable. If quantification is all that is required, then data on human-error probabilities at *any* level is acceptable, and the higher the level at which the data exists or can be synthesised, the less effort is required to fulfil the PSA's objectives.

There are, however, two major problems with this approach. The first is that if error reduction is required, there may not be enough detail on the human errors to identify and tackle the root causes of the error rather than addressing it superficially, according to its overt manifestation. For example, if the failure to respond to an alarm happens to be a significant contributor to the level of risk, error-reduction efforts might suggest the use of a second, higher-priority alarm to reduce the error. However, if this failure to respond is a problem, instead, of *bounded rationality*, then a second, higher-priority alarm may not help the operators; in fact it might make even things *worse* by increasing the level of stress on the operators. Error-reduction measures based on a Level-B decomposition would fare more favourably, while those based on Level C would be the *most* likely to succeed in reducing errors. The danger with insufficient modelling is therefore that of being able only to make *superficial* error-reduction recommendations whose degree of efficacy remains indeterminate.

A second problem relates to error-recovery and dependence. If a failure to respond is modelled, then a natural recovery path would be the response of a *second* operator. But if the bounded rationality in this case is endemic to the operators' training, then a second operator would also be unlikely to recover the error; and this means that either such a recovery step should not be modelled or it should be completely dependent on the first error.

These two potential disadvantages of using a relatively high level of decomposition are essentially pitfalls associated with undermodelling the human contribution to risk. This is a complex problem that is not easily solved. Techniques of quantification differ in their 'preferred' level of decomposition. Some – for example, HEART (human-error reduction and assessment technique, Williams, 1986) and Influence Diagrams (Phillips *et al*, 1986) – involve quantification at a task-based level which might encompass a range of human errors at Level A – which would now be called *Level 0*. THERP is at the other end of the spectrum, and although it does not entertain PEMs, it does decompose errors down to a high resolution. Other techniques such as APJ (absolute probability judgement, Seaver & Stillwell, 1983) and SLIM (success-likelihood-index method, Embrey *et al*, 1984) are also fairly flexible, dealing with any level of decomposition that the experts (these techniques both rely on the judgement of experts) are confident to quantify.

In practice, there are two interim solutions to the dilemma posed by the question of how far human errors should be decomposed. The first is to model errors in detail at least *qualitatively* during the error-analysis phase, so that even if the PSA makes use of Level A (or Level 0), the assessor responsible for making ERA decisions will be aware of the reasoning underpinning the external error modes in the trees. Similarly, if there *has* been a detailed qualitative modelling, the assessor will be judicious with the use of error-recovery modes, probabilities and dependences. The second solution, if a detailed qualitative modelling is not possible given the available resources, is to utilise conservative human-error probabilities and to ensure an adequate use of HPLVs so as to account for 'modelling uncertainty'. If this latter approach is adopted, then, if error-reduction measures are required, further qualitative modelling will have to take place.

In summary, the level-of-decomposition problem is something that assessors should be aware of. In most cases this level will be determined by PSA expectations, the Human Reliability Quantification (HRQ) technique selected and the amount of resources available. In real PSAs, HRAs rarely go beyond Level B in the logic-tree representation, because it is not always clear how, then, to aggregate the various PEMs at Level C (i.e. some could be additive, others mutually exclusive – but this is not always clear). But in detailed HRAs, on the other hand, the qualitative modelling underpinning the representation at Level A or B *is*, in essence, a Level-C modelling. This is a complex area requiring further research and development, and in the meantime the assessor will play a large role in fixing the required level of decomposition. As for the amount of resources made available for determining the level of qualitative modelling in a HRA, and as for the benefits accruing from such modelling, these will depend on the perceived usefulness of qualitative information to the managers of the HRA/PSA project.

Summary of representation issues

This section has elucidated on the use of fault and event trees for HRA purposes, and has dealt with the modelling, and quantified inclusion, of two types of dependence in the context of risk assessments, and test and maintenance error representation issues have been considered. Several screening methods have also been put forward for large and resource-constrained HRAs, and it has been emphasised that all screening methods run the risk of under-assessment if not applied with skill and judgement. The difficult and unresolved area of the *level* of decomposition required has been discussed to make the reader aware of the potential problems it raises, and this issue will be returned to in Section 5.9, during a consideration of PSAs that require Level-0 HRA-inputs. In summary, methods for representing human errors of many varied forms and combinations *do* exist and *can* be used relatively robustly, as long as the effects of dependence are also taken into account. The next section describes how the human errors, having been represented in the logic trees, can now be quantified.

5.5 *Human Error Quantification*

This section firstly and briefly considers the historical perspective in this field, to show what types of approaches have developed over the past three decades, and then considers the main techniques available today. Of these, six are described in detail (SLIM, THERP, HEART, APJ, PC, and HRMS). In Appendix III, three critical evaluations of techniques for Human Reliability Assessment (HRA) are detailed which will facilitate the selection of Human Reliability Quantification (HRQ) techniques for a particular purpose/project. The results of Appendix III are also included in a set of tables in Section 5.10, and at the end of this section, to facilitate the selection of HRQ techniques. The aim of this section is therefore to describe the techniques in sufficient detail so that the reader can appreciate how they are applied in practice.

In two cases, the techniques cannot be given a full description: THERP is itself a comprehensive technique, too large to contain within a single section, while the human-reliability-management system (Kirwan, 1990) is not in the public domain. The former technique is described as well as possible within this book because it is one of the most popular techniques available. The HRMS approach, too, is described, because, though not generally available, it represents state-of-the-art HRQ technology, being a fully computerised system dealing, in an integrated fashion, with all ten aspects of the HRA process.

5.5.1 Historical perspective

The following sketches the history of HRA in terms of the predominant techniques and developments over the past three decades. For more information, see Topmiller *et al*, 1984, and Dougherty & Fragola, 1987. The historical perspective is not necessary for the practitioner, but is included for interest as it shows the trends that have driven human reliability over the past three decades.

The origin of human-reliability analysis is usually traced back to the American Institute for Research's (AIR's) database of human-error probabilities (Munger *et al*, 1962). This involved a set of HEPs that covered both operable equipment and system-design variables. It was Swain, in 1967, who later showed this approach to be inadequate mathematically. The first formal attempt to include task/environmental variables, along with human-engineering-design characteristics, for estimating HEPs was made by Swain, *et al*, with the Sandia human-error-rate bank (SHERB) (see Rigby, 1976), which used PSFs (arranged on a 7-point scale) to guide the analyst in deriving the human-error rate for a particular task. In the late 60s and early 70s, furthermore, human-error data concerning input errors in naval operations was collected (the operational recording and data system (OPREDS), Coburn, 1971). Other significant attempts at data-bank-driven approaches included the Bunker-Ramo tables (Mills and Hatfield, 1974) and the technique for estimating personnel-performance standards (TEPPS, Blanchard *et al*, 1966).

Thus, during the first decade of HRA, there was a desire to create *human*-error data-banks, by way of a parallel to those being successfully created for hardware components. There was also a realisation of the need to consider performance-shaping factors (PSFs). By the end of the first decade of HRA, however, it was being realised that the data-bank approach was not working.

In 1975, the WASH-1400 Reactor Safety Study (Rasmussen, 1975) was compiled (the prototype of the Probabilistic Risk Assessment) which included a still-used human-error database. At this time, practical examples of the THERP technique were beginning to emerge (Swain, 1976). In parallel, in the early 70s, Siegel *et al* (1974) developed digital simulation methods for estimating the reliability of man-machine systems: these included a consideration of human errors created as a result of inadequate performance.

Askren & Regulinski (1969) considered human error to be dependent on time, and they based their approach on the use of the traditional reliability ratio of mean time to failure (or rather, *error* in this case).

In the 70s, therefore, three strands of HRA research were evolving. The first involved a continuation of the *database* approach in the form of the THERP system, which, although it is a highly decompositional approach by modern standards, was far more integrative than the earlier data-bank approaches, dealing far more with *tasks* than with elemental behavioural units. The second strand was the *simulation* approach, which offered stochastically to simulate the human operator's reliability by using distributions of performance times combined with a Monte Carlo-style simulation. The third strand was the *time-reliability* approach, again a parallel development to approaches to reliability assessment, involving the modelling of the 'mean-time-to-failure' ratio for hardware components. All three strands were destined to survive, in modified forms, until the present, and they are still in use.

In 1979, the Three Mile Island accident shook the nuclear-power world to its core, and interest in HRA surged. In 1981, the draft THERP handbook was made available by Swain & Guttman (the finished version was published in 1983), and the use of THERP for PRAs/PSAs soared, THERP being the most effective approach at this time of crisis in the nuclear-power industry. In 1982, the first glimpses of a Human Cognitive Reliability (HCR) approach were beginning to appear (Hall *et al*, 1982); the HCR technique was arguably a logical descendant of the Askren & Regulinski approach of the 70s. This in turn led onto SHARP (systematic human-action-reliability procedure, Hannaman *et al*, 1984) which was the second total framework for HRAs, from task analysis right through to quantificantion (the first being THERP).

In 1984, the SLIM-MAUD approach was published as part of a USNRC Nureg report (Embrey *et al*, 1984). This represented a *psychologically* dominated approach, as opposed to engineering- and reliability-based approaches such as THERP and HCR. Shortly after SLIM-MAUD came the SHERPA system (the systematic human-error reduction and prediction approach, Embrey, 1986) – the third 'total' HRA framework, along with SHARP and THERP (although all three were also heavily *quantification*-biased). The APJ and paired-comparison techniques were also fully documented for use in HRAs and PRAs (Seaver & Stillwell, 1983). In 1986, the human-error assessment and reduction technique appeared (HEART, Williams, 1986) which was heavily based in the ergonomics domain, and which actually encompassed error-reduction guidance. The technique of Influence Diagrams, the first one capable of considering aspects such as safety culture, was also starting to be used in PRAs (Philips *et al*, 1983).

Although Three Mile Island was beginning to be written off as a one-off accident, and the need for HRAs was being questioned, the subsequent occurrence of accidents such as Bhopal (1984), and then Challenger and Chernobyl (both 1986) ensured that the HRA stayed firmly a part of the PSA process. There was now a flourishing range of techniques. Moreover, the pre-dominance of *engineering* approaches was now going into reverse, and greater credibility was being given to more psychological and expert-opinion-based quantification approaches.

In 1988, the HRAG document (Kirwan *et al*) presented the first full, de-tailed peer review of HRA techniques, to be followed a year later by Swain's own exhaustive review (Swain, 1989). Validations and comparative experi-ments began to appear in the early 80s (Kirwan, 1982; Brune *et al*, 1983; Comer *et al*, 1984; Rosa *et al*, 1985; Kirwan, 1988; Waters, 1989), and are still continuing.

At present, there are one or two new or hybrid techniques that involve unusual (from a historical perspective, at least) partnerships – for example, that between the SLIM and HCR (Bley *et al*, 1988). However, today, efforts are beginning to be directed away from quantitative HRAs and towards *qualitative* HRA insights instead.

It can be seen from the above that, originally, HRAs attempted, somewhat unsuccessfully, to develop data-banks. Later on, HRA practitioners resigned themselves to using expert-judgement techniques such as the SLIM and APJ, or else techniques which were a mixture partly of data and partly of expert judgement (i.e. that of the technique's author), such as THERP, HEART and, latterly, HRMS. Recently, however, there has been a renewed interest in the data-bank concept, most notably with the NUCLARR project (Gertman *et al*, 1988), as well as with a few others (e.g. Kirwan *et al*, 1990; Taylor-Adams & Kirwan, 1994). Currently, such data-banks are not themselves ready to be used for a direct HRQ, but their development rate is such that this situation may change in the near future. Appendix II shows some examples of the types of data that are available, not primarily for HRQ purposes, but for the purpose of demonstrating human-performance-failure probabilities.

The remainder of this section will consider some of the more useful and available HRQ techniques themselves. For other reviews, the reader may consult Swain (1989), and Kirwan *et al* (1988).

5.5.2 Human Reliability Quantification techniques overview

Human Reliability Quantification techniques all involve the calculation of a the human-error probability (HEP), which is the measure of Human Reliability Assessments. The HEP is defined as follows:

$$\text{HEP} = \frac{\text{The number of times an error has occurred}}{\text{The number of opportunities for that error to occur}}$$

Thus, if, on average, *tea* is accidentally purchased 1 time in 100, when one is trying to buy *coffee* from a vending machine, the HEP is taken as 0.01. (It is somewhat educational to try and identify HEPs in everyday life that have a value of less than 1 in 1000 opportunities – or even one as low as 1 in 10,000.)

In an ideal world, many studies based on industrial incidents would exist in which HEPs were recorded. In reality, there are very few such studies. The ideal source of human-error 'data' is industrial studies of performance and accidents, but at least four reasons can be deduced for the lack of availability of such data:

- Difficulties involved in estimating the number of opportunities for error in realistically complex tasks (the so-called *denominator problem*).
- Confidentiality.
- An unwillingness to publish data on *poor* performance.
- A lack of awareness of why it would be useful to collect such data in the first place, and hence a lack of financial resources for such data-collection.

There are a number of other potential reasons too (*see* Kirwan *et al*, 1990). The net result is a sparsity of HEP data.

Other sources of information include simulator data (e.g. data on exercises using high-fidelity simulators) and data derived from experimental, laboratory-based studies, as reported in human-performance literature. Two problems exist with respect to simulator studies, the first being that such simulators are used almost exclusively either for training purposes or else in connection with operator re-certification as in the US-nuclear-power field. Hence, personnel on the simulator scene are highly motivated and frequently know already what is on the training curriculum – i.e. they *know* which scenarios to expect. Secondly, it is not clear how realistic is an emergency faced in a simulator compared with the real thing – the *cognitive fidelity* issue.

Human-performance literature has a similar problem in that studies in this field are usually highly controlled, often looking at one or two independent variables (unlike in industry where there are many different PSFs, changing and interacting), and using reasonably motivated subjects for a short period of time. Generalising from such studies to complex, industrial, multi-personnel situations is not easy, and is indeed a questionable strategy.

Overall, therefore, there is a definite 'data problem'. And even where adequate data *do* exist, e.g. for a chemical-plant operator's failing to respond to an alarm in Scenario X, how can this then be generalised so as to cover Scenario Y, or the same Scenario, X, but on a *different* chemical plant with a *different* operating regime, or even a *nuclear-power* plant as opposed to a chemical plant? What, in other words, defines the 'generalisability' of data?

The existence of such difficult and as-yet-unresolved issues as these have led to the development of non-data-dependent approaches, and in particular to the use of expert opinion. This is by no means necessarily a bad thing: expert opinion has been used successfully in other areas (Murphy & Winkler, 1974), and is in any case used at least occasionally in Probabilistic Safety Assessments (Nicks, 1981), where similar problems often exist. Also, as will be seen, expert judgements can often penetrate further than any set of data when it comes to pronouncing on the importance of PSFs, etc. (e.g. when using the SLIM). For relevant data examples, see Appendix II.

The following subsections outline the main techniques currently in use in HRAs. Those that are expanded upon later in detail are only briefly described below (see also Swain, 1989; Kirwan *et al*, 1988).

5.5.2.1 APJ: absolute probability judgement

The APJ technique involves the use of experts for the direct generation of human-error probabilities. It may occur in various forms, from the single expert assessor to a large group of individuals who may work together – or whose estimates may be mathematically aggregated. APJ firstly requires experts who must know in depth the area they are being asked to assess. A second requirement is that the experts must also have some *normative* expertise – i.e. they must be familiar with probability calculus – for otherwise they will not be able to express their expertise in a coherent quantitative form. It is preferable, if experts are meeting and sharing their expertise, and discussing their arguments in a group, to make use of a *facilitator*. The facilitator's role is to try to prevent both biases arising as a result of personality differences within the group and biases arising during the making of expert judgements from distorting the results.

5.5.2.2 Paired comparisons

The technique of *paired comparisons* (Hunns, 1982) is borrowed from the domain of psychophysics (a branch of psychology). Like the SLIM, it is a means of defining preferences between items (human errors). Unlike SLIM, however, it also asks the experts to make judgements, albeit of a relatively simple variety. Each expert compares all possible pairs of error descriptions and decides, in each case, which of the two errors is more probable. For n tasks, each expert makes $n(n-1)/2$ comparisons. When comparisons made by different experts are combined, a relative scaling of error likelihood can then be constructed. This is then calibrated using a logarithmic calibration equation which, as for the SLIM, requires that the HEPs be known for at least two of the errors within the task set. Paired comparisons are relatively easy for the experts to carry out, and the method usefully determines whether each expert has been consistent in the judgements he or she has made; inconsistency of judgement would suggest a lack of substantive expertise.

5.5.2.3 HEART: human error assessment and reduction technique (HEART)

HEART (Williams, 1986), which is a relatively quick technique to use, is based on a review, by its author, both of the literature on human factors and, in particular, of experimental evidence showing the effects of various parameters on human performance. The technique defines a set of generic error probabilities for different types of task; these are the starting point for a HEART quantification. Once a task is thus classified, it is then determined by the analyst whether any error-producing conditions (EPC) – as defined in HEART – are evident in the scenario under consideration. For each EPC evident, the generic HEP is multiplied by the EPC multiplier, which thus increases the HEP. The technique also has a set of practical error-reduction strategies which can be used to reduce the impact of errors on the system, or even to prevent them entirely.

5.5.2.4 THERP: technique for human error rate prediction

This technique basically comprises a database of human-error probabilities, Performance Shaping Factors (such as stress) which affect human performance and can be used to alter the basic human-error probabilities in the database, an event-tree-modelling approach and a dependency model. It is often referred to as a 'decompositional' approach in that its descriptions of tasks carried out by human operators have a higher degree of resolution than many other techniques. Having said that, it is also a most *logical* approach, and one which puts a larger degree of emphasis on error recovery than do most other techniques.

5.5.2.5 SLIM-MAUD: success likelihood index method using multi-attribute utility decomposition

SLIM-MAUD is a computerised technique originating from the world of decision analysis. The SLIM, on the one hand, is essentially a method of defining preferences amongst a set of items – in this case, a set of human-error tasks. MAUD, on the other hand, is a sophisticated approach which helps to ensure that the expert group that is used to define the preferences do so without letting biases affect their judgements (Tversky & Kahneman, 1974). SLIM-MAUD, as a single technique, defines the preferences representing the relative likelihoods of different errors as a function of various factors which can affect human performance (Performance Shaping Factors: PSF, such as the level of training, the quality of the procedures, the time available or the quality of the interface with the operator). SLIM-MAUD creates a relative scale representing the likelihoods of different errors, called the success-likelihood index (SLI). This index can be 'calibrated' so as to allow the generation of human-error probabilities via a logarithmic relationship based both on experimental data and on a theoretical argument presented on behalf of the paired-comparisons technique (Hunns, 1982).

5.5.2.6 Influence Diagrams approach (STAHR: sociotechnical approach to Human Reliability Assessment)

The Influence Diagrams approach (or STAHR, Phillips *et al*, 1985) is the only technique, to date, which can explicitly model some of the more pernicious PSFs, e.g. those to do with safety culture, or with the influence of management on reliability/safety. It was developed for use in HRA by Embrey, Phillips and Humphreys, and also comes from the field of decision analysis. It takes PSFs or influences which are non-independent and uses experts to define how these PSFs interact to influence the success or failure probability for a task. The approach estimates probabilities directly, and does not utilise calibration as the SLIM and the paired-comparisons methods do. The technique has rarely been used, but if the practitioner is called on to assess organisational/motivational parameters on human performance, Influence Diagrams should be considered (see also Kirwan *et al*, 1988).

5.5.2.7 HCR – human cognitive reliability

The HCR approach has had many variants since its inception as cited in early works by Hall *et al* (1982) and Wreathall (1982). Essentially, the HCR model assumes that, in diagnostic tasks, the probability of a non-response in an emergency is greatly influenced by the time available for a diagnosis, although recent variants of this approach have allowed other PSFs to have marginal effects on the HEPs derived. The HCR technique is important for two reasons. Firstly, it is an approach which can easily be integrated with other engineering-based assessments of transient behaviour and modelling; and secondly, it has received a considerable amount of funding in the USA. It has, however, a number of disadvantages and associated theoretical problems, which are discussed in more detail by Kirwan, Embrey & Rea (1988). In particular, the validity of the approach has been seriously undermined recently, both on theoretical and empirical grounds. On theoretical grounds, the main arguments are that, whilst time is undoubtedly an important PSF, it is in many cases unlikely to be dominant; nor is it likely to be monotonic in its predictive relationship to human performance (see Kirwan, 1992). On empirical grounds, there are two independent sets of simulator trials that have, to a large extent, disconfirmed the predictive relationships put forward in the HCR approach (Kantowitz and Fujita, 1990; Dolby, 1990). This method is therefore not currently in vogue. For a fuller review of the various HCR approaches, see Dougherty (1987), and Kirwan (1992).

5.5.2.8 Accident sequence evaluation programme (ASEP)

ASEP (Swain, 1987) is a shortened version of THERP which can be used with less training and with less of a need for HRA expertise to produce screening estimates for the PRA. It consists of pre-accident screening and a nominal HRA, together with a post-accident screening and nominal HRA facilities. The screening approach essentially uses more conservative values than would be obtained by using THERP, both for HEPs and for reactions to events occurring both prior to and after the occurrence of an incident/accident ('pre-accident' and 'post-accident' respectively). With the ASEP-THERP combination, ASEP can be used as a fine screening approach, while THERP can be used for those errors or tasks that are found to contribute significantly to the level of risk involved in the system.

ASEP is far quicker to carry out than THERP, and can be computerised (this has been done by at least one group; Whittingham, 1989). For more information, see Swain (1987, 1989).

5.5.2.9 Systems analysis of integrated networks of tasks (SAINT)

SAINT is not an HRA tool in itself but a simulation language which brings together both a task analysis and a Monte Carlo simulation of the performance of the operator (Siegel & Wolf, 1969), thereby allowing a network of tasks to be 'run' dynamically via the simulation. Therefore, as an example, a nuclear-power-plant accident sequence could be broken down into the constituent tasks and sub-tasks that are required to handle the scenario successfully. If both data on task and sub-task completion times (and on distributions of those times, or the parameters for those distributions) and the error-recovery probabilities can be estimated, then the simulation can be run. The resulting output will give either the number of actions completed within the scenario's time-scale or the average time taken to complete all the tasks, with minimum and maximum times for the potential range of operating crews.

The SAINT approach, and others like it, offer a potentially very powerful approach to the HRA, as long as the task analysis is comprehensive, and as long as time/performance data can be derived or estimated. It is this latter requirement that is the most difficult to satisfy, since, as already explained in this section, there is such a paucity of data (whether in a HEP format or in a performance-timing format) for the generation of realistic process-control task-combinations. Nevertheless, SAINT has been the subject of two feasibility studies (see Swain, 1989) which have demonstrated that it has potential PSA application (e.g. see Kozinsky *et al*, 1984).

5.5.2.10 Maintenance personnel performance simulation (MAPPS)

MAPPS (Siegel *et al*, 1984; Kopstein & Wolf, 1985) is particularly concerned with maintenance-related human reliability, and therefore occupies a special niche in HRA, since, as discussed near the end of Section 5.4, PSAs often do not model maintenance errors. Like SAINT above, MAPPS is a computerised, stochastic, task-oriented model of human performance. Its function is to simulate a number of human 'components' to the system – e.g. the maintenance mechanic, the instrument and control technician, the electrician, and the quality-control supervisor – together with any interactions (in the form of communications, instructions, etc.) between these people and the control-room operator. MAPPS is a simulation approach, but in particular is partly driven by PSFs (such as the workplace, the layout, fatigue) in its efforts first to predict the levels of performances for sub-tasks and then to quantify the level of performance for the *whole* task. For more information on MAPPS, see Swain (1989).

5.5.2.11 Human reliability management system (HRMS)

The HRMS (Kirwan & James, 1989; Kirwan, 1990) is a fully computerised HRA system with deals with all aspects of the HRA process. Its quantification module is based on actual data, which is supplemented by the author's own judgements on how data can be extrapolated to new scenarios/tasks, according to six major PSFs (the time-scale involved; the quality of the interface; training/experience/familiarity; the degree of adequacy of procedures; how the task is organised; and the degree of complexity of the task). Like the SLIM, the system can also carry out a PSF-based sensitivity analysis to determine how to reduce the likelihood of error, and like HEART, the system provides error-reduction mechanisms, but the latter, in the HRMS, are fairly task-specific. The HRMS has only been used extensively in one major PSA, and is currently not commercially available. It is included in this book (and expanded upon shortly) merely to show an example of how a fully integrated system may appear.

5.5.2.12 Justification of human error data information (JHEDI)

JHEDI (Kirwan, 1990) was developed for the same reason as that for which, presumably, ASEP was developed – namely, to provide a faster screening technique than that of its 'parent', the HRMS approach, of which it is a derivative. The HRMS is resource-intensive, and requires a significant amount of training, whereas JHEDI is quick (for one thing, it involves the same amount of resources as HEART), and requires little training. (Also, the JHEDI database is a more conservative version of that utilised in HRMS.) The HRMS-JHEDI combination, as with the THERP-ASEP partnership, offers the practitioner a resources-flexible approach to HRAs, which is particularly useful in large PSAs/HRAs.

There are other techniques that have been utilised in HRA (see Swain, 1989), and others are doubtless in the throes of development. The following subsections detail six of the above techniques which the assessor can get hold of (with the exception of HRMS) and apply, and which all have a fairly healthy track record of multiple applications in real assessments. There are a number of consultancy- or company-based training courses available on the use of these techniques, though the major requirement with all of them is *practice*. With THERP in particular however, training *is* strongly recommended, especially as the simple subsection in this book cannot hope to encompass the entire THERP methodology, which is considerable.

5.5.3 Detailed review of techniques

Having defined briefly the main HRA techniques available, six of the more influential techniques are defined in more detail below.

5.5.3.1 Success likelihood index method (SLIM)

Methodology
The SLIM (see Embrey *et al*, 1984) can best be explained by means of an example of a Human Reliability Assessment – in this case, an operator decoupling a filling hose from a chemical (chlorine) road tanker. In this situation, the operator may forget to close a valve upstream of the filling hose, which could lead to undesirable consequences, particularly for the operator – i.e. the operator could get a rather nasty and possibly fatal dose of chlorine. The human error of interest here is the 'Failure to close valve V0101 prior to decoupling the filling hose'. In this case, the decoupling operation is simple and discrete, and hence the failure occurs catastrophically rather than in a staged fashion.

The 'expert panel' required to carry out the SLIM exercise would typically comprise, for example, two operators with a minimum ten years' operational experience, one human-factors analyst and a reliability analyst who is familiar with the system and who also has some operational experience.

The panel is initially asked to identify a set of Performance Shaping Factors (PSFs), defined as any factors relating to the individual(s), environment or task which affect performances positively or negatively. The expert panel could then be asked to nominate the most important or significant PSFs for the scenario under investigation. In this example, it is assumed that the panel identify the following major PSFs as affecting human performance in this situation:

• Training
• Procedures
• Feedback
• The perceived level of risk
• Time pressures involved

PSF RATING

The panel are then asked first to consider other possible human errors arising in this scenario (e.g. mis-setting or ignoring an alarm), and then to decide to what extent each PSF is optimal or sub-optimal for that task in the situation being assessed. The 'rating' for whether a task is optimal or sub-optimal for a particular PSF is made on a scale of 1 to 9, with '9' as optimal. For the three human errors under analysis, the ratings obtained are as follows:

Errors	Training	Proc's	Feedback	Perc'd risk	Time
V0101 open	6	5	2	9	6
Alarm mis-set	5	3	2	7	7
Alarm ignored	4	5	7	7	2

PSF WEIGHTING

If each factor were equally important, one could simply add each row of ratings and conclude that the error with the lowest rating-sum (i.e. alarm mis-set) was the most likely error. However, the expert panel in this example, as with most panels, does not feel that the PSFs are all of equal importance. In this particular case, it feels that the *perceived risk level* and *feedback* PSFs are the most important, and are in fact *twice* as important as *training* and *procedures*, which are, in turn, one and a half times as important as *time pressure* (in this case, as it is a routine operation, time is not perceived by the panel to be particularly important). Weightings for the PSFs can be obtained directly from these considered opinions, normalised so as to add up to unity, as follows:

Perceived risk level	– 0.30
Feedback	– 0.30
Training	– 0.15
Procedures	– 0.15
Time pressures	– 0.10

Sum = 1.00

Both the SLIM and the decision-analysis technique on which the former is based, the *simple multi-attribute rating technique* (Edwards, 1977), propose simply that the degree of preference can be worked out as a function of the sum of the weightings multiplied by their ratings for each item (task error). The SLIM calls the resultant preference index a *success-likelihood index* (SLI). This is illustrated below with a table that shows the weightings (W) multiplied by the ratings (R), such that SLI = (sum) WR.

Weighting	PSFs	V0101	Alarm mis-set	Alarm ignored
(0.30)	Feedback (0.3×2) = 0.60		0.60	2.10
(0.30)	Perc'd risk	2.70	2.10	2.10
(0.15)	Training	0.90	0.75	0.60
(0.15)	Procedures	0.75	0.45	0.75
(0.10)	Time	0.60	0.40	0.20
	SLI (total)	5.55	4.30	5.75

In this case, the lowest SLI is 4.3, suggesting that 'alarm mis-set' is still the most likely error. However, due to the weightings used, the likelihood-ordering of the other two errors has now been reversed (a close inspection of the figures reveals that this is because the feedback PSF is held to be important, and because there is ample feedback for 'alarm ignored' but not for 'V0101 open'). Clearly, at this point, a designer would realise that increased feedback to the operator about the position of V0101 might be desirable.

The SLIs are not yet probabilities, however. Rather, they are indications of the relative likelihoods of the different errors. Thus the SLIs show the ordering of likelihoods of the different errors but do not yet define the *absolute probability values.*

In order to transform the SLIs into human error probabilities (HEPs), it is first necessary to 'calibrate' the SLI values. (Note: the paired-comparisons technique also requires this calibration, and it uses the same basic formula.) Two earlier studies on calibration (Pontecorvo, 1965; Hunns, 1982) have both suggested a logarithmic relationship of the form:

$$\text{Log p(success)} = a\,(\text{SLI}) + b$$

If two tasks for which the HEPs are known are included in the set of errors which are being quantified, then the parameters of the equation can be derived via simultaneous equations and the other (unknown) HEPs can be quantified. If, in the above example, two more tasks, X and Y, were assessed which had known HEPs of 0.5 and 10^{-4} respectively, and were assessed SLIs of 4.00 and 6.00 respectively, then the equation would be derived as follows:

$$\text{Log (HEP)} = a\,(\text{SLI}) + b$$

$$\text{Log (HEP)} = -1.85\,(\text{SLI}) + 7.1$$

The HEPs would then be:

V0101   = 0.0007
Alarm mis-set = 0.1400
Alarm ignored = 0.0003

This is the body of the rationale underlying the SLIM. In practice, this method is in fact more complex, and is computerised so as to promote ease of use, as well as to prevent biases, often found in the elicitation of expert opinions, from influencing the results. The computerised version, known as SLIM-MAUD (SLIM using multi-attribute utility decomposition), will, due to the mathematics in the software which is present partly to avoid such biases, produce slightly different HEP values from those arrived at via the hand-calculated method above. In particular, the simple summary of weightings and ratings is refined in several ways (beyond the scope of this book), according to the more detailed mathematical requirements of multi-attribute utility theory.

It should also be noted that several forms of the calibration equation exist. In the original report, the equation used is as follows:

$$\text{Log } p(\text{success}) = a \text{ (SLI)} + b \qquad \text{(Equation 1)}$$

Alternatively, the complement of the SLI, called the failure-likelihood index (FLI), can be assessed (see Zimmerlong, 1992), as follows:

$$\text{Log (HEP)} = a \text{ (FLI)} + b \qquad \text{(Equation 2)}$$

Since the official document is the definitive version of the SLIM approach, it is suggested that this version (Equation 1 above) be utilised. (The earlier equation in the example is, however, easier to calculate and demonstrate, and hence was used for the example.)

PRACTICAL GUIDANCE

The following discussion attempts to give practical guidance on the application of the SLIM and SLIM-MAUD approaches, based on the author's experiences in the application of the technique. The guidance should be treated as recommendations rather than mandatory rules to be followed, since in many cases (e.g. how many PSFs should be used, and what is the minimum/maximum number of PSFs?) strict guidance cannot be given. Anyone planning to carry out a SLIM exercise should also first read the original source reference (Embrey *et al*, 1984), and, preferably, consult the Human Reliability Assessor's Guide (Kirwan *et al*, 1988).

The above has shown the general rationale of the SLIM, and enables the reader to understand more easily how the technique works. More formally, the SLIM procedure goes through the following stages:

1. The selection of the expert panel.
2. The definition of situations and subsets.
3. The elicitation of PSFs.
4. The rating of the tasks on the PSF scale.
5. The ideal-point elicitation, and scaling calculations.

6. Independence checks.
7. The weighting procedure.
8. The calculation of SLIs.
9. The conversion of SLIs into probabilities.
10. The uncertainty-bound analysis.
11. The sensitivity analysis for error-reduction analysis purposes.
12. The documentation process.

Each of these is briefly dealt with below from a practical viewpoint (see also Kirwan *et al*, 1988, and Embrey *et al*, 1984).

The selection of the expert panel
The selection of experts is critical when carrying out SLIM assessments. There are three basic requirements for any expert panel:

• Substantive expertise
• Normative expertise
• Group cohesion

Substantive expertise means that the experts must be both knowledgeable and experienced in the subject matter of the HRQ exercise. Generally, 10 years of experience is a minimum requirement for the main substantive experts, and generally, again, the more experienced the better. Normative expertise requires that the experts be familiar with probabilities, and be able to appreciate the magnitude of the differences between them (e.g. the difference between 1E–3 and 1E–4; they must appreciate just how remote 1E–4 is).

Group cohesion is perhaps the wild card when selecting a panel, since often the assessor running the SLIM session (called the facilitator) has little control over which experts are in the panel, and has little or (usually) no prior knowledge of their personalities until they arrive for the session. Almost by definition, some experts are likely to have very strong opinions about the scenarios and their probabilities, etc., and may be unreceptive to the views of other experts. This is where the assessor's role as a facilitator and mediator of the session becomes critical. This role is discussed in more detail in the subsection on APJs later in this section. If things really do get out of hand, e.g. if expert-panel members stop talking to each other, or start heated arguments, the assessor must be prepared to abandon the session, or at least adjourn it. Things rarely get quite this bad, however (see Section 5.5.3.4).

A workable group would have three main substantive experts, one human-factors professional, one safety assessor and one facilitator, who should remain unbiased throughout the session and not give any quantitative or qualitative judgements, e.g. on PSF weights or ratings.

The definition of scenarios and subsets
It is critical that all members in the panel share the same mental model of the scenarios. To this end, the assessors gather as much information as possible

on such scenarios, including, where possible, information on likely PSFs, prior to the panel session. This should include diagrams and pictures, where relevant. Once the panel meets, sufficient time should be allotted for the discussion to ensure that *all* members of the group understand and agree to the definitions of the different scenarios. Often such a discussion will yield additional material which the assessor was unaware of, or else it may lead to a correction of some of the original scenario descriptions. However, the assessor should beware of the 'denial' phenomenon, in which, at the outset, the experts are dismissive of the scenario, saying that it is simply impossible and unrealistic and would never happen. That may be so, but the assessor's job is to quantify the scenario; and in any case, he or she can always incorporate the experts' conceptions about its low likelihood into the assessment. The assessor must also beware of the panel's changing the description and conditions/assumptions of a scenario to such a great extent that, whilst they are then *more* happy to assess it, it is no longer relevant to the PSA in question. Once the scenarios have been explored by the group, the assessor can then group these scenarios into subsets. The rationale for doing this is not usually based upon the number of scenarios but rather has to do with the degree of *homogeneity* of the PSFs affecting them. Whilst PSF ratings can be evaluated differently for different scenarios, PSF weightings will remain the same once allocated for each subset of scenarios. Subsets are therefore defined by a process of considering which scenarios or tasks are believed to have not only common PSFs but also common PSF *weightings*. Since, at this stage, the PSFs will not have been decided, nor weighted, the grouping of scenarios into subsets may prove an iterative process. Often, scenarios are grouped into categories such as skill-based, rule-based, knowledge-based or rule-violating, but this is only an approximation, and it is the commonality and homogeneity weighting of the PSFs that must be the determining factor for the subset. In terms of numbers in each subset, there should be less than 10 scenarios (including calibration tasks), and six or seven is a good number of tasks on which to carry out a SLIM exercise.

The elicitation of PSFs
The assessor (or the computer, if the SLIM-MAUD program is being used) will ask the panel to consider three of the scenarios, and to consider whether one of the tasks differs from the other two in a way which will differentially affect human performance. Usually, there is little problem with this approach, though the facilitator should have some suggestions ready in case the panel has difficulties. Typical PSFs used are as follows:

- The time pressure or stress levels
- The quality of information or of the interface
- The quality of procedures
- Consequences as perceived by the operator
- The level of complexity of the task

- The amount of teamwork required
- Whether or not there is adequate training, or an adequate level of competence

As with the Influence Diagrams approach, what the SLIM requires are the *dominant* PSFs and not *all* the possible PSFs which could have an impact on the task (there could be close to 100, after all). The SLIM is therefore looking for a 'requisite' number of PSFs – usually about six, though as few as four and as many as seven may occur, depending on the assessment scenarios.

The rating of the tasks on the PSF scale
The rating of each PSF for each task is usually done on a scale that runs from 1 to 9, whether on computer or on paper (by the facilitator). It is usually preferable to consider one PSF at a time for each task since this helps the panel to make relative judgements about how the rating for a particular PSF should alter for each successive task. However, if the experts want to do *all* the relevant PSFs for each task in turn, this is also possible.

The assessor/facilitator must remember that the PSF scales are linear in nature, and must be rated accordingly. Sometimes, experts actually rate logarithmically or exponentially, so that, for example, the following anchored ratings might be derived for a task in which *the quality of the procedures* figured as a PSF:

1.0 – Almost perfect procedures, with checksheets, highlighting of important information, etc.
5.0 – Average procedures
8.0 – Poorly written procedures, and no highlighting of warnings
8.5 – Procedures 10 years out of date; inaccuracies present
8.8 – Procedures incomplete, inaccurate; much cross-referencing, etc.
9.0 – No procedures, even though the task demands them

Such a set of ratings is logarithmic, such that the difference between, say, 5 and 6 is *not* the same as the difference between a rating of 8 and 9. In general practice, 1 and 9 should represent, respectively, the best and the worst actual practices likely to be encountered in the industry under consideration, with most experts' assessments therefore falling within these boundaries, e.g. between 2 and 8. Therefore, the assessor must ensure that the SLIM PSF ratings are linear rather than logarithmic.

The ideal-point elicitation, and scaling calculations
The ideal point on each PSF scale must then be derived. This will usually be 1 or 9. Preferably, always use one end of the scale, e.g. 9, to represent success-inducing levels of the PSF – in practice, it is easy to make errors when carrying out the SLIM if the ideal rating oscillates arbitrarily between 1 and 9. In some cases, the ideal point may lie between the two extremes of the

scale, e.g. in the case of the PSF of stress: too little stress can have negative effects on performance, as, also, can too *much* stress. A mid-range value (e.g. 6) can be selected by the experts to represent the ideal value of such PSFs.

The MAUD part of SLIM-MAUD will then rescale all PSF ratings in terms of their distance from the ideal point, and will produce a rating value, for each task's PSFs, of between 1.0 and zero. If carrying out these calculations by hand, there will only be a problem if there is an ideal point which does not lie at one or the other end of the scale. The simplest way to get round this problem is to define such PSFs (e.g. stress) in a way that allows the ideal point to be placed at one or the other of the poles of the scale e.g. 9 is optimal stress, 1 is very sub-optimal stress (since 'Low stress' is not necessarily optimal). Once this has been done, rescaling becomes relatively unnecessary since the ratings (with all PSF-scale ideal points lying at a value of 9) can be multiplied directly by the PSF weightings, as described below.

Independence checks
The computerised system will warn of any PSF ratings that correlate highly with the way in which another PSF was rated for the same tasks, a situation which suggests that these PSFs are not independent, and actually may be representing the *same* PSF – or a more fundamental PSF common to *both* the correlated PSFs. (Time pressure and stress are often highly correlated, and rarely independent.) The facilitator must then ask the experts if this is the case. If the group decides that they are different, and the facilitator agrees, the computer will continue – that is, it will not prevent continuation. This point is important, since the facilitator must make an executive decision as to whether the two PSF are correlated or not. In any case, if such a warning occurs, the facilitator should make a record of the warning, the decision made and the justifications for that decision.

If the procedure is being carried out by hand, the facilitator should monitor the PSF-rating process and watch out for any correlated ratings, although this is, of course, easier said than done, and is one example of where the computerised SLIM-MAUD system is advantageous.

The weighting procedure
The computerised version has a sophisticated weighting procedure for determining the PSF weights (see Embrey *et al*, 1984). These decisions are, in practice, quite difficult to make, and the process is best done iteratively, i.e. by carrying out the weighting procedure for all PSFs reviewing the PSF weights derived and, finally, letting the experts decide if they are happy with the results or if they wish to re-do them.

When calculating these by hand, it is best, firstly, to get the panel to agree on a rank ordering for the PSFs, i.e. to determine the most important through to the least important. A process of estimating ratios for PSFs can then be utilised. For example, *training* may be considered the most important PSF, and it may be considered 10 times more important than *perceived consequences*, the least important PSF. The second most important PSF, *time pressure*, may be considered half as important as training, but two and a half times as important as procedures. If 'perceived consequences' is given a value of 1, this means that corresponding values for the other PSF would be as follows: Training (10); Time Pressure (5); and Procedures (2). Together with the value for perceived consequences, these values sum to 18. The weighting is then derived by dividing each PSF value by the sum of PSF values, leading to the following PSF weightings, which sum to unity:

Training Weighting $= 10/18 = 0.56$
Time Pressure Weighting $= 5/18 = 0.28$
Procedures Weighting $= 2/18 = 0.11$
Perceived Consequences Weighting $= 1/18 = \underline{0.05}$
$$\sum_{n}^{1} = 1.00$$

This can be represented linearly as follows:

```
10 ----- 9 ----- 8 ---- 7 ----- 6 ----- 5 ----- 4 ----- 3 ---- 2 ---- 1
 !                               !                       !    !
Training                   Time pressure      Proced's  Perc'd
                                                        Con's
```

Training is given a value of 10 since it is the most important PSF and is 10 'units' from perceived consequences. Since time pressure (5) is half as important as training, it must be five times more important than perceived consequences, while procedures (2) must be twice as important as perceived consequences.

These values are summed thus: $10 + 5 + 2 + 1 = 18$.

Normalised weighting values are then calculated directly:

Training $\qquad$ $10/18 = 0.56$
Time Pressure $\qquad$ $5/18 = 0.28$
Procedures $\qquad$ $2/18 = 0.11$
Perc'd Conseq's $\qquad$ $1/18 = \underline{0.05}$
$\qquad\qquad$ Sum $= 1.00$

The ratios between each of the PSF weightings can then be checked to ensure that the calculation process has been carried out correctly.

The calculation of SLIs

The SLI formula has already been demonstrated in the earlier example, and is quite straightforward. More formally, it is as follows:

$$SLIj = SUM\ (Rij \times Wi)\ for\ i = 1\ to\ i = x$$

Where:

SLIj = Success-likelihood index for task j
Wi   = Normalised importance weighting for the ith PSF
Rij = (Scaled) rating of task j on the ith PSF
x   = The number of PSFs considered

Conversion of SLIs into probabilities
The calibration equation as noted earlier is as follows:

$$\text{Log (probability of success)} = a\,(SLI) + b$$

where logs to base 10 are used, and a and b are constants that can be derived either by the computer system or by the process of simultaneous equations, as long as at least two calibration probabilities have been assessed within *each* task subset (preferably, *three* calibration points should be used, for reasons described below).

Figures 5.5.1a & b show a hypothetical worked example of the calibration equation that uses two calibration points. SLIM-MAUD will allow the user to investigate the 'tightness of fit' of the calibration – i.e. the predictive validity of the results as a function of the degree of variance accounted for by the regression line produced by the calibration. In practice, however, the most robust way to consider the predictive validity of the results is, as already mentioned, to use at least *three* calibration points. If two points are being used and the calibration is invalid, a line can always be fitted between them, but if three points are being used and the calibration is invalid, this invalidity will become evident in the scatter-plot, as shown in Figures 5.5.2a & b. The assessor must also be wary of a *calibration-point inversion*, in which the panel actually transposes the two calibration points and then believes that the higher HEP calibration point is actually lower in probability than the lower HEP calibration point – as also illustrated in Figures 5.5.2a & b. If this occurs, erroneous results will be produced and the session data will have to be abandoned. Inversion implies either that the calibration data is invalid for the particular subset or that the experts have misinterpreted the data, or that the data themselves are invalid (or at least not fully described or accounted for by the PSFs chosen), or finally that the experts are *not* experts when it comes to these data. An inversion is a rare occurrence, however.

In general, it is therefore best to have three calibration points, and to have real calibration data relevant to the subset scenarios. In practice, such data are often unavailable, and the assessor may have to resort to expert-judgement techniques such as APJ to calibrate SLIM. This is not a particularly satisfactory state of affairs, but it will hopefully be improved as databases develop.

The uncertainty-bound analysis
Uncertainty bounds can be estimated via APJ confidence bounds around the resultant SLIM-based HEPs (see APJ subsection 5.5.3.4).

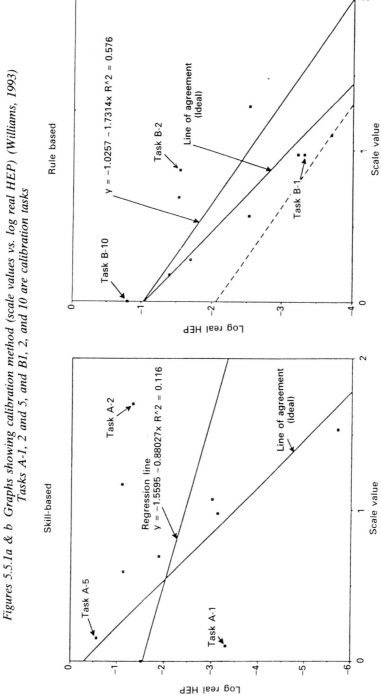

Figures 5.5.1a & b Graphs showing calibration method (scale values vs. log real HEP) (Williams, 1993) Tasks A-1, 2 and 5, and B1, 2, and 10 are calibration tasks

Figures 5.5.2a & b Scatter plots showing good and poor calibrations (log real HEP vs. log estimated HEP) (Williams, 1993)

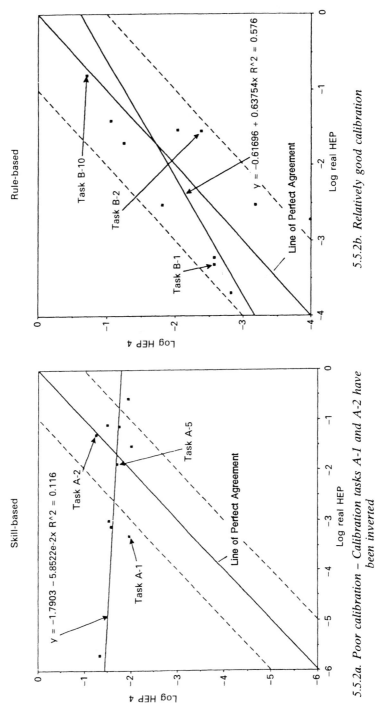

Skill-based

$y = -1.7903 - 5.8522\text{e}{-}2x \ R^2 = 0.116$

Task A-2

Task A-5

Task A-1

Line of Perfect Agreement

Log real HEP

Log HEP 4

5.5.2a. Poor calibration – Calibration tasks A-1 and A-2 have been inverted

Rule-based

Task B-10

Task B-2

Task B-1

$y = -0.61696 + 0.63754x \ R^2 = 0.576$

Line of Perfect Agreement

Log real HEP

Log HEP 4

5.5.2b. Relatively good calibration

Figure 5.5.3 Sensitivity analyses

Results   **SLIM-MAUD**

Weighting	PSE	Original	1	2	3	4	5	6	7	8	9
.1	Training	6	6	3	3	1	6	3	3	3	6
.1	Procedures	8	2	8	6	8	8	2	2	8	2
.2	Interface	6	6	6	6	6	3	3	6	3	3
.3	Time	5	5	5	5	5	5	5	5	5	5
.3	Distraction	8	8	8	8	8	8	8	8	8	8
Σ = 1.0											
	SLI	.22	.54	.46	.50	.40	.34	.96	.80	.60	.70
	HEP	.10	.03	.07	.04	.02	.09	.001	.005	.01	.007
	Reduction factors	*	3	1.4	2.5	5	.1	100	20	10	14

* = Indeterminate reduction factor

Sensitivity analysis and error reduction analysis
SLIM and SLIM-MAUD are particularly useful for determining which PSFs will be the most effective at reducing the HEP. In particular, the program SLIM-SARAH can carry out various sensitivity analyses which can be used to explore the effects of altering PSF ratings. An example of the type of output that can be derived from such analyses is shown in Figure 5.5.3. It is important to ensure that any such analysis does not become merely 'a numbers game' in which, via a manipulation of the PSF ratings, the desired HEP is achieved. It must be found out how to achieve the required change (i.e. improvement) in the PSF rating(s) (see also Section 5.7 on error-reduction analyses). Ultimately, if the PSF-rating changes are 'operationalised' in terms of defining, for example, what an improvement in the adequacy of procedures from a rating of 3 to a rating of 7 actually requires, then SLIM can offer the risk manager or designer a powerful cost-effectiveness-investigation tool, so that the latter parties can decide how best to spend money on improving human performance.

The documentation process
SLIM and SLIM-MAUD/SARAH need to be highly documented, which is a fairly automatic process with the computerised systems. What are also important, however, are the assumptions and justifications that the panel make during the session(s), which will have to be recorded manually or taped. If such

information are not recorded, then the assessor will not be able to use the data for future related assessments, e.g. with a different panel, who may have different assumptions about the system's operations and about human performance. If all assumptions are properly recorded, however, such data can potentially be used as calibration data in future assessments.

ADVANTAGES AND DISADVANTAGES OF SLIM

Advantages

- The SLIM system is generally a highly plausible approach for the assessors, regulators and experts who participate, and is undoubtedly sophisticated and well-developed. It has also already been used, more in the USA than in the UK, in a number of different assessments, and is thus a fairly acceptable approach. Not surprisingly, therefore, the evaluation by Rosa *et al* (1985) is fairly complimentary.
- It allows gross cost–benefit evaluations to take place (as in the earlier example, wherein improvements in feedback would have improved the level of human reliability). These will be important if it is likely that design recommendations will be required, and if an ergonomist is involved in the assessment, then these recommendations can also be made fairly specific.
- In the assessment of complex scenarios (e.g., abnormal events or emergencies), it is important to obtain accurate results. SLIM's level of accuracy, as with other techniques is, due to the lack of data, indeterminate. However, its theoretical validity *is* at a reasonably high level, and if the HEPs are calibrated with other, *known* HEPs, these are likely to fall within the right 'ballpark'. In complex scenarios, where many factors are affecting human performance, it is plausible that a technique which analyses and re-synthesises those factors will be more successful than one which does not do these things but instead merely demands a global estimate.
- SLIM is a flexible technique, one able to deal with the total range of human-error forms, and it does not require a detailed decomposition of the task – as, for example, does THERP.

Disadvantages

- The logarithmic calibration relationship could well do with further verification.
- The choosing of PSFs is currently a somewhat arbitrary and therefore unsatisfactory affair, and at the very least a set of PSFs should be provided for selection by the judges. There is the very fundamental question as to why experts (e.g., operators) should be expected to know accurately what factors affect them, and to what degree, and why, even if it is reasonable to assume that they have such knowledge, such knowledge should be assumed to be reliable.

- There is a chronic lack of data, which in itself begs the question as to how one can calibrate the SLIs derived using SLIM. The principal author of SLIM has resorted to using APJ to calibrate the SLIM, but this is somewhat in contradiction with the very reasons for which SLIM was created in the first place.
- SLIM requires a small panel of experts, but as with any expert group there will be the problem of defining what an expert is (and then verifying this expertise), and the not inconsiderable task of assembling several experts for a few days to make the judgements. These considerations, plus the cost of SLIM and the training required, therefore means that SLIM puts a fairly heavy burden on resources. However, if SLIM is utilised often, a database can be built up, and this has the result that the amount of resources required per datum diminish.
- SLIM's PSFs are fairly global, in comparison to the more specific and perhaps more useful PSFs found in a technique like HEART.

5.5.3.2 Human error assessment and reduction technique
(HEART: Williams, 1986)

This technique is of particular interest to ergonomists as it is based on human-performance literature. It has been designed by its author as a relatively quick method for Human Reliability Assessments, and to be simple to use and easily understood. Its fundamental premise is that in reliability and risk equations, one is interested only in those ergonomics factors which have a large effect on performance, e.g. which cause a decrement in performance by a factor of 3 or more. Thus, whilst there are many well-studied ergonomics factors, and consequent guidelines, which are urged by ergonomists themselves (e.g. lighting recommendations), many of these factors actually have (in reliability terms) a negligible effect on the operator's performance. HEART therefore concentrates on those factors which *do* have a significant effect.

This point is important because, in part, it underlines something of a communications gap between engineers and ergonomists. Engineers and designers designing a plant cannot spend unlimited funds on the optimization of ergonomics aspects, and they often ask how important (in quantitative terms) an ergonomics recommendation is. Frequently, the ergonomist is unable to give the engineer an answer to this fundamental question. HEART in particular, as well as some of the other techniques (e.g. SLIM), allow the human-reliability analyst to answer this question quantitatively.

The first function of the HEART assessment process is to refine the task in terms of its generic, proposed, nominal level of human unreliability, as shown in Table 5.5.1. Thus, the task is first *classified*, so as to allow a nominal human-error probability to be assigned to it (it is classified as a complex task, or a routine task, etc.). The next stage is to identify error-producing conditions (EPCs) which are evident in the scenario and would have a negative

Table 5.5.1 HEART generic categories (after Williams, 1986)

Generic task	Proposed nominal human unreliability (5th–95th percentile bounds)
(A) Totally unfamiliar, performed at speed with no real idea of likely consequences	0.55 (0.35–0.97)
(B) Shift or restore system to a new or original state on a single attempt without supervision or procedures	0.26 (0.14–0.42)
(C) Complex task requiring high level of comprehension and skill	0.16 (0.12–0.28)
(D) Fairly simple task performed rapidly or given scant attention	0.09 (0.06–0.13)
(E) Routine, highly-practised, rapid task involving relatively low level of skill	0.02 (0.007–0.045)
(F) Restore or shift a system to original or new state following procedures, with some checking	0.003 (0.0008–0.007)
(G) Completely familiar, well-designed, highly practised, routine task occurring several times per hour, performed to highest possible standards by highly-motivated, highly-trained and experienced person, totally aware of implications of failure, with time to correct potential error, but without the benefit of significant job aids	0.0004 (0.00008–0.009)
(H) Respond correctly to system command even when there is an augmented or automated supervisory system providing accurate interpretation of system stage	0.00002 (0.000006–0.0009)

influence on human performance. Table 5.5.2 shows the major EPCs found in HEART.

A hypothetical example of how these are used to quantify a human-error probability for a task (Kirwan, Embrey & Rea, 1988) is shown below.

EXAMPLE

A safety, reliability or operations engineer wishes to assess the nominal likelihood of an operator's failing to isolate a plant bypass route following strict procedures. Unfortunately, the scenario not only necessitates that a fairly inexperienced operator apply an *opposite* technique to that which he normally uses to carry out isolations, but also involves a plant of whose inherent major hazards he is only dimly aware. It is assumed that the man could be in the seventh hour of his shift, that there is talk of the plant's imminent closure, that his work may be checked and that, finally, the local management of the company is desperately trying to keep the plant operational, despite a real need for maintenance, because of its fear that a partial shutdown could quickly lead to total, *permanent* shutdown.

Table 5.5.2 HEART EPCs (Williams, 1986)

Error-producing condition	Maximum predicted nominal amount by which unreliability might change, going from 'good' conditions to 'bad'
1. Unfamiliarity with a situation which is potentially important but which only occurs infrequently, or which is novel	× 17
2. A shortage of time available for error detection and correction	× 11
3. A low signal-to-noise ratio	× 10
4. A means of suppressing or overriding information or features which is too easily accessible	× 9
5. No means of conveying spatial and functional information to operators in a form which they can readily assimilate	× 8
6. A mismatch between an operator's model of the world and that imagined by a designer	× 8
7. No obvious means of reversing an unintended action	× 8
8. A channel capacity overload, particularly one caused by simultaneous presentation of non-redundant information	× 6
9. A need to unlearn a technique and apply one which requires the application of an opposing philosophy	× 6
10. The need to transfer specific knowledge from task to task without loss	× 5.5
11. Ambiguity in the required performance standards	× 5
12. A mismatch between perceived and real risk	× 4
13. Poor, ambiguous or ill-matched system feedback	× 4
14. No clear, direct and timely confirmation of an intended action from the portion of the system over which control is to be exerted	× 4
15. Operator inexperience (e.g. a newly-qualified tradesman, but not an 'expert')	× 3
16. An impoverished quality of information conveyed by procedures and person–person interaction	× 3
17. Little or no independent checking or testing of output	× 3
18.* A conflict between immediate and long-term objectives	× 2.5
19. No diversity of information input for veracity checks	× 2.5
20. A mismatch between the educational-achievement level of an individual and the requirements of the task	× 2
21. An incentive to use other more dangerous procedures	× 2
22. Little opportunity to exercise mind and body outside the immediate confines of a job	× 1.8
23. Unreliable instrumentation (enough that it is noticed)	× 1.6
24. A need for absolute judgements which are beyond the capabilities or experience of an operator	× 1.6
25. Unclear allocation of function and responsibility	× 1.6
26. No obvious way to keep track or progress during an activity	× 1.4

*18–26. These conditions are presented simply because they are frequently mentioned in the human-factors literature as being of some importance in Human Reliability Assessments. To a human-factors engineer, who is sometimes concerned about performance differences of as little as 3%, all these factors are important, but to engineers who are usually only concerned with differences of more than 300%, they are not very significant. The factors are identified so that engineers can decide whether or not to take account of them after the initial screening.

Using a simplified HEART, the safety reliability and operational engineer's assessment could look something like this:

Type of task			Nominal human unreliability
F			0.003

Error-producing conditions	Total HEART effect	Engineer's assessed proportion of effect (from 0 to 1)	Assessed effect
Inexperience	× 3	0.4	$((3 - 1) \times 0.4) + 1 = 1.8$
Opposite technique	× 6	1.0	$((6 - 1) \times 1.0) + 1 = 6.0$
Risk misperception	× 4	0.8	$((4 - 1) \times 0.8) + 1 = 3.4$
Conflict of objectives	× 2.5	0.8	$((2.5 - 1) \times 0.8) + 1 = 2.2$
Low morale	× 1.2	0.6	$((1.2 - 1) \times 0.6) + 1 = 1.12$

Assessed, nominal likelihood of failure:

$$0.003 \times 1.8 \times 6.0 \times 3.4 \times 2.2 \times 1.12 = 0.27$$

Time-on-shift effects would be ignored since there is no indication of monotony.

Similar calculations may be performed, if desired, to work out the predicted 5th and 95th percentile bounds, which in this case would be 0.07–0.58. This is done by inserting the lower and upper confidence bounds for the task type (i.e. F) alternatively in the final equation above, instead of the median HEP for task type F. Since the total probability of failure can never exceed 1.00, if the factors, once multiplied, take a value above 1.00, the probability of failure has to be assumed to be 1.00 and no more.

The relative contribution made by each of the error-producing conditions to the amount of unreliability modification is as follows (derived from the final equation by dividing each multiplicative factor in that equation by the sum of all the multiplicative factors):

	% contribution made to unreliability modification
Technique unlearning	41
Misperception of risk	24
Conflict of objectives	15
Inexperience	12
Low morale	8

Table 5.5.3 A subset of HEART remedial measures (Williams, 1986)

1. Technique unlearning (× 6)	The greatest possible care should be exercised when a number of new techniques are being considered that all set out to achieve the same outcome. They should not involve that adoption of opposing philosophies.
2. Misperception of risk (× 4)	It must not be assumed that the *perceived* level of risk, on the part of the user, is the same as the *actual* level. If necessary, a check should be made to ascertain where any mismatch might exist, and what its extent is.
3. Objectives conflict (× 2.5)	Objectives should be tested by management for mutual compatibility, and where potential conflicts are identified, these should either be resolved, so as to make them harmonious, or made prominent so that a comprehensive management-control programme can be created to reconcile such conflicts, as they arise, in a rational fashion.
4. Inexperience (× 3)	Personnel criteria should contain experience parameters specified in a way relevant to the task. Chances must not be taken for the sake of expediency.
5. Low morale (× 1.2)	Apart from the more obvious ways of attempting to secure high morale – by way of financial rewards, for example – other methods, involving participation, trust and mutual respect, often hold out at least as much promise. Building up morale is a painstaking process, which involves a little luck and great sensitivity.

Thus, a HEP of 0.27 (just over one in four) is calculated, which is a very high predicted error probability, and probably warrants error reduction. In this case, technique unlearning is the major contributory factor to this poor performance, and so clearly if an error reduction were required, then either some form of retraining or a redesigning process, to make the isolation procedures consistent across the plant as a whole, must be considered. HEART, however, goes further than other techniques in error reduction, since for each EPC it gives corresponding, suggested error-reduction approaches; for example, for the task above, the remedial measures in Table 5.5.3 would be proposed. The following six sections give some practical guidance on this technique.

1. HEART SCREENING PROCESS

In addition to the mechanics of HEART demonstrated in the example above, there is also a HEART screening process. This is a simplified qualitative set of guidelines for identifying the likely classes, sources and strengths of human error for a particular scenario. These guidelines afford the user qualitative insights into the types of error that might be observed, and can be used by the assessor as a screening device for deciding what errors to quantify. Unfortunately, the author of this book has no experience of the use of these guidelines, nor any examples, and so cannot comment on its practical application (see Williams, 1986).

2. HEART GENERIC CATEGORIES

The HEART generic categories have a critical influence on the HEPs that will be derived. Since HEART's error-producing conditions generally increase the nominal HEP, once the generic category is set, this acts as a ceiling that human reliability will not rise above. In practice, assessors can differ considerably in terms of how they select the generic category. This is because the descriptions of the different categories are, by nature, generic and hence open to interpretation by the assessor. What is deemed a 'complex task' (category C) by one assessor may be deemed the 'restoration of a system to its original state' (category F) by another, the difference in the nominal HEP here being roughly two orders of magnitude. In a validation exercise (Kirwan, 1988), it was noted that if an assessor considered that a task could fit into two categories, then the lower category (i.e. the one indicating the most reliable performance) should be selected. This improved the accuracy of the technique, and prevented it from being used to produce overpessimistic results.

It is hoped that in the near future more guidance will be given, by HEART's author, on the selection of generic categories, by way of examples of actual tasks being categorised. It would also be useful if another category were created to deal with *alarm-handling*, which can currently fit into several categories. Work on this front is currently being undertaken by the author of the technique.

3. HEART EPCS

Error-producing conditions (EPCs) are also highly significant in determining the final human-error probability (HEP). In the aforementioned validation exercise, it was noted that using only a *small* number of EPCs generally produced more accurate results. If too many were utilised, the nominal HEP would rapidly increase towards unity, even if a low generic category were being utilised. Therefore, it appears that only those EPCs that *are* evidently present in the scenario should be utilised by the assessor, so as to avoid his or her being over-pessimistic.

Another aspect involved in the use of EPCs is the need to avoid including EPCs which are already accounted for in the generic categories. For example, an EPC related to *time pressure* is *not* required if category A, which itself has to do with time pressure, has already been chosen. At present, this danger must simply be avoided by the assessor, but it is hoped that further guidance will be given on this aspect as HEART develops.

4. ASSESSED PROPORTION OF AFFECT

This is sometimes perceived by assessors as the most difficult part of the HEART process, since the selection of generic categories and EPCs is a relatively structured affair. There is little published guidance on how to decide the degree of effect of an EPC, and assessors vary in their approaches. The inclusion of EPCs that have a relatively small effect (e.g. a value of less than 0.1) on a scenario, i.e. of small 'proportions of affect' (POAs), as they are called, would appear to go against the basic philosophy of HEART, since such POAs would have a marginal effect on the overall probability. Similarly, the use of spuriously accurate POAs (e.g. '0.43') would also seem out of place in a HEART assessment.

The use of large POAs, for example those greater than 0.8, or even equal to 0.9, should perhaps be given careful consideration, since in many cases involving EPCs such a use would signify a very poor system or task design, possibly, even, the worst that might be encountered in industry (which ties in with the PSF range used in the SLIM, where at the end of the rating scales are to be found the worst conditions likely to be encountered). If a plant had scenarios with a large number of EPCs, all assessed with high POAs, it would be the sort of plant the assessor may well want to shut down! The 'binary' EPCs, such as those related to a stereotype violation (e.g. where one valve turns according to an convention opposite to that governing all other valves: EPC 9) would not, of course, contribute to such a situation. Such EPCs, if present, will often have a value of 1.0, since any intermediate value may be meaningless, given what the EPC is trying to represent. As with generic categories and EPCs, it is hoped that as HEART develops, more structured guidance will become available on how to assess the 'proportion of affect'.

5. REMEDIAL MEASURES

Remedial measures are useful, but in many cases will be unlikely to go far enough towards deriving error- or task-specific error-reduction mechanisms (ERMs) – as with other aspects of HEART, they are intentionally generic rather than system-specific. The assessor is therefore likely to have to make his or her own interpretation, to a considerable extent, of the remedial measures, or in fact to generate new ones based on the qualitative analysis – for example, the error or task analyses – performed prior to the quantification. Nevertheless, if the risk managers or designers want merely an indication of the type of remedial measures that could apply, those in HEART may well suffice. The assessor, however, should not resist the urge to add any ERMs or error-reduction strategies that arise from the qualitative analyses, and to place the weighting on these as he or she sees fit.

6. DOCUMENTATION

If many HEART calculations are being carried out, it is important that the assessor's assumptions be adequately recorded, particularly with respect to

Figure 5.5.4 Example of HEART calculation format

Fault/event tree	Emergency-shutdown device		
Operator action identifier	ESDI		
Operator action description	Operator fails to operate ESD when required		
Category	Trip and shutdown failure		
Generic task	E: 0.02		
Error-producing conditions	Maximum affect	Assessed proportion	Assessed factor
(1) Unfamiliarity	× 17	0.1	(17 − 1) × 0.1 + 1 = 2.6
(2) Time pressure	× 11	0.1	(11 − 1) × 0.1 + 1 = 2
(7) Irreversible action	× 8	0.2	(8 − 1) × 0.2 + 1 = 2.4

Assessed probability of failure

$0.02 \times 2.6 \times 2 \times 2.4 = 0.25$

Assessment assumptions and notes

This assessment assumes that the requirement to operate the ESD is relatively clear; it also assumes both that the ESD must be operated very quickly and that the demanding event is severe and may surprise/stress the CCR operators, which could cause a delay. This HEP may seem pessimistic, but behaviour during such a severe and rapid event is difficult to predict, and must be assessed conservatively.

Dependence relationships

N/a

Assessor's comments

A clear and rapid emergency operating procedure/flowchart should exist. Simulator training for the operation of this emergency shutdown device should be implemented.

the assessed proportion of affect (POA). An example of a tabular format for a HEART calculation is shown in Figure 5.5.4. Quality assurance can be important in ensuring that EPCs and POAs have been applied consistently, and for this reason some of the space on the table in Figure 5.5.4 has been used to show that this aspect (as well as the rudimentary calculation procedure itself) has been checked.

HEART'S ADVANTAGES AND DISADVANTAGES
HEART offers a quick and simple human-reliability-calculation method which also gives the user (whether engineer or ergonomist) suggestions on error reduction. Its strengths are the small degree of use of resources and its error-reduction potential. HEART does not, however, model dependence, and its further weaknesses include the lack of justification for such a simple

multiplicative model which sees human error as a function of EPCs, and its *varied* use by different assessors. The first problem would be resolved by further theoretical validation – or, more likely, by empirical validation showing that the model adequately predicts HEPs. It is difficult, however, to counter the theoretical argument that some EPCs (or PSFs) will *interact with each other* (e.g. time and complexity), and yet such interactions are not modelled by HEART (see Williams, 1992, for a discussion of some of these issues). In practical situations, HEPs derived are very sensitive to the initial classification stage, as well as to the exact number of PSFs (and their POAs) that the assessor includes as relevant to the scenario. More guidance is needed for the assessor on both of these critical HEART steps or else different assessors may generate significantly *different* HEPs. Having said all this, the popularity of the HEART technique is currently increasing, undoubtedly due in large part to its level of resource-use and its versatility.

5.5.3.3 Technique for human error rate prediction (THERP)

The THERP (Swain & Guttmann, 1983) is in itself a total methodology for assessing human reliability that was developed over a significant period of time by Swain, its main author. The THERP handbook is a large and highly useful document to which the brief treatment given in this book cannot hope to do justice. Only a flavour for this influential technique can be given here, and the reader/practitioner wishing to pursue HEART is advised to read the handbook itself.

The THERP, being a total methodology, also deals with task analyses (e.g. documentation reviews and walk/talk-throughs), error identification (as briefly reviewed in Section 5.3.4) and representation (the HRAET and THERP dependence models were outlined in Section 5.4), as well as the quantification of HEPs. The treatment of the THERP in this section will focus primarily on its application as a quantification technique, and will also briefly revisit the dependence model used in with the THERP.

THERP quantification process
The quantification part of the THERP comprises the following:

- A database of human errors which can be modified by the assessor so as to reflect the impact of PSFs on the scenario.
- A dependency model which calculates the degree of dependence between two actions (e.g. if an operator fails to detect an alarm, then the failure to carry out appropriate corrective actions reliably cannot be treated independently of this failure).
- An event-tree-modelling approach that combines HEPs calculated for individual steps in a task into an overall HEP for the task as a whole.
- The assessment of error-recovery paths.

Figure 5.5.5 Outline of a THERP procedure for HRA (adapted from Bell, 1984)

Phase 1: Familiarisation
  – Plant Visit
  – Review Information From System Analysts

Phase 2: Qualitative Assessment
  – Talk- or Walk-Through
  – Task Analysis
  – Develop HRA Event Trees

Phase 3: Quantitative Assessment
  – Assign Nominal HEPs
  – Estimate the Relative Effects of Performance Shaping Factors
  – Assess Dependence
  – Determine Success and Failure Probabilities
  – Determine the Effects of Recovery Factors

Phase 4: Incorporation
  – Perform a Sensitivity Analysis, if Warranted
  – Supply Information to System Analysts

The basic THERP approach is shown in Figure 5.5.5 (Bell, 1984). Probably because of its relatively large human-error database, and its similarities with engineering approaches, the THERP has been used to a greater extent, in industrial applications, than any other technique. This small section can only cover the rudiments of the technique; for more information, the reader is referred to the source references above (see also Kirwan, Embrey & Rea, 1988).

An example of the type of basic event data given in THERP is shown in Table 5.5.4 (Webley & Ackroyd, 1988), and an example of the type of 'raw data' that is derived from the THERP database itself is shown in Table 5.5.5.

A human-reliability-analysis event tree (HRAET, as shown in Figure 5.5.6) can be used to represent the operator's performance.

Alternatively, as shown in Figure 5.5.7 (Whittingham, 1988), an operator-action event tree can be used. In each case, the event tree represents the *sequence* of events, and considers possible failures (of omission, commission, etc.) at each branch in the tree. These errors are quantified, and error-recovery paths are then added to the tree, where appropriate. Figure 5.5.8 shows the PSFs which can be used in a THERP study. Often, in such THERP studies, only one or at most a few of these are utilised quantitatively (e.g., stress).

It is important to model recovery, particularly when investigating highly proceduralised sequences, since an operator will often be prompted by a *later* step in the procedures to recover from an earlier error in a previous step. For example, if an operator omits a step to turn the power on, and then a second

Table 5.5.4 Task analysis: initiation of flow via stand-by train (Webley & Ackroyd, 1988)

Task identifier*	Task of interest	Error identifier	Human errors	Human error probabilities	
A	Identify loss of flow via duty train	A_1	Fail to identify loss of flow	$10^{-4} - 0.25$	Alarm response model
B	Start correct procedure	B_1	Fail to start procedure	$10^{-3} - 10^{-2}$	Procedure used
				$10^{-2} - 5 \times 10^{-2}$	Procedure not used
C	Roving operator opens correct valve	C_1	Error of omission – verbal order	10^{-3}	
		C_2	Error of commission – selecting incorrect valve	$10^{-3} - 10^{-2}$	
D	1st operator starts stand-by pump via remote control	D_1	Error of omission – written procedures available	$10^{-3} - 10^{-2}$	Procedure used
				$10^{-2} - 5 \times 10^{-2}$	Procedure not used
		D_2	Error of commission – selecting incorrect control	$10^{-3} - 10^{-2}$	
E	Supervisor checks 1st operator	E_1	Error of omission – written procedures available	$10^{-3} - 10^{-2}$	Procedure used
				$10^{-2} - 5 \times 10^{-2}$	Procedure not used
		E_2	Error of commission – selecting incorrect valve	$10^{-3} - 10^{-2}$	
G	Shift Manager checks activation	G_1	Fail to initiate checking function	$10^{-3} - 10^{-2}$	
			Fail to identify errors	$5 \times 10^{-2} - 0.2$	

* Task F – no errors identified for this step.

*Table 5.5.5 Extract from THERP human-error database (adapted from Swain & Guttmann, 1983): Initial-screening models of estimated HEPs and EFs for diagnosis, by control-room personnel and within time T, of abnormal events annunciated closely in time**

Item	T (minutes after $T_0{}^+$)	Median joint HEP for diagnosis of a single, or the first, event	EF	Item	T (minutes** after $T_0{}^+$)	Median joint HEP for diagnosis of the second event[++]	EF
(1)	1	1.0	—	(7)	1	1.0	—
(2)	10	0.5	5	(8)	10	1.0	—
(3)	20	0.1	10	(9)	20	0.5	5
(4)	30	0.01	10	(10)	30	0.1	10
(5)	60	0.001	10	(11)	40	0.01	10
(6)	1500 (= 1 day)	0.0001	30	(12)	70	0.001	10
				(13)	1510	0.0001	30

* 'Closely in time' refers to cases in which the annunciation of the second abnormal event occurs while control-room (CR) personnel are still actively engaged in diagnosing or planning (or both) responses to cope with the first event. This is situation-specific, but for the initial analysis, use 'within 10 minutes' as a working definition of 'closely in time'.
   Note that this model pertains to the CR crew rather than to one individual.

[+] T_0 is a compelling signal of an abnormal situation, and is usually taken to indicate a pattern of annunciations. A probability of 1.0 is assumed for observing that there is some abnormal situation.

[++] Assign a HEP of 1.0 for the diagnosis of the third and subsequent abnormal events annunciated closely in time.

step is attended to, which involves checking certain power-supplied instruments, the operator will rapidly recover the first error.

If recovery opportunities for such proceduralised tasks are not identified, then the human-error factor may be overestimated. THERP is, in fact, the only technique which emphasises error recovery in this way. Error-recovery paths are shown on the operator-action event tree in Figure 5.5.7 as dashed lines, and in this particular tree, these largely involve recoveries of one person's error by another person.

THERP also models dependence relations between human errors. If, for example, an operator, in a high-stress situation, is trying to carry out a procedure quickly, or is demotivated or fatigued, etc., then a whole subsequent section of procedures, as well as recovery steps within these procedures, may be carried out wrongly. This is an example of a 'human-dependent failure'. An important example of dependency modelling concerns one operator checking another operator. It is unlikely that the action on the part of operator 1 and the subsequent check by operator 2 will be completely independent, since the first operator may assume that the second will be trying to detect

Figure 5.5.6 Example of human-reliability-analysis event tree

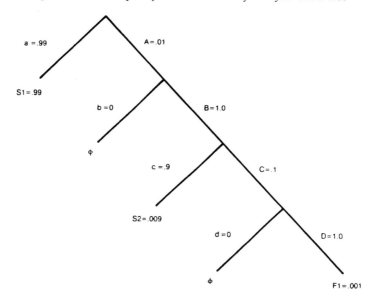

A = FAILURE TO SET UP TEST EQUIPMENT PROPERLY
B = FAILURE TO DETECT MISCALIBRATION FOR FIRST SETPOINT
C = FAILURE TO DETECT MISCALIBRATION FOR SECOND SETPOINT
D = FAILURE TO DETECT MISCALIBRATION FOR THIRD SETPOINT
φ = NULL PATH

any faults, while the second may assume the first *did* do the job properly. THERP is one of the few techniques which actually quantitatively models this relation of dependence between actions/errors. It uses a simple five-level model of dependence: from zero dependence (i.e. total independence), through low, medium and high dependence levels, to complete dependence (considered in Section 5.4). The effects of these levels of dependence are mathematically calculated in the HRAET/OAT. The following five sections give practical guidance on the THERP technique.

1. BASIC THERP PROCEDURE
    (i)   The error which requires quantification must be defined.
    (ii) Once the error has been defined, the user must use the flowchart in Chapter 20 of the handbook to decide which human-error data table to consult (see Table 5.5.6). There are tables for most, if not all, error types (e.g. the response to alarms, moving switches or push-buttons, etc.). The basic or normal HEP is selected from the table that most closely resembles the error being assessed.

Figure 5.5.7 Operator-action event tree (Whittingham, 1988)

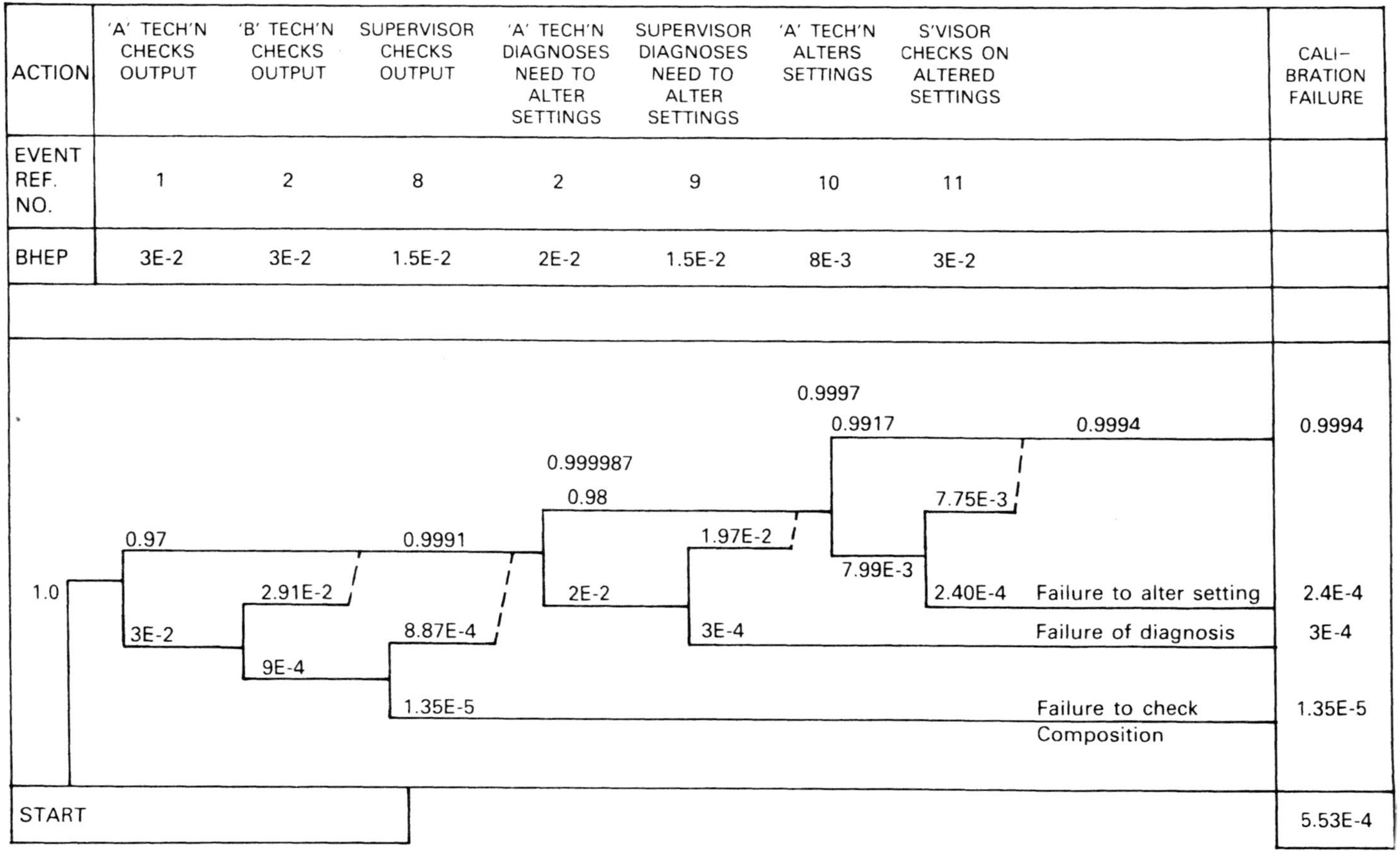
ACTION
'A' TECH'N CHECKS OUTPUT
'B' TECH'N CHECKS OUTPUT
SUPERVISOR CHECKS OUTPUT
'A' TECH'N DIAGNOSES NEED TO ALTER SETTINGS
SUPERVISOR DIAGNOSES NEED TO ALTER SETTINGS
'A' TECH'N ALTERS SETTINGS
S'VISOR CHECKS ON ALTERED SETTINGS
CALI– BRATION FAILURE
EVENT REF. NO.
1
2
8
2
9
10
11
BHEP
3E-2
3E-2
1.5E-2
2E-2
1.5E-2
8E-3
3E-2
1.0
0.97
3E-2
2.91E-2
9E-4
0.9991
8.87E-4
1.35E-5
0.999987
0.98
2E-2
1.97E-2
3E-4
0.9997
0.9917
7.99E-3
7.75E-3
2.40E-4
0.9994
Failure to alter setting
Failure of diagnosis
Failure to check
Composition
0.9994
2.4E-4
3E-4
1.35E-5
START
5.53E-4

Figure 5.5.8 *Performance Shaping Factors from the THERP system*

External PSFs	Stressor PSFs	Internal PSFs
Situational characteristics: Those PSFs general to one or more jobs in a work situation	Psychological stressors: PSFs which directly affect mental stress	Organismic factors: characteristics of people resulting from internal and external influences
Architectural features Quality of environment: Temperature humidity Air quality, and radiation Lighting Noise and vibration Degree of general clean liness Work hours/work breaks Shift rotation Availability/adequacy of special equipment, tools and supplies Manning parameters Organisational structure (e.g. authority, responsibility, communication channels) Actions by supervisors, co-workers, union representatives and regulatory personnel Rewards, recognition, benefits	Suddenness of onset Duration of stress Task speed Task load High jeopardy risk Threats (of failure, Loss of job) Monotonous, degrading or meaningless work Long, uneventful vigilance periods Conflicts of motives about job performance Reinforcement absent or negative Sensory deprivation Distractions (noise, glare, movement flicker, colour) Inconsistent cues	Previous training/experience State of current practice or skill Personality and intelligence variables Motivation and attitudes Emotional state Stress (mental or bodily tension) Knowledge of required performance standards Sex differences Physical condition Attitudes based on influence of family and other outside persons or agencies Group identifications
Task and equipment characteristics: Those PSFs specific to tasks in a job	Physiological stressors: PSFs which directly affect physical stress	
Perceptual requirements Motor requirements (speed, strength, precision) Control-display relationships Anticipatory requirements Interpretation Decision-making Complexity (information load) Narrowness of task Frequency and repetitiveness Task criticality Long- and short-term memory Calculational requirements Feedback (knowledge of results) Dynamic vs step-by-step activities Team structure and communication Man–machine interface Factors: Design of prime equipment, test equipment, manufacturing equipment, job aids, tools fixtures	Duration of stress Fatigue Pain or discomfort Hunger or thirst Temperature extremes Radiation G-force extremes Atmospheric-pressure extremes Oxygen insufficiency Vibration Movement constriction Lack of physical exercise Disruption of circadian rhythm	
Job and task instructions: Single most important tool for most tasks		
Procedure required (written or not written) Written or oral communications Cautions and warnings Work methods Plant policies (shop practices)		

Table 5.5.6 *Table-selection flowchart from chapter 20 of THERP handbook (adopted from Swain & Guttmann, 1983)*

	Diagnosis	1		
SCREENING				
	Rule-based actions	2		
	Nominal diagnosis	3		
DIAGNOSIS				
	Post-event control-room staffing	4		
	Written materials mandated			
	Preparation	5		
	Administrative control	6		
	Procedural items	7		
ERRORS OF OMISSION				
	No written materials			
	Administrative control	6		
	Oral-instruction items	8		
	Displays			
	Display selection	9		
	Read/record quantitative	10		
	Check-read quantitative	11		
ERROR OF COMMISSION				
	Control and MOV selection and use			
	Locally operated valves	12		
	Valve selection	13		
	Stuck-valve detection	14		
	Tagging levels	15		
PSFs	Stress/experience	16		
	Dependence	17	18	19
	Other PSFs (see text)			
	Estimate UCBs	20		
UNCERTAINTY BOUNDS	Conditional HEPs and UCBs	21		
	Errors by checker	22		
RECOVERY FACTORS	Annunciated cues	23	24	
	Control-room scanning	25	26	
	Basic walk-around inspection	27		

(iii)   PSFs related to the error and to the conditions of the scenario should then be identified and considered. These should then be noted as assumptions on the part of the assessor. The error factor associated with the basic HEP may then be applied by the assessor in such a way as to modify the basic HEP (in much the same way as the HEART uses the assessed 'proportion of affect'). When doing this, the assessor should follow any relevant guidance in the THERP manual (there is a considerable amount of discussion on the impact of certain PSFs in the handbook).

(iv)   The value of the HEP should be modified according to the dependence model, if dependence relations are present. If there is no dependence, then *zero dependence* should be noted. Assumptions about dependence should also be noted and documented by the assessor.

(v)   External recovery paths should be identified. HEPs associated with these should be dealt with as specified in steps i–iv above. In particular, it should be ensured that dependence relations between errors and successive recovery steps are adequately accounted for.

(vi)   The errors and error recoveries should be represented in a HRAET.

(vii) The evaluation of the tree should be carried out (and the overall level of reliability of the task calculated).

(viii) A sensitivity analysis, if this is required, should be performed.

(ix)   Error-reduction recommendations, if these are required, should be generated.

(x)   The analysis should be documented.

A short case study using the THERP is presented in detail in Appendix VI. This endeavours to show the basic mechanics of the THERP approach. It should be noted that this case study does not come from the THERP authors themselves, and therefore should only be taken as *one way* to carry out a THERP analysis, and not necessarily *the* way to apply THERP. However, the case study does involve a *real* application of the THERP by a UK assessor.

2. MODELLING

Despite the highly structured nature of the THERP methodology, there has been considerable variation in the way different assessors have utilised the approach. This variation is most evident in the modelling part of the THERP, i.e. the error-identification and error-representation stage. Different assessors tend to model the same scenario in different ways (see Brune, 1983; Waters, 1989). This became particularly evident in an independent THERP-validation study (Kirwan, 1988) in which two assessors differed significantly in the depth of decomposition of the same scenarios. One assessor broke down the task into more separate elements than the other, and this left the assessor in question with more detailed and complicated event trees. Ironically in this study, however, the overall task-failure probabilities derived by these two assessors actually tended to converge to the *same* result. Despite this fact, there still appears to be considerable latitude for interpreting how scenarios should be modelled. This is one of the reasons why training in the usage of the THERP is recommended (training courses are run in the USA by the originators of the THERP).

3. USAGE OF PERFORMANCE SHAPING FACTORS (PSFS)

A number of practitioners make only limited use of PSFs, or else *no* use at all. Undoubtedly in some cases this will be warranted, but equally there will be other cases in which PSFs really should be modelled. In these latter cases, assessors may simply use THERP's basic-HEP error factor (EF: usually, a factor of 3, 5 or 10 is used to increase the basic or nominal HEP) which is shown in Table 5.5.5. There are two issues related to this non-use or indirect use of PSFs. Firstly, whilst little structured guidance is given in THERP handbook on the quantitative inclusion of PSFs in the HRA, it is clear that it is intended by THERP authors that PSFs should be both considered and quantitatively accounted for (via the error factor) if they are deemed likely to have an impact on performance (which would be more likely to be the case if THERP were being used in non-nuclear-power contexts; most of the data assumes a nuclear-power-type context, but the reliability levels quoted could be expected to differ, for example, for an offshore platform). The assessor *should*, therefore, duly consider PSFs when carrying out THERP assessments, and should preferably record what PSFs are underlying the use of error factors. Many assessments simply refer to a single PSF of 'stress'. Whilst this PSF may well be present in post-trip scenarios, other PSFs could be equally or even more important.

The second issue related to the non-consideration or indirect consideration of PSFs is that of error reduction. If PSFs are not being considered explicitly, then error reduction will be limited to engineering-based solutions to human error (such as the automation or implementation of interlocks), and to the insertion of recovery points into the task (e.g. through the use of either diverse personnel checks or checksheets, as part of the procedures). Such error-reduction approaches may work in many situations, but they do, also, tend to limit the system designers' ability to carry out a full error-reduction analysis as pointed out in Section 5.7. In particular, error reductions via other PSFs such as 'interface design' may be overlooked, or may require a separate qualitative assessment.

THERP was originally developed when the emphasis in HRAs was almost exclusively quantitative rather than qualitative (see the historical perspective, earlier in this section, and so it would be useful if the THERP were further developed to provide more guidance on the structured use of PSFs in assessments, with a view to later error-reduction analyses. It would also be useful if THERP could be updated so as to take into account control rooms which are dominated by computerised displays rather than the hardwired panels and annunciators which the current database has grown up around.

4. DEPENDENCE

THERP's dependence model, as noted in Section 5.4, is currently the only dependence model in use in HRAs (as far, at least, as the author is aware).

Any practitioner will become quickly aware of the effects of modelling dependence. In particular, one HEP which is highly dependent on another HEP will be modified to a value of approximately 0.5, whatever its initial (independent) value. Given that in many cases such an increase will be warranted, the assessor must take care in assigning dependence levels. Incorporating dependence is not a matter of fine-tuning: it can have a great impact, at a *basic* level, on the assessed human contribution to risk. The assessor must therefore utilise the dependence model with a good deal of consideration and judgement.

5. DOCUMENTATION

THERP, by its very detailed nature, lends itself to extensive documentation. What are important, and are sometimes omitted, are the detailed assumptions underpinning the assessment – for example, those relating both to PSF usage and to dependence considerations. If these are not recorded, then, although calculations can be checked, an assessment, on the other hand, cannot be properly audited.

ADVANTAGES AND DISADVANTAGES OF THERP

THERP is an overall methodology, one founded on the database included in Chapter 20 of the THERP handbook but tempered by its use of PSFs, by dependence modelling and by the analyst's further use of recovery-modelling. Both THERP and its recent offspring, the accident-sequence evaluation programme (ASEP, Swain, 1987), can be used to model most human-error situations, including those involving diagnostic failures.

THERP's strengths are that it has been well-used in practice and that, as the product of two decades of work by Swain, it offers a powerful methodology which can be made auditable by the assessor. In the comparative-evaluation study cited in an earlier section (Kirwan, 1988), THERP performed well in terms of accuracy, although its resource-use level was variable and its usefulness in achieving error reduction was also limited. THERP's disadvantages are that it can be resource-intensive and that it does not offer enough guidance in modelling both scenarios and the impact of PSFs on errors. As a result, some users make extensive use of PSFs in determining impacts upon HEPs, whereas others use only a nominal effect of 'stress' in some cases. Another disadvantage of THERP is that it is relatively psychologically opaque, dealing only with external error modes rather than psychological error-mechanisms; it is difficult to adapt THERP in such a way as to allow it to consider cognitive modelling. And finally, its *consistency* as a technique has been questioned (Brune *et al*, 1983; Waters, 1989).

THERP, as with many of the techniques being discussed, is what the assessor makes of it. Too few people actually read all of the handbook, which, for the practitioner, is a welcome and rich source of useful human-reliability perspectives. For proceduralised tasks, and for well-structured problems for which quantification only is required (i.e. no error reduction involved), THERP is adequate as a methodology, and should definitely be a strong candidate in the selection of techniques. Although Swain (1989) reviewed the THERP–ASEP partnership as being the most powerful in the HRA world today, only time will tell whether or not the THERP paradigm has seen its greatest day and is due to be overshadowed by more psychologically-based methodologies that focus more on error-reduction modelling.

5.5.3.4 *Absolute probability judgement (APJ)*

Note: the subsections on APJs and paired comparisons (PCs) draw their material, to an extent, from the examples and formats utilised in the *Human Reliability Assessor's Guide* (Kirwan *et al*, 1988), of which this author was the lead author (particularly for the APJ and PC sections). The guide was produced by the Human Reliability Assessment Sub-Group of the Human Factors in Reliability Group. Some of the mathematical material on APJ and PC is also drawn from Seaver and Stillwell (1983), which defined the mathematics involved in a range of expert-judgement techniques, and Hunns and Daniels (1980). The reader or practitioner requiring more information is directed to these works, as well as to both Goossens *et al* (1989), who have developed sophisticated computer-modelling tools that facilitate the use of expert opinion in HRAs and in PSAs, and Comer *et al* (1984), for their validation exercises.

INTRODUCTION

The APJ approach is conceptually the most straightforward HRQ approach. It simply assumes that people can either remember or, better still, estimate directly the likelihood of an event – in this case, a human error. When it comes to risk assessments for existing plants or systems, it is arguable that the more experienced personnel will have a reasonable memory of their own errors, as well as of other operators' errors and their 'rates' of occurrence (particularly where supervisors and trainers are concerned). Such memories will naturally suffer some erosion, and will be subject to distortions that occur either at the time or in the intervening period before such errors are recollected. But having said that, there is an obvious desire, nonetheless, to tap such a potential source of 'quasi-data', particularly if such recollections can be corroborated.

Additionally, even where personnel do not have direct experience of the event in question, if the event is similar to one for which an expert *does* have knowledge and experience, then such an expert, or group of experts, may be able to construct a reasonably accurate estimate of the likelihood of occurrence of such an event.

In both these cases, a critical assumption is being made that the 'expert' has useful and accurate knowledge of the problem domain being investigated. This is known as having *substantive expertise*. Without substantive expertise, any HEP-estimation procedures that are carried out on the basis of the judgement of experts will have no predictive validity – the so-called 'garbage in, garbage out' situation. However, if there *is* substantive expertise (and in some cases, arguably, there must be), the next question is: can it be tapped accurately? The answer appears to be yes – in at least some cases, and to a certain degree of precision (see the appendix on validation). The next logical question then becomes: how can it best be tapped? This section, and the one on paired comparisons (PC), attempt to answer this question, which is of the most interest to the practitioner.

First, it is necessary to restrict the number of potential APJ approaches, since there are many such approaches. And then, it is necessary to consider three more semi-technical terms.

The first is that of an *expert*. There has been much debate about what defines an expert, but one pragmatic definition that has been preferred, and is easily measurable, is that an expert is someone who has been practising in a particular field or job for at least 10 years (Rea, 1986). This definition does not always hold true, but nevertheless, even when someone with 10 years' experience in a power plant turns out not to be a useful expert (because their predictions are inconsistent or just plain wrong), this person is still probably the most useful starting point in the search for experts. Finding good (so-called 'well-calibrated') experts is the hardest part of any APJ or PC exercise, and little more help for the practitioner can be offered at this time.

The second term is that of *normative expertise*. This, basically, is the extent to which the expert can deliver his or her expertise accurately, in the form of probabilistic statements which are mathematically consistent with his or her recollections or aggregations. Essentially, therefore, this term refers to the degree to which the expert is familiar with probability calculus, and with probabilities in general, so that, for example, the difference between 1E–3 and 1E–4 would be truly understood by this expert. There are various ways of overcoming a lack of normative expertise in cases where the expert has some very useful knowledge but has never worked with probabilities, or is not particularly numerate; and for a review of these, see Stael von Holstein & Matheson (1979). For the purposes of this section, however, an adequate degree of normative expertise is assumed in the descriptions of the approaches. The determination of the level of adequacy of the experts' normative expertise is usually straightforward, and is based on their background and their typical duties, or on a number of probability questions (or both). If there is any doubt as to their normative expertise, then measures such as those described in the above reference (Stael von Holstein & Matheson, 1979) should be consulted.

The third term relates to the distortions of memory, or the distortion of the expert's mental estimation processes, and is known simply as *bias*. Towards the end of this section, a number of biases will be considered. The magnitude of the effect of biases should not be underestimated: the author himself has witnessed two such biases cause, in the process of estimation, a shift from the 'original' value by a factor 1000, i.e. by three orders of magnitude. This is, to a large extent, the rationale behind the development of 'indirect' expert-judgement methods such as the SLIM and the Influence Diagrams approach.

In terms of scope, this section deals with group approaches to APJs. The adage of 'two heads are better than one' is nowhere more true than in APJ, and in the light of this, single-expert approaches are not considered further (the author's own bias). There are several ways of aggregating several experts' opinions: they can estimate alone, with their opinions then aggregated mathematically; or they can estimate alone but have limited discussions for clarification purposes; or they can meet as a group and discuss their estimates until they reach a consensus. This section will briefly consider all these methods and then focus on group-consensus methods in particular. This approach is partly a reflection of the author's own preference, but it is also a reflection of the fact that when experts are using the APJ approach to calculate HEPs (as opposed to using it to estimate hardware reliabilities or some other dimension), it is essential that a discussion occurs of *all* the factors leading to the error, for usually, one expert will forget something, or be unaware of some critical piece of information. In the author's own experience, experts who take part in a group-consensus APJ session usually (whether they like it or not) come away from the session richer for their experience, and usually the HEPs derived are felt to be closer to reality than those derived in the absence of any discussions. However, it is occasionally not possible to have group-consensus APJ sessions. In such cases, mathematical aggregations must be carried out; some basic formulae for such procedures are also, therefore, given in this section. Additionally, it should be noted that there are APJ methods within the mathematical-aggregation approach which can test the expertise of the expert (i.e. how well-calibrated the expert is). The reader is referred to Goossens *et al* (1989) for a consideration of these methods.

The four group-APJ approaches are described briefly below:

1. AGGREGATED INDIVIDUAL METHOD

This method entails that the experts do not meet but make estimates individually. These estimates are then aggregated statistically by taking the geometric mean of all the individuals' estimates for each task (i.e. all the expert's probabilities are multiplied together, and the 'nth' root of the product is then taken). The principal advantages of this method are firstly that it does not require experts to be located together in one place, and secondly that it avoids the kind of personality conflicts which can occur in a small group, and which may bias group estimates away from their 'true' value. The principal disadvantage, on the other hand, is that this method does not allow the experts to share their expertise, and to point out fallacies and biases in other experts' judgemental processes.

2. DELPHI METHOD

The Delphi method (Dalkey, 1969) is another 'anonymous' technique. Experts make their assessments individually, and then all the assessments made are shown to all the experts. The individuals then consider this information and reassess the HEPs. These HEPs are then statistically aggregated (i.e. the geometric mean of the estimates is calculated), as above. This method may be considered an improvement over the first method, since here the information is *shared*. However, this method still does not allow experts to discuss and resolve their different perspectives.

3. NOMINAL GROUP TECHNIQUE (NGT)

The NGT is similar to the Delphi method, except that here some limited discussion is allowed between experts for clarification purposes. This method thus enhances the sharing of information, but still largely manages to avoid the problem of individual personalities in groups dominating the assessments, since, after the group discussion in question, each expert makes his or her own assessment. These assessments are then statistically aggregated.

4. CONSENSUS-GROUP METHOD

In a consensus group, each member contributes to the discussion, but the group as a whole must then arrive at an estimate upon which all members of the group agree. This method maximises information-sharing, but at the same time it necessitates that the experts all be brought together in one place. Also, with consensus groups, personality variables and other biasing mechanisms (see later) may come into play, affecting the group estimation process. And lastly, if a 'deadlock' arises (i.e. if a consensus simply cannot be reached), the group must resort to a statistical aggregation of the individual estimates.

A more detailed description of problems encountered in group approaches where individuals meet (problems such as personality clashes and group pressures) is given in Eils & John (1978).

The following general recommendations (which are sometimes difficult to adhere to) apply when any of the interacting-group methods, such as the consensus-group method or the NGT are being used:

- Ensure that each member is committed to the assessment task.
- Encourage 'free speech'.
- Avoid 'hot' discussions.
- Use a group small enough for everyone to have a fair say.

In summary, the overall preference for group methods is as follows, based on the maximisation of information-sharing:

1. Consensus-group method
2. NGT

3. Delphi method
4. Aggregated-individuals method

However, it will be up to the practitioner carrying out the study to decide which method to use by assessing the requirement for information sharing and accuracy in the estimates made against the possible practical difficulties involved in co-locating and 'managing' a group of experts.

APJ PROCEDURE
The overall APJ procedure is as follows:

1. Select subject-matter experts.
2. Prepare the task statements.
3. Prepare the response booklets.
4. Develop instructions for subjects.
5. Obtain judgements.
6. Calculate inter-judge consistency.
7. Aggregate the individual estimates.
8. Estimate uncertainty bounds.

These steps are now detailed below (see also Comer *et al*, 1984, and Seaver & Stillwell, 1983).

STEP 1: SELECT SUBJECT-MATTER EXPERTS
The experts making the judgements must be familiar with the tasks to be assessed. If the tasks involve, for example, control-room operators in a nuclear-power plant, the experts must have an in-depth knowledge of plant systems, operations and control-room procedures. Control-room operators are thus an obvious choice for experts, as indeed are supervisors (preferably those who have previously been operators). If a response to rare events is being reviewed, *simulator instructors* may also serve well as additional experts.

The number of experts needed to make the required judgements cannot be stated unequivocally. As many experts as practicable should try to partici-pate. Seaver & Stillwell (1983) suggest six experts would be sufficient for a direct estimation, although more would be preferable. In practice, however, financial and other constraints often lead to the use of a smaller group of only three or four experts. In general, if a group consensus is aimed for, a group of 4–6 people is preferable, since problems are likely to occur with larger groups. If the method of aggregation is being utilised, then a *larger* size of 'group' is more advantageous (e.g. greater than 10) since this will compensate for the fact that the experts are not allowed to meet; and the averaging process should be able to counter the effects of individual biases. If many experts are available, then it may prove advantageous to utilise two inde-pendent consensus groups, which would thus lead to two independent sets of estimates. Where there is an agreement between the two groups, this will lend credibility to the estimates derived. Where there is disagreement, on the other hand, this will raise doubts over the validity of the results. In the latter case, a re-meeting of the groups, or the use of an alternative method, should be suggested.

STEP 2: PREPARE THE TASK STATEMENTS

Well-defined task statements are a critical aspect of the APJ-estimation procedure. The more fully the tasks are specified, the less they will be open to individual interpretation by the experts when they are making their judgements. The level of detail of task definitions will vary according both to the nature of the task itself and to the final use of the human-error-probability (HEP) estimate.

The degree of clarity of the task definitions involved should be examined in a pre-test setting. A few experts should be consulted so as to ensure that the level of detail of the task definition in question is sufficient, and that the stated assumptions for a task or group of tasks have been made clear and explicit. Specifically, as Stillwell, Seaver & Schwartz (1981) delineated:

- The role of Performance Shaping Factors (PSFs) in the task should be defined.
- Tasks not currently under consideration but which might be confused with the task to be assessed should be clearly separated from the task that *is* currently under consideration.
- Larger sets of tasks to which this task belongs should also be described.

Illustrations may also be usefully included, such as diagrams either of the instrumentation relevant to a particular task or of the control-room layout, or photographs, or video recordings. In some cases, there will be an advantage in carrying out the APJ assessment within the plant enclosure itself, e.g. in a special conference room, so that the experts can always visit the plant itself to gain more factual data if required. Obviously, this is only possible with existing and, to a limited extent, simulated systems.

STEPS 3 AND 4: PREPARE THE RESPONSE BOOKLETS AND DEVELOP INSTRUCTIONS

A key consideration when using the APJ approach is the type of scale on which experts will indicate their judgements. A scale such as the one in Figure 5.5.9 may, for example, be used. Other types of scales have been described by Stillwell, Seaver & Schwartz (1981). It is important that the chosen scale be of sufficient detail to allow the degree of sensitivity of the expert to individual differences to be indicated. The scale values must also reflect the estimated range of the true error probabilities of the tasks, where these are known. If they are not known, then a range of 1E-0 to 1E-6 is sufficient (unless extreme rule violations/sabotage/suicide attempts, which are outside the scope of this book, are being considered).

Figure 5.5.9 APJ estimation scale, incorporating example of completed direct estimate

(THIS END OF THE SCALE IS FOR INCORRECT ACTIONS WITH A HIGH LIKELIHOOD OF OCCURRENCE)		
	Probability	**Chances of occurrence**
	1.0	1 chance in 1
	.5	1 chance in 2
	.2	1 chance in 5
	.1	1 chance in 10
UPPER BOUND	**.05**	**1 chance in 20**
	.02	1 chance in 50
ESTIMATE	**.01**	**1 chance in 100**
	.005	1 chance in 200
LOWER BOUND	**.002**	**1 chance in 500**
	.001	1 chance in 1,000
In bold: the estimated	.0005	1 chance in 2,000
chances that an operator	.0002	1 chance in 5,000
will read information from	.0001	1 chance in 10,000
a graph incorrectly. (What	.00005	1 chance in 20,000
assumptions were made	.00002	1 chance in 50,000
that influenced the final	.00001	1 chance in 100,000
estimates?)	.000005	1 chance in 200,000
	.000002	1 chance in 500,000
	.000001	1 chance in 1,000,000
(THIS END OF THE SCALE IS FOR INCORRECT ACTIONS WITH A LOW LIKELIHOOD OF OCCURRENCE)		

Having prepared the task statements and the scale design, the response booklets can now be prepared. Instructions, assumptions for the task set and sample items should appear first in this booklet (see Comer *et al*, 1984, for examples).

If the purpose of the study is not adequately conveyed, experts will draw their own conclusions as to the reasons for the study, and if they suppose it to be for negative reasons, or for reasons with which they do not accord, their estimates of human reliability may, as a result, be biased (this is known as a motivational bias, and it may or may not occur consciously).

It is therefore important that the reasons for the study are accurately conveyed to the experts, especially, for example, if they are to assess the design of a system, or to perform cost–benefit analyses of human-related improvements to the system.

STEP 5: OBTAIN JUDGEMENTS

Experts are asked either to work through their booklets or, when operating in a group-consensus mode, to discuss each task in turn and arrive at a consensus estimate. It may be useful, when the consensus mode in operation, is to let the experts review all the tasks and start on one they feel confident that they can assess: starting with a *difficult* one will not help to encourage group

Table 5.5.7 APJ-derived human-error probabilities

Expert (m)	Event (n)			
	1	2	3	4
1	0.013	0.07	0.0025	0.068
2	0.56	0.13	0.33	0.063
3	0.005	0.01	0.00017	0.00033
4	0.005	0.00033	0.0017	0.0025

Table 5.5.8 Log HEPs

Expert (m)	Event (n)				Total
	1	2	3	4	
1	−1.90	−1.15	−2.60	−1.17	−6.82
2	−0.25	−0.90	−0.48	−1.20	−2.83
3	−2.30	−2.00	−3.77	−3.48	−11.55
4	−2.30	−3.48	−2.77	−2.60	−11.15
Total	−6.75	−7.53	−9.62	−8.45	−32.35
Average	−1.69	−1.88	−2.41	−2.11	

cohesion, and may even cause a morale problem, leading to unreliable estimates or even to no estimates at all.

If a consensus group is being utilised, a facilitator will be required to overcome any personality/group problems, and to prevent potential biases, discussed later, from negatively affecting the judgements.

STEP 6: CALCULATE INTER-JUDGE CONSISTENCY

Individual (i.e. not group-consensus) HEP estimates should only be used if there is a reasonable level of agreement between the experts. The following procedure for calculating inter-judge consistency is based on Seaver & Stillwell (1983). For a general description of the rationale underlying the analysis-of-variance (ANOVA) technique, which is described below in its application to the calculation of inter-judge consistency, see Chatfield (1978) or Meddis (1973).

An example set of human-error probabilities (HEPs) are shown in Table 5.5.7.

To make subsequent calculation easier, the set of HEPs obtained from the experts (either directly, as HEPs, or as odds converted into HEPs) are then translated into their logarithmic equivalents, resulting in the set of figures in Table 5.5.8.

The computational instruction for a two-way analysis of variance are set out below (adapted from Meddis, 1973):

1. Calculate the column totals (t): -6.75, -7.53, etc., above.
2. Calculate the row totals (r): -6.82, -2.83, etc., above.
3. Calculate the grand total (T): -32.35 above.
4. Calculate the correction term (C): $C = T^2/mn$, therefore

$$C = (32.35)^2/16$$
$$= 65.41$$

5. Calculate the sum of the squares (x^2) of the raw scores:

$$(-1.9)^2 + \ldots + (-2.60)^2 = 83.05$$

6. Calculate the total sum of the squares (TSS):

$$83.05 - 65.41 = 17.64$$

7. Calculate the 'between column sum of squares (t^2)' (Col SS):

$$\frac{(-6.75)^2 + (-7.53)^2 + (-9.62)^2 + (-8.45)^2}{(n =)4} - 65.41$$
$$= 66.55 - 65.41$$
$$= 1.14$$

8. Calculate the 'between row sum of squares (r^2)':

$$\text{Row SS} = \frac{(-6.82)^2 + (-2.83)^2 + (-11.55)^2 + (11.15)^2}{(m =)4} - 65.41$$
$$= 78.06 - 65.41$$
$$= 12.65$$

9. Calculate the 'residual sum of squares':

$$\text{Residual SS} = \text{TSS} - \text{Col SS} - \text{Row SS}$$
$$= 17.64 - 1.14 - 12.65$$
$$= 3.85$$

10. Enter the appropriate degrees of freedom ('df's) into the summary table:

Columns differential df = number of columns – 1   = 3
Rows differential df   = number of rows – 1   = 3
Total differential df   = number of scores – 1   = 15
Residual differential df = total df – col. df – row df = 9

11. Calculate the variance estimates by dividing each of the sums of squares by the appropriate degrees of freedom:

Column (events) variance = 1.14/3   = 0.38
Row (expert) variance   = 12.65/3 = 4.22
Residual variance   = 3.85/9   = 0.43

12. Calculate the F ratios

> F (columns) = (column variance)/(residual variance)
> F (rows)   = (row variance)/(residual variance)

Therefore:

> F (columns) = 0.38/0.43 = 0.88
> F (rows)   = 4.22/0.43 = 9.81

13. The last step is to determine the intra-class correlation coefficient, according to the following formula:

$$r = \frac{F - 1}{F + (n - 1)} = \frac{0.88 - 1}{0.88 + 3} = -0.03,$$

where F is the ratio for the events factor. The resulting value in this example is very low, suggesting that the experts are inconsistent. With results such as these, the personnel running the study may well consider giving feedback to the individual judges, and then either get them to make new estimates or form a consensus group and obtain consensus estimates. Alternatively, a *new* group might have to be assembled.

Table 5.5.9 Summary of analysis-of-variance (ANOVA)

Source	Sums of squares	df	Variance	F ratio
Events (columns)	1.14	3	0.38	0.88
Judges (rows)	12.65	3	4.22	4.22
Residual (error)	3.85	9	0.43	
Total	17.64	15		

In this example, the judges' F ratio is significantly high, indicating that there is a significant difference between these different experts' estimates, and thus also verifying the lack of agreement indicated by the low intra-class correlation coefficient.

STEP 7: AGGREGATE THE INDIVIDUAL ESTIMATES
If a consensus group is not used, and if, also, the level of agreement between judges is adequate, it will next be necessary to aggregate the different individuals' estimates for each HEP. This is achieved by taking the geometric mean of the individual estimates. In the example above, the geometric means for the four tasks (events) would be 0.02, 0.013, 0.004 and 0.008, respectively.

STEP 8: UNCERTAINTY-BOUND ESTIMATION
Uncertainty bounds are calculated using the following formulae adapted from Seaver & Stillwell (1983).

Upper and lower uncertainty bounds are equivalent to:

$$\text{Log HEP} + 2 \text{ s.e.}$$

where s.e. = standard error = $\sqrt{\dfrac{V\,(\log \text{HEP}_i)}{m}}$,

$$V\,(\log \text{HEP}_i) = \frac{\left(m \displaystyle\sum_{j=1}^{m} (\log \text{HEP}_{ij})^2 \right) - \left(\displaystyle\sum_{j=1}^{m} \log \text{HEP}_{ij} \right)^2}{m\,(m-1)},$$

m = number of experts
and n = number of events

For example, for Event 1 in Table 5.5.7 (which has an average log HEP of -1.69):

$$V\,(\log \text{HEP}_i) = \frac{4 \times 14.25 - (-6.75)^2}{4 \times 3}$$
$$= 0.953$$

Thus, the s.e. $= \sqrt{\dfrac{0.953}{4}}$
$$= 0.49$$

Therefore, the 95 per cent confidence range is log $(-1.69) \pm 0.94$

This produces, by taking the antilogarithms, the following confidence interval for Event 1:

$$2.1 \times 10^{-3} < 2 \times 10^{-2} < 1.9 \times 10^{-1}$$

This range therefore spans two orders of magnitude: the variation in the judges' estimates and the small number of experts used. It would generally be hoped, however, to achieve a smaller range, i.e. one spanning only *one* order of magnitude.

If a group-consensus approach is being used, uncertainty bounds can be estimated via the APJ method itself. Each expert is asked to estimate these bounds (see Step 5 above), which are then aggregated statistically. There is a danger that a bias towards 'confident' estimates (see next section) may serve to shrink the confidence interval, so that the upper bound (i.e. the highest HEP value) is underestimated while the lower bound, on the other hand, is overestimated. This effectively means that the judges are overconfident, and that the uncertainty bands are therefore narrower than they should be. The analyst should be aware of this possibility.

BIASES IN EXPERT JUDGEMENT

All forms of APJ, whether group or individual, are prone to particular biases which can detract from the accuracy of the experts' assessments. If APJ is being used, it is therefore important to be aware of these potential problem areas. One of the major ways of reducing the problem of biases in expert judgements is to employ a 'facilitator' during the experts' group session. The primary function of the facilitator (who will generally 'manage' an expert-panel session without himself or herself providing any quantitative estimates) is to try to overcome these biases. A few of the more well-known biases are described below: for further details, see Tversky & Kahneman (1974) and Stael von Holstein & Matheson (1979).

With any group process, there is the risk of group-dynamic effects. A secondary function, therefore, of the facilitator is to overcome any personality conflicts, or other problems, which may occur in small groups, and influence the assessment process. A personality conflict may, for example, arise in a group where there is a individual who takes a domineering approach and yet who is *not*, in fact, the most authoritative expert in the group. In such a scenario, the facilitator must try to ensure that the other experts have a fair say. The prevention of such problems requires skills of diplomacy and arbitration, amongst others – especially when emotive topics are being dealt with in the assessment exercise. The skill of facilitation is not within the scope of this guide, and so the reader is referred to Eils & John (1978) for more information on this subject. The following deals only with some of the more well-known biases, and not the various personality conflicts which can occur.

The *overconfidence bias* (also referred to as 'conservatism' though throughout this book 'conservative' estimates are held to be erring towards pressimism) has already been mentioned. This causes uncertainty bands to be too narrow; i.e. it generally causes the *under*estimation of very high failure probabilities, on the one hand, and the *over*estimation of very low failure probabilities on the other hand.

Another bias is the *availability bias*. Here, the value of the estimate made by the experts involved is affected by the ease with which they can bring to mind previous occurrences of the events in question. Thus, an expert may be (erroneously) influenced by a disproportionate amount of exposure to a particular event or piece of media reporting, or may find it difficult to imagine a particular type of error and thus *under*estimate its likelihood. As an example, if, in a case where several experts are considering the likelihood of certain errors made in the cockpit of a commercial aircraft, there has recently been a well-publicised air disaster, such a disaster is likely to sway the experts' judgements so that such human errors are considered *more* likely than they are in reality, or are given a greater value than they would have been given had no such disaster taken place.

In the case of an *anchoring bias*, an expert, or expert group, starts with some initial value, suggested by one member, and adjusts this value so as to derive the best estimate, frequently failing, however, to adjust it to a sufficient extent. In a consensus group, therefore, it is important first to discuss the task qualitatively without initially quantifying and thus anchoring the HEP. Then, when each individual member has written down his or her estimate – before anyone is 'anchored' by anyone else – the facilitator can 'reveal' the estimates, and a discussion can begin towards reaching a consensus.

One of the more difficult biases to deal with is the *motivational bias*. This occurs when an expert, or group of experts, have a *vested interest* in obtaining probabilities of a certain value – or else, for example, low probabilities versus high ones (or vice versa). One of the experts may actually have been personally involved, for instance, in the design of the system being operated, and may therefore feel responsible for its reliability. Such an individual may suffer a conflict of interests if the group generates *high* HEP estimates, and as a result, he or she may try, consciously or unconsciously, to produce relatively *low* HEPs, while also encouraging others to do the same. This kind of bias is very real, and is difficult to deal with once such a member has already been brought into a consensus group. It is therefore advisable to try to avoid its occurrence in the first place by carrying out a careful screening of experts so that any vested interests that may exist can be detected during the expert-selection process. Even with such screening attempts, however, such a bias *can*, sometimes, get into the group, and the author himself has had to abort at least one session, as well as weeks of preparation, due to someone with a motivational bias, and acting contrary to instructions, anchoring the group far from their intended group-median estimate.

SUMMARY OF ADVANTAGES AND DISADVANTAGES OF APJ
The principal advantages of the APJ approach are as follows:

- The technique has also been shown to provide accurate estimates in *other* fields (e.g. weather forecasting; see Murphy & Winkler, 1974).
- The method is relatively quick to use, and yet it also allows as much detailed discussion as the experts think fit; this kind of discussion, if documented, can often itself be qualitatively useful.
- Discussions can also be turned towards a consideration of how to achieve error reductions. In such a situation, the group becomes like a HAZOP group, and can develop some highly credible and informed suggestions for improvements. This development is also beneficial where the group members are themselves operational staff, since this fact would improve the chances of such recommendations being accepted and then *properly* implemented.

The principal disadvantages of the APJ approach are as follows:

- The APJ technique is prone to certain biases, as well as to personality/group problems and conflicts, which, if not effectively countered (e.g. by a 'facilitator'), can significantly undermine the validity of the technique.

Figure 5.5.10 Continuum of perceived error-likelihoods, combined with a paired-comparison scale

				Judgement continuum		
Low error probability					**High error probability**	
E	F	D	A	B		C
0	10	20	30	65		100
				Relative scale values		
				(A–F are human errors for particular actions)		

- Since the technique is often likened to 'guessing', it enjoys a somewhat low degree of apparent, or 'face', validity.
- The technique is critically dependent on the selection of appropriate experts, but there are few really practically useful criteria for the selection of 'good' experts.

5.5.3.5 *The method of paired comparisons*

RATIONALE

The paired-comparisons method is a scaling technique based on the idea that expert judges are far better at making simple comparative judgements, i.e. at comparing one item with another and determining which is higher or lower on some sort of scale, than they are at making *absolute* judgements. In other words, it is easier for an expert to say 'a car crash is more likely than a plane crash' than it is for him or her to say 'a car crash will occur every x minutes, and a plane crash will occur y times a month'. Thus, a comparative judgement, generally speaking, may be an easier way of eliciting an expert opinion than an absolute judgement. Judgements arrived at by means of the paired-comparisons method can be transformed into an interval scale which, as with SLIM, can then be calibrated to give actual HEPs, provided that at least two tasks whose probabilities are known to the assessors but not to the experts have been 'seeded' into the task set being assessed.

Thus, when many experts make the same paired comparisons, an interval scale can be derived (as based on the theory of psychophysics: Thurstone, 1927). An interval scale is a quantitative scale (e.g. from 0 to 100) upon which different items or events can be placed relative to each other according to some specified dimension (in this case, the likelihood of error). The 'location' of an action on such a scale signifies its likelihood of occurrence relative to all the other actions on the small scale (see Figure 5.5.10). The scale is thus a numerical representation of a 'psychological continuum' of perceived human-error likelihoods drawn from the experts' 'communal mind' through the process of statistical aggregation. At no time are the experts asked to make numerical judgements. For a detailed description of the theory, and the assumptions, underlying the development of this scale, see Torgerson (1958) or Hunns & Daniels (1980).

The next step is to turn the relative likelihoods of human error into *absolute* human-error probabilities. A method for doing this was suggested by Blanchard *et al* (1966) and used by Hunns & Daniels (1980). Essentially, it rests on the premise that there is an exponential relationship between the scale values and the absolute probability of failure, of the form log HEP = a.S + b, where S is the paired-comparison scale value and a and b are constants. This premise is based both on an experimental study by Pontecorvo (1965) and on a theoretical argument put forward by Hunns & Daniels (1980).

Once such a relationship has been assumed, the 'calibration' procedure for converting the scale into an *absolute* one is as follows. When selecting the actions or tasks to be evaluated by the paired-comparisons procedure, two actions for which the actual probabilities of failure are known (ideally from empirical data) must be included in the set. These calibration tasks should preferably have a high and a low failure probability respectively, and should ideally lie *at*, or at least *near*, the two opposite ends of the scale. An equation is developed which is based on the relationship between the known failure probabilities for the calibration tasks, on the one hand, and their corresponding positions on the paired-comparison scale on the other. This calibration equation is then used to convert the scale values for the unknown tasks into failure-probability estimates for these tasks. It is possible to derive a general calibration equation for converting paired-comparison scale values into HEPs, while using two calibration HEPs (see Figure 5.5.11). However, it is usually simpler to use *simultaneous equations* to derive a specific calibration equation for a particular study. And if possible, it is better to use *more* than two calibration HEPs, since this allows a least-squares regression equation to be generated between the scale values and the reference HEPs. With such an equation, the goodness of fit of the logarithmic assumption can now be tested.

In the example in Figure 5.5.11, six tasks have been used as a part of a paired-comparisons procedure to produce a scale of rankings. Tasks F and C were 'calibration' tasks, for which the probabilities of failure were known. By means of equation (1), the failure probabilities for the other tasks can be computed.

PAIRED-COMPARISONS PROCEDURE

The complete paired-comparisons procedure for estimating human-error probabilities is as follows (primarily based on descriptions by Hunns & Daniels (1980) and Seaver & Stillwell (1983):

1. Define the tasks involved.
2. Incorporate the calibration tasks.
3. Select the expert judges.
4. Prepare the exercise.
5. Brief the experts.

Figure 5.5.11 Calibration method for generating human-error probabilities (example taken from Hunns & Daniels, 1980)

	E F D A	B	C
	S_1 S_x		S_u
	P_1		P_u

(1) Log $P_x = aS_x + b$
(2) Log $P_u = aS_u + b$
(3) Log $P_1 = aS_1 + b$

From (2) and (3):

$$(1)\ P_x = P_1 \left(\frac{P_u}{P_1} \right)^Z$$

where $Z = \dfrac{S_x - S_1}{S_u - S_1}$

A–F	= tasks
F, C	= calibration tasks
P_1, P_u	= known failure probabilities for F (lower) and C (upper) calibration tasks respectively
S_x	= scale value for task 'x' (called, in this case, task 'D')
S_u, S_1	= scale values for upper and lower calibration points respectively
P_x	= unknown probability

6. Carry out paired comparisons.
7. Derive the raw frequency matrix.
8. Derive the proportion matrix.
9. Derive the transformation X-matrix.
10. Derive the column-difference Z-matrix.
11. Calculate the scale values.
12. Estimate the calibration points.
13. Transform the scale values into probabilities.
14. Determine the within-judge level of consistency.
15. Determine the inter-judge level of consistency.
16. Estimate the uncertainty bounds.

DEFINE THE TASKS INVOLVED

The tasks should be defined simply, unambiguously and comprehensively, to ensure that every expert has the same knowledge of the task being assessed. The tasks should also be homogeneous – for example, they should all be maintenance tasks, or emergency actions – since, otherwise, experts will be trying to compare fundamentally different items on one single scale, and this would be likely to lead to invalid results. If there are several different types of HEPs being assessed, the whole set should be broken into homogeneous *subsets*, as with the SLIM.

INCORPORATE THE CALIBRATION TASKS

At least two (or preferably more) task descriptions for which the HEP values are known should be included in the task set. These two HEP values should preferably define the range within which the other HEPs will fall, and it is therefore preferable for them to lie, respectively, at the two extreme ends of the scale developed. Otherwise, there will be an element of uncertainty involved in the prediction of the other tasks (for example, if, in Figure 5.5.11, the calibration tasks were in fact D and A, then the HEP estimates for B and C would be subject to a good deal of uncertainty since these two tasks lie outside the range for which the calibration equation would be optimal).

It is also important that the calibration or reference tasks be homogeneous with the rest of the tasks being assessed. If, for example, the user calibrates a set of maintenance HEPs alongside two known values for the degree of reliability of an operator in a control-room emergency, the results may well be meaningless since the judges will find it difficult to compare two such dissimilar kinds of task.

SELECT THE EXPERT JUDGES

The judges must have relevant experience of the tasks being assessed (i.e. so-called *substantive expertise* – see the APJ description). Unlike the APJ technique, the paired-comparisons technique does not require the experts to have any great amount of expertise, either in probability theory or in the use of mathematical/statistical concepts.

The number of experts should preferably be more than 10. Although a smaller number of experts can be used, the calculations become more difficult, and the derived scale is less robust.

PREPARE THE EXERCISE

For n tasks, there are $n(n-1)/2$ possible paired combinations. For example, for four tasks, A, B, C and D, there are $4(3)/2 = 6$ possible pairs (AB; AC; AD; BC; BD; CD). Thus, for 10 tasks, 45 pairs must be considered by each expert, and for 20, 190 pairs.

As the number of tasks increases, the amount of inconsistent judgements is also likely to increase, and the judges are likely to get tired and thus make more mistakes. There are, however, methods for reducing the number of comparisons (Seaver & Stillwell, 1983), such as choosing a small number of 'standards' and then comparing each item to each standard. If m is the number of standards and n is the total number of tasks, then the required number of comparisons becomes $(mn - (m(m + 1)/2))$. Therefore, with 30 tasks and four standards, 110 comparisons are needed (as opposed to 435).

Each pair of tasks should be presented on its own, so that the expert only considers one pair at a time (this is necessary since the paired-comparisons (PC) approach assumes that each decision being made is independent from any other decision). Preferably, each task is described in a brief paragraph, with associated pictures or diagrams if these are helpful. Thus, a small PC exercise booklet can be developed, with each page covering one pair of task descriptions.

BRIEF THE EXPERTS

It is of critical importance that the experts fully understand the nature of the tasks being assessed, as well as the PC procedure itself. It will therefore be useful to discuss all the tasks involved with the experts, identifying possible causes of error, as well as factors affecting performance, and ruling out similar, confusible tasks which experts might be concerned with but which are not actually being assessed. Discussions with the experts on topics such as these may well lead to changes in the task descriptions, since the experts may know more about the task than the analyst who has prepared such descriptions.

The experts should then be briefed both on the purpose of the study and on the nature of their task. Each expert should be instructed to consider each pair on its own, to decide which human-error probability is more likely and to 'tick' or otherwise highlight the most likely human error. It should be explained to the experts that there is an assumption of independence between comparisons, so that experts should not refer back to previous pairs to see what judgements they made before. Lastly, experts must be urged to make a comparison of *all* pairs, even when this is difficult. It should be explained that if they are merely guessing randomly, then the technique will identify this, but if they omit to make certain difficult decisions, then *no* results will be obtainable. It is important that experts are motivated to carry out this procedure conscientiously.

CARRY OUT PAIRED COMPARISONS

Where possible, the analyst should be present while the experts are making the paired comparisons. This is useful where subjects request the clarification of certain task descriptions – which is a common occurrence. The analyst can also be present to remind experts not to refer back to previous comparisons (which often happens if participants are left alone).

The following example of a calculation is based on an example taken from Hunns & Daniels (1980).

DERIVE THE RAW FREQUENCY MATRIX

A matrix is constructed with (n × n) elements, where n is the number of tasks being assessed. Each cell entry in the matrix shows the number of judges who thought that the event at the top of the matrix was more likely than the corresponding event at the side. For example, in Table 5.5.9, the cell in the first row and second column tells us that 6 of the 20 experts thought that B was more likely than A.

Table 5.5.9 Raw frequency matrix

	A	B	C	D
A	–	6	7	11
B	14	–	12	20
C	13	8	–	14
D	9	0	6	–

Table 5.5.10 Proportion matrix

	A	B	C	D
A	–	0.30	0.35	0.55
B	0.70	–	0.60	1.00
C	0.65	0.40		0.70
D	0.45	0.00	0.30	–
Total	1.80	0.70	1.25	2.25

DERIVE THE PROPORTION MATRIX

The next stage is to convert each element in the above matrix into a proportion of the experts holding a particular opinion. For example, 11 out of 20 experts means that 55 per cent of the experts believed D more likely than A. This conversion 'normalises' the raw frequencies above. The corresponding proportion matrix is shown below (Table 5.5.10).

DERIVE THE TRANSFORMATION X-MATRIX

The next step is to convert these proportions into their equivalent unit normal deviate, using tables covering the area under the normal distribution (see Table 5.5.13). The columns are also put into rank order, according to the column totals shown in the proportion matrix in Table 5.5.10. In other words, they are represented in order of increasing probability. The resulting scheme is called the transformation X-matrix.

If all the judges agree, or if all disagree, on a particular paired comparison, this produces a proportion of 1 or 0, and a corresponding unit normal deviate of plus or minus infinity. These can be resolved as shown in subsequent steps.

DERIVE THE COLUMN-DIFFERENCE Z-MATRIX

As shown in Table 5.5.12, a column-difference matrix is derived from the transformation X-matrix by means of a simple calculation of the differences between the adjacent column values shown in Table 5.5.11 (e.g. for D and A: $0.13 - 0 = 0.13$; ? $- 0.52 = $?; $0.52 - 0.39 = 0.13$; $0 - (-.13) = +0.13$). Each element in this matrix corresponds to a pair of tasks, and represents an estimate of the scale separation between the two tasks.

Table 5.5.11 Transformation X-matrix

	D	A	C	B	(increasing probability)
A	0.13	–	−0.39	−0.53	
B	?	0.52	0.25	–	
C	0.52	0.39	–	−0.25	
D	–	−0.13	−0.52	?	

Table 5.5.12 Column-difference Z-matrix

D–A	A–C	C–B
0.13	0.39	0.14
?	0.27	0.25
0.13	0.39	0.25
0.13	0.39	?
0.13*	0.36	0.21*

* Average based on 3, not 4, estimates. An alternative calculation method which avoids the problem of obtaining indeterminate values in the column-difference matrix is shown in Figure 5.5.12.

CALCULATE THE SCALE VALUES

The scale values are calculated by taking the average column differences from Table 5.5.12 (in this case, 0.13, 0.36 and 0.21) and converting these directly into a linear scale. To achieve this, the most preferred task (in this case, 'D') is first set to zero, representing a high probability of error. Then, the next most preferred task is set at the distance of its average column difference (in this case 13 units) away from D. The next task is similarly put as (36 + 13), or 49, units away from D, and the least likely human error, which here occurs task 'B', is set (21 + 36 + 13), or 70, units from D.

The final scaled ranking is therefore represented diagrammatically as follows:

D   A   C   B

0 ----------------------- 13
   ---49
   -- 70

The values 0, 13, 49 and 70, which represent scale values corresponding to the statistically aggregated judgements of the experts regarding the likelihoods of errors, are not yet absolute probabilities. The next step is therefore to transform these into such probabilities by the process of calibration.

Table 5.5.13 Table of normal deviates

P	01	02	03	04	05	06	07	08	09	10
z	-2.33	-2.05	-1.88	-1.75	-1.64	-1.55	-1.48	-1.41	-1.34	-1.29
P	11	12	13	14	15	16	17	18	19	20
z	-1.23	-1.18	-1.13	-1.08	-1.04	-.99	-.95	-.92	-.88	-.85
P	21	22	23	24	25	26	27	28	29	30
z	-.81	-.77	-.74	-.71	-.67	-.64	-.61	-.58	-.55	-.53
P	31	32	33	34	35	36	37	38	39	40
z	-.50	-.47	-.44	-.41	-.39	-.36	-.33	-.31	-.28	-.26
P	41	42	43	44	45	46	47	48	49	50
z	-.23	-.20	-.18	-.15	-.13	-.10	-.08	-.05	-.03	-.00
P	51	52	53	54	55	56	57	58	59	60
z	+.03	+.05	+.08	+.10	+.13	+.15	+.18	+.20	+.23	+.2
P	61	62	63	64	65	66	67	68	69	70
z	+.28	+.31	+.33	+.36	+.39	+.41	+.44	+.47	+.50	+.53
P	71	72	73	74	75	76	77	78	79	80
z	+.55	+.58	+.61	+.64	+.67	+.71	+.74	+.77	+.81	+.85
P	81	82	83	84	85	86	87	88	89	90
z	+.88	+.92	+.95	+.99	+1.04	+1.08	+1.13	+1.18	+1.23	+1.29
P	91	92	93	94	95	96	97	98	99	99.
z	+1.34	+1.41	+1.48	+1.55	+1.64	+1.75	+1.88	+2.05	+2.33	+2.5

Figure 5.5.12 Alternative calculation method that avoids indeterminacies in matrices

	Raw frequency matrix			
	A	**B**	**C**	**D**
A	–	6	7	11
B	14	–	12	20
C	13	8	–	14
D	9	0	6	–
Total	36	14	25	45
Divided by 20 (no of judges)	1.8	0.7	1.25	2.25
Plus .5 (for comparison of A with A, B with B, etc.)	2.3	1.2	1.75	2.75
Divided by 4 (no of comparisons)	0.575	0.3	0.4375	0.06875
Z scores	0.19	-0.52	-0.16	0.49
Transposition to a 0–100 scale	0	36	70	100
	B	C	A	D

Note: this alternative method has the advantage of avoiding the indeterminacies involved in both the transformation-X and column-difference matrices. It will also, however, yield slightly different values for the HEPs, values which may or may not be more conservative than those derived by means of the other calculation method. It is therefore left to the discretion of the analyst to decide which calculation method will be used.

ESTIMATE THE CALIBRATION POINTS
If possible, the error probabilities for the calibration tasks should be estimated from frequencies obtained via actual observations of the tasks in the field. Alternatively, the absolute-probability-judgement (APJ) approach can be used if field data is unobtainable.

TRANSFORM THE SCALE VALUES INTO PROBABILITIES
If, for example, it is known that the HEP for D is 0.5, and that the HEP for B is 10^{-4}, the following HEPs for A and C can be derived via the method of simultaneous equations, and by using the logarithmic relationship log HEP = $ax + b$ cited earlier:

$$\text{Log } (10^{-4}) = a(70) + b$$
$$\text{Log } (0.5) = a(0) + b$$

$$a = -0.053$$
$$b = -.301$$

Therefore, the calibration equation is:

$$\text{Log } p(x) = -0.053(x) - 0.301$$

Table 5.5.14 Circular triads

Event	1	2	3	4	a_i	a_i-a
1	–	–	1	0	1	-0.5
2	1	–	0	0	1	-0.5
3	0	1	–	0	1	-0.5
4	1	1	1	–	3	1.5

Substituting scale values for C and A, we obtain the following HEPs:

$$p(C) = 1.3 \times 10^{-3}$$
$$p(A) = 0.10$$

DETERMINE THE WITHIN-JUDGE LEVEL OF CONSISTENCY

This procedure is adapted from Seaver & Stillwell (1983). Experts may ex-
hibit internal inconsistencies that manifest themselves as 'circular triads', such
that a judge says A is greater than B, B is greater than C, and C is greater
than A – thus effectively saying A is greater than itself! It is necessary,
therefore, first to determine the number of circular triads (c) put forward by
the expert, as shown below:

$$c = \left(\frac{n \times (n^2 - 1)}{24} \right) - \frac{T}{2}$$

where n is the number of events, $T = (a_i - a)^2$, $a = (n-1)/2$ and the a_i values
are the number of times that an event a_i was judged to be more likely than
any other event. Table 5.5.14 shows an example of a calculation involving a
small set of tasks (in this example, n = 4; a = 1.5).

Therefore, $c = \dfrac{4(15)}{24} - \dfrac{3}{2} = 1$

In the above example, the triad is 3>1>2>3 (see the numbers underlined in
Table 5.5.14).

The coefficient of consistency (K) is then found by means of the following
formula:

K = 1 – (24c/n(n² – 1)) if n is odd, and
K = 1 – (24c/n(n² – 4)) if n is even
Therefore, K = 1 – (24 × 1/4 × (12)) = 24/48 = 0.5

If n is small, K can be treated as a *correlation coefficient*. If, on the other
hand, K is large enough to be statistically significant (according to correlation-

coefficient tables such as those found in Siegel, 1956) then the results should be rejected. For inter-judge consistency analysis see Seaver and Stillwell (1983).

ESTIMATE THE UNCERTAINTY BOUNDS

Seaver & Stillwell (1983) provide a statistical procedure for calculating uncertainty bounds. The reader should consult this reference if estimates of statistical uncertainty bounds are required.

DISADVANTAGES OF THE PAIRED COMPARISONS APPROACH

The paired-comparisons (PC) method relies on certain assumptions which can be violated during a Human Reliability Assessment. The first of these relates to the *complexity* of the event being considered. The method was originally developed to consider relatively *simple* perceptual phenomena, such as tastes, the perceived level of comfort, etc., and these are usually *uni-dimensional* phenomena. If the expert is asked to compare two very complex operator-error scenarios, such as two nuclear-power-plant emergencies, then a simple comparison may prove difficult due to the multidimensional natures of the two situations. The expert may have to consider a number of specific Performance Shaping Factors in each situation, as well as then aggregating their degrees of impact on an error before making comparisons. This is arguably a much more complex mental exercise than the method was designed to involve. Paired comparisons may not, therefore, be appropriate for complex human-error predictions.

The second assumption, underlying the PC approach, which may be violated in certain applications of the technique, is that of the *homogeneity* of the events or tasks being considered (this has already been alluded to in the text). The tasks being considered should be homogeneous, or of a similar type, or else the comparison process becomes very difficult and may, therefore, be subject to error. One way to overcome the heterogeneity problem is to create *subgroups* of homogeneous tasks.

The third assumption, which is easy to violate and difficult to control, is that of the *independence* of each comparison made. The reasons for the independence requirement are too complex to document in detail here, but essentially, if an expert mentally compares the possible results of a comparison with those of previous comparisons, then a *dependence* relation starts to be established which will create a bias effect that in turn will distort the results. Unfortunately, there *is* a natural tendency to compare one comparison with previous ones. In order to mitigate this problem, the independence assumption should be explained to the experts, and they should be encouraged to consider each comparison as if it were the only comparison being made. In particular, when using workbooks, the analyst must prevent experts from re-reading what they decided on previous comparisons. This tactic can be facilitated by carrying out the comparisons on a computer where the computer controls the presentation of the pairs, and allows no 'back-tracking'.

The disadvantages of the technique, therefore, arise from the fact that its stringent underlying assumptions are easily violated in the context of a Human Reliability Assessment. These disadvantages can be summarised as follows:

- The tasks being considered may be too complex to allow an easy comparison.
- The tasks may not be homogeneous.
- The comparisons made may not be independent of each other.
- If the number of comparisons is large, the judges may become tired and, therefore, carry out later comparisons differently from earlier ones.

ADVANTAGES OF THE PAIRED COMPARISONS APPROACH

The advantages of the PC approach may be summarised as follows:

- Subjective knowledge *can* be profitably extracted from comparative judgements, provided that the assumptions of the methods are upheld and the value of the human judgement proves greater than the value of any confusion arrived at via direct numerical assessments.
- The technique makes it possible to determine whether or not individual judges are qualified to make judgements about a particular datum or data set.
- Since the technique can work with a minimum of two empirically estimated HEP values, this enables the most effective use to be made even of scarce amounts of empirical data.
- Even without calibration, the technique provides a useful means of deriving a measure of the relative importance of difference human errors or human events. This technique was used, for example, to prioritise different types of ship-collision threats to offshore platforms in the UK sector (Technica, 1985).
- With a small number of tasks and a set of readily available experts, the technique can be applied fairly quickly, especially when carried out on a computer (see Hunns & Daniels, 1980).
- Experts do not have to carry out the comparisons as a group. Not having to do so will, of course, eliminate the the logistic problems of bringing experts together for the comparison.

5.5.4.7 Human reliability management system (Kirwan & James, 1989; Kirwan, 1990)

The following sections describe the modules of the human-reliability-management system in terms of what functions they achieve and how, broadly, they work. An overall schematic presentation of the HRMS has already been given earlier in Section 5.3, and since the task-analysis and human-error-analysis (HEA) parts of the HRMS have already been described, these will not be repeated here.

REPRESENTATION MODULE

The purpose of this module is to ensure that errors are represented or described in a way which can be accurately quantified both by the quantification module (PHOENIX) and by the fault-event-tree methodology when integrated into the Probabilistic Safety Assessment (PSA) and evaluated within the logic-tree format. A small library of typical fault and event trees, covering human errors and recovery paths, exists (at present, only for reference purposes) within this module. The user may refer to these for three purposes: to note aspects involved in the construction of fault and event trees, particularly with respect to dependence; for guidance on the use of HPLVs; and to note preferences for event over fault trees or vice versa.

The primary role of this module, however, is to determine which of the errors identified are to be represented and quantified in the assessment. Five categories exist which may be removed from further assessment at this stage: errors which have no consequence; errors already identified; errors which are incredible; errors with a very high (virtually certain) chance of recovery; and errors which may be 'subsumed' under other HEPs, i.e. aggregated within a higher-level datum (see also pages 112–114).

QUANTIFICATION MODULE

The quantification module is based primarily on the best available data from actual historical measurements, from simulator studies and from experimental research, as well as on certain derived data. The derived data comes from two sources, the first being a generic HEP guideline database (itself based largely on WASH-1400 and on judgements on the part of assessors) and the second being the BNFL experiment (Kirwan, 1988) involving several validated techniques (THERP, APJ, HEART). With regard to this second source, if five estimates, made using THERP, APJ and HEART, all yielded approximately the same HEP for a particular scenario, then this *convergence* tended to suggest that the HEP was probably valid, given also that these same techniques (and the respective assessors and experts) were proved accurate when tested against the actual tasks involved – i.e. given that they were effectively *well-calibrated*.

However, it is unlikely that such a limited database would suffice for all safety cases. Furthermore, such a database would yield little, if any, useful information on error reduction during an assessment. As a result, the quantification system also utilises six operationally defined Performance Shaping Factors, derived both from SLIM analyses carried out during various experiments, and from a review of several other techniques and assessments. These six PSFs are:

- Time
- The quality of the operator interface
- The quality of procedures
- How the task is organised

- Training/competence/familiarity
- The level of complexity

Each of these PSFs involves various levels, and any scenario can be described by reference to a particular PSF. In addition, any scenario can also be described from the point of view of any one of the PSFs; thus, each datum has a particular *profile* for each of these PSFs, as well as an attached probability of human error.

The system of quantification essentially allows the user to refer to a particular type of task, via a simple task or behaviour taxonomy, after which the user's task is matched with tasks in the database in such a way as to bring the user to the datum which most closely matches his or her own scenario. The user then qualitatively 'rates' the scenario against the PSF in question. If the ratings are identical, then the HEP for the datum itself is generated for the purposes of use in the assessment. If, on the other hand, the PSF profile is different (as is likely for at least some of the PSFs), then the datum on the HEP is modified appropriately by a series of PSF multipliers and the user is presented with a new HEP.

The essential philosophy underlying PHOENIX is to start from good data and extrapolate to the particular datum required. As the database becomes larger, the system becomes more robust, and the level of uncertainty decreases. The PSF matrices make no assumption of independence between different PSFs, an assumption which is difficult to support. Dependences, or interactions, between *time* and *complexity* (which are very likely), for example, are quantified by the PSF matrix approach.

SENSITIVITY ANALYSIS

From the description given in the above section on the quantification module, it is apparent that for any given error, the relative importance of different PSFs can be calculated. If an HEP is derived which is too high, then the sensitivity-analysis capability within HRMS will allow the user to see how much of a change in probability could be brought about by a modification of the degree of importance assigned to the various PSFs. This is one method of error-reduction analysis.

ERROR REDUCTION MODULE

A set of guidelines are available on how to reduce errors of various types. These guidelines, influenced by the above-named PSFs, are practically oriented, in terms of design and operational parameters, and are aimed at reducing the root causes of error, at increasing a system's level of error tolerance, at enhancing error recovery or else at generally improving the standard of human performance. Guidelines also exist on how to feed error-reduction assumptions through the quantification process, and how, then, to ensure that they are implemented. Specifically, the facilities provided as part of the HRMS are as follows:

- Reducing the consequences of an error, or blocking an error's pathway: a system change which removes the operator from danger, or removes the consequences of the error. Usually, error-reduction mechanisms (ERMs) in this module are hardware-oriented, and are most appropriate in cases involving a severe human-reliability problem.
- Error-recovery enhancement: the introduction of error-recovery loops into the logic system dealing with failure, either via procedures or other forms of task organisation.
- PSF-based error reductions: altering the task's PSFs in such a way as to improve the HEPs.

The first two mechanisms affect the event- or fault-tree logic, either by making certain paths less dominant or by *removing them altogether* from the analysis, while the PSF-based ERMs, on their part, allow the user to consider the effects of, for example, inserting a separate hardwired alarm, or of instituting refresher training, on the HEP. Furthermore, the PSF-based error-reduction module helps the user to decide which PSF would be most effective, once considered and improved, at reducing an individual HEP by automatically carrying out a basic sensitivity analysis that uses the same algorithms as those used to calculate the original HEP. Any improvement made to a PSF has to be justified and then logged. Ultimately, all ERMs have to be validated, via a process of, for example, noting which design drawings were changed (all such changes are logged together with the scenario analysis, and become auditable assumptions underpinning the safety case).

DOCUMENTATION

The system is largely self-documenting, via print-outs, occurring at various stages in the program, which can be appended to safety cases and recorded in the HRMS 'library', as well as being stored on computer disk. The system also documents the user's identity and details of the safety case, as well as relevant dates, etc.

QUALITY ASSURANCE

All documented assumptions are passed on to respective design and operations departments and followed through until sanctioned and, ultimately, satisfactorily implemented. If one or more assumptions are not 'cleared' in this way, a reassessment of the safety case will be carried out.

Changes made to the operational design of the system during its operational lifetime can be checked to see if they affect any of the safety cases or safety-case assumptions. In addition, any incident information about an incident or incidents that is relevant to the safety cases can be analysed to determine whether the HEPs are accurate, and the HEPs can be adjusted if necessary. In this way, the database would become more realistic as the plant ages.

A sample of HRMS extracts is given in the following figures, which cover a screening/representation-analysis output, tabular task analysis outputs, and error-reduction-analysis outputs. These figures, 5.5.13–5.5.17, also show the level of auditability, as well as the level of self-documentation, of the system.

SUMMARY OF HRMS

HRMS is currently an 'experts' tool, requiring a good deal of human-reliability experience. It is a highly in-depth approach, focusing both on quantitative modelling and on PSF-based justifications. Its error-reduction module is a logical but more powerful and explicit extension of similar formats found in other PSF-based techniques (see Kirwan, 1990). The HRMS's strong point is its integrated nature and its modelling power: it is an exhaustive system. However, this same advantage also means that it makes an extensive use of resources. This problem has led to the development of a faster but far less detailed system (JHEDI), described in section 6.0, used for the majority of assessments, with HRMS only utilised for 'important' errors.

HRMS has yet to be validated, and it has not yet 'closed the loop' between those PSFs implicit in the questions posed in the HEI module and those contained in the quantification module. More research is required on this front, as well as on error reduction via root-cause analyses. All in all, the HRMS offers a powerful and coherent HRA framework; but only time will tell whether it is a useful and used advance over other available techniques.

DISCUSSION AND CONCLUSIONS CONCERNING HUMAN RELIABILITY QUANTIFICATION

A conclusion made in the *Human Reliability Assessor's Guide* (Kirwan *et al*, 1988) was that there were already techniques available for carrying out the quantification of *most, if not all*, types of scenario; but also that the practitioner needed to be flexible in selecting techniques appropriate to each situation. Swain, in his (1989) review, was less generous. Whilst noting that there *were* techniques capable of carrying out HRAs, he lamented the lack of R&D funding in this field, and pointed out, furthermore, that until a proper database was set up and more evaluations were carried out, Human Reliability Assessments would continue to remain the least credible part of the Pro-babilistic Risk Assessment (PRA) process. It is interesting that Swain, whose THERP data-base is the most widely used data-bank in existence, is still arguing for the generation of *real* data as opposed to 'synthetic' or 'simulator' data (the present author sides entirely with Swain in this respect), and there is, in fact, a resurgence of interest in collecting usable data. There is also an increasing interest in the carrying-out of validations of HRQ techniques, both to see how accurate they are in practice and to allow the techniques themselves to be improved.

In the meantime, before databases are constructed and techniques have been fully validated, the practitioner does have plenty of tools from which to choose. Some selection guidelines, based on the HRAG assessments, but updated following a validation experiment (Kirwan; 1988) are shown in Figures 5.5.18 and 5.5.19. It is hoped that these will aid the practitioner in selecting techniques for real applications (see also Section 5.10).

Figure 5.5.13 HRMS outputs: human-error-screening report

Assessor: B I KIRWAN
Task: DOG-FAN-FAILURE SCENARIO
Safety case:
File name:

Step	No. 1	DETECT AND ACKNOWLEDGE VDU ALARMS
Error	No. 01090	CRO fails to respond to all signals
Error mechanism		Cognitive/stimulus overload
Recovery		SS/HARDWIRED ALARM/OTHER ALARMS
Dependency/exclusivity		
Screening		
Comments		REC HIGHLY LIKELY. TO BE QUANTIFIED AS CRO CONFUSION DUE TO HIGH DENSITY ALARMS

Step	No. 2	RESPOND TO HARDWIRED ALARM
Error	No. 07090	CRO fails to detect hardwired alarm amongst others
Error mechanism		Discrimination failure
Recovery		
Dependency/exclusivity		
Screening		SUB
Comments		HARDWIRED ALARM VERY SALIENT, 'COPIED' TO SS CONSOLE. SUB 01090: ALARM-QUANTIFICATION MODULE QUANTIFIES TOTAL RESPONSE TO ALARM SET

Step	No. 3.2	CRO suggests fan failure as scenario
Error	No. 13010	CRO or SS (or both) fail to identify the scenario in time
Error mechanism		Cognitive/stimulus overload
Recovery		SS, EMERGENCY PROCEDURES
Dependency/exclusivity		
Screening		
Comments		THIS ERROR REFERS TO A FAILURE TO IDENTIFY THE OVERALL GOAL OF PREVENTING BOILING BY ACHIEVING RECIRCULATION (BY CHANGING VALVE STATUS)

Figure 5.5.14 HRMS outputs: Tabular Task Analysis

Task Goal	Time	Personnel	System Status	Information Available	Decision/Action/ Communications	Equipment/ Location	Feedback	Distractions/ Other Duties	Comments and Operator Expectations
1. Normal Operations		CRO, SS	Normal operating envelope			CCR		Busy on activities	Single CRO, SS not busy on another fault. Assume beginning of each sequence (worst-case timescale)
2. Respond to first alarm	T.10s	CRO	Both DOG fans fail for a cause other than CNI failure	VDU alarm (yellow) on low DOG fan shaft speed	CRO targets 'fetch alarm': goes to level 4 mimic.	VDU, DOG console	Flashing yellow alarm message	Previous tasks: other alarms begin to occur	
3. Respond to further alarms	T.70s	CRO	Further alarms occur due to: (i) low DP across fans (ii) low flow in DOG (iii) low DP across dissolver (iv) low pressure drop across columns	VDU alarms: – yellow – red – red – red	CRO targets 'next alarm': goes to level 4 mimic. CRO goes to group alarm listing. Looks at printer.	VDU, printer	Flashing audible messages	Other alarms	
4. CRO identifies fan failure & calls supervisor (acknowledge alarms)	T.60	CRO/SS	Alarms on: (i) Low flow DOG duct at stack monitoring room (ii) Hardwired in DOG duct manifold	VDU alarms: – red – Hardwired	CRO monitors level 3 mimic / CRO acknowledges alarms, calls supervisor / SS detects alarm from own console	DOG, VDU / DOG, VDU & BUP / SS, VDU console	Audible & visual / Audible & visual	SS involved in other tasks	

Figure 5.5.15 Error reduction menu

--> Introduction

  Selection of Error Reduction Mode

  Consequence Reduction / Error Path Blocking

  Increasing Error Recovery

  Reducing Error Likelihood

  Root Cause Reduction

  Sensitivity Analysis

  Print Error Reduction Analysis Report

  Return to Main Menu

Use Up-Down cursor keys to select . . . Confirms with <Enter>

Figure 5.5.16a PSF Reduction Menu-Prior to Reduction

Step: 1   Go to sample point

Error: 02310   Goes to wrong sample point

Original HEP: 1.0E-1   Current HEP: 1.0E-1

	Original PSF	Current PSF	Target PSF	HEP Result
Time	4	4	2	1.0E-001
Quality of information	4	4	2	1.0E-001
Training/experience	0	0	0	1.0E-001
Procedures/instructions	4	4	2	1.0E-001
--> Task organisation	8	8	6	2.0E-002
Task complexity	2	2	0	1.0E-001

  Initialise Current to Original
  Return to Error Reduction Menu

Figure 5.5.16b PSF Sensitivity Analysis-Based Error Reduction: after reduction

Step: 1   Go to sample point

Error: 02310   Goes to wrong sample point

Original HEP: 1.0E-1.0E-1   Current HEP: 1.0E-002

	Original PSF	Current PSF	Target PSF	HEP Result
Time	4	4	2	1.0E-002
Quality of information	4	4	2	1.0E-002
Training/experience	0	0	0	1.0E-002
Procedures/instructions	4	4	2	1.0E-002
--> Task organisation	8	2	0	1.0E-002
Task complexity	2	2	0	1.0E-002

Initialise Current to Original
Return to Error Reduction Menu

Figure 5.5.17 Error Reduction Mechanisms Derived

Assessor:
Task:
Safety Case:
File name:
Task Step: 1   Go to sample point
Error: 02310   Goes to wrong sample point
Error Mech.:   Topographic misorientation   Event Ref: OGTWSP

CONSEQUENCE REDUCTION & ERROR PATHWAY BLOCKING

Description: Implementation of auto-sampling system (10)
Comments: Removes error from tree

INCREASING ERROR RECOVERY

Description: auto-sampling prompts operator with sample point (1)
Comments: adds error recovery step into tree

REDUCING ERROR LIKLEHOOD

Description: Remove distractions
Comments: Move this sampling operation under the auto-sampling system.

Description: Built in self-checking
Comments: Use a tick-sheet procedure which notes the sampling point to be
             sampled and the code of the actual sampling local valve.

Figure 5.5.18 Summary of evaluations of techniques (adapted from Kirwan et al, 1988, and updated to account for later evaluations)

	APJ	PC	THERP	HEART	IDA	SLIM	HCR
Accuracy	Moderate/High	Moderate/Low	Moderate/High	Moderate	(Low) Undetermined	Moderate	(Low)
Validity	Moderate/High	Moderate/High	Moderate	Moderate	Moderate	Moderate	Low
Usefulness	Moderate	Low	Moderate/Low	High	Moderate/High	High	Low/Moderate
Effective use of resources*	Moderate/Low	Moderate	Moderate	High	Low	Low	Moderate/High
Acceptability	Moderate	Moderate/High	High	High	Moderate	Moderate/High	Low
Maturity	High	Moderate	High	Moderate	Low/Moderate	High	Moderate/High

* A rating of high on this criterion means that the technique makes few demands on available resources.
The HRMS and JHEDI are not included in this table since they are currently unavailable for public evaluation.

Figure 5.5.19 Selection matrix (adapted from Kirwan et al, 1988, and updated)

	APJ	PC	THERP	HEART	IDA	SLIM	HCR
Is the technique applicable to:							
Simple and proceduralised tasks?	Y	Y	Y	Y	Y	Y	N
Knowledged-based, abnormal tasks?	Y	Y	Y*	Y	Y	Y	Y
A misdiagnosis which makes a situation worse?	Y	Y	N	N	Y	Y	N
Sociotechnical factors?	Y	N	N	Y	Y	Y	N
Are qualitative recommendations possible?	Y	N	Y	Y	Y	Y	Y
Is a sensitivity analysis possible? (i.e. for error reduction purposes)	N	N	Y	Y	Y	Y	Y
Are specific, PSF-based ergonomics recommendations possible?	N	N	N	Y	Y	Y	N
Does the technique have:							
Only minimal training requirements?	N	Y	N	Y	N	N	Y
Requirements for calibration data?	N	Y	N	N	N	Y	N
Requirements for experts (judges)?	Y	Y	N	N	Y	Y	N

* Via its nominal diagnosis model, or via ASEP.

5.6 Impact assessment

Once the human-error probabilities (HEPs) have been assigned to the various events in the fault or event trees (or both), these trees will be mathematically evaluated, i.e. the top-event frequencies will be calculated for the fault trees and event-sequence-outcome likelihoods will be calculated for the event trees. It is at this point that the relative contributions of individual human errors to accident frequencies, as well as the contribution of human error as a whole, are determined.

In fault trees, it is possible to calculate directly the relative importance of each cut-set or 'basic event' quantified in such trees, and indeed many fault-tree software packages do this automatically. This calculation gives a measure of the degree of 'sensitivity' of the top event to each basic event or cut-set. It is then clear which events, and *combinations* of events, are determining, to the greatest extent, the accident's likelihood of occurrence. Similar approaches can be used for event trees, but usually an inspection of the tree itself is enough to determine which events play a major part in determining risk.

An example of the type of output that can be obtained from a fault-tree program has already been shown earlier in Figure 5.4.4 and Table 5.4.1. These illustrated, respectively, the fault tree and the cut-sets in order of probability, as well as the relative importances of each event in connection with the top event. These examples show which human-error probabilities are contained in the dominating cut-sets and which individual events contribute most significantly to the level of risk, which is calculated throughout the cut-sets. Whilst the latter kind of information may appear, at first sight, the most useful for impact assessments, the cut-set ordering is also of importance, and often it is the combinations of events found in cut-sets that must be considered, rather than individual events (treated independently) if the level of risk is to be reduced.

When the trees have been thus evaluated, their calculated accident frequencies will be compared against predefined accident criteria. If these frequencies are well within (i.e. below) the acceptable risk levels for the industry, then it may be that nothing more needs to be done to the analysis besides documentation and quality assurance. If, on the other hand, the frequencies are violating the criteria (i.e. they are higher than the allocated risk allowed), then the Probabilistic Safety Assessment (PSA) must now determine which individual events (be they hardware failures or human errors) or combinations of events (as represented by cut-sets) make a great impact on the accident frequencies. It is these high-impact events or event sequences that must, in some way, be reduced. If high-impact or sensitive *human errors*, exist, these will be targeted for error reduction (as discussed in the next section). If error-reduction measures are implemented, then the level of risk must be recalculated accordingly, until the required levels of acceptable risk are achieved, or until the risk level is as low as is reasonably practicable/achievable in cases where such error-reduction targets are being actively pursued.

In practice, an impact assessment is an iterative process. Initially, if the risk level is too high, or unacceptably close to the criteria levels, the risk analysis will undertake an initial exploration of ways to reduce the risk level. Such an investigation will naturally focus on major risk-generating events. On the hardware side of the PSA analysis, ways may be sought to improve reliability, e.g. by increasing the number of back-up ('redundant') machines, by increasing the *diversity* of equipment used or by using a more reliable component.

On the human side, attempts will be sought from the HRA analyst to decrease the HEP, usually by at least a factor of 10 – but sometimes only by a factor of three. Such aspects as training or the redesign of the operator's interface may also be considered by the HRA analyst (see the next section). And the analyst may even be asked to calculate the likelihood of failure of the new enhanced human-operating system being developed, as well as be asked to estimate the cost of error-reduction measures. It is important, as was pointed out in Section 5.7, that the HRA analyst neither *over*estimates the reliability of the enhanced system nor *under*estimates either its cost or the time-scale required for its implementation.

Once these factors have been estimated, the PSA analysts will be in the best position for deciding whether the hardware system, the human system or a *combination* of both should be targeted for improvements that would reduce the level of risk inherent in the system, or whether, in fact, the level of risk *cannot* be sufficiently reduced to meet the criteria. In the case of the latter situation, the project will either have to be reconsidered, or scrapped, or put 'on hold' until some new design innovation can be used drastically to reduce the risk level.

This section has been brief, since little has been written on this subject, and the subject itself is not particularly formalised. Thus, whilst the mathematics and methods of impact assessments are fairly robust (thanks, for example, to fault-tree-cut-set analysis), the decision-making processes used to decide what to implement to reduce the impact of human error are still dependent both on the industrial context and on the company or companies involved. The case studies in Appendix VI will, nevertheless, show the type of impact assessments that are carried out, as well as an idea of the decisions that are sometimes made on what to do about unacceptably high levels of risk.

5.7 Error reduction analysis

This section discusses a formalised approach to error-reduction analysis, based on the results of a quantified Human Reliability Assessment (HRA), which can be applied irrespective of which techniques of error identification and quantification are being utilised. This section therefore discusses ways of identifying error-reduction measures (ERMs) and incorporating them into the system design in such a way that they will 'take root'. The actual nature of the ERMs, which can be gleaned from related discussions in other sections (see Appendix V; and also the HEI section), really requires that the reader investigate other human-factors/ergonomics materials outside of the scope of this book; useful and accessible references are given in Appendix IV.

5.7.1 Error reduction analysis in HRAs

Error-reduction analysis has already been mentioned in several parts of this book, under the following headings:

- The use of task analysis to identify and reduce errors (tabular task analysis: see Table 5.7.1); e.g. via enhanced feedback, or training.
- The use of Human Error Identification methods (e.g. SHERPA, or the human-HAZOP approach or HEIST) to identify errors and derive error-reduction approaches (see Table 5.7.2).
- The use of PSA sensitivity-analysis methods (e.g. fault-tree-cut-set-importance analysis, or a review of event trees) to identify 'sensitive' errors which can be targeted for error reduction (see earlier Section 5.6.).
- The use of quantification methods with built-in error-reduction strategies (see the HEART, and other examples, in Section 5.5).
- The use of quantification methods with sensitivity-analysis capabilities (see the SLIM, HEART and HRMS in Section 5.5).
- The use of quantification methods with error-reduction-analysis capabilities (e.g. HRMS: Section 5.5, and the discussion below).

As well as the above methods within the HRA, there is the entire vast domain of human-factors engineering, which aims to design for optimal or maximal performance from the start. This area could be made to follow the task-analysis approach, rather than the HRA process (see Kirwan & Ainsworth, 1992; Wilson & Corlett, 1990). There are, therefore, many ways to improve error-reduction measures aimed at quantified estimates of risk. Since, however, this book is dealing only with HRAs carried out as part of a Probabilistic Safety Assessment (PSA), it will be restricting itself to the HRMS approach, an approach which is, to the author's knowledge, one of more advanced of the formal HRA approaches to error reduction.

The first step in an error-reduction analysis (ERA) is the PSA impact or sensitivity analysis. This determines which scenarios are sensitive and, within those scenarios, which human errors are important. If no such errors are found, then an ERA is unlikely to be required since, in this case, human errors will ostensibly impact on the risk level. However, it *is* possible for an element of human intervention to be introduced into a sensitive scenario which previously had no such element of human involvement (e.g. the introduction of a human-fault-detection-and-recovery procedure into a previously hardware-controlled, 'closed' system-control loop which is liable to failure).

Once the critical error (or set of errors) in a sensitive scenario has been identified, the following error-reduction approaches can be applied.

5.7.2 HRMS error reduction approach (Kirwan, 1990)

The HRMS error-reduction approach has four basic components, the first two of which focus on reducing the risk level by directly decreasing the system's level of vulnerability to human error, and the second two of which aim to decrease the human error's likelihood of occurring, or of its *continuing*. These are explained below.

Table 5.7.1 TTA derived Error Reduction Analysis (Reed, 1992)

Step no.	Description	Displays	Required action	Feedback	Comms.	Notes
A	RACK FROM STORAGE TO PLATE					THE SUCCESSFUL COMPLETION OF EACH SUB-TASK IS ASSUMED THROUGHOUT THE TASK
A1	RACK HANDOVER					CS1 situated on a thoroughfare. Therefore, frequent distractions and possibility of accidental control activations
1	Accept control of transfer machine from storage at Control Station (CS1)	CONTROL OFFERED light illuminated	ACCEPT push-button	CONTROL OFFERED light goes out, and CONTROL ACCEPTED is illuminated	Confirmation of control with Storage Control Room	CONTROL ACCEPTED light should stay illuminated throughout CS1 control of the transfer machine
2	Read rack number	CCTV monitors	Camera controls			Possible poor quality of CCTV system. Possibility of incorrect reading/forgetting of number, rely on information given by Storage – which may be incorrect
3	Input rack number to computer for checking	Previous reading Computer prompt	Input number using keyboard	Computer response – continue or re-enter		Input error not detected by computer check
A2	DELIVER RACK TO C1					
4	Open Gate A and the building door		2 push-buttons	None		No Gate A Building Door status indicator lights
5	Release travel locks		RELEASE LOCKING BOLT push-button	LOCK OPERATING light will go out		No RELEASED/ENGAGED indicator lights, therefore bolt status not shown
6	Travel to C1		TRAVEL TO PLANT STOP push-button	OPERATING LIGHT illuminated		No indication of transfer machine location
7	Lower rack		LOWER HOIST push-button	Wait for FULLY LOWERED to be illuminated		
8	Release rack		RELEASE PINTLE GRAB	Pintle grab OPERATING illuminated		Two OPERATING lights and labelling does not specify what is operating
9	Raise hoist		RAISE HOIST push-button	FULLY RAISED illuminated		

Table 5.7.2 HEA derived ERA (Reed, 1992)

Error No.	Design inadequacy	Errors	Possible consequences	Recovery points	Remedial actions		
					Design	Training	Procedure
1	The required updating of the pond contents on return to computer control	Incorrect input of the location of equipment, racks, etc.	Potential risk of equipment collisions	Notice error via observation of monitors	Improved feedback		Recording & checking of information
2	Absence or insufficient positional information of RH 'Long travel', 'Cross travel', 'Grab height', 'Grapple drive tube' and 'Slew (Direct control station)	Incorrect movement operation of crane (e.g. travel or slew when not at transport height, travel when crane slewed incorrectly, crane not fully lowered when load released, collide with removal machine	Plant and equipment design	Emergency stop	Numerous additional displays		CCTV checks
3	'Slew raise' beneath 'Stem lower' on joystick	Slew move up instead of down or vice versa	Equipment damage. Loss of throughput		Alternate 'raise' and 'lower'		
4	No auditory or visual alarms (Direct control station)	Fail to realise mechanical fault	Plant, fuel and equipment damage. Loss of throughput	Machine safety interlocks	Provision of basic alarms		
5	Little personnel restriction to direct control station	Various	Various – risk of collision by inexperienced personnel	Various			
6	No indicator lights for Gates B and C	Crane collides with gates	Equipment and plant design. Loss of throughput	CCTV check	Addition of indicator lights		
7	No indicator lights to show status of Gate A and the building door (CS1 – transfer machine)	Collision of transfer machine with gate	Equipment and plant damage. Loss of throughput	Visual check	Addition of indicator lights		
8	Inadequate communication system, subject to delays and interference (all control stations)	Various	Various – including collisions		Additional intercoms – telephones		
9	No prioritisation of alarms (CS1 and CS2)	Overlooking of important alarms due to distracting, less significant alarms	Various (e.g. delays)	Various	Discrimination of alarms by colour coding and location		

Figure 5.7.1 Barrier-analysis structure

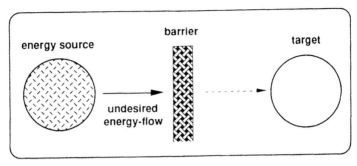

5.7.2.1 Consequence reduction

The consequence-reduction process is related to *barrier analysis* (Haddon, 1973; Trost & Nertney, 1985), whose basic structure is shown in Figure 5.7.1. Essentially, to prevent the level of risk from increasing, the target (in this case, a human being) must be protected from a potentially harmful energy source. This may be achieved in a number of ways, but principally there is a barrier either on the source, between the source and the target or on the target itself. This simple approach has led to a useful set of preventive design measures that can be considered and then implemented when designing a hazardous system, or when carrying out an error-reduction analysis. These are shown in Table 5.7.3.

Some of the approaches outlined in Table 5.7.3 would involve costly design changes, and hence these approaches are more appropriate to the *early* stages of the design process (e.g. the conceptual and preliminary-engineering phases). Others are more amenable to consideration at various different typical HRA stages, such as the detailed-design stage, or a later stage in the system-design life cycle.

Protecting the operator from exposure to adverse conditions or effects is best achieved by (further) isolating the source from the individual, or else, for example, by making the task remote instead of local, which may be possible via new hardwired or 'soft-wired' control systems. *Reduced occupancy* may be a worthwhile additional measure for consequence-reduction purposes, but this is unlikely significantly to affect PSA results since PSAs usually make conservative or even pessimistic estimates of occupancy levels.

A second approach to reducing the consequences of human errors is to substitute the human operator with a mechanised/computerised piece of technology – i.e. *automate* the process by making use of programmable electronic systems, etc.

A third approach is to use safer substitute materials, e.g. synthetic, *non-toxic* alternatives. This may not always be feasible, however, and would moreover require significant research and development to ensure that the

Table 5.7.3 Barrier-analysis examples (Koorneeff, 1990)

1. Limit energy	Speed, size Safer solvents Smaller quantities Lower-voltage tools
2. Prevent build-up	Fuses Sharp tools Gas detectors
3. Prevent release	Storage, mounting, Testing, tool rest Insulation Life line Guard rail/toe board
4. Provide slow release	Rupture disc Safety valve Seat belt
5. Channel away	Exhaust Electric earthing Aisle marking Restricted area
6. On source	Guard Filters Noise treatment Sprinklers
7. Between	Glass shield Fire doors Welding shields Interlock guards
8. On target	Goggles Protective equipment
9. Raise threshold	Acclimatisation to heat/cold Calluses Selection Damage-resistant material
10. Ameliorate	Emergency showers Job transfer Rescue Emergency medical care

alternatives were in fact safer to use and added no new, *more* dangerous failure paths to the logic trees that deal with risk.

A fourth approach to error-consequence reduction is that of *personnel protection*, via protective equipment. This is not as powerful as the other approaches, however, simply because it is not always used by operators.

5.7.2.2 Error pathway blocking

Error-pathway blocking involves redesigning the system in such a way that the error can no longer occur; a process which will involve hardware/software changes. Thus, for example, the error of failing to respond to an alarm can be made less significant by introducing an automatic trip into the system. If this is done and the alarm still occurs, the human's role is now to ensure that the trip has occurred, and the human action now involves a recovery path rather than amounting to a direct error that makes a significant impact on the system.

A second approach often utilised is the implementation of *interlocks*, i.e. mechanical devices which prevent an action from taking place (e.g. the door of a washing machine, which locks itself automatically when the wash cycle is in progress, thus preventing the error of opening the door during this cycle). Interlocks are used a great deal in the nuclear-power, chemical and offshore sectors of industry.

One danger with interlocks is that if they are perceived to interfere with the job, operators may be inclined to disable them. If a case does occur in which the interlock must (for safety reasons) be overridden, then it must be ensured that this can be done in a safe fashion – perhaps, for example, via the supervisor's password, which grants access to the relevant device. Most experienced operators will know a way of defeating an interlock if this becomes necessary, though the methods will often be unorthodox – and, in some cases, rather ingenious.

Error-pathway blocking may require the functional reorganisation of a system, and can even lead to an alteration of a given process at a fundamental level, e.g. via its *automation*. Such changes require a guarantee, that they will not lead to *new* kinds of maintenance error.

5.7.2.3 Error recovery enhancement

Error-recovery steps may be introduced into the fault or event tree if this is feasible. These may already have been identified in the Human Error Identification (HEI) phase or the task-analysis phase. An extra level of supervision, if properly implemented and rendered independent to as great an extent as is possible with human systems, may be expected to reduce the level of error by a factor of up to 10, for example. On the other hand, extra procedural checks, or the use of 'check-off' sheets, may be expected to have a reduction factor of only about 3. Training for 'self-recovery' is worthwhile in real systems, but in PSAs it will often be given little credit.

If a process of error-recovery enhancement is used as a major error-reduction strategy, the dependence factor must be given strong consideration. In many cases, the dependence relation between an error and a recovery step must be assumed, in the first instance, to be strong, until it can be shown that the relation is in fact a weak one. If the dependence relation between the two factors does remain strong, however, then a net reduction factor of only 2 will be gained.

In a PSA-driven error-reduction analysis, where the amount of risk due to human error is too great, a process of error-recovery enhancement will often not be enough to achieve the desired risk-reduction effect. In such cases, it will be used instead to 'fine-tune' the risk-reduction process. In real terms, however, the use of an error-recovery-enhancement process should always be given serious consideration by the HRA analyst, since, as a technique, it is often very easy to implement – for example, by slight modifications to procedures, and via team training. Since one of the main objectives in the field of hazardous industries is to reduce risk to as low a level as is reasonably practicable, error-recovery-enhancement measures will, in many cases, be advisable. Error-recovery-enhancement methods should therefore be a prime consideration even in cases where risk levels are satisfactory.

5.7.2.4 *PSF-based error reduction*

The SLIM, HEART, HRMS and THERP systems all quantify performance as a function of Performance Shaping Factors (PSFs), which can be reviewed to see which of them appears most important from the point of view of the quantitative analysis. If a HEART PSF is found to be important, then there will be a specific error-reduction mechanism (ERM) which can be implemented, and the analyst will then judge how far this ERM will reduce the EPC value in the HEART equation. If the SLIM or THERP are being used, the PSF of importance can be reviewed to decide what possible ERMs could be utilised to reduce this PSFs impact on performance. This all requires a certain amount of judgement and, possibly, some human-factors expertise. Once again, the analyst must judge how much of the reduction in the error's impact can be attributed to the ERMs being developed. With HRMS, the ERMs can be derived and specified by means of the PSF quantification questions, and the impact of any ERM is directly calculable via the HRMS quantification system, and thus does not require judgement. Furthermore, the ERA module in HRMS actively prompts the analyst to consider specific ERMs according to the PSF profile for that error.

5.7.3 Other approaches (based on Hale and Glendon's 1987 framework)

The following human-factors/safety-science approaches to improving reliability in process-control situations are built upon the framework by Hale and Glendon's (1987). Whilst there are many ways to improve the level of reliability, depending on the task's context, these four aspects are particularly useful for improving the level of human performance in relation to the control of systems in both normal and abnormal system situations.

5.7.3.1 *Increasing predictability*

Increasing the ability of the operators to *anticipate* problems is a useful additional recovery strategy, and one which may involve the implementation of predictor displays, early alarms, trend displays, etc. All such devices not only allow system disturbances to be avoided in the first place, but also allow the operators to perform one of their most useful functions for detecting impending system abnormalities: that of *pattern recognition*. Predictor and trend displays, in particular, also serve the function of keeping the operator alert and 'in the loop' rather than in a state where he or she is simply passively waiting for a system disturbance to occur, seeing if the system adequately copes with the abnormality.

5.7.3.2 *Enhancing detectability*

The detectability of situations can be increased in two main ways: by enhancing the ability to pinpoint an abnormality, and by enhancing the ability to diagnose what that abnormality is – by means of problem-solving and hazard-identification training prior to the processing of information. Both methods allow the operator to detect effectively the event at an *earlier* point in time.

Detectability may be enhanced via the provision of event-specific alarms or displays, or via generic displays which show that the system is moving (or has moved) outside of its specified operating modes. The latter types of display are most clearly illustrated in the critical-function monitoring systems (CFMSs) that have been implemented in the nuclear-power industry (e.g. see Carnino, 1988). These systems, rather than presenting detailed and hence complex pictures of the parameters governing the system, instead present a simple but powerful picture of the health of the *primary* safety parameters. In this way, the CFMSs can be seen to involve a very high-level safety display. One of the advantages of this type of display is that it can give operators a considerable amount of feedback in cases where, for example, they have misdiagnosed a scenario, and the scenario is now deteriorating rather than recovering as a result of their misdiagnosis. The CFMS concept (and a similar version called the safety-parameter display system) was developed in response to the type of misdiagnosis which occurred and persisted, despite contradictory feedback, at Three Mile Island.

5.7.3.3 *Increasing controllability*

Increasing the level of controllability involves increasing the power which operators have to recover systems. For example, the ability to avoid system-dynamics designs which lead to sudden, catastrophic and essentially non-recoverable kinds of failure, or the use of 'buffers' in the process cycle to render systems less tightly coupled, both make systems more controllable. As with error blocking and some barrier methods, this approach usually entails redesigning the system, which may be seen as a prohibitively expensive process once the plant or system is under construction. Sometimes, however, a system that proved hitherto difficult to control may be made more controllable via improved training and procedures. In such cases, it may be that a task-analysis study is required to determine how better to control the system – as has occurred in at least one case study (Rycraft *et al*, 1992). In other cases, it is likely that new control-systems engineering may be required, which will also require a new or modified control–display interface. In such cases, qualified human-factors expertise will almost certainly be required.

5.7.3.4 Increasing competence

Increasing the competence and confidence of operators at reacting to and controlling abnormalities can be achieved via training, and in particular *simulator* training. (In many cases, full-scope simulator training will not, however, be required (see Ball, 1991), and in some cases the training will simply amount to structured talk-throughs, or low-fidelity simulations or mock-ups.) Simulator training can be particularly useful in the context of three different types of training: refresher training (e.g. training for rare events, where the subjects should be refreshed every six months); team training (focusing on encouraging control and coordination, cooperation, compliance and competent responses to situations – see Ball, *op. cit.*); and problem-solving training (e.g. training in novel scenarios both to build up problem-solving strategies and to familiarise operators with the stress of uncertainty that can occur when they are faced with events at the edge of the design 'envelope' or beyond).

Such training methods are potentially cheaper alternatives to design solutions, but all three require more research (see Wilson & Kirwan, 1992). It may be that by failing fully to utilise training solutions to deal with problems of reliability, systems designers and operations groups are squandering the most powerful resource inherent in the system.

The above four methods are not so easy to include as a factor in PSAs, and in particular, the effect of an increase in competence may be difficult to take into account in PSA terms, unless a PSF-based Human Reliability Quantification (HRQ) technique is being used which actually includes training itself as a PSF.

One other important 'systemic' error-reduction mechanism is the feed-forward of PSA information into the training system, so that operators can be trained to deal with scenarios which, though infrequent, are often the focus of PSAs. The subjects involved can, in addition, be 'refreshed' periodically – for example, once per half year.

5.7.4 Error reduction strategies

A final but important point of interest concerning error-reduction analysis is the distinction between error-reduction mechanisms (ERMs) and error-reduction *strategies* (ERSs). The former address an individual error, while the latter addresses a single *ERM* that is applicable to a number of different scenarios within the scope of the PSA. An ERS concerned, for example, with procedures, would require the reorganisation and redevelopment of an *entire set* of operational procedures rather than just a single procedure.

Where an error-reduction analysis is being carried out for a number of scenarios, and for many errors, it is likely that an ERS approach may be more cost-effective than a host of individual ERMs. This is not a rule of thumb, merely a fairly important (in the cost sense) consideration for the system-project management sponsoring the PSA.

Summary

This section has briefly covered the subject of error-reduction analysis in HRA. The error-reduction process may base itself on one or more of several stages involved in the HRA process, once the impact assessment has highlighted errors for reduction via the task-analysis, error-analysis and quantification phases. As the need for error reduction increases, i.e. where the impact of human error is great and leads to an unacceptable level of risk, the error-reduction phase is more likely to to be applied to design factors, as well as procedural and training factors. In most cases, the recommendations arising out of the task-analysis, error-analysis and quantification stages will tend to converge, so that a coherent picture of what kind error-reduction technique is required can usually be built up as the analysis proceeds. If this does not occur, the analyst must either use his or her own judgement to decide what error-reduction mechanisms to recommend or seek the advice of a human-factors professional.

In many HRAs today, the error-reduction approach taken is based on the assessor's individual judgement in the wake of an impact assessment and a sensitivity-analysis of the results. In many cases, this is acceptable, since many practitioners are *also* human-factors specialists. However, a more formalised approach, as suggested in this section, does guarantee the auditability of the analysis, as well as its documentability (so that someone else can implement them if, for example, the analyst is only a consultant, and will not be involved in their implementation). These last two factors are dealt with in the next section, on quality assurance and documentation.

5.8 Quality assurance and documentation

Since quality assurance often relies on the checking of documentation, these two stages of the HRA process are clearly interrelated, and thus are dealt with together in this section.

5.8.1 Quality assurance

There are two major aspects to quality assurance (QA) in a HRA. The first is the assurance that a quality HRA has been carried out, i.e. that the objectives have been achieved within the scope of the project and without error, and the second is the assurance that error-reduction assumptions relating to improvements in human reliability are, and remain, warranted. These two aspects are dealt with below.

5.8.1.1 Quality assurance in the HRA

QA, when applied to an HRA, involves three general objectives:

- Ensuring that the HRA achieves its goals within their intended scope.
- Ensuring the accuracy of the qualitative aspects of the analysis, as well as the validity of assumptions made by the assessor.
- Ensuring the accuracy and consistency of the quantitative aspects of the HRA.

The first of the above cannot be dealt with in this book since it has to do with project management, and will not only be influenced by many factors but will also be highly context-dependent. (Nevertheless, it is usually the aspect of QA taken most seriously by management; some technical issues related to the first objective *are* therefore dealt with, in Section 5.9.)

The second and third objectives concern the validity, and the degree of realism, of the qualitative and quantitative analyses which cover the following components of the HRA process (depending both on the depth of the Human Error Identification (HEI) process and on which Human Reliability Quantification (HRQ) technique(s) are utilised):

- Problem definition
- Task-analysis information
- Identified human errors
- Screening
- Representation (involving logic trees and dependence relation)
- Quantification
- Impact assessments and error-reduction procedures
- Assumptions on the part of the assessor

Each of these (except the first, which is dealt with later, in Section 5.9) is considered below in terms of what types of QA problems could arise and what practical measures can be taken to overcome such problems.

QA AND TASK ANALYSIS

The task-analysis phase is a critical one since it forms the basis for all subsequent analyses, and so its importance should not be underestimated. A typical QA weakness which may occur during the task-analysis phase involves the development of a task description which is fine in theory but is not an accurate representation of how the task *is really carried out.* This drawback often manifests itself in cases where the assessor has no access either to the real system or experienced operators, and as a result conducts the analysis at a distance using only procedures and other such material. Such an eventuality should be avoided wherever possible, and all observations should be supplemented by interviews, reviews of incidents, etc.

A second problem (which applies to detailed task analyses only) concerns a task description that is inadequate, i.e. one which does not properly specify how the task is actually to be carried out. In the context of a hierarchical task analysis (HTA), this may lead to deficiencies in the plans, or in the redescriptions utilised. The former can best be avoided firstly by getting an expert both to carry out another review of the task domain and to comment on the HTA, and secondly by having another assessor see if he or she can understand how to do the task by following the plans rigorously.

QA AND HUMAN ERROR IDENTIFICATION

The most difficult problem in HEI is trying to be *comprehensive* with respect to critical errors. If this problem is to be avoided, multiple techniques, multiple independent assessors or group HEI approaches like HAZOP will have to be used. Using a tabular human-error-analysis (HEA) approach also enhances the degree to which an independent assessor can check the degree of effectiveness of a Human Error Identification exercise, by showing what errors were identified for each task step in the analysis. Ensuring the validity and comprehensiveness of the process of identifying misdiagnoses and rule violations will, however, prove far more difficult.

A common problem in HEA, and one that is hard to eradicate, is the biasing effect created by the assessor's own beliefs about the operational culture of a system, about what errors might occur and about what errors would seem incredible. Thus, for example, an assessor may, during the early stages of an assessment, regard certain errors as entirely incredible, and to amount to extreme rule violations, only to find out later that they *do* actually happen. This amounts to a mismatch between the assessor's perceptions of the system, on the one hand, and the system as it really is, on the other. In some cases, it is not uncommon for assessors to consider themselves as operators, and to decide if they themselves could ever commit that error. Such an approach is usually not good enough. It is better to study a representative range of operators, and always also worthwhile both to implement Critical Incident Techniques (CITs) and to carry out incident reviews/interviews, to determine more realistically what the operational culture is actually like. Some operators will denounce proposed errors as impossible and insulting, whereas others may tell you, aside, that such things have actually happened. Do not, therefore, rely on the testimonies of single operators alone.

SCREENING

Screening is also critical, since if any errors 'screened out' from further analysis are, in fact, *critical* errors, then the resulting risk calculations may underestimate the level of risk. Whilst coarse screening will be fairly 'safe' in this respect, fine screening must be given more scrutiny, as a fine-screening value for a skill-based error, for example, could underestimate the HEP if the PSFs in that scenario happen to be particularly unfavourable. The assessor should use his or her own judgement to decide if the fine-screening values *are* in fact pessimistic, and, if uncertain about the situation, should also apply a quick check (e.g. via the HEART) so as to gain more confidence in the matter. The person carrying out a QA function on the screening process should obviously focus on the numerical process underpinning screening, and should also check to see if the effects of dependence have been properly accounted for in the screening process.

REPRESENTATION

The question of the validity of the logic trees centres round several issues that are worth considering in a QA context. The first is whether the logic within fault trees is itself correct. An error which is not uncommon (and usually rapidly detected by an independent assessor) is that of using an 'AND' gate instead of an 'OR' gate, or vice versa. Since the former is the most potentially damaging to the validity of a risk assessment (since it leads to an *under*estimation of the risk level), the assessor and independent assessors carrying out QA should double-check that all the events under AND gates in the tree in question *do* need to occur together before the top or intermediate event above them can take place. Many fault-tree assessors have, at some stage, made this simple but dramatic mistake (and have then corrected it themselves, or had it corrected for them).

A second issue concerns dependence in logic trees, which is without doubt one of the more difficult areas to get right. It is best, when considering each pair of potentially dependent events, first to ask the simple question 'Does event A have anything to do with event B?' and then consider the validity of the following equation with respect to the two events being compared:

$$P(B/A) = P(B)?$$

Dependence in *event* trees is usually easier to detect, because event trees contain information in a *sequential* form. When looking for dependences in fault trees, therefore, it is useful to convert a cut-set, or part of a fault tree, into event-tree notation. Dependences often then become apparent. If, on the other hand, there are in fact no dependences, the assessor may well find it difficult to construct the event tree, and it is this, indeed, which tends to indicate the *absence* of dependent relationships. This is not a *guaranteed* indicator, however, since in cases where all events are highly independent but non-sequentially related, the same difficulty in event-tree construction will arise.

The use of HPLVs or other CMF-limiting-value or beta-factor methods should be checked by an assessor carrying out a QA function. The assessor should beware of generating logic structures which result in an HPLV dominating the tree (suggesting too much uncertainty in the analysis and the need for more detailed modelling). Similarly, having *two* HPLVs in the same cut-set, which literally compounds uncertainties and renders the resultant validity of the cut-set somewhat indeterminate, should also be avoided.

QUANTIFICATION

Whatever quantification method is utilised, an independent assessor must check all the calculations involved in the HRQ procedure. It is also highly useful to consider the level of consistency among experts wherever possible – for example, formally, within the APJ and PC approaches, or informally, within the SLIM. In such expert-based approaches, it is useful to have the experts involved generate confidence statements about the resultant HEPs.

In the context of non-expert techniques, the QA-checking assessor must look for inconsistencies in the application of nominal HEPs, as well as in PSFs/EPCs. This involves reviewing both the PSF information and the assessor's assumptions underpinning the quantifications and then comparing this information with the data governing the application of the nominal HEPs and PSFs/EPCs in each HEP assessment. The assessors themselves, when carrying out quantifications, should beware of doing too many assessments in one day, since there may be a shift in the way in which, for example, EPCs are applied as the day wears on, and as the assessor grows tired. The QA assessor can look out for such shifts, which may manifest themselves in the form of a decreasing number of EPCs being considered, or a decrease in the number of assumptions and comments on the part of the assessor that are being noted. Two ways to avoid this are to limit the quantification period to, for example, two 2-hour periods in a day (as is sometimes done with HAZOP studies, which can be mentally exhausting) and to carry out each calculation on a pre-prepared calculation sheet, noting both the time the calculation was commenced and the time it was completed. This gives the assessor direct feedback about his or her own progress. What's more, any speeding up or slowing down of procedures will be noticed. This same fatigue/time-variability aspect also affects groups of experts. In such cases, the facilitator should not drive them on but instead introduce a break.

Another point of interest concerning the QA aspect in relation to the SLIM and PC techniques has to do with the calibration equation. It is preferable to utilise at least *three* calibration points in connection with either technique, so that any deficiencies in the calibration process become clear.

Lastly, an independent assessor may also wish to ratify a number of the HEPs. This can be done in two ways. The first is by actually carrying out a set of independent HEP calculations, preferably without knowing what the original assessor's estimates are. The second is simply to compare all the HEPs with the generic guideline data (specified earlier, in a table in Section 5.5) to see if any HEPs appear significantly out of line with the guideline figures (e.g. a complex alarm response with a HEP of 1E–4 is suspicious enough to warrant a QA investigation).

IMPACT ASSESSMENT AND ERROR REDUCTION

The impact-assessment phase is primarily driven by a quantitative analysis of the logic trees, which will obviously be checked by the PSA assessors. The interpretation of the results is a difficult business, and deciding what strategies to adopt for reducing the level of risk should be a group process, and an *iterative* one at that. This is because, often, as assessments are carried out, ERMs become apparent, or are suggested by operators, which may not, however, be seen as the most effective after all the calculations have been completed and all the events and event combinations involved in the impact assessment have been evaluated. It is therefore important not to become biased about which ERMs should be implemented. Often, PSA results contain a few surprises: e.g. the error everyone was concerned about turns out not to be as dominant as expected, or an error thought to be of minor importance actually occurs in so many cut-sets that it becomes a prime target for an ERA. Sometimes, also, even when a large and detailed HRA has been carried out, the human-error factor turns out not, in the end, to have any great influence on the level of risk, so that no ERA is necessary. Naturally, if the results are a total surprise to all analysts, then this clearly needs investigating, as it suggests something has gone wrong in the logic-tree-evaluation process.

The error-reduction QA process involves two activities, namely the selection of an effective ERM to combat the error, and the effective implementation of that same ERM (or ERS). (The latter part is dealt with in the next subsection.) The determination of an effective ERM is often not an easy task. The QA of ERMs, i.e. the business of guaranteeing their effectiveness, is an important task. The QA assessor should ask two basic questions for each ERM:

- How does this ERM prevent the error (does it tackle its root cause?)?
- Does this ERM introduce any *new* error potential?

The QA assessor should also have documentation on each ERM concerning how it was derived and from which part or parts of the HRA process it originated (ERMs which appear to have popped out of thin air should be treated with suspicion). The next subsection considers a major aspect of QA, that of ensuring that ERMs/ERSs are effectively implemented.

THE ASSESSOR'S ASSUMPTIONS

Throughout the HRA process, the assessor is likely to need to make a large number of assumptions, particularly if the system is still in the design stage.

It is important both that all such assumptions be documented, and, particularly when they underpin sensitive (i.e. risk-sensitive) HEPs, that ways of confirming these assumptions are sought – which can be done either by trying harder to track down factual data that would substantiate the assumptions or by gathering together, in a formal setting, a group of experts whose opinions will help to ratify such assumptions.

If assumptions are not recorded, then the assessor carrying out the QA function will find it difficult, if not impossible, to do more than a superficial job. The QA assessor who finds him or herself in such a position should go back to the original assessor and demand more information, or even demand that a new assessment be carried out. In either case, the QA assessor should *not* 'sign off' the QA as having been carried out until more information has been received that does full justice to the QA job process. Although the QA assessor does not, of course, review the whole analysis in depth (though every single calculation *should* be checked), it *should* be possible, on the other hand, to review individual *segments* of the analysis in depth.

5.8.1.2 *Quality assurance in the implementation of ERMs*

This subsection gives an outline of some of the various aspects of human-work-systems design which need to be implemented if it is to be ensured that an error-reduction procedure will not merely be theoretical but will achieve its intended impact.

Having identified ERMs (and, perhaps, ERSs), the next step is to ensure that their implementation will have the desired effect on the level of performance of the system. This may not be as straightforward as one would at first sight imagine, because interventions in systems designs occur not only in a technical environment but also in a psychosocial one. Thus, for example, there is no guarantee that new error-reduction technology will be successfully integrated into an existing system, for a number of reasons: retraining will be required; personnel's responsibilities and degree of perceived autonomy will shift, possibly in directions they do not desire; a lack of consultation with personnel affected may lead to feelings of distrust and resentment, etc. Such problems can seriously undermine the assumption that an ERM will work.

Furthermore, ERMs and ERSs may have an initial impact which subsequently subsides without achieving the expected level of improvement in performance. Some ERMs/ERSs do, indeed, appear to be shorter-term than others, and this is due to the fact that the changes created by, for example, training, selection, or motivational interventions in human-systems designs or redesigns can often be remoulded, or can often simply 'decay', back to their original state as a result either of pressure from the environment or, simply, of a decline in motivation. This is why, in particular, safety motivational programmes tend to have a short-lived impact on performance. Once the novelty has 'worn off', the standard of performance slides back to its pre-intervention levels. Many companies, in response to this problem, have invoked the use of *continually changing* safety-awareness programmes, as a means of keeping a high level of safety awareness while preventing the novelty from wearing off.

Certain ERMs/ERSs may aim to achieve a level of performance which is beyond maintainable limits. For this reason, it is quite unfair to expect a *100 per cent* performance, even in cases where tremendous motivational programmes have been implemented (as was the case with the early USAF 'Zero Defects' programme, which failed to achieve *this* goal). Similarly, training programmes aimed at overcoming an inherently difficult design-display problem will also fail, as will asking the human operator to perform beyond human-speed/-accuracy/-discrimination capabilities. Furthermore, an exhortation to adopt safe working practices (e.g. the use of protective clothing) which either interfere with the task itself or contravene social variables such as status, pride or even 'machismo', will make only a limited impact, and will generate conflict within the system's organisation which could even lead to *new* problems and errors worse than those which were the target for reduction.

The solution to this problem is to adopt a 'systems' approach to error reduction. What this approach does is convert any error-reduction mechanism into an error-reduction *strategy* which affects other human-performance areas as well as the one which is the focus of the mechanism (for example, the changing of a display, even when seen as an error-reduction *mechanism* rather than an error-reduction strategy, will still also have an effect on both training and procedures systems. The impact on the task's design structure, and hence the organisational structure around the task, must then be investigated, and altered if appropriate. If the changes made to the nature of the task are significant, then a process of reorganisation, retraining and possibly even reselection may be required.

Thus, although an original ERM intervention, derived from an analysis of the HRA's results, may appear to occur in only one of the 'human system' design dimensions, human performance is such that *all* potentially related aspects must be considered.

In addition to the considerations of parallel effects in other human-system-design dimensions, psychosocial variables must also, as was pointed out earlier, be taken into account. An intervention is more likely to be effective if those who will be affected have been consulted and hence know that they will be involved in the forthcoming changes, and even in the choosing of the method of intervention. If this is the case, the involvement of any other psychosocial variables associated with the intervention will have most likely been detected. Thus, for example, concerns about the impact of new technology on traditional roles and skills (and hence on the matter of *status*), will already have been voiced, and then rendered acceptable by whatever means.

The easiest way to achieve this systems approach to error reduction is to involve operations personnel right at the outset, even to the extent of granting them a degree of involvement in the HRA process itself (as could be achieved, for example, by using expert-judgement approaches for the identification or quantification of errors, as well as for task analyses). This is not a particularly revolutionary idea, as operations personnel are likely to have already been involved in other parts of the PSA, e.g. the HAZOP and the detailed PSA stages; and their involvement in the HRA itself would be quite logical, given its subject matter.

The quality assurance of ERMs/ERSs is thus a critical business, since, if it is not carried out, the assumed benefits will either never materialise or *will* do so but only briefly.

The impact of reduction measures on human performance, and human-system performance in general, must also, furthermore, be monitored, throughout the life of the system, via an incident-reporting system and via periodical PSAs. The area of incident investigation and analysis (which is outside the scope of this book) is an important QA approach, since it gives real feedback on the effectiveness of the ERMs/ERSs, as well as on the validity of the PSA as a whole. This area is considered further in the next subsection, and in Section 5.11.

5.8.2 Documentation

The documentation of the HRA should satisfy the following requirements:

- The HRA should be auditable by, and understandable to, an independent assessor.
- All assumptions made by the assessor during the assessment should be documented.
- The HRA should be repeatable.

Auditability

Auditability is becoming of increasing importance in highly regulated industries such as the nuclear-power industry. It is therefore essential that assessors ensure that they leave an 'audit trail', so that other, independent assessors or regulators can check their work and ensure its quality. This applies even in cases where expert-judgement techniques are being used. A non-auditable assessment currently carries little weight in many, if not most, of the hazardous industries.

Assumptions

Assumptions have already been discussed in the previous subsection, and the importance of their documentation is well-known to experienced practitioners. Assessors tend to underestimate how much of the system's detail they have actually come to understand and know during the assessment process. Many important assumptions on how the system works in theory, and more importantly in practice, become unconscious, or taken for granted once acquired, and as a result, when such assessors come to document their rationales of assessment, they forget how much they did not know at the outset of the assessment process, and so do not record everything that was acquired during the assessment. This causes great problems for another assessor who is trying to apply QA to the work, or who is trying to repeat it five years later when the original assessor is no longer available for consultation. The documentation of assumptions therefore best proceeds by having an independent assessor review the documentation as it is produced, asking clarification points as it proceeds, and thus enlarging the set of assumptions.

Assumptions should also clearly be things that are *assumed*, rather than factual data. Otherwise, there is a danger (in PSA and engineering as a whole, not just the domain of HRA) that assumptions made about operational practices, which may be necessary in, for example, the design phase, are not queried or updated later, when operational experience has actually been gained. It is therefore necessary that assumptions be literally documented under a heading 'assumptions'.

Repeatability

A further reason for ensuring that the HRA is fully documented is so that later PSAs can utilise the results and interpretations of human-error analyses without having to start 'from scratch'. Thus, whilst future PSAs/HRAs for a system do not, of course, want simply to copy and superficially update already existing HRAs for a system (because new information will have been gained on operational experience, and hardware/organisational/operational changes will have occurred), such future PSAs *should* endeavour to reap the benefits of hindsight: seeing what was predicted and understanding *why* it was predicted, as well as why other scenarios were thought unrealistic, etc.

Living PSAs/HRAs

In addition, the assessment should ideally be seen by the operations personnel as a document whose use will extend over the whole lifetime of the system itself, rather than as a document that is simply put in the archives once its immediate purpose has been served. For example, information about both task-performance PSFs and expected error probabilities (and how to reduce them) may be useful not only to those who have to analyse incidents but also those who wish to improve performance levels. The central concept here is that of the 'living PSA'. Such a PSA, not to mention the HRA itself, acts as an information source, for operations staff and for management, during the whole of system's lifetime, and in addition allows the *overall* course of the system to be charted against PSA predictions, and the PSA to be regularly updated as the system get older. A PSA represents one of the most detailed investigations of system-performance problems ever to be undertaken for a system, and it is sensible, given this fact, to utilise such an investigation to the full.

When the living-PSA concept, or simply the concept of making a HRA useful for future users, is adopted, documentation becomes not merely a chore, a mere means of making an 'audit trail', but rather a way of installing information into the 'corporate memory' of the company. After all, it is likely that a PSA-predicted accident, should it occur, will probably take place many years after the PSA/HRA analysts (and the original designers) have moved on to other pastures: the recipients of their legacy will probably need all the help they can get. And what's more, if good documentation is maintained, then when such incidents do occur, PSA/HRA predictions *can* then be validated (or invalidated), which not only makes the predictive process more accurate but also perhaps even allows the generation of useful human-error-probability data (see Appendices II and III).

In summary, QA and documentation are critical phases, ones that make the difference between a HRA that is merely a table-top exercise and a HRA that is realistic, and remains useful and accurate throughout its predictive lifetime.

5.9 Problem definition revisited

It is difficult to explain some of the difficulties involved either in problem definition or in the setting of the scope of the HRA, without first getting the reader to understand the techniques, and their data requirements, used in HRAs generally, as well as the limitations of available HRA technology. Therefore, although the problem-definition phase logically *precedes* the HRA, there are certain aspects of HRA problem definitions that can more clearly be discussed *after* the techniques involved in a HRA have been reviewed. As a result, problem-definition issues have been split, in this book, into general problem definition issues, dealt with in Section 5.1, and, on the other hand, issues particularly relevant to large PSAs and certain HRA techniques – which are dealt with in this section. The preceding chapters have therefore assumed that the scope of the HRA has been fixed, and that, in general, a fairly detailed level of HRA is being described. This section discusses, in particular, HRAs that quantify performance at a *task*-based level – a kind of HRA that was alluded to in the section on HRQ techniques and which is utilised in some large PSAs, but one which also has ramifications, in particular, for detailed error-identification and error-reduction processes. This problem-definition issue, and two subsidiary ones, both of which affect the scope of the HRA, are dealt with below.

5.9.1 Task-based HRAs/PSAs

If the HRA is being carried out as part of a large Probabilistic Safety Assessment, then one major consideration may be whether there is a requirement for a detailed task analysis and a detailed Human Error Identification procedure to ensure that all relevant errors have been appropriately identified and can be incorporated into the risk assessment. In many cases, the PSA or Probabilistic Risk Assessment (PRA) will have identified areas of potential plant or system vulnerability, extracted the human-involvement factors from these areas and then proposed these same areas for a HRA. The HRA can start by describing the tasks and telling the operators what they have to do to perform adequately, for system-success purposes. Then, instead of carrying out a detailed error-identification procedure, the HRA can assess the entire task in terms of its success/failure likelihood.

What makes all this possible (at least in theory) is the fact that, as was seen in the chapter on quantification, some techniques (e.g. the HEART, APJ and SLIM approaches) are able to quantify human performance at a relatively high task level, whereas others instead break down the task into discrete operations and individual errors, quantifying how each of these affects the total success/failure likelihood for each task. The latter approach is known as a *decompositional* approach, the former a *holistic* approach. The main advantage of the decompositional approach is that the contribution of every potential error is individually considered. The primary disadvantages with this approach, on the other hand, are firstly that it is resource-intensive and secondly that many of the errors that are identified may make little or no impact on the system's assessed level of risk. The advantage of the holistic approach to HRAs is that it is potentially less resource-intensive. The major disadvantages, on the other hand, are firstly that there is always the risk that a significant error which could make a great impact on the system's risk level may be missed out of the analysis, and secondly that, as a result of this first drawback, the error considerations will not be detailed enough to act as a base for error-reduction measures/strategies. Some would also question whether such a holistic approach, wherein a single task HEP would aggregate the effects of *many* undecomposed HEPs, could be said to be accurate at all.

The normal approach for a holistic HRA, driven by a PRA, is therefore to carry out a task analysis and quantify the task's likelihood of success directly. In addition, the analyst who carries out such a task-based PSA would also carry out an analysis of the task in question in order to identify any significant errors, such as errors of commission or latent failures or misdiagnoses, that would not ordinarily be associated with task-failure considerations, and then take these into consideration when quantifying the task's success-likelihood – even quantifying their likelihood, if necessary. Furthermore, if the system's level of risk is found to be particularly sensitive to one particular task, then the task analysis may be rendered more detailed, and a more detailed error-identification procedure may thus be carried out.

The holistic approach to the HRA is therefore primarily PRA-driven, entailing task analyses and the direct quantification of task success-likelihoods, and taking into account any significant error opportunities. If the risk element is particularly sensitive to *human* errors, then a more detailed task analysis

Figure 5.9.1a The holistic HRA approach to human-error descriptions

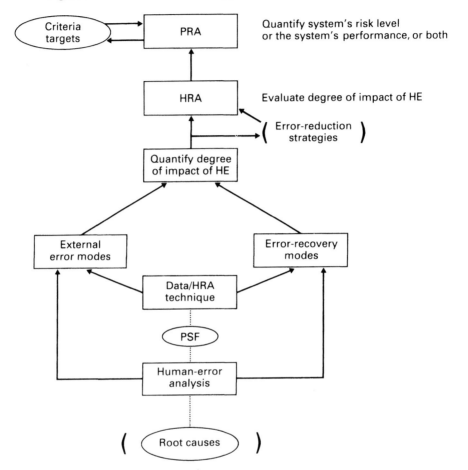

and a more detailed error-identification procedure may be carried out to ensure that the risk level is accurately assessed. The differences, described earlier, between the holistic approach and the decompositional approach are shown schematically in Figures 5.9.1a and 5.9.1b. It is part of the problem-definition phase of a HRA to decide which approach will be taken, and the decision criteria here usually includes the system's level of vulnerability to human error, the size of the PSA (larger PSAs will tend to go for task-based approaches so as to avoid massive fault trees and event trees which may be difficult to evaluate and interpret), the extent to which the HRA is PRA-driven and the amount of resources available.

Some HRAs may be so PRA-driven, however, that only a brief task analysis or consideration of significant human errors occurs, the HRA instead simply quantifying the likelihood of success or failure of those tasks identified

Figure 5.9.1b The decompositional approach to human-error descriptions

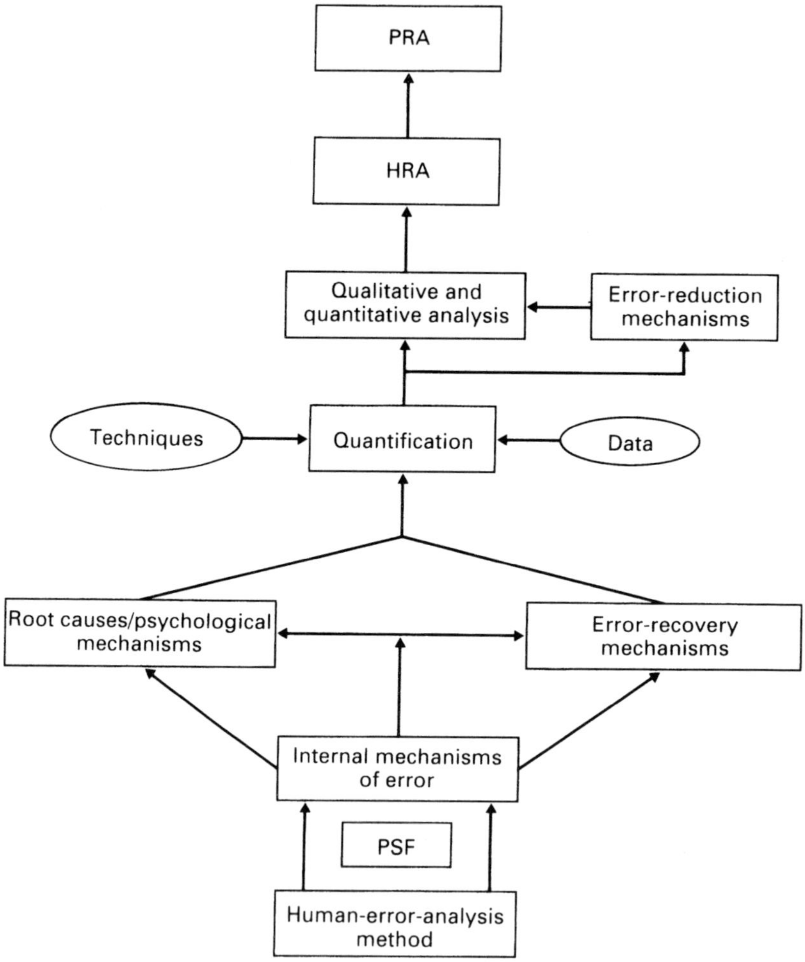

by the PRA as involving important human actions. In such cases, the HRA assessor is really not being asked to identify significant involvements or errors on the part of the operator, nor to carry out a task analysis of those that have been identified in the PSA. This kind of approach was particularly evident a decade ago in the chemical and petrochemical risk-analysis areas, when the HRA was not seen as central to risk analysis, and was largely associated only with human-error/-performance quantification, and not with task analyses, critical-task identifications, or error identifications. The problem with this approach is that there may be other important human involvements which will not be identified via this route because of their complex interactions with other hardware and software failures. Risk analysts developing their fault and event trees, and reviewing operational experience, may well, on the other hand, succeed in identifying many, even most, of the critical human impacts upon the level of risk. Even in the latter case, however, a certain percentage of critical human involvements may still nonetheless be omitted from the analysis, and the full potential impact of human error in a PRA-identified critical human task may still therefore be underestimated, in the absence of a proper task analysis.

The decision of whether to take the *PSA* perspective on the important human errors as the *final* perspective, thereby reducing the importance of the task analysis and the error-identification process, is not one on which guidance can be given here. The scope set for a HRA is a critical determinant of its degree of accuracy, its usefulness and the kind of impact it makes, and it is likely that if a system is potentially hazardous, complex or novel, and has a high degree of human involvement, then it will be beneficial to carry out a *broader* initial task and error analysis, to ensure that all significant errors are ultimately captured in the assessment. If resources allow, then a decompositional approach is to be preferred here (note that the HEART, SLIM and APJ approaches can operate in either mode), since this technique involves a more detailed modelling and understanding of the human contributions to risk, and of how to alleviate them.

5.9.2 Qualitative HRA outputs

There are a number of potential HRA outputs, and the desire for these affects the final scope the HRA approach that is adopted. The principal potential outputs are:

- Human-error probabilities
- Critical operator tasks (and associated task analyses)
- Significant errors
- Timings for the operator's handling of events
- All potential errors
- Error-reduction mechanisms or error-reduction strategies (or both)
- Information to be fed into training and procedures systems
- Information to be fed into a performance-monitoring system

The above have been listed in approximate order of availability according as the HRA adopts a more qualitative basis. Thus, for example, a resource-restricted, PRA-driven HRA for a non-novel system with a low level of vulnerability to human error may only yield HEPs, whereas a less resource-constrained one will also yield both critical operator tasks and significant errors. A detailed HRA will also be able to identify *all* potential errors concerned, and will be more likely to yield error-reduction mechanisms or error-reduction strategies or both, as well as information which can feed forward to become part of the training and procedures content. And a full, or top-level, HRA can also feed forward information to a performance-monitoring system.

Therefore, if error-reduction measures are desired, more emphasis will be put on the need for qualitative analyses (e.g. detailed task analyses, or error analyses – or both) within the HRA, and the HRA will take on less of a purely quantitative, PRA-driven character.

5.9.3 Comprehensiveness, risk and the scope of the HRA

The main problem facing risk and human-reliability analysts may be explained by using the analogy of a spoked wheel (Ackroyd, 1992; personal communication). In this analogy, the accident scenarios and sequences are seen as the spokes spreading out from the centre of the wheel, while the area covered by the spokes represents the risks assessed. Now ideally, the analysis would cover the *whole* of the area inside the wheel, but this is rarely possible because of problems and limitations connected with tractability, assessment technology and resource levels. It is therefore adequate if the spokes cover enough space for *most* risks to be considered, and if, furthermore, that risks that are omitted are essentially quite close or similar in nature and frequency to those considered. If this is the case, then the spokes in the wheel will have a fairly close and even arrangement, which will make the wheel stable. If, however, the wheel either does not have enough spokes, or else has them unevenly distributed (which suggests that certain risk-contributing areas are being omitted from the analysis), then the wheel may collapse, which equates to the risk analysis grossly underestimating the system's failure-frequency.

An additional problem for risk and human-reliability analysts is that there are many assessable errors or failures in performance which, because of the design or because of the level of error-tolerability inherent in the system, will have *no*, or a *negligible*, effect on the system's inherent risk level. Such areas of human involvement may serve to distract resources away from the errors that are *more* significant, in risk terms. Many such significant errors or critical operator tasks will be readily identifiable, and probably will already have been identified by the PRA analysts. However, a number of significant errors are unlikely to be detected without a detailed task *or* error analysis (or both). Furthermore, there will be errors which may interact in complex ways with hardware failures or environmental events (or both) to form new failure paths which prove very difficult to predict. Lastly, there may be unusual kinds of error or rule violation which either will not be identified, because they are seen as involving aberrant behaviour, or will be thought to be too infrequent to be worthy of consideration. These latter two types of potential human contribution to risk levels will be difficult to identify, and so will tend to be absent from the risk analysis. It is therefore up to the individual assessor to endeavour to assess even these types of errors, to ensure the comprehensiveness and accuracy of the assessment.

Human behaviour is effectively infinite in the forms it can take, whereas resources and imagination in the context of risk assessments are always limited. All that can be done with respect to these latter two types of error form is to keep assessments erring on the conservative (slightly pessimistic) side, and to make judicious use of HRA approaches to quantifying modelling uncertainties (such as HPLVs). In addition, if the PRA and the HRA provide readily available means of reducing the risk factor, such measures, if reasonably practicable, should always be adopted, even if they are not technically required according to the results of the PRA. This is because PRAs and HRAs are inevitably never as accurate as risk assessors would wish, and as a result it is better to install defences and reduce vulnerability levels than rely solely on figures, which involve a fair level of uncertainty in their make-up.

The regulatory bodies for an industry may also have a particular requirement or set of requirements concerning PRAs or HRAs (or both) which will influence the level of HRA adopted. In general, such bodies are likely to specify guidance notes on the best or the minimum acceptable practice – or at least a critical comment on the different approaches available (e.g. ACSNI, 1991). In specific cases, however, they may stipulate particular requirements for the PSA/HRA for a specific system design. Such general guidance and specific instructions will both have to be adhered to by those orchestrating the risk assessment. As a result, regulatory opinion, formal or informal, is potentially a strong constraining factor when it comes to determining the scope of the HRA.

Other human-error impacts that can, potentially, be included in a risk analysis are the effects of management action on the risk level, on the one hand, and sociotechnical errors on the other. These have been defined elsewhere, and are beyond the scope of current HRAs, though approaches do exist for assessing their gross potential impact (see Section 5.11). If it is desirable to include these in the scope of the HRA in question, then, at present, implementing a performance-monitoring system will prove the most effective way of mitigating their effects.

Three 'levels' of HRA are defined in Table 5.9.1, namely a full in-depth analysis of all the operators' interactions, a task-based PSA-driven approach and a purely qualitative approach. This table may help assessors and PSA project managers to decide the level of HRA to implement. In an ideal world with unlimited resources, a full detailed analysis would be carried out in every PRA, and this would ensure a maximum degree of modelling and understanding of the human involvements in a system, with the least chance of omission of any significant human-error contribution to risk. Nobody, however, lives in such an ideal world.

5.10 A practitioner's framework

This section attempts to summarise and merge the results of preceding sections into a practical framework for the human-reliability practitioner. It builds on the foregoing sections, focusing on particular, previously presented tables which are useful for deciding what approach(es) to adopt for a particular type of application. Two new tables are also presented, the first giving a quick summary of the main factors to consider, plus examples of techniques to

Table 5.9.1 Level of HRA

Level 1: PSA-driven task-based HRA
This level of HRA is the most basic. There is no task analysis, and the required human-error inputs are derived solely by means of the PSA. These human-error inputs are basically task failures (e.g. the failure to achieve recirculation) that have not been decomposed into individual actions (e.g. 'Fails to close valves XYZ1, ABC2, etc'). A task-quantification method may be used (e.g. the HEART, SLIM or, APJ), and very few error-reduction procedures will be implemented. Level 1 is included simply because it exists in practice, rather than because it is necessarily adequate. This author recommends that a Level-2 analysis, at the very least, be carried out for any PSA work.

Level 2: PSA-driven task-based HRA with targeted task analysis
This level of HRA is similar to Level 1 but with one significant exception: for a set of pre-identified scenarios involving important human error contributions, task analysis is now carried out. The scenarios themselves will still be identified by the PSA, but the task analyses will give far more meaning to the quantifications carried out, and moreover will make error reductions more achievable and more effective. The critical ingredient in this level of HRA is the correct choice of scenarios for in-depth qualitative analyses. The scenarios chosen should be representative of the total set of human actions being assessed in the PSA.

Level 3: HEA-driven HRA within PSA
This level of HRA starts from the problem-definition phase and proceeds through the HRA process. Unlike Levels 1 and 2, it is the HRA itself that decides what errors should be quantified, via HEA, and quantification generally takes place at the *action* rather than at the task level (the THERP and other similar techniques can therefore be used here). The advantages of a Level-3 analysis are its degree of comprehensiveness when it comes to error identification and the subsequent degree of effectiveness of the error-reduction mechanisms and strategies derived and then prioritised via the PSA. One negative aspect, however, is its extensive use of resources. When evaluating a novel system with a high risk potential and with human involvements, this is, however, the level to aim for.

Level 4: Qualitative HRA
This level usually occurs when a human-performance problem has been highlighted by a particular incident, or by production problems, in cases where, moreover, a PSA framework is felt to be unnecessary, or even undesirable. The HRA process is utilised, except for the quantification and impact-assessment stages, and the representation stage is also usually implemented, since it is the identification both of the causes of the problem and of the error-reduction potential that is the objective of the study. The quantification of HEPs can also occur, as a way of prioritising ERMs.

---

apply, at each stage of the HRA process, and the second table adding the dimension of a system-design-life-cycle stage to the practitioner's selection of approaches to decision-making.

It should be stated at the outset that this section can only give *approximate* guidance and not hard and fast rules on exactly which techniques to use. This is because of both the large range of tools available and the inherent flexibility of the HRA process, and also because not all of the possible permutations of techniques have yet been tried out. In any event, the practitioner or company, adopting a HRA seriously as a long-term approach, will try different techniques, and will develop preferences for particular techniques or combinations of techniques. Such preferences will hopefully fit into the framework outlined in this book and in this section, but if they do not, that does not necessarily mean they are incorrect.

The following therefore reconsiders (in brief) the main aspects involved in the different stages of the HRA process, relating back to previous tables where appropriate, or to the two new tables in this section.

5.10.1 Problem definition

It is of primary importance to define the problem being evaluated, as well as, in particular, what is 'driving' it – i.e. whether it is part of a PSA (in which case, scenarios may be pre-identified by the PSA for the HRA), whether it is part of a vulnerability analysis (in which case, in-depth techniques capable of error reduction should be used) or whether it is a particular human-performance problem – one targeted for analysis. The most useful table for this purpose is Table 5.9.1, which summarises four 'levels' of HRA.

Other considerations include whether the top events generated by the PSA adequately account for the impact of human error on the system-risks being assessed (if not, new top events or event sequences must be identified) and whether maintenance errors are being included in the analysis (they should be, generally). If new top events need to be identified, reviews of past performance can usefully be undertaken.

It is also generally advisable, if carrying out a risk assessment in the field of engineering, to work with *other* risk assessors (RAs) as this hybrid approach will make the identification and definition of different scenarios a more comprehensive process. Although some degree of independence is necessary, if the HRA analyst is to be seen to be extending the scope of the PSA, this may require the HRA analyst to look at the PSA.

5.10.2 Task analysis

The flowchart in Figure 5.2.15 is the most useful summary of the task-analysis (TA) techniques available for HRAs. It shows what techniques are appropriate, given the objectives of the analysis. Where several techniques are available, the practitioner will then have to decide which particular technique to utilise, on the basis possibly of resources considerations, ease of learning or application, etc. Whatever the approaches selected, the practitioner will need to apply techniques of data collection and representation as a minimum for HRA (or any other) purposes.

A data-collection task analysis can be based on procedures, on the observation of personnel carrying out tasks, on walk-throughs or, in a top-level way, on the PSA itself. These approaches can then be used to develop, as a minimum representational format, the linear, 'nested-loop' task analysis, or they can be expanded to create a more detailed task analysis in a hierarchical, tabular or tabular-scenario-analysis format (in increasing order of both resource-use levels and the level of detail involved).

It may be useful at this stage to consider which tasks are skill-, rule-, or knowledge-based, especially when using techniques such as the SLIM for quantification purposes. The modelling should be at a level that reflects the degree of human involvement in the scenario, so that a complex problem-solving situation is not treated simply as a single act but is modelled in more detail, showing the depth of problem-solving required, and also whether the situation is ambiguous, unexpected, etc. Thus, if there is a *knowledge-based* potential for error, the more *cognitive* task-analysis formats should be considered.

If the HRA is primarily qualitative in nature, then a tabular human-error analysis, as well as a prior HTA, will probably dominate the representational task-analysis formats utilised.

5.10.3 Human Error Identification

The first question in this phase, if this question has not already been answered during the task analysis, is whether there is a cognitive-error potential, and the flowchart in Figure 5.3.16 should be used to help answer this question. The next question is what level of human-error analysis (HEA) should be adopted, and to answer this question, the onion framework, shown in Figure 5.3.14, should be consulted. Then, it is necessary to review the human-error-analysis modes and interactions in Tables 5.3.16 and 5.3.17 to see what techniques are available for each level of analysis. These tables can then be used to decide which specific technique(s) will be used, based on the level of resources available and other criteria.

In general, when dealing with skill/rule-based behaviour, the predominant kinds of behaviour when it comes to HRA, THERP can be used to identify external error modes (EEMs) fairly quickly; otherwise, the SRK approach, or the SHERPA system, can be used. A somewhat underestimated approach is the group, or human-HAZOP, approach. Whichever of the above are utilised, a HEA table should be the final result, on which consequences and potential-recovery opportunities can usefully be recorded as the HEA progresses. Human-factors (HF) deficiencies and potential ERMs can also be noted at this stage; they may save time later on in the HRA.

5.10.4 Representation

The Probabilistic Safety Assessment (PSA) can be a determinant of which of the analyses, i.e. the fault-tree or the event-tree analysis, to use, if the nature of the scenario does not suggest the use of one of these in particular.

If a quick method of representation is required, then the analyst can first identify EEMs in fault trees or event trees and then simply 'fill in' the PSA gaps by quantifying the EEMs during the next stage of the HRA process and add in human-performance limiting values (HPLVs). If a more intensive approach is being undertaken, then the HRA will drive the human part of the PSA, and may also generate sub-trees whose top events or end probabilities feed into the PSA 'gaps'. In addition, a misdiagnosis may be considered, even if not previously identified in the PSA, and actions making the situation worse can be introduced into cut-sets and event sequences.

Table 5.4.2, which summarised the relative advantages of the major methods of representation (fault, event and HRA-event trees), can be used for selecting an approach or group of approaches. In general, HPLVs should be utilised if cut-sets are below 1E−3 in value. The only currently accepted model of inter-event dependence is the THERP dependence model.

It is useful, when carrying out representation, to consider how far down in the fault tree the errors occur. If they are low down, they are often not of importance, quantitatively speaking. If this is the case and yet the scenario also has a high degree of human involvement, consider if there are any errors missing from the upper levels of the tree. Also, check that 'AND' gates *are* really such and should not, in fact, be 'OR' gates. Look for dependence relations between errors; and bearing in mind all the above constraints and exhortations, retain a representation which is at least moderately scrutable, or else the 'wood will not be seen for the trees'.

5.10.5 Quantification

The PSA, which we are assuming is quantitative in nature, will require the process of Human Reliability Quantification (HRQ). Figures 5.5.18 and 5.5.19, which evaluate the major techniques in use, help the practitioner select the right technique for a particular application. In brief, the HEART and JHEDI are quick methods, while the APJ/PC, SLIM, THERP and HRMS approaches are all more in-depth.

If the objective is not to produce a quantitative output, the HRQ techniques can still be used to rank the errors/events in order of importance. Such a ranking procedure is relatively robust. When the HEP values have been calculated, it is useful to consider whether they conform to reasonable expectations or whether they appear as something of a surprise. If they *are* surprising, attempts at gaining corroborative evidence by applying another approach (e.g. one of the 'quick' methods) should be considered.

If experts are involved, their degree of expertise may be tested by means of the paired-comparisons method. If a facilitator is being used (in connection with the SLIM or APJ approaches) it must also be ensured that the facilitator has had adequate training: the maxim of 'garbage in, garbage out' is highly apt in the context of HRQs that make use of expert judgement.

5.10.6 Error reduction analysis

The PSA sensitivity analysis and the impact assessment can define the importance of HEs, as well as show their degree of impact on the level of risk. An error-reduction analysis (ERA) is only strictly necessary if the PSA is close to, or has breached, the limits of its own scope. Having said that, an ERA can be utilised on an 'as low as reasonably practicable (or even achievable)' design or safety-assessment basis. This is negotiable between the company and the regulatory body.

Quick (but not necessarily inexpensive) and effective methods include error-pathway blocking and consequence reduction, as well as other forms of design change. The HEART can also offer quick guidance on ERAs themselves. Otherwise, the SLIM and HRMS approaches can carry out PSF-based error reductions.

If a significant error reduction is required (>>10), it is the design-change option that is recommended, backed up by any other changes, in other human-performance dimensions, which would properly support such a design change. The rule is to check that the changes made do not merely replace old errors/psychological mechanisms with new ones. One must also ensure that a psychosocial protocol is followed, otherwise the changes made may have no effects, or may even end up having a negative effect.

The HEIST system and Appendix V may be considered in relation to an ERA. If the ERA required is considerable, or complex and far-reaching, the services of an ergonomics practitioner must be sought.

5.10.7 Quality assurance and documentation

Recommendations and assumptions that are made as part of the assessment process must remain true and effective, or else the PSA/HRA becomes irrelevant. Ultimately, the HRA and the PSA should be periodically repeated, and their insights used during the whole lifetime of the system. This is called (as has already been mentioned) a 'living PSA', and it ensures that all problems that arise remain in the system's 'corporate memory'.

The importance of documentation is often underrated. All assumptions and calculations should remain scrutable, auditable and justifiable throughout the lifetime of the system. Aim to render the study repeatable by hiring the services of another, independent analyst.

Summary tables

Table 5.10.1 considers the main steps involved in the HRA process. It is divided into five vertical sections, as follows:

- HRA step: one of the 10 steps involved in the HRA process, from problem definition to quality assurance.

Table 5.10.1 A practitioner's framework

HRA step	System-risk perspective	HRA tools			Comments
		Quick	Exhaustive		
Problem definition (see Table 5.9.1.)	What drives HRA? - PRA? - Vulnerability? - ERMA? Focus - Top event? - Maintenance?	Review PRA	Interview - assessors - operators Review - incidents - near-misses		- Collaborate with RAs but maintain independence
Task analysis (see Fig. 5.2.15)	- Procedures - PRA - Observation - Walk-through	Linear TA ITA CIT/OER Documentation review	HTA Tabular/timeline scenario analysis/TSA/WT/TT Link analysis/DAD		- Classify tasks as S, R or K - Model complexity appropriately
Human Error Identification (see Figs 5.3.16, 5.3.14, and tables 5.3.16 and 5.3.17)	- HEs in PRA - Operator's actions, in sequences - Maintenance of key safety systems - EOC	THERP EEMs only	- SRK - SHERPA/GEMS - Human HAZOP/ EOCA - HEA table		- Note consequences and recovery potential - Note HF deficiencies - Note ERMS
Representation (see Table 5.4.2)	- PRA bias - FTA/ETA - CCF	Fill in PRA-gaps HPLVs THERP dependence model	HRA defines parts of HRA Exacerbators/EOCs Misdiagnosis THERP dependence model		- FTs: are errors near to TOP? - Gates AND vs OR? - Retain meaning in representation

Table 5.10.1 (Continued)

| HRA step | System-risk perspective | HRA tools | | Comments |
		Quick	Exhaustive	
Quantification (see Figs 5.5.18 and 5.5.19)	– HRA techniques better than simple engineering judgement	– HEART – (JHEDI) – ASEP	– APJ/PC – HEART – SLIM – THERP – (HRMS)	– If non-quantitative, rank errors in order of importance – Are the HEPs reasonable? – Multiple approaches – If using experts, test them – Facilitator critical for group techniques

Error-reduction-mechanism analysis (ERMA) (see Appendix V)	– Sensitivity analysis
	– HE importance
	– Impact on risk
	– Proximity to criteria
		– Error-pathway blocking
		– Design-based system change
			– PSF-based change
			  . SLIM
			  . HRMS
			  . HEART
			– Psych-mechs based (see Appendix V and HEIST)
				– If require >> 10 error reduction, use error-pathway blocking plus PSFs
				
				– If blocking pathway, do new errors arise?
				
				– If PSF-ERM, ensure implemented adequately
				
				– If Psych-mech ERM, are other psych-mechs created?

HRA step	System-risk perspective	HRA tools		Comments
		Quick	Exhaustive	
Documentation	PRA	– Document all assumptions	– Document sufficiently for someone else to recreate the study	– Auditability – Note *all* assumptions
Quality assurance	Living PRA	– Check all calculations	– QA the methodology – was the right level of HRA selected? – Do the HRA tools fit together? – Carry out a sample of independent assessments using a different HRA technique – how much agreement is there?	– Ensure recommendations are implemented adequately – Ensure the implications of human error stay in corporate memory

Table 5.10.2 _Earliest life-cycle stage for HRA-technique application – summary table_

HRA process stage	Concept	Preliminary phase	Detailed phase	Commission	Operations and Maintenance
1. TA (see note)	HTA	Documentation review Link Analysis ITA	TTA/TLA/TSA WT/TT	DAD	OER/CIT
2. HEA	HAZOP	SHERPA THERP (EEMs)	GEMS/HRMS Murphy/SRK HEIST	CES	CMA IMAS/CADA FSMA EOCA
3. Representation		Fault/event	HRAET		
4. HRQ	APJ	PC/SLIM HEART	THERP IDA HRMS JHEDI		
5. ERA	Layout	Interface, staffing	Training procedures	Management	

Note: see also Kirwan, Ainsworth & Pendlebury (1992) for an expanded version for 25 task-analysis techniques.

- System-risk perspective: this lists those critical aspects which the practitioner should consider if the HRA is embedded within a PSA (some are relevant even if *not* embedded within a PSA).
- HRA tools – 'Quick': this column lists the more cost-efficient techniques appropriate for this stage of the HRA process.
- HRA tools – 'Exhaustive': this column notes the more in-depth and powerful techniques appropriate for this stage of the HRA process.
- Comments: these are the author's own comments on each of the various approaches that help to render the relevant stage of the HRA process easier/more fruitful.

Note that, with the exception of the table row 'Quantification', the 'exhaustive' column generally incorporates whatever is also in the 'Quick' column: i.e. the exhaustive approach assumes that the 'quick' approaches have also been carried out.

Table 5.10.2 considers the usually accepted earliest life-cycle stage, at which some of the more predominant techniques can be applied, via the conceptual, preliminary-design, detailed-design, commissioning and operations-and-maintenance stages.

5.11 Management and organisational boundaries in HRA

This section explores a current frontier of research in the field of Human Reliability Assessment (HRAs): the effects of management actions on the safety level of a plant. The recognition that management and organisational (M&O) influences can contribute significantly to risk and human-error level has only formally occurred in the past 10 years. Arguably, it was the Bhopal disaster of 1984 that was the first accident to be placed in this category – or rather, the first to demand this category, which formerly had not been recognised. In this accident, it was allegedly commercial pressures that had led to the disabling of safety systems and the subsequent erosion of Bhopal's protective systems, to a point at which the plant appeared, virtually, to be *waiting* for an accident to happen. The problem was that a HRA carried out during the design stage of the plants life cycle would not necessarily have addressed such a source of influence in its assessments, and would probably have judged the plant as acceptably safe. Such assessments would clearly have been wrong.

Two years later, the Chernobyl accident occurred in the nuclear industry, shattering any complacency that may have built up since Three Mile Island and amazing risk assessors with a sequence of events that, at the time, no serious HRA or PSA could have predicted. In the same year, the Challenger Space Shuttle disaster also occurred, throwing serious doubts over the adequacy of safety management found in the project, both in the months (even years) leading up to the lift-off and, in particular, in the early morning hours just prior to the lift-off. In both cases, a management/organisational influence was involved as an antecedent to the accident: in Chernobyl, pressure was put on the reactor personnel, who were regarded as at the top of their 'league', to resolve the experimental problem they were wrestling with (ironically, a problem concerned with increasing safety); and in the Challenger disaster, considerable delays to the Space Shuttle programme had been experienced, and there was intense pressure to get Challenger launched.

Since 1986, there have been two more notable accidents in the M&O category: the capsizing of the Herald of Free Enterprise and the Kings Cross station fire in London (see Reason, 1990). The problems created by leaving the bow doors open while sailing had been reported but not acted upon, and similarly, requests for bow-door indicators on the bridge had been ignored. The Kings Cross fire similarly involved factors which had been ignored by the very authority empowered to render this transport system safe.

The problem of M&O influences clearly does exist, and perhaps has always existed. The HRA practitioner faces three main hurdles when trying to deal with it. Firstly, it is not yet properly understood what factors give one company a good 'safety culture' and what factors give another an inherently dangerous one, nor is it properly understood what exactly a good safety culture is. Secondly, the ways in which to change a poor safety culture into a good one are not yet clear; and expert judgement currently suggests that any such shift could take as long as a decade. And thirdly, the practitioner has no quantitative models of the effects of the safety culture on the level of risk: all he or she has is the perception that such effects might be more complex than the representations of traditional PSA methods, such as fault and event trees, would suggest.

A failure to account for M&O effects on risk could lead to an underestimation of risk by perhaps as much as several orders of magnitude, making M&O an incredibly important common mode to consider in a PSA. This common mode could have a general effect on error rates, or a destabilising effect on the integrity of the safety system, or an increasing effect on an event-initiating frequency, or finally an increasing effect on the generation of both rule violations and errors of commission (extraneous errors).

The assessment problem is therefore quite daunting, but it is nonetheless one that is central to the risk-assessment industry. Three possible ways of dealing with the M&O problem are briefly explored in this section:

- Developing inherently safe industrial cultures.
- Assessing M&O effects on risk levels, and altering PSA predictions accordingly.
- Setting definite M&O boundaries for risk assessments so that the deterioration of the safety culture will be signalled by the PSA rather than assessed directly.

Each of these potential avenues is explored below.

5.11.1 Inherently safe industry cultures

(Note: this particular subsection, 5.11.1, draws heavily from the section on MORT in *A Guide to Task Analysis* (Kirwan & Ainsworth, 1992); this original section was co-written by Kirwan and Whalley.)

An 'inherently safe' company culture would be one in which safety was a primary concern at all levels of the management and the workforce, in which all procedures would have been assessed so as to allow robustly safe procedures and practices to be created, and in which the safety managers would continually be actively trying to improve the safety of the plant in question. At first sight, such an approach might seem to be going too far. However, with complex systems the routes to failure are extremely diverse, and as systems grow and items wear out, and as workforce members become complacent about incidents, the conditions and attitudes that allow accidents to happen start to develop. It appears, therefore, that for complex systems such as nuclear-power plants, a high-profile and *proactive* approach to safety is essential.

Some companies do appear to have been successful in achieving good safety cultures, and these same companies have simultaneously reaped the commercial benefits of *loss prevention* (Dupont, for example, is widely believed to be in this category).

One of the main problems for an industry wishing to secure a good safety culture is that of trying to find out what is actually required, besides a set of platitudes on safety management which are, to say the least, operationally undefined. One of the more useful models to have arisen out of the need to tackle this problem is the *management oversight risk tree* (MORT: Johnson, 1980), which is based on a normative model of how safety should be managed in a company.

The management oversight risk tree (MORT) technique was developed by Johnson from the mid and late 1970s through to the 1980s for use by the US Department of Energy. The technique investigates the adequacy of safety-management measures, and as a result it can be used either to ensure that adequate safety-management functions are in place or, in an accident investigation, to determine how such functions have failed. MORT is one of the very few techniques in existence that directly addresses management effects on safety.

The principal output of the MORT technique is an outline of the causes which have led to an event/accident (if it is being used for an accident analysis), together with an identification of areas where the safety-management structure appear vulnerable.

MORT can be applied late in the design stage, prior to the operation and commissioning of a specific plant, or at any subsequent point during the life of a system. It is also particularly appropriate for accident investigations. MORT can be applied at any venue, provided that the assessor can gain access to company personnel to confirm any aspects that are unclear. The major usage of the technique to date has been for accident-investigation purposes, e.g. on behalf of the US Department of Energy. Its secondary use, along with a host of extra (but MORT-related) techniques, has been for a consideration of the degree of operational readiness of a system, i.e. whether a sufficient safety analysis has been carried out, and whether there are enough safety-management systems in place for the plant to go safely into operation. It is this secondary potential role of MORT which is of concern to us in this section.

MORT is a diagrammatic representation of potential management weaknesses, and one which gives an easily accessible and comprehensive overview of the situation. The MORT chart is based on the fault-tree concept, with the undesired event (accident) or potential for loss at the top of the tree, and the logic of causal events beneath. MORT's logic is basically the same as that used in fault trees – i.e. it uses AND or OR gates to define its structure – but its approach is not a quantitative one. The top level of the MORT tree and an example of a sub-tree are shown in Figures 5.11.1 and 5.11.2 respectively. Since the top level of the MORT tree refers to an accident, this will be briefly dealt with before the focus is moved onto its normative management structure.

The event box at the top of Figure 5.11.1 summarizes the extent of the accident's associated losses, and postulates any future undesirable events. Once the extent of the accident has been established, the user arrives at the first logic gate, which is an OR gate. Only those risks which had been identified, analysed and accepted at the appropriate management level can be deemed *assumed* risks; unanalysed or unknown risks become oversights or omissions by default. It is important to remember that because mistakes could have been made when a risk was initially accepted, an assessment *should* still be applied.

The next major subdivision separates 'what happened' from 'why'. The 'what happened' section of the tree considers the specific control factors that should have been in operation, whilst the 'why' section of the tree considers general management-system factors. It is the 'what happened' branch of the tree which forms the major assessment route during an accident analysis, whilst management-system factors, on the other hand, become of primary importance when company's safety programme is being assessed.

The major parts (sub-trees) of the MORT tree are as follows:

- Technical information systems, incorporating, for example; knowledge about the type of incident; performance-monitoring systems; data-collection systems; and hazard-assessment systems. Why did these not predict the accident?
- Maintenance: did, for example, the maintenance plan contribute to the event?
- Inspection: did, for example, the inspection plan aid the detection of the factors causing the build-up to the accident?

Figure 5.11.1

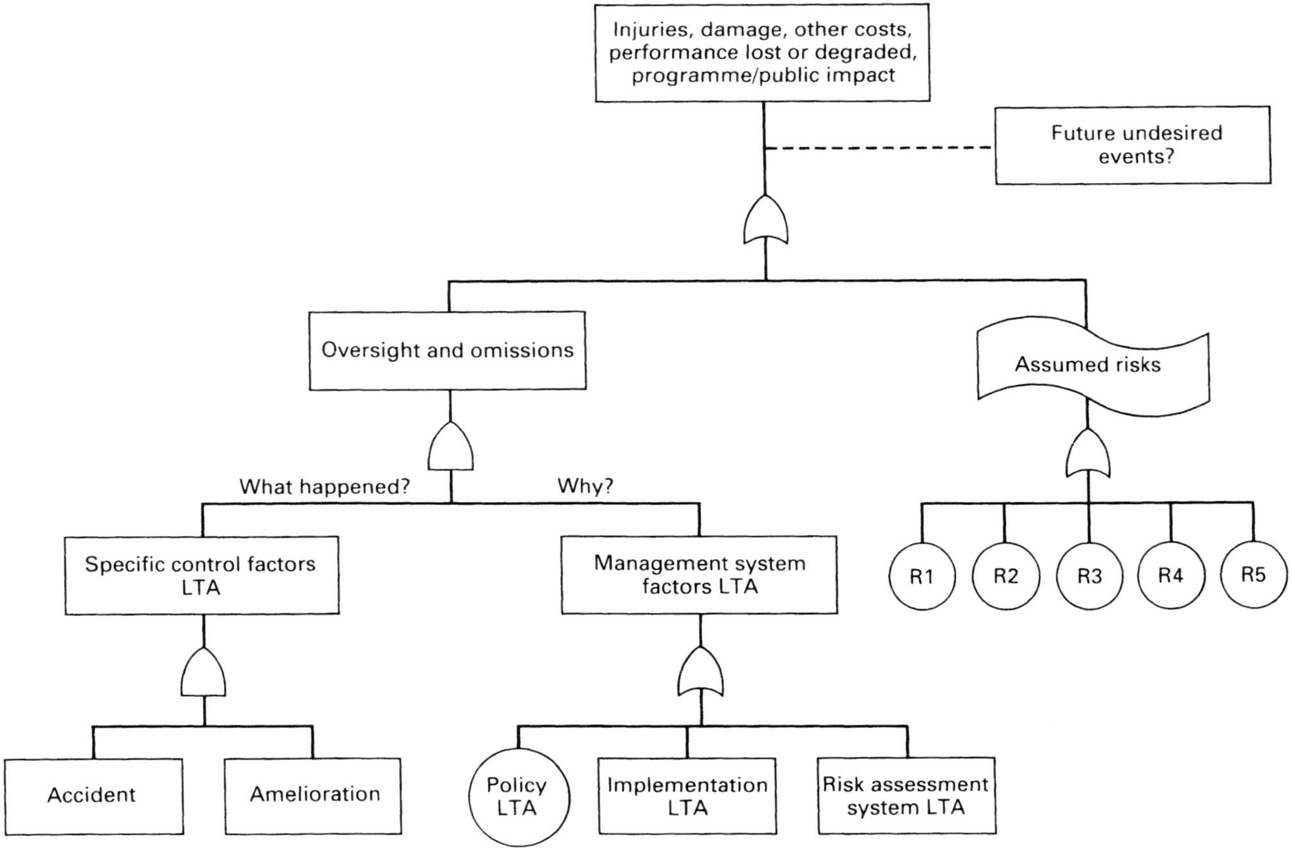

Injuries, damage, other costs, performance lost or degraded, programme/public impact
Future undesired events?
Oversight and omissions
Assumed risks
What happened?
Why?
Specific control factors LTA
Management system factors LTA
R1
R2
R3
R4
R5
Accident
Amelioration
Policy LTA
Implementation LTA
Risk assessment system LTA

Figure 5.11.2

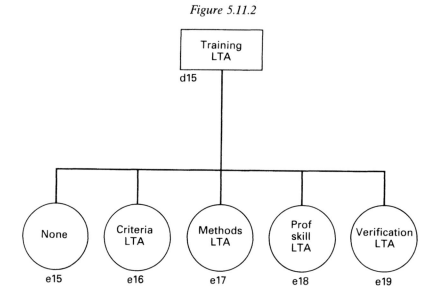

- Safety-analysis-recommended controls: were, for example, the worksite controls found on the equipment or on personnel (or both) inadequate, and thus a factor contributing to the accident?
- Services: did, for example, a lack of resources, of research and fact-finding activities, or of standards and directives contribute to the accident sequence?
- Human Factors: were, for example, the tasks allocated to machines and to the operator Less Than Adequate (LTA), or was it a failure to predict errors that contributed to the accident?

In addition, other major tree subheadings in MORT are:

- Task-performance-errors LTA: for example, for personnel selection, training, motivation, procedures, etc.
- Amelioration LTA: for example, for rescue, emergency responses, etc.
- Supervision LTA: for example, for performance, higher supervision, etc.
- Management-systems LTA: for example, for policy, policy implementation, risk assessments, etc.
- Concepts-and-requirements LTA: for example, for the definition of goals and concepts of risk, safety criteria, life-cycle analyses, etc.
- Design-and-development-plan LTA: for example, for energy control procedures, independent review methods, general design processes, etc.

In addition to indicating management factors specific to the analysed accident, it is also possible to identify more fundamental management-system aspects which could cause problems elsewhere. Each event included in the

MORT tree is separately defined in an accompanying handbook, with associated questions to consider during the analysis. MORT relates the different sub-trees to different sections in the MORT handbook. These sections provide methods and suggestions for improving inadequate situations.

As an example, when considering the degree of adequacy of the training given to the individual or individuals involved in the incident, the following types of question must be answered (see Figure 5.11.2):

d15 Training LTA – is the personnel training adequate?

e15 None – is the individual trained for the task?

e16 Criteria LTA – are the criteria used to establish the training programme adequate in scope, depth and detail?

e17 Methods LTA – are the methods used in training adequate to meet the training requirements?

e18 Professional skills LTA – is the basic professional level of skill of the trainers adequate for the training programme?

e19 Verification LTA – is the verification of the person's current trained status adequate? Are the training and requalification requirements of the task defined and reinforced?

Negative answers to any of the last five questions above (e15–e19) mean that an LTA response is generated for d15. Since these questions or 'events' are connected to the intermediate event 'Training' via an OR gate, this means that if any *one or more* of these five questions is 'LTA' then the superordinate event 'Training' is also 'LTA'.

In summary, the MORT tree comprises a normative model of safety management. Its model is based on approximately 20 years of research and accident investigation, the latter often as part of judicial proceedings. The MORT approach can be utilised to investigate the level of adequacy of safety management in a company or plant – in particular, it can be used to determine how well incident reports are made use of as accident-protection information. In practice, few cases exist of this kind of use of MORT, its use being virtually wholly directed towards accident *investigations*. Two research projects are in progress, however, that are investigating the development of MORT as a potential safety-management design tool, with a view to generating a blueprint for an inherently safe culture. The MORT approach therefore appears to offer a long-term route for dealing with the M&O problem facing the HRA practitioner: by moving it from the HRA domain over to the management-design domain. If such a MORT-type (or a philosophically/functionally similar type) system of safety management were developed, then a good safety culture would naturally follow, and such a system would continue to monitor its own safety progress, and make appropriate responses. However, until such an approach develops, the assessor must seek alternative means of dealing with the M&O problem.

5.11.2 Techniques for assessing M&O effects on PSA predictions

There are a range of qualitative tools for carrying out safety-management audits. These are not based on any normative model of safety management. Rather, they are based on many accumulated years of operational and safety experience. They have also been helped by the identification of 'Performance Indicators' (PIs), which assessors believe correlate with general safety performance. Such indicators may range from safety policies to the extent of supplies of back-up maintenance parts in the stores. Usually, such audit tools organise their indicators into groups, and have a means of aggregating or scoring their ratings so as to derive a total-safety index rating. In one system, the International Safety Rating Scheme (ISRS), it is possible to achieve a five-star rating, which would suggest a very good safety culture (see Eisner & Leger, 1988, for a critical review).

The principal difference, therefore, between a normative model and a PI-based system is that the former is explicit and essentially deterministic (i.e. if all safety management functions are addressed satisfactorily, then the level of safety will be high), whereas the latter approach is *correlational* in nature (i.e. if all PIs are satisfactory, then the level of safety is probably high; but there may, of course, be unknown PIs, not in the model, that would threaten to reduce the level of that safety). The PI approach does not rely on a normative model; instead, it must merely be accepted or rejected by potential users.

These safety-audit tools have been developing for some time, and have already been used both for safety-management audits on plants and for making qualitative recommendations, if judged desirable by the auditors, or if the scoring achieved by the company appears to warrant improvement. It is only recently that these techniques have been linked to PSAs in a *quantitative* fashion. This is being investigated at present primarily with two techniques, the above-mentioned ISRS approach, and the MANAGER (management-safety-systems-assessment guidelines in the evaluation of risk, Pitblado *et al*, 1990) audit tool.

With both of the above systems, an auditor carries out an extensive, on-site audit of the company via observation, the evaluation of documentation, interviews and inspection. To give an example of the scale involved, the ISRS, the larger of the two systems, can often require answers to as many as 600 audit questions. The MANAGER tool, on its part, consists of approximately 114 questions, divided into 12 areas:

- Written procedures
- Safety policy
- Formal safety studies
- Organisational factors
- Maintenance
- Emergency resources and procedures

- Training
- The management of change
- Control-room instrumentation and alarms
- Incident and accident reporting
- Human factors
- Fire-protection systems

The ISRS has 20 categories, which, though they overlap somewhat with those of the MANAGER approach, are sufficiently different in orientation to warrant a separate listing below:

- Leadership and administration
- Management and training
- Planned inspections
- Task analysis and procedures
- Accident/incident investigation
- Task observation
- Emergency preparedness
- Organisational rules
- Accident/incident analysis
- Employee training
- Personal protective equipment
- Health control
- The programme-evaluation system
- Engineering controls
- Personal communications
- Group meetings
- General promotion
- Hiring and placement
- Purchasing controls
- Off-the-job safety

It can be seen that the ISRS elements are more biased towards organisational factors. This in part underlies the ISRS philosophy that safety management and general management are inextricably bound: good management implicitly means the inclusion of good safety management.

The MANAGER technique was the first to consider linking up ratings on its audit questions with PSA results. The logic was as follows: members of Technica's team of assessors noted a range of three orders of magnitude, in terms of accident frequency, between inherently good plants and unsafe plants. This meant that the maximum difference between the PSA results of two otherwise identical plants, where one had a good safety culture and the other a poor one, came to a factor of 1000. It therefore remained to create an expert-judgement-derived equation that would relate answers given to its assessment questions to a PSA-top-event *modification factor* (MF). Thus, a plant can be

audited using the MANAGER technique, and then its PSA-top-event frequency modified, accordingly, by up to a factor of 10 – in terms of improvements on the original PSA predictions – or by up to a factor of 100 – in terms
of higher risk estimations on the part of the PSA. The partitioning of the
total potential effect, 1000, into, on the one hand, 10 for improvement and,
on the other hand, 100 for increased risk is a fairly arbitrary strategy, but still,
however, quite a conservative one. ISRS could be used in a similar way
(Smith, 1992).

The approaches adopted by both techniques however involve some fundamental drawbacks:

- The factor of 1000 is fairly judgemental, and should be further substantiated.
- The audit tools themselves, although *PI* tools rather than normative models, and although correlational in nature, are nonetheless treated more as
  probabilistic/stochastic models. In reality, however, the degree of uncertainty involved in these models is not only unknown but, in the absence of
  empirical validation or a comparison with a normative model, it is virtually
  unknowable.
- The technique of deriving impact levels on the basis of risk estimations is
  crude to say the least: it would be perhaps more correct and more useful
  to calculate an MF for different types of error/failure, or for related cut-
  sets in the fault/event trees, or even to introduce new events, related to
  M&O effects, into the PSA (Smith, 1992).
- If these two methods are being used, the reduction of the risk level can
  only be logically achieved by improving the status of the various negatively
  assessed indicators. But these indicators *are only indicators*, and are not
  necessarily the source of the problem. This leaves open a dangerous route
  wherein one correctly assesses a plant as unsafe but then deals in an *incorrect*
  way with undesirable superficialities, as a result of which one subsequently
  mistakenly declares the plant as safe.

The brief review in this section does not as yet endorse this second, PI-
based approach to dealing with the M&O problem because of the disadvantages raised above. This does not mean that these approaches are not worth
investigating, merely that their potential theoretical and practical drawbacks
must be recognised. Such methods should not be endorsed until substantial
independent validations of their adequacy have taken place. Such independent
validations largely have yet to occur.

5.11.3 Delineation of PSA M&O boundaries

A third approach is to recognise the problem but to confine the PSA to its
domain of expertise and, rather than trying to quantify what may well be

Figure 5.11.3

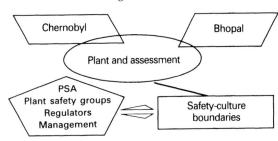

DSR has an implicit safety-culture envelope

unquantifiable, to define the M&O boundaries of the PSA instead. This means in practice that indicators will be set up at the periphery of the PSA to mark where the M&O assumptions underpinning the PSA could be violated. This strategy is represented in Figure 5.11.3, which shows that, originally, Bhopal's and Chernobyl's PSAs must have been adequate for predicting the level of risk inherent in those systems. However, as pressures acted upon the management of these systems, they gradually moved outside their respective PSA domains, so that the PSAs no longer represented the true states of either plant. What is needed when using this model is a set of warning indicators telling the plant management, the safety groups, and the regulatory authorities, unequivocally that the safety culture has begun to degrade – i.e. that the PSA 'sell-by date' has expired.

There would have to be a mixture of general indicators and specific indicators for the plant in question, some potential examples being:

- Higher rates of unplanned maintenance
- Maintenance-schedule delays
- Incident rates
- Operating-rule-violation frequency
- Risk-taking occurrences
- The occurrence of errors of commission
- Error/failure frequencies higher than those found in the PSA
- An error/failure occurrence not identified in the PSA

To monitor such events and contrast them with the relevant predictions would require a performance-monitoring system operating independently of the plant-management process. The advantages of this approach are that it does not require a reliance either on an undeveloped normative model or on unproven correlations. Nor does it negate the value of the PSA – instead, it only tries to establish the PSA's true domain of accuracy. As with the other approaches, however, it requires development.

5.11.4 Summary

M&O effects upon the level of risk is a new and difficult area, but one that is critically important in the risk-assessment of real systems. Three approaches to dealing with this area have been outlined, all of which require some development. The first and third approaches are the preferred ways forward, as far as this review is concerned. It is likely that over the next five years, there will be much development in all three of these areas – as well as, unfortunately, more accidents involving this genre.

6

Future directions in HRA

This section reviews current and likely future trends in the field of HRA, and focuses on four areas: methodology, resource-efficiency in Human Reliability Quantifications (HRQs), new technical-application areas and new industrial-assessment areas. This section cannot hope to be comprehensive, but it attempts to give a flavour of current directions in HRA, showing likely areas of expansion and likely developmental trends for the near future as regards the techniques available for HRAs.

6.1 *Methodological trends in HRA*

6.1.1 Human error analysis

This is still one of the underdeveloped areas of HRA, in terms of techniques based on sound theoretical models, empirical data, and techniques which have been positively validated. Two areas which are currently receiving attention are those of Human-HAZOP-technique development and error-of-commission analyses (Gertman *et al*, 1992). These approaches, in particular, will be likely to lend more credibility to HRAs carried out as part of a Probabilistic Safety Assessment, since they will allow more of the types of error that actually occur, but are difficult to predict, to be included in risk assessments. Such error types will include rule violation and maintenance contributions to the level of risk, currently not fully addressed in many, if not most, PSAs.

The area of cognitive-error analysis is also being investigated, with simulations such as the cognitive environment simulation (CES, Woods *et al*, 1987) and the cognitive simulation model (COSIMO, Cacciabue *et al*, 1989) representing the leading edges in this area, as far as potential new tools are concerned. If these simulations are successful, it is likely that smaller and cheaper/more generic models will be developed to allow predictions to be made for a wider industrial base (both CES and COSIMO are nuclear-power-plant-based).

6.1.2 Dependence analysis

There is a growing realisation that the effects of dependence can be highly significant in human-error analysis. Unfortunately, the tools available for modelling such effects are comparatively crude at present (e.g. Swain's five-level direct-dependence model, and Kirwan's HPLVs). What are needed are more discriminative and prescriptive models and tools, both for deciding different levels of dependence and for defining common-mode dependence levels for different types of system and different types of scenario involving operators. This avenue of research is currently being investigated in the UK as part of a Human Factors in Reliability Group (HFRG) extended research project (see further information section at the end of the book).

6.1.3 Databases

In the section on HRQ, there was a review of early attempts at databases. The conclusion arrived at in that review was that these attempts were not particularly successful, for a number of reasons. In the past half-decade, however, there has been a resurgence of interest in database development. The most notable of these developments has been the NUCLARR project (Gertman *et al*, 1988), a US database system. Other projects have also recently begun (see Taylor-Adams & Kirwan, 1994) which aim to produce usable and robust data, whether for comparative purposes, for the validation of HRQ techniques, for the calibration of techniques such as the SLIM and PC approaches, or for direct application themselves.

6.1.4 Validation

There have been a number of validation exercises in the history of HRA, with positive results for some techniques and negative results for others (see Appendix III). Not enough comparative validations have taken place, however, that make use of real, empirically robust and industrially relevant data (some validation methods, for example, have involved using the THERP's database data to validate other techniques' predictions). Clearly, there is a need for further independent validation exercises, especially since the techniques available for HRQ are sufficiently developed and mature to undergo rigorous testing. Such validations will in themselves lend a greater degree of credibility to the HRA field, and should enable the fine-tuning of the techniques as a means to improving their accuracy.

6.1.5 'Super-techniques'

The 'super-technique' is a term coined by Williams (1990) for HRA techniques which effectively deal with a large part, if not the whole, of the HRA process. The HRMS (Kirwan & James, 1989; Kirwan, 1990), is one such example, SCHEMA (Embrey, 1990) being possibly another. The advantage of such computerised approaches is their integrative nature, and the ease with which they lend themselves to documentation and quality assurance. Their disadvantages usually include both their development costs and a potential reduction in flexibility when it comes to selecting different tools within the HRA framework. Nevertheless, for a company that has a large amount of HRA work to complete in a systematic and consistent fashion, the use of such an integrative HRA modular system probably has more advantages than disadvantages, even if they develop their own integrated approach, rather than using an already-developed or 'bespoke' system.

6.1.6 Safety management and safety culture assessment

As noted in Section 5.11, this is currently a growth area, and the main issue here is whether PSAs and HRAs include quantitative assessments of the degree of adequacy of the safety culture or safety-management procedures in their predictions, or whether such assessments will remain as a *qualitative* aspect of the PSA. A basic theoretical problem in this area is that not enough is known about human errors in fairly *simple* contexts, let alone contexts involving complex management and cultural aspects. Although it is understood that the prevailing culture obviously has an effect on behaviour, current psychological and sociological modelling theories are still a long way away from being predictive even for general patterns of behaviour, let alone operators' tasks, and individual errors and violations. Nevertheless, there is a large amount of funding currently going into this area, so it remains to be seen what develops.

6.1.7 Standards

There is currently ongoing international work on the development of two standards of Human Reliability Assessment, one that forms part of a more general reliability-technology standard (BS5760 in the UK) and one that acts as a standard in its own right (IEC/TC/56/WG11). These standards, which are likely to appear in the next half-decade, will aim to give guidance to industry on recommended HRA practices. It is likely that such standards will be advisory, rather than mandatory, at least for some time.

6.2 Resource-efficiency in HRQ

As HRA (or indeed any) technology matures, the issue of resources-efficiency becomes increasingly important. PRAs and HRAs can be very labour intensive, and yet may yield little in the way of beneficial results. Screening was mentioned earlier as a concept, and recently, the ASEP procedure has been developed by Swain which allows the HRA to deal more sensitively with significant human-error contributions to risk. The likely reasoning behind ASEP's development is simply that to verify that many of the errors identified actually pose little or no threat to safety may cost significant amounts in resources if a full THERP analysis takes place.

A parallel philosophy has recently been developed with respect to the HRMS (Kirwan, 1990). As has been seen, HRMS is a potentially powerful but resource-intensive technique. A faster and cruder version has therefore been developed as a screening tool. Its application, together with the provisional criteria for calling upon the larger HRMS, are described below.

The approach utilises two systems, the first is the justification-of-human-error-data-information (JHEDI) system, and the second is the HRMS itself. Both systems are computerised, both involve a database and both are highly auditable. The following description of JHEDI is given not to show the technique itself, but rather to show how such a screening partnership can be implemented (e.g., such a partnership could be developed between HEART and SLIM, or HEART and THERP).

6.2.1 The JHEDI system

This system comprises the following modules/functions:

- Scenario description: a free-text description.
- Task analysis: linear, 'nested loop' architecture.
- Human Error Identification: a rapid Human Error Identification module.
- Quantification: the use of a HRMS database/alarm-handling algorithms and JHEDI-specific algorithms (eight databases in total).
- PSFs/assumptions: documentation of the status of various important Performance Shaping Factor (PSF) parameters, as well as documentation of additional assumptions underpinning the assessment.
- Approval/audit by HRA manager: the HRA manager checks all the JHEDIs produced.

A case-study example using the JHEDI system is shown in Figures 6.1–6.5 at the end of this section. This example (Kirwan, 1990) involves the assessment of a case wherein operators in a plant were required manually to undertake a sampling operation prior to transferring the contents of a vessel to its next location (called *sentencing*).

The scenario descriptions, the task analysis, the human-error output and an example of the PSF output, together with the derived HEP (chosen from the database by the assessor), are all shown. These are actual print-outs from the JHEDI system, all of which are input by the assessor, who may have had

Figure 6.1 JHEDI task-analysis example

FILE NAME:

ASSESSOR:

DATE:

SAFETY CASE:

SCENARIO

The enrichment monitoring and blending sequence reaches the point at which a daily sample of UN is required. The CCR operator requests that a local operator go to the plant room and take a sample from the correct tank. The local operator does this, and despatches the liquor to the laboratory.

In the laboratory, the analyst performs the test, records the result and transmits it back to the CCR. Here, the CCR operator enters the result into the DCS. If this result differs markedly from the on-line measurement, then the DCU draws this to the operator's attention. The operator then makes a decision on sentencing the liquor, and must obtain a Management Authorisation before proceeding. It is assumed that this supervisor's check should reveal any mistake made by the CCR operator when he or she was inputting the result, as well as any difference between this result and a corresponding on-line measurement which has not been noticed by the CCR operator.

TASK ANALYSIS

q1. INITIATE SAMPLING
1.1 CCR operator recognises need for daily UN sample
1.2 CCR operator communicates with local operator
1.3 Local operator goes to plant room and takes sample
1.4 Local operator despatches sample for test
2. ANALYSIS
2.1 Analyst carries out test
2.2 Analyst records results
2.3 Analyst transmits results
3. INPUT OF RESULT INTO DCU
3.1 CCR operator enters lab result into DCU
3.2 DCU warns operator of discrepancy between measurements
3.3 CCR operator obtains Management Authorisation
3.4 CCR operator sentences liquor

some limited formal training in HRA theory/practice. The database is not shown. The assessor also has access to alarm-handling-performance algorithms, which give the assessor an HEP, either for a simple alarm response (e.g. a failure to respond to a high-level alarm) or for more complex cases involving a diagnosis, based on the answers the assessor gives to a number of PSF questions. An extract from the alarm-quantification module found in the JHEDI system is also shown. It should be noted that the PSF questions in the HRMS and JHEDI systems are not judgemental – i.e. they do not ask the assessor to make relative judgements but instead ask for answers appropriate to the factual conditions involved. This can be seen from the questions and answers used in the PSF assessment. Both the HRMS and the JHEDI systems

Figure 6.2 JHEDI HEI output

Assessor: Date: Safety case: File name:	
1.2	CCR operator communicates with local operator
03120	Communication error between CCR and local operator
1.3	Local operator goes to plant room and takes sample
04120	Local operator takes wrong sample
2.1	Analyst carries out test
07120	Analyst makes mistake whilst undertaking test
2.2	Analyst records results
08120	Analyst records result wrongly
3.1	CCR operator enters lab result into DCU
11120	CCR operator enters wrong value into DCS
3.2	DCU warns operator of discrepancy between measurements
12120	Operator doesn't react to discrepancy warning
3.3	CCR operator obtains Management Authorisation
13120	Supervisor's check fails

ask the assessors only what they can be expected to answer – not, for example, such complex questions as 'How would you rate the interface in this scenario on a scale of 1 to 10?', which even a human-factors specialist would have difficulty in answering, let alone a safety assessor with no ergonomics experience. The fault tree for this sampling/sentencing scenario (not printed out by the JHEDI system) is also shown (courtesy of RM Consultants, Warrington, using the LOGAN fault tree system).

The JHEDI system is relatively rapid, and may take up to half a day for a scenario with several HEPs, assuming that the assessor knows the task requirements reasonably well. Answers concerning training and procedures, if not known by the assessor at the time of assessment, can be documented as assumptions and then included, for future implementation, in the training/ procedures systems used by the operations group.

This system is therefore auditable, has a certain degree of validity (thanks to its database and is relatively fast compared with several other techniques. Its database is also fairly conservative, so that it is more likely to err on the pessimistic rather than the optimistic side.

Figure 6.3 JHEDI quantification sheet

FILE:

STEP: 2.1 Analyst carries out test

ERROR: 07120 Analyst makes mistake whilst undertaking test

TIME

The time available is several hours.
The available time is significantly more than required.

QUALITY OF INFORMATION

The task step takes place in an area other than a control room.
Is information presented as a written/typed document? YES
Are diverse hardware- and software-based alarms available? NO
Is the information available in the situation clear and unambiguous? YES
Is feedback given promptly on all control actions, confirming that
  that they have been carried out effectively? NO

EXPERIENCE/FAMILIARITY

The event is very frequent (at least weekly).

PROCEDURES/INSTRUCTIONS

Written procedures will be available.

TASK ORGANISATION

The personnel involved is a single operator.
Do you think the operator would understand the task's significance,
  i.e. the importance of reliable performance and the cost/consequence
  of errors? YES

TASK COMPLEXITY

The task involves some interpretation or straightforward calculations in a basically
  non-complex procedure.

Probability chosen: 3E–3   HED reference: HE6

Authorised by:

6.2.2 Resources comparison of JHEDI and HRMS

Both systems have been computerised for ease of use, and are largely self-training and self-documenting, which allows such resources as training time and the reliance on 'experts' to be cut down. At present, only certain personnel are designated users of the HRMS system. The level of resource-use varies dramatically. The JHEDI system may take up to half a day, from start to finish, and in some cases only an hour. The HRMS system, on its part, may

Figure 6.4 JHEDI alarm-response quantification

Assessor: Task: Safety case: File name: Error:	
RESPONSE TO ALARMS – SIMPLE	
Is the alarm response based in the Central Control Room?	YES
Is the alarm an ESD signal, both hardwired and displayed on the VDU, and occurring both audibly and visibly?	NO
Is the situation unambiguous, requiring a simple and straightforward response?	YES
Is the situation high priority?	YES
Are there diverse hardwired and software-based alarms?	YES
The response time available is about 60 minutes.	

Assessor: Task: Safety case: File name:		
1	RESPOND TO ENR DISCREPANCY ALARM	
	CRO fails to respond to discrepancy alarm	
DATA POINT	Alarm response – simple	1
HEP	0.010000	

take two weeks to assess a scenario covered by a design project, because a good deal of time is required to collect information at the task-analysis phase (the process involves discussions with designers/operators, reviews alarm and instrumentation schedules, etc.) Once the task analysis has been completed, then, unless only a trivial problem is involved, the Human Error Identification, Representation and Quantification phases will take one or two days, depending on the complexity of the scenario. The error-reduction process, if this is required, can involve several iterations: the analyst must not only consider making improvement on the PSF front but must also discuss, with operational/design personnel, the improvement's feasibility.

From purely the resources perspective, HRMS may seem extravagant, but it does offer several advantages which, under certain circumstances, may justify the amount of resources used. Firstly, the scenario is analysed in great depth, and this does make the people involved feel confident that the human-error inputs have been properly and comprehensively modelled. Secondly, the system is based on information of a far more specific variety, due to the

Figure 6.5 JHEDI values in fault tree

Event ref	Data ref	Event description	Failure rate	Trepair/ Ttest	Probability	R/U
CE	HE6	Communication error – wrong sample taken			3.0E-03	
LOF	HE5	Local operator takes sample from wrong point			1.0E-02	
AE	HE6	Analyst error			3.0E-03	
ARWR	HE7	Analyst records wrong result			1.0E-03	
CCRORW	HE7	CCR operator enters wrong value into DCU			1.0E-03	
SCF	HE3	Supervisor's check fails			1.0E-01	
OFEM	HE5	Operator fails to react to measurement discrepancy			1.0E-02	
DCUFE	50	DCU fails to detect discrepancy between measurements	8.8E-04	1.0E-00	4.4E-04	U
OCF	75	Operator console fails to display alarm message	1.0E-02	1.0E-00	5.0E-03	U
SCF	HE3	Supervisor's check fails			1.0E-01	

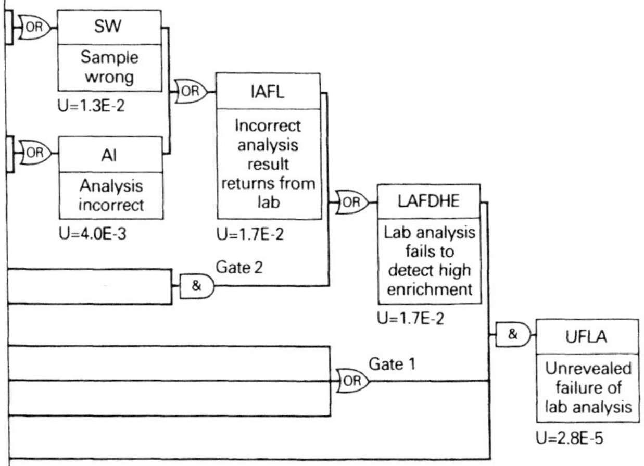

Generated by LOGAN v2.14 on (m-d-y) 01-10-1990 – Logan owned by RM Consultants, Warrington, UK.
Logic data file:
Event data file:

detailed PSF consideration, i.e. a specific scenario in a specific plant is assessed rather than a generic scenario in a general plant (the latter is often the case in the context of Probabilistic Risk Assessments). This may or may not ultimately prove more accurate in an absolute sense, but the specificity of the assessment does enable a clearer recognition of the relative importances of different errors, according to their PSF profiles. And thirdly, the error-reduction module allows a proper analysis of the different ways of improving human performance, as well as allowing the effects of error-reduction mechanisms to be quantified in the assessment.

6.2.3 Selection of appropriate method

In any PSA, the estimated risk level is compared against formal risk (frequency- and consequence-based) criteria to decide whether or not the design is acceptable. In the light of this, it is useful to relate the decision as to which technique is required to any such risk criteria, which will vary from industry to industry but will all tend to have the same general format; e.g. the criterion that the predicted frequency of an event which gives rise to particular consequences should not exceed 1E-6 per year.

The JHEDI and HRMS systems, if they are to be regarded as resources-flexible approaches, require a mechanism which will enable them to be selected according to specified criteria, which themselves should be based on the degree to which the PSA risk estimates impinge on the plant's allocated risk criteria. It was noted earlier that the JHEDI system was utilised for all assessments in the first instance. The criteria which can lead, on the other hand, to the use of the HRMS system are:

1. If the risk calculation is within a factor of 10 of the allocated risk criteria (or if it is, in fact, *above* those criteria), AND if a tenfold increase in a single HEP or group of dependent HEPs causes at least a doubling of the calculated risk element.
2. If the event-outcome frequency is sensitive to one or more HEPs, such that a tenfold increase in a given HEP or group of HEPs would yield a corresponding tenfold increase in the event-outcome frequency, AND if the outcome frequency is within two orders of magnitude of its allocated risk criteria.
3. If a particular scenario requires a more in-depth assessment as a result of a desire to encourage a greater degree of confidence in that assessment (especially, for example, with high-consequence scenarios), or in order to improve the level of scenario's reliability, either as a consequence of a request on the part of the assessor, designer or operator or as a means to demonstrating the integrity of an operator–machine system.

The rationale for these criteria is as follows. If the risk calculated is within a factor of 10 of the criteria, then a more in-depth assessment is warranted, given potential inaccuracies or omissions of critical error-pathways. Even in cases where the risk calculation is only two orders of magnitude away from its target, if the event frequency is directly sensitive to the human-error input, the changes made to several HEPs could ultimately come too close to the allocated criterion. And secondly (but of no less importance), if certain personnel believe that a scenario is worth investigating in depth, then this should be reason enough to carry out an investigation: there is a danger, in any 'art' such as that of safety assessment, of ignoring a basic lack of confidence in an assessment – as well as the subsequent desire to model it more fully. Ultimately, any assessment is only as good as the assessor carrying it out. If, therefore, the assessor has a low degree of confidence in an assessment, then it should, at the very least, be repeated by another assessor – or, preferably, modelled in more depth until more confidence is gained, or a problem is found which undermines the original assessment and was probably also underlying the assessor's concern.

These criteria, which can be used to decide when the HRMS should be utilised, rely on the monitoring of the degree of impact of safety cases on their associated risk criteria, as well as on sensitivity analyses of safety cases.

6.2.4 Summary concerning resources efficiency in HRA

A resources-flexible approach to HRA has been postulated for dealing with human-error scenarios in PRAs, and it is based on both a quick but auditable method and an in-depth HRA system. The advantage of this approach is that it allows resources to be allocated according to the degree of severity, or potential severity, of the human-error impact. It has been demonstrated via two specific techniques, but other combinations of techniques can also be used.

Resource-flexible approaches (HRMS-JHEDI; THERP-ASEP) are a provocative development in the field of HRA, suggesting a certain maturity in the subject, even if only in the quantification domain. As noted earlier, although this maturity is laudable enough, the focus of attention ought really to be deflected away from quantification (though without giving it up!) and back to the earlier and arguably more critical phase of the HRA, namely the qualitative modelling phase. It is here that the successful future of HRA lies.

6.3 New technical application areas

Since the management's impact on the level of risk has already been addressed in the previous subsection, it will not be reconsidered here (though it could logically fit into either this or the earlier section). Two particular technically challenging areas are outlined below. Whilst these are both currently only to be found in the nuclear-power domain, this particular domain does tend to set the pace for other industries when it comes to applied HRAs, and is often a harbinger of problems and solutions that eventually crop up in other industries.

6.3.1 Outage PSAs

There has been a recent surge of interest in assessing risk levels for plants that exist in a shutdown or 'outage' state, which prevails whenever reparative maintenance, refuelling or some other such activity is going on, such that the plant is not in its normal running state. In many cases, a plant undergoing maintenance may have safety mechanisms reduced so that safety devices and interlocks, etc., may be circumvented for maintenance purposes. For some systems, plants will be effectively 'emptied' of hazardous materials prior to such maintenance, but for others, the hazards may remain, and in this case they must therefore still be monitored and guarded against. Often, such protection measures are in effect dependent on administrative procedures and controls, which are highly human-dependent. For this reason, the risk factor brought into play during the shutdown and maintenance of some nuclear-power and chemical facilities cannot be assumed to be negligible; in fact, it is conceivable that risks incurred during such outages could be more hazardous than those incurred during normal running conditions (due to the lack of in-depth protection). Errors made during outages may also become latent failures for the operational phases of the plant.

6.3.2 Advanced light water reactors (ALWRs) – the role of the operator

The next generation of nuclear-power plants – i.e. those that will be built in the next century – will have a far more inherently safe design. They will, for example, make use of many more passive safety-barrier and defence systems (e.g. natural-circulation and gravity-feed mechanisms), better (and simplified) reactor-system and protection-system designs and electrical back-up supply systems that will last 72 hours (Nobile, 1992). Such systems as those currently in the conceptual stage in the USA, Japan and Italy have significant implications for the role of the human operator, as well as for HRA methods.

The first implication for the operator is that in many scenarios, action will not be required for 72 hours – a far cry from the current situation, wherein most operators will have to make at least an initial response within a 30-minute time-frame. This questions the very role of the operators, since it puts them in more of a supervisory and monitoring role than they occupy at present, both during normal, and now during abnormal, situations. This will have significant ramifications for such human-factors issues as training and procedures, displays, whether to 'lock-out' the human operator during initial phases of incidents, and the impact this shift in role emphasis will have on job satisfaction, morale and motivation.

The second implication is for HRA methods themselves. The quantification techniques have evolved to consider responses made in a short time-frame. Excessively long time-frames (e.g. 72 hours) cannot be assumed to go hand in hand with asymptotic levels of high human reliability, especially since responses or non-responses made within such time-frames must straddle several shifts of staff. Furthermore, in future, there will probably be a higher component of risk from two particular sources: maintenance errors and errors of commission. As noted earlier, these two sources are themselves underdeveloped methodological areas. The ALWR concept thus raises a new challenge for HRAs. There is a sufficiently long time-frame for the development of requisite tools, provided that development begins now.

6.4 New industrial assessment areas

6.4.1 Low technology risk

The HRA field has mainly concerned itself with the high-risk, high-technology industry sector, including nuclear-power plants, chemical-plants, etc. There also exist, however, a large number of other, lower-technology sectors, e.g. mining, which also often involve high-risk factors as a result of a large number of 'small' accidents (e.g. 1 or 2 fatalities), in contrast to the situation found in high-technology industries where the high-risk factor is instead defined by a probability value which, though very small, is attached to a *large-consequence* accident. This is clearly an area where applied HRA methods should be able to help lower the level of risk. Some workers have already begun to use HRA approaches in such areas (e.g. Collier & Graves, 1988).

6.4.2 Space

Since the Challenger Space Shuttle disaster in 1986, there has been an acute awareness in the space industry of the importance of human reliability. The European Space Agency has commissioned various research and development projects as a means to developing its own approach to HRA (Rosness *et al*, 1992).

6.4.3 Pharmaceutical and medical research

Industries associated with the production of drugs and medicines have very high standards to meet, and yet such industries are also ultimately dependent on human reliability. This is therefore a likely growth field for HRA methodology – which can be used by these industries to ensure that they succeed in attaining such high standards. Some kinds of medical research deal with highly contagious viruses, etc., which could have catastrophic effects in the event of an accident. This is an area of research in which risk-assessment and human-reliability methods could help to assess such accidents and then prevent them from occurring.

6.4.4 Service and business sectors

HRAs have been used in a limited number of studies in the service sector (e.g. for postal-service reliability) and the business sector. As the drive for quality services increases, and as the costs of business failures – caused by errors or decision failures – rise, the services offered by an HRA-style approach are likely to become more attractive. Doubtless, the methodology and data would have to change significantly so as to reflect the contexts involved, but the basic HRA objectives would still be applicable: identify what errors, oversights or decision failures increase the element of risk or lower the level of quality, prioritise them in terms of order of consequence and order of likelihood and then determine defences against them. The stock exchange, for example, would seem to be one suitable area for developmental research in business-sector HRA methods.

6.4.5 Sociopolitical decision-making

The ultimate use to which human-reliability approaches could be put would be guidance for sociopolitical decision-making, i.e. avoiding political errors. This is not a new idea: the Influence Diagram approach (or rather, a variant of it) has been used in at least one instance to predict a political-decision failure (unfortunately, this remains confidential). Several of the HRA techniques (in particular, the SLIM, APJ and Influence Diagram Approaches) are in fact highly relevant to the field of decision-making, since they are concerned with the structuring of expert opinions. Whether HRA tools or philosophies *are* ever developed and applied to politics remains to be seen, but this would undoubtedly be the most significant challenge for the HRA approach, one unlikely to develop until the next century.

6.4.6 Summary concerning contemporary issues

The above has attempted to give a broad-ranging pot-pourri of likely developments and possibilities in the field of applied Human Reliability Assessments. It must never be forgotten that predicting human behaviour is always an *approximate* process, since there are always so many variables. Ultimately, HRA will survive only as long as it produces useful outputs – and as long, of course, as it continues to be applied.

7

Conclusions

In the preceding sections, the various techniques involved in the HRA process have been outlined in detail. These sections, together with the appendices, should hopefully be of use to practitioners and students, as well as (see the case studies in Appendix VI) to inexperienced assessors. As a whole, this book has tried to give the user a relatively comprehensive framework, and a flexible toolkit with which to apply HRA methods to a wide range of human-reliability contexts and scenarios. What this book cannot do is take the place of real experience of HRA methods applied in real assessments. The prospective user simply has to start trying out such methods.

There are several overall conclusions that serve to summarise the position and philosophy which this book has tried to convey. These are as follows:

- There are sufficient tools available for assessing many, if not most, human-error scenarios and contributions to the given level of risk.
- These tools fall into the reasonably simple framework that is the HRA process. Within this framework, there are usually multiple techniques for each stage of the process, so that the assessor enjoys some degree of flexibility in deciding which tools to implement according to his or her criteria.
- In a number of case-study applications, and in many real PSAs/HRAs, HRA methods have proven useful in first calculating and then reducing the risk element incurred as a result of human error.
- The HRA approach still requires significant amounts of skill and judgement on the part of the individual assessor, and there is still no substitute for experience. In this way, HRA is still as much an art as a science, though not necessarily a difficult art to learn. What this means is that books such as this one are limited in the degree to which they can be prescriptive.
- On the theoretical-and empirical-validation front, HRAs have a fair way to go. Nevertheless, HRA methods look better when seen in an applied setting than they do in a theoretical one, so it is more constructive, therefore, to judge a HRA by its *practical* merits rather than by its theoretical standing.

- The HRA method offers a useful overall approach to modelling the degree of human impact on systems in a system context. And this approach could easily, moreover, be extended to other spheres of human involvement. The degree to which this occurs remains to be seen.

Appendix I

Human-error forms in accidents

This appendix reviews a set of 12 system accidents and one incident (Davis-Besse, 1985), which have occurred over approximately the last two decades. These brief reviews, based on or adapted from published analyses, are exemplary, showing the different types of error and combinations of events that have occurred in real accidents.

It should be noted that it is always difficult to reconstruct what actually happened in an accident, and the descriptions in this appendix are based on other published accounts of what allegedly happened. The brief descriptions here inevitably focus only on the human-error aspects, and so, in a more general sense, may not give the kind of balanced perspective that would be appropriate for an accident-enquiry report. However, since the intention here is merely to highlight the *human-error* aspects of accidents, these brief descriptions *are*, for our purposes, appropriate. For other descriptive references to a range of accidents, see Bignell *et al* (1978), Bignell & Fortune (1984) and Reason (1990).

Year	Event	System domain
1966	The Aberfan disaster	Mining
1972	The Crash of the BEA Trident 1	Aviation
1973	The Paris air disaster	Aviation
1974	The Flixborough disaster	Chemical
1975	The Browns Ferry fire	Nuclear power
1975	The Dutch States Mines explosion	Chemical
1976	The Seveso incident	Chemical
1977	The Ekofisk Bravo blowout	Offshore
1978	The Bantry Bay disaster	Petrochemical
1979	The Three Mile Island accident	Nuclear power
1984	The Bhopal catastrophe	Chemical

| 1985 | The Davis Besse incident | Nuclear power |
| 1986 | The Challenger Space Shuttle disaster | Space |

1. The Aberfan disaster (1966)

The Aberfan disaster on 21 October 1966 led to 144 deaths in a small mining village. The disaster was caused by a coal-mining rubbish tip falling down onto the town of Aberfan (see Bignell, 1978, for a precise chronology of the accident). Howland (1981) cites the major causes of this accident to have been:

- The neglect of many prior warnings, including reported tip-movement six months before the disaster.
- An inadequate disaster plan.
- An inadequate tip-inspection procedure.
- Human inertia, ignorance and ineptitude, a failure in communications and inadequate safety training at all levels (Howland, 1981).

A principal factor in the latter-cited communication failure was, according to Bignell (1978), a disagreement between two members of the reporting hierarchy. This type of social problem, which can occur in a technical environment (hence the phrase 'sociotechnical'), is rarely cited in accident case histories, though this may be due to a reporting bias. The sociotechnical error is a threat to any system, and one which can probably only be dealt with via effective management and organisation. The other main causes that Howland cites fall into the general categories of inadequate safety management and inadequate preparedness/emergency planning.

2. The crash of the BEA Trident 1 (1972)

The crash of the BEA Trident 1 at Staines on 18 June 1972 incurred a death toll of 118. Specifically, the aircraft stalled during take-off, due to the pilot's failing to maintain sufficient air-speed. Howland (1981) cites three main causes for this accident:

- The absence of a mechanical interlock preventing the retraction of slats at too low an air-speed.
- A lack of experience/training, on the part of the crew involved, for this type of emergency.
- The captain's ill health during take-off.

All three component causes are important. The use of interlocks is an important way of avoiding certain types of potentially fatal, or at least hazardous, errors. Interlocks may be omitted where it is felt that personnel have enough experience not to do 'incredible' things, but this strategy can still clearly fail – as in this case, where an inexperienced team was involved. However, too *many* interlocks may lead to operators or maintainers trying to use various methods to overcome them, for example, an incident occurred in which one maintenance crew overcame an interlock in a nuclear-reactor plant. This led to an incident in which, after they had gone to lunch leaving the interlock disabled, another operator attempted to operate the system, unaware of the status of the system and of the illegal interlock-override that had been carried out.

The lack of experience and training for a particular type of emergency is an often-cited problem. In any kind of emergency which causes high stress levels, it is somewhat optimistic to expect operators to handle an event which may be complex and totally new to them. A case in point was the Kegworth air crash (1989 – see AAIB, 1990), which presented the pilots with a complex scenario for which training and their interface had left them ill-prepared. This perception accounts for the large amount of simulator training which pilots traditionally undertake.

Illness – or worse, the incapacitation of one member of a crew – is a potential problem which a high-risk plant must address via its staffing policy.

3. The Paris air disaster (1973)

In 1973 a Turkish Airlines DC-10 crashed in a forest shortly after take-off from Paris, as a result of a mishap involving a pressurised cargo door. The cargo door opened outwards, and its accidental opening was prevented by means of locking pins (Williams, 1987). The 'operators' knew that the door would be secure if a handle, connected to the locking-pin mechanism, could be put in the closed position, hence allowing pressurisation. However, whilst the operators believed that the successfully closed handle indicated a fully secured door, what they were unaware of was the case where, due to a source of resistance elsewhere (the door not properly closed), a rod in the mechanism could buckle and allow the handle to appear closed without actually locking the pins home – the situation which occurred on this day, leading to a tragic loss of life (via sudden depressurisation during the late stages of take-off-the door flew open at altitude). The primary cause, aside from a simple 'design error' (the operator could not have foreseen this event), was, as Williams states, the violation of a human-factors principle, namely:

- The feedback mechanism should have given *direct* (and hence reliable or true) feedback to the effect that the pins were locked home, which it did not.

This lack of direct feedback recurred in the Three Mile Island incident, discussed later. This type of problem has also been mentioned with respect to the activation of a control component. For example, the operator should know not only that a control input to an actuator has been sent, or even simply that it has been received, he or she should in fact know that the component has moved, or fired, or achieved whatever its mission was.

4. The Flixborough disaster (1974)

On 1 June 1974, a pipeline rupture at a Flixborough plant led to the explosion of a cyclohexane vapour cloud, causing 28 deaths and a very large amount of collateral damage to the surrounding populated area (1821 houses and 167 shops were damaged). The cause of this accident was primarily an inadequate design modification on Reactor Number 5, in the form of a bypass pipe. This modification was made in response to damage to the reactor, namely a 5′ 6″-long split, whose cause was not fully investigated. The team appeared to assume that the installation of the bypass line was a relatively routine plumbing job, failing to realise the necessity for a proper technical assessment of safety considerations. There was no pressure testing carried out for the modified system, and there was also a failure to site the explosive material away from the control room, or adequately protect the control room from any potential blast, despite the fact that the HMFI had apparently already suggested the installation of shatter-proof windows in the control room. The three main factors involved were therefore:

- The inadequate design/safety-testing of the bypass pipeline (latent failure).
- The inadequate training/engineering experience of the staff involved (Howland, 1981).
- The inappropriate siting of explosive material or a lack of protection facilities for the control room (or both).

Another contributory factor in this disaster appears to have been the desire to get the system back into service as soon as possible. This possible production bias recurs in later accidents discussed in this appendix, such as Bhopal.

5. The Browns Ferry fire (1975)

On 22 March 1975 there occurred an operator-induced fire in the cabling duct of the Browns Ferry nuclear-power plant. The operator was leak-testing (i.e., testing containment integrity) using the naked flame of a candle, and the polyurethane foam in the cabling duct caught fire. As most of the control cabling for the plant ran through this duct, this caused both operational problems (in particular the disabling of the core cooling system) and damage to the plant. The main human errors were:

- An inadequate design, which rendered virtually all the plant's safety systems vulnerable to even a relatively small fire (common mode failure).
- An inappropriate leak-testing method.

This event in particular highlights the vulnerability of systems to (without hindsight) 'incredible' human errors, or to human-induced dependent failures. In this case of Browns Ferry, the operators *did* manage to regain control of the plant, albeit by unconventional methods – which at least demonstrates the human capacity to recover from failure or errors.

6. The Dutch State Mines explosion (1975)

Early in the morning of 7 November 1975, at the DSM plant at Beek in the Netherlands, a vapour cloud of approximately 5.5 tonnes of hydrocarbons (mainly propylene) was released, and shortly thereafter exploded. The explosion killed 14 and injured 104 on the site, as well as three outside it.

The focus of interest in this accident is not the cause of the accident but the *response* to it. The large vapour cloud was apparently detected visually by personnel on the plant, although it appears *not* to have been detected by any instruments in the control room. Personnel on the plant warned the control-room personnel, and the order to raise the fire alarm was given. In the albeit-brief two minutes between the first visual sighting of the cloud and the explosion (the control room was not blast-proofed, and was severely damaged in the explosion) the fire alarm was *not*, however, sounded, little effective mitigating action was apparently taken and the operators were killed (it has been commented that perhaps there was very little that *could* have been done for the control-room operators). The principal human-error form of interest here is therefore:

- An inadequate immediate emergency response.

One possible interpretation of this unfortunate event highlights the potent effect of very high levels of stress upon rational performance, especially in acutely short time-frames. Consideration must therefore be given both to enhancing performance levels in emergencies (via emergency training and the optimising of crew functioning) and to ensuring that alternative (hardware-based) systems will be able to detect and initially react to severe and sudden onsets of emergencies. This accident therefore warns against practitioners' making optimistic assumptions about the reliability of human performance in the first few minutes of a scenario, particularly when the danger is severe (see also Berkun's Studies, cited in Williams, 1987).

7. The Seveso incident (1976)

On 10 July 1976, a shutdown reactor pressurised at a chemical plant in Seveso, Italy, releasing a large cloud containing a mixture of toxic chemicals, including dioxin. Aside from the death of a large amount of livestock, the release apparently resulted in 30–40 abortions and a decontamination bill of approximately $50m. The principal human-error forms that contributed to the incident were:

- The failure of the operator to follow explicit procedures – in particular, to: distil off 50 per cent of the ethylene glycol (only 25 per cent was distilled off); add 3000 litres of water in order to cool the reactor mix; observe the temperature recorder, and in this case turn it off, so as to ensure that a safe reactor temperature prevailed.
- A failure to remain with the reactor until it had reached a safe temperature region (i.e. 0–60°C).
- A failure on the part of the designers to ensure that the reactor products were adequately contained within the plant's boundaries.

If the procedural failures noted above were put together in a single fault-tree branch, each coming together under an 'AND' gate (i.e., all of them being necessary for the 'Top' event to occur), it is very likely that these individual errors would all have been given low probabilities, with the resultant prediction that this accident effectively could not have happened. Clearly, this is an example of a form of human-dependent failure, the result possibly of inadequate procedures or a lack of understanding of the hazards involved in the system, as a result of which there was a need both for regular checks and for an adherence to procedures. This incident also clearly shows the need for supervision/management to have come in on this task, since *any* accident-modelling fault tree which contains a set of single-operator errors that are so closely linked to the top event can only indicate the lack of an adequate human-system design. It also suggests that the temperature recorder should have been capable of giving off an alarm.

Lastly, this incident, and some other incidents, raise the question of whether risk-analysis information (e.g. predicted failure paths) should form part of the basis for safety-training procedures given to operators. After all, if particular feasible scenarios *are* identified, and if, moreover, they are safety-assured, on the grounds of adherence to procedures, by well-trained operators, then why not give this information to the operators? As a prime example, the basic form of the TMI incident had been predicted in Wash-1400, and a similar incident had in fact already occurred in the USA six months before TMI, though here without such disastrous results. If information regarding feasible scenarios were more effectively transmitted, more incidents and accidents would be averted.

8. The Ekofisk Bravo blowout

The Ekofisk Bravo blowout incident, which occurred in 1977, caused a large degree of collateral damage. A blowout developed while an offshore well was being worked on. A blowout *preventer* should have been able to prevent this, but this had unfortunately been incorrectly installed. The events that led to the accident occurred over approximately a 36-hour period, although some of the deeper, root causes appear to relate to high-level failures to impose proper safety controls. The significant human-error forms elicited from this incident were:

- The failure of information to be properly transferred across shift boundaries.
- The fact that *one* individual, in control of the operation, had been on shift almost continuously for nearly 36 hours. In addition, a critical safety component was inserted upside down, making it ineffective. This mistake could have been rendered impossible by the design of the system; nor was the mistake obvious to the operators, since particular visual information that would have indicated which way up it should be inserted was missing.
- The failure on the part of government departments to impose high-level controls on the safety aspects of particular operations, and controls such as blowout preventers.

Of particular interest is the failure of important safety information to be transferred effectively across the shift boundary, as a result of inadequate shift-handover procedures. Some symptoms of an impending blowout *had* occurred (signs of a 'kick' developing), but, none of the shift workers appeared to make any effort to tell any of the the oncoming shift's workers about these signs. There were problems with the operation, and one of the key personnel had been on shift for an exhausting amount of time. The lack of visual feedback as to the correct orientation or nature of a component is not uncommon in maintenance, and yet here we see the total disabling of a major safety barrier (preventing an oil/gas blowout) via one simple maintenance-type error.

9. The Bantry Bay disaster (1978)

On 8 January 1977, the oil tanker *Betelgeuse* caught fire at the Bantry Bay terminal at Whiddy Island, Eire, killing 50 people (Howland, 1981), mostly by drowning (or asphyxiation) rather than as a direct result of a fire or explosion. There were two main causal factors of interest in this accident:

- The badly corroded centre part of the ship apparently split in half during ballasting. The owners of the ship had been aware, since 1977, of the developing corrosion problem, but had not yet implemented measures to deal with it.
- Prompt emergency-response action was *not* taken during the first 10 minutes after the outbreak of the fire.

The first cause suggests a lack of effective safety management. The second is related to several specific observations made during the investigation. Firstly, of the two tugs that should have been available and in sight of the tanker, only one was available. This tug, which in any case was anchored *out of sight* of the tanker, was called too late for effective action, other than that carried out in order to prevent the spread of fire to the tank farm, to be taken. There was also no automatic fire hose on the jetty. Other erroneous factors included the failure (to start) of two of the three emergency Landrovers, which had not been maintained in a truly ready state, the fact that the foam tender in the fire station was inoperative, and the breaking-down of another vehicle along the route. Accidents like this seem to suggest that when things go wrong, they go wrong in a *big* way. Clearly, *preparedness* is something which must be maintained permanently – routinely tested and periodically assessed. The Alexander Kjelland tragedy (1976; see Bignell and Fortune, 1984) also showed a lack of adequate preparedness, as well as a failure to account for the 'coupling' of an accident with adverse environmental conditions. In view of the tendency of real accidents occasionally to occur in non-benign conditions, perhaps preparedness requires the further consideration of worst-case scenarios.

10. The Three Mile Island accident (1979)

At 4.00 am on 20 March 1979, a serious accident occurred at the Three Mile Island Unit 2 nuclear-power plant near Middletown in Pennsylvania. The accident was initiated by mechanical malfunctions in the plant, and then exacerbated by a combination of human errors committed in response to these malfunctions. Of particular interest is the fact that over the next few days the extent and seriousness of the accident was not made clear either to the managers of the plant, to state officials or to the general public. Its impact upon the nuclear-power world has, however, been quite dramatic. The essential fact that gave rise to the accident was that the reactor was running at 97 per cent of its normal power capacity when the feedwater pumps tripped, and this led, shortly thereafter, to the scramming of the reactor (i.e. the rapid insertion of the control rods) on the part of the operator. An erroneous action prevented emergency feedwater from reaching the reactor, and at the same time caused water to drain from the reactor, thus exacerbating the overheating problem. The essential error forms were that:

- The operator failed to notice two lights indicating a valve closed on each of the two feedwater lines (one light was covered by a maintenance tag). Seeing high pressure in the system, the operators deduced that there was too much water pressure in the reactor, and thus switched off the emergency feedwater line reaching the reactor.
- A light on the control panel indicated that a PORV (pilot-operated relief valve) had re-seated itself, when in fact it was stuck open (the signal saying it had shut was indirect; what in fact this signal said was that the power had been cut from the PORV opening mechanism).

- Water leaking into the polisher's control system during maintenance may have led to the tripping of the feedwater pumps, which was what initiated the incident. This had happened once before at TMI.
- The error recovery from switching off the emergency feedwater 2 minutes into the incident occurred after eight minutes into the incident when some-one 'discovered' that no emergency feedwater was reaching the steam generators.
- The operators, further to the last point, nevertheless failed to interpret the following signals which suggested they had an 'open PORV' and a loss-of-coolant accident on their hands.
  - An alarm signalling high water (over 6 feet) in the containment sump (a clear indication of a leak).
  - The neutron-measuring instruments showed a higher count than normal inside the core.
  - A rupture disk on the drain tank burst as the pressure in the tank rose.
  - The temperature and pressure indications rose rapidly, and although the operators did respond to these indications by initiating cooling fans, they failed to realise that the rising temperature and pressure had been caused by a LOCA.
  - Four reactor-coolant pumps vibrated badly, due to the fact that they were pumping *steam* as well as water.

Shortly after 6.00 am, a tele-conference was held. At 6.22 am, two hours and twenty-two minutes after the PORV had opened, the operators closed the block valve upstream of it, this essentially terminating the progress of the accident – though the core damage had already been done.

This accident *rocked* the nuclear-power world sharply, and perhaps irre-trievably undermined the public's confidence in nuclear-power safety. In par-ticular, it emphasised (see also the BEA Trident crash) the need to give operators *direct*, and not indirect, feedback (the operators in this scenario believed the PORV was closed, when it was not). This incident is also often cited as an example of a fixed 'mind set', or of 'tunnel vision'. These terms refer to the inability to draw back from an incorrect diagnosis once has been made, and they suggest the need for diverse diagnostic facilities, either in the form of diverse personnel (e.g. the nuclear-power world's 'shift technical advisor') or in the form of some kind of computerised decision-support system – or else a rigorous proceduralised system. A more problematic consideration is whether physically to prevent operators from switching off emergency safety systems – e.g. by the use of interlocks, or some other means – in the first stages of an incident situation.

The Human Factors Society, after reviewing the TMI accident (Hopkins *et al*, 1982), derived four main contributory factors, which in turn yielded four areas of work and research in the drive to prevent such incidents from recur-ring. These were:

- Control-room design
- Procedures
- Training
- The operator's qualifications

Some specific factors in this accident appear to have been largely classic human-factors omissions:

- Inadequate labelling
- Excessive and confusing information
- An inadequate indication of critical controls
- A failure to put important related controls together
- Inconsistent colour coding

Finally, the PORV in question had stayed open many times in the past, and this fact should have been integrated into the operators' diagnostic processes. This particular device should also clearly have been fixed earlier by maintenance, which itself highlights the problem of ineffective maintenance scheduling.

11. The Bhopal catastrophe

The Bhopal catastrophe occurred on 3 December 1984, killing more than 2500 people and injuring and permanently affecting many more (over 200,000 sought medical treatment, and over 70,000 were evacuated). This accident involved a major release of MIC (methyl-isocyanate), a highly toxic compound, which quickly spread over the surrounding population. The release was primarily caused by an ingress of water, leading to an exothermic reaction with the stored MIC. In common with the Mexico City fire (also in 1984), the number of houses being built next to the plant was getting out of control. Some of the specific failures in operation and maintenance appear to have been as follows:

- The existence of a 'jumper line' between the vent headers.
- A failure to respond to a sudden increase in pressure in the MIC storage tank from 2 psig to 30 psig.
- Allowing the MIC refining still to operate at a higher-than-normal temperature.
- Incorrectly transferring off-spec MIC to Tank 610.
- Three safety features had been off-line for 6 months.
- A failure to plan maintenance in accordance with safety goals.
- A failure to slip-plate lines during maintenance.
- The removal of refrigeration.

There also appeared to be economic pressures on the plant, and these led to short-cuts which in turn affected safety (Bellamy, 1986). These short-cuts appeared to involve:

- A decrease in manning costs
- A reduction in equipment (including, for example, the switching-off of refrigeration)

There also appeared to be some mix-up concerning the safety roles performed by the following (Bellamy, 1986):

- The operators
- The supervisors and management
- The US parent company
- Local central government
- The Indian government

The Bhopal incident emphasised the importance both of providing *automatic safeguards* – rather than relying totally on both manual responses and the activation of safety systems – and of installing safety interlocks on critical systems. In this incident, it appears that the operator was given little support, in the above and other respects, by the design of the plant (Bellamy, 1986).

Safety management procedures at Bhopal also appeared to have failed dramatically, in a number of major ways. For example, there were failures:

- To ensure that the relevant procedures were followed.
- To ensure that personnel at all levels had enough knowledge and training concerning the plant and its hazards.
- To ensure that the plant's design was safe.
- To maintain the plant effectively, and to evaluate modifications.
- To provide information about the high degree of toxicity of MIC, as well as information about the risks this factor involved both for the surrounding population and for the operators themselves.

This accident also suggested there may have been an inadequate level of communication between shifts. And there were furthermore several questionable design decisions – most notably the decision to *bulk-store* MIC, which in fact can be produced in smaller but sufficient quantities *on-line* as part of the process in which it is used.

12. *The Davis Besse incident*

On 9 June 1985, an incident involving a total loss of main and auxiliary feedwater occurred at the Davis Besse nuclear-power plant, while the plant was operating at 90 per cent power. The event involved 12 malfunctions, including several dependent failures, two maintenance errors and two operator errors, including a misinterpretation and the overriding of certain other personnel's advice. The event lasted 30 minutes and, frankly, can be seen as either a comedy of errors, a patent example of Murphy's Law (i.e.: if something can go wrong, it will) or a heroic recovery. The significant error forms are noted below, some of them most poignant since this incident occurred well after both the TMI incident and the subsequent recommendations on human-factors issues which that incident generated:

- The actuation of a safety system was prevented by a human error that arose either as a result of an inadequate design for a panel, or as a misconception by the operator.
- Four common-mode failures occurred, thus enhancing the confusion created by the incident.
- Four maintenance-related actions complicated the operators' tasks.
- The shift supervisor overrode procedures (arguably correctly), and then waited for the successful resurrection of the auxiliary feedwater system, rather than resorting to the more costly 'feed and bleed' procedure advocated.
- The operators involved, having exhausted all operational possibilities available at the CCR, sent personnel to open valves, etc., on the plant. These personnel quickly had to get past locked doors, via keys and plastic security cards, as well as go down steep stairs resembling ladders, and other such 'obstacles'.
- A PORV pilot-operated relief valve failed to close, as happened at TMI, but the operator *ignored* the warning given by the PORV-open indicator (placed in the CCR after TMI).
- Several kinds of failure occurred during this incident which had already occurred previously and yet which had still not been corrected by maintenance.

The Davis Besse incident shows the high degree of complexity that can occur in a real incident, as well as, of course, the ingenuity of the human operators in overcoming it safely. As with TMI, the importance of making the right diagnosis in a complex, fast-reacting system is paramount, and given this fact, the operators must have sound, clear information. Since the chronology of this incident exemplifies many important points already mentioned, the chronology in question is detailed below, followed by a summary of specific human-factors implications.

Time	Narrative	Malf'n	CMF	Operator
01:35:00	Reactor trip occurs. Operator detects event. Primary/secondary side operators work controls in accordance with procedures. STA (Shift Technical Advisor) contacted.			
01:35:37	Main steam safety valves open: MSIVs (main steam isolated valves) had closed. Operators detect abnormality.	1,2	1	
01:40:00	Main feedwater terminated.			
01:41:08	CRO (Control Room Operator) incorrectly actuates (manually) the SFRCS, and in so doing prevents it from coming on automatically.	(3)	2	1
01:41:44	Both ancillary feedwater pump turbines trip after going too fast.			
01:42:00	Isolation valves AF-599 and AF-608 fail to reopen themselves. (This error is totally dependent on the SFRCS error occurring at 01:41:08.)	6,7	4	
01:44:00	Operators exhaust CCR options for restoring feedwater, and as a result find themselves in a highly unusual and stressful situation. Many uncertainties about the likely success of future actions. Primary Operator directs equipment operators to go to the AFWP (auxiliary feedwater pump) room to determine what is wrong:			
	(1) Two equipment operators sent to AFWP manually to restore AFW pump back into service. (2) Assistant shift supervisor (ASS) goes to make SUFP (start-up feedwater pump) available.			

Time	Narrative	Malf'n	CMF	Operator
	(3) Two equipment operators sent to OTSG (once-through steam generator) AFW isolation valves AF-599 and AF-608 to open them.			
01:4?:??*	Problems experienced in controlling main steam pressure, which further decreases steam inventory. Operators in various locations race past locked doors via keys and plastic cards.			8
01:4?:??*	Assistant SS (Shift Supervisor) decides to place SUFP in service to provide feedwater to the generators. Has to get four fuses, since maintenance staff had taken it (and left it) off-line last January. ASS takes 4 minutes to perform a 12-minute task.			
01:4?:?*	Two operators in AFWP room unsuccessful in their task (neither had ever performed it before). Third, more experienced operator joins them and helps them to carry out the task. Difficulties in this task are eventually overcome.			
01:50:??	Operator leaves primary-side control station to reset the isolation signal to the SUFP values. Operator also asked to reset atmospheric vent valves. Operator absent from panels from 2 minutes.			
01:48:49	Pressuriser PORV (pilot-operated relief valve) opens and closes (01:50:09) twice without operator's knowledge.			

* Exact time unclear

Time	Narrative	Malf'n	CMF	Operator
01:47:??*	SPDS (safety parameter display system) inoperative. Trend information not available and difficult to glean from the instrumentation. Operators do not know that both OTSSs have boiled dry. This condition requires (procedurally) MU/HPI (make-up/high promare injection) or feed-and-bleed (FAB) cooling.			
	Secondary-side operator recommends FAB when hot-leg reaches 591°F, and operations superintendent tells SS that if an AFWP is not providing cooling facility within 1 minute, SS should initiate FAB.			
	SS decides to wait for AFWP to become available, on account of:			
	• Economic disadvantages of FAB • Breach of radiological barriers • Fact that cold shutdown would be delayed and holds FAB as a last-resort option.			
01:51:49	PORV fails to close.	9		
01:51:49	Operator ignores TMI PORV acoustic monitor. Equipment operators in AFWP room experience difficulties caused by communication and hardware problems.			2
01:53:00	Feedwater reaches OTSGs, and heating of reactor coolant system ends (peaks at 592°, then goes down to 540° in 6 minutes). Excessively rapid cooling leads to overcooling transient. Operators align system to prevent safety-features activation system from activating.			

* Exact time unclear

Time	Narrative	Malf'n	CMF	Operator
01:58:00	No 1 AFW pump suction transferred spuriously from the condensate storage tank to the service-water system. This action had occurred before, and not been corrected.	10		M*
01:59:00	A source range nuclear instrument becomes inoperable. This had occurred before, and had not been properly repaired. Control-room ventilation system trips into its emergency-re-circulation mode, which had also previously occurred.	11 12		 M*
02:01:??*	Operator terminates the AFW flow. Conditions become stable.			

(M* = maintenance failure)

* Exact time unclear

Some of the more important implications of this and other incidents are briefly listed below:

- Cognitive errors (related to diagnosis and decisions) are important, and must be prevented or protected against.
- Maintenance scheduling must adequately prioritise the fixing of controls and instrumentation. All too often, hardware failures in incidents have *already* occurred prior to the incident.
- Crew functioning and interrelationships are critical in real accident sequences. This crew, overall, reacted well, whereas less effective crews may have utterly failed to cope with the complexity of this incident.
- If an instrument fails, the operators probably need to know about it.
- The design of the panel is critical, particularly in highly stressful situations.
- Important alarms must gain the attention of the operators, and must be reliable, or else they will be ignored.
- Experience and adequate training are of high importance in emergency situations.
- If important safety events occur while the operator is absent from his normal position, some means either of drawing him or her back or of saving but still highlighting the information must be used.
- The communications hardware must be robust, and diverse media should be available.

- The control room should have diverse sources of instrumentation for critical parameters.
- There is a need, via training, to prepare operators in such a way that, in the event of unforeseen problems, they can successfully implement knowledge-based strategies to get the plant out of trouble.
- There is a need for a risk-management system which, in an unusual (non-proceduralised), situation would enable the correct executive decisions to be taken at the highest level of control, decisions which would determine the correct kind of response to different events.

13. The Challenger Space Shuttle disaster (1986)

On 28 January 1986, the Space Shuttle Challenger was destroyed in an accident during take-off, together with its full crew of seven astronauts. Essentially, the accident was caused by a low-temperature-induced fuel-seal (o-ring) failure on the booster rockets. Fuel leaking out eventually reached the ignition source at the bottom of the firing rocket boosters, whereupon the boosters catastrophically failed and much of the rocket disintegrated, the shuttle itself plunging back towards the ocean.

Watson and Oakes (1986) reviewed the accident and found the following major causes:

- An ineffective 'silent safety program' within NASA (this was the principal cause).
- The lack of a clear corporate safety organisation.
- The lack of adequate communications down the line and between lines (e.g., between project managers and flight-operations staff).
- An increased flight rate combined, on the other hand, with *decreased* resources.
- Maintenance-management problems.

Overall, the Challenger disaster appeared to lack a clear and adequate safety-management structure, and this may reflect the difficulties that a largely technical body such as NASA experienced in trying to cope with commercial pressures with which it was not familiar. This accident also highlighted the use of risk analysis in determining operational safety in a pressurized situation. According to the inquiry accidents documented in Rogers *et al* (1986), the company producing the boosters was concerned at the intention to carry on with the launch following very cold temperatures, since they could *not* verify the safety of the o-rings (the devices which failed) following such low temperatures. Thus, the engineers claimed that this matter fell outside of their design basis, and hence that they could not say if it was safe to launch. A management decision then overrode these concerns by arguing that the engineers could not at the same time prove that it was *unsafe* to launch – with hindsight a most unsatisfactory decision stance to adopt. This shows the effect of pressures on the management of technical systems, and demonstrates how good safety advice can sometimes be ignored.

Appendix II

Human-error data

Introduction

This appendix briefly reviews problems and major issues in data collection and then presents a series of tables of human-error data. These tables are not intended for use in HRA quantification exercises but are instead intended to give the assessor an insight into actual recorded incident rates and error probabilities. In each table, the source of the data/datum is also stated, as well as the type of data.

Data collection issues

(Note: the following three sections, but not the data tables themselves,* draw from a review of human-error data collection (Kirwan *et al*, 1990).)

There are two major types of human-error data which can be collected:

- Qualitative data: this information provides both general error-reduction strategies, based on human-factors experimentation, and also specific error-reduction guidelines, based on feedback from operational experience.
- Quantitative data: this information can be in the form either of relative data, e.g. 'the probability of error A is half that of error B', or of absolute data, e.g. 'the probability of error A is 0.1'.

Both types of data are useful in the context of human-reliability assessments, but there is in particular a need for the collection of absolute quantitative data for use in Probabilistic Risk Assessments (PRAs). These human-error-probability (HEP) estimates can then be used either in the validation of techniques which have been developed to quantify human error or more directly, for quantification purposes, if enough useful data exists.

---

* Data in the tables are drawn from Kirwan (1982), Kirwan *et al* (1990), and the database used in the Kirwan (1988) validation experiment.

Whilst this paper is primarily concerned with the derivation of (absolute) quantitative data in the form of HEPs, qualitative data are also of fundamental importance, and this kind of data can be collected at the same time as quantitative data. Such qualitative data is of great use, e.g. with respect to incident follow-ups and the determination of means of preventing incidents from recurring. As will become apparent later in the paper, qualitative data enriches the meaningfulness of the HEP, and ultimately enhances the determination of its range of applicability to various PRA scenarios.

Three potential sources exist for the collection of data suitable for the generation of HEPs. These are:

- Data derived from relevant operating experience
- Data derived from experimental research
- Data derived from simulator studies

Ideally, all data collected would be taken from relevant operating experience, or from sufficiently robust and industrially relevant experiments. Unfortunately, very little data have been collected from such sources, and thus recourse has had to be made to data from other sources such as the judgement of experts. The main reason for this is that there are a number of serious difficulties associated with the collection of operational experience data; these are discussed below.

Problems in the collection of operational-experience data

The following are major problems associated with the collection of (qualitative) operational-experience data in cases where such information is utilised for the generation of quantitative HEPs.

- If a reporting scheme is instituted whereby disciplinary action may be taken against those who commit an error, there will be a natural reluctance for events to be reported, either wholly or accurately. Error-reporting schemes should therefore preferably be based on no-blame or anonymous approaches such as have been adopted by the aviation industry. Schemes which are not based on this approach will usually lead to an incomplete database.
- A major prerequisite for the implementation of a successful data-collection scheme is the need to assign to specific individuals the responsibility for investigating incidents and collecting error data (see Lucas & Embrey, 1989). Implicit within this recommendation is a firm management commitment to data collection and data analysis, as well as a belief in the benefits of having dedicated personnel carry out such tasks.

Despite such difficulties, a number of data-collection schemes do exist. However, there are further technical problems associated with deriving actual HEPs from such information, and these are discussed in the next section.

Technical problems associated with HEP data derivation

Three major technical problems exist in relation to the generation of HEP data, even when an operational-experience data-collection system is in place:

- Human errors which do not lead to the violation of a plant's technical specifications are unlikely to be reported (even though they may lead to a violation of the limits of acceptability defined in connection with the errors under consideration). If this occurs, then the database, consisting of the number of errors collected, will be incomplete.
- Human errors which are recovered almost immediately, i.e. especially those recovered by the person who committed the error, are unlikely to be reported, and thus, again, the database will be incomplete.
- An analysis of operational experience may not yield sufficient information either on low-probability errors or on errors which may only occur during low-probability *event scenarios* – such as a fault diagnosis following a plant transient. This is because an insufficient number of events would be discovered during the data search that would prove statistically significant. Thus, for these types of error, recourse has to be made to other sources of data, such as simulator studies, or expert judgements.
- Current error-reporting schemes seldom carry information on the *root causes* of events, such as inadequate procedures, a poor working environment, ambiguous information-feedback to the operator, etc. Usually, only the consequences or observable manifestations of the error (called the external error modes – EEMs) are reported, such as 'Valve left open following test.' For an analysis of errors contained within a database, it is useful and arguably essential to understand the actual error mechanisms (how the error occurred in terms of operator functioning), as well as their associated Performance Shaping Factors (PSFs). Otherwise, errors may be aggregated which in fact are only superficially similar, which means that error-reduction measures may be inefficient or even ineffective.

These technical difficulties mean that the HEPs derived may involve a degree of uncertainty concerning their accuracy, in addition to the statistical uncertainty which will be attributed to them by the data-generation process (based on the number of observed events, etc.). Some attempts have been made to initiate data-collection schemes by the use of licensee event reports (LERs; for example, see Metwally *et al*, 1982), and these attempts have tried to overcome the technical problems listed earlier. The LER system is a mandatory reporting scheme which is controlled by the US Nuclear Regulatory Commission (USNRC). Any event which occurs at a US nuclear-power station in violation of its technical specifications must be reported as an LER, as part of the licensing requirement on the plant. Thus, this scheme provides a large and comprehensive database, of which human error forms a part. And attempts to generate HEPs from an analysis of this database have proved encouraging. However, the LER database gives practically no information on the particular PSFs that are associated with each event, and as a result its usefulness is limited. Other schemes which have attempted to address some of these critical aspects of data-collection are the human-performance-evaluation system (HPES; see, for example, Paradies & Busch, 1988), which is run by the Institute for Nuclear Power Operations (INPO), and the nuclear computerised library for assessing reactor reliability (NUCLARR, Gertman *et al*, 1988), but few useful outputs, in terms of data for HEPs in PRAs, have as yet emerged from these databases.

In view of the above history of data collection and its associated problems, it is appropriate to review the requirement for HRA data in line with current trends and theories in this area.

As noted earlier, the human-error probability (HEP) is defined simply as follows:

HEP = Number of errors observed/Number of opportunities for error

For a long time, this was the major, if not the sole, form of human-error data in PSAs. Thus, early attempts at database generation took the form either of performance-reliability statements (e.g. 'The operator can achieve the turning of a valve in the *right* direction "X" times out of a thousand') or of general guidelines concerning the probabilities associated with a generic type of error (e.g. 'The operator omits a step in a procedure (i.e. any procedure) "X" times out of a thousand, or "Y" times out of a thousand if the step is an "isolated" step at the end of the task.')

These two types of data can be utilised in PRAs, and *have* indeed been used extensively, but there are limitations and disadvantages associated with both of them. The first problem is concerned with the degree of specificity inherent in the data given for the plant undergoing a PRA. Large variations exist between different plants in terms of the way they are operated, their training and procedural facilities, the safety-management culture, the ergonomical level of adequacy inherent in the equipment's design, etc. In PRAs, either of these types of data are likely to be applied indiscriminately to widely different types of plant. However, although this may not be an ideal solution to the problem of how to quantify the human-error impact on a specific plant, it could be counterargued that the differences in HEPs for specific plants (given the same human error in question) would be relatively small in PRA terms, and so would not lead to a markedly wrong prediction for either plant, especially given the large degree of uncertainty already inherent in the PRA process. The resolution of the argument concerning the need for more plant-specific data, or even for more data which is sensitive to a plant's individual characteristics, is only likely to be conclusively achieved if various amounts of data are derived from different plants and then compared. Such comparisons are unlikely in the near future.

However, there is a second and more immediate problem, which concerns the usefulness of the data for error-reduction purposes. The types of data mentioned above do not give information on how to improve human reliability in those cases where it is found during the PRA evaluation that the plant is not satisfying the risk criteria (due to the human-error impact). In this case, if there is no easy way of improving human reliability, then other approaches (interlocks, automation, extra safety systems, etc.) may have to be considered. This may not be the most cost-effective error-reduction strategy. There are many factors known to influence human performance (the quality of the operator interface, the extent of training, etc.) which could be used to raise the standard of human reliability. The types of data described above, therefore, are not 'useful' in the sense of providing means of error reduction, should such means be required.

A third problem with purely quantitative data, one related to the second problem, is that such data only states the external form, or observable manifestation, of the error (i.e. the external error mode – EEM). Returning to the above-mentioned example of turning the valve the wrong way, this error could be due to a momentary aberration on the part of the operator – called a 'slip'. It could also occur as a result of the operator's experience on *other* plants where valves had to be turned in the *opposite* direction. The point here is that the operator involved in the second situation is far more likely to make the error, due to a 'population stereotype', than he or she would if the error were due purely to a 'slip'. And the associated HEP could also differ dramatically. The external error mode is the same but the actual root cause, or psychological error mechanism (PEM), is different. There are two main implications connected with failure to identify PEMs during a PRA;

- If data is generated in the form of EEMs only, and it has not been determined what are the psychological error mechanisms underlying the EEMs, then the application of such data in a PRA for a plant may lead to inaccurate results because of the further existence, in the plant scenario, of dominant PEMs which were not accounted for in the original data. Therefore, the data may be inaccurately applied.
- Although PSFs may be used to reduce the impact of human error, this may not be as effective a strategy as eliminating the root cause of the error. Furthermore, the recommended change in PSF terms may in certain circumstances go against the more appropriate solution suggested by the root-cause problem; for example, more training might be provided in a case where the root cause of the problem is in fact *overlearning*. Therefore, the data may not be effective in error reduction.

The above conclusions concerning the use of data in HRAs carried out within a PRA framework can now be summarized. The original data believed to be required comprised HEPs for associated generic task types or error types. However, it has been argued that such data will largely be applied indiscriminately, will not be useful in error reduction and may also be inaccurately applied in certain situations. It appears, therefore, that in modern PRAs, what is required is not merely data in the form of EEMs but data with associated PSFs and PEMs. Such data could, in theory, be applied more accurately in the assessment of human-error contributions, and could prove highly useful in error reduction.

The study reported in Kirwan *et al* (1990) details one recent successful approach to deriving data. The study was based on the investigation of incidents at BNFL Sellafield, using the EEM/PEM/PSF classification embedded in the HRMS. Out of the 70-plus incidents reviewed, 34 HEPs were derived. Although these are still being researched, in order for their uncertainty bounds, and other aspects, to be determined, the study did at least show that data concerning PSFs, PEMs and EEMs can be successfully complied.

Data tables

Tables II.1–II.4 show some data available from a range of sources.* These forms of information comprise:

- Generic data: typical judgement-derived kinds of data that nevertheless provide acceptable guidelines for HRAs
- Data from operational plants
- Data based on ergonomics studies
- Data from simulator studies

It is not intended that these be used directly. Rather, they are presented to give the practitioner a 'feel' for error rates, as far as this is possible. Furthermore, for most of the following data presented, there is little information on PEMs and PSF. This is the state of the art, though research is ongoing at present to improve matters (Taylor-Adams and Kirwan, 1994).

The practitioner may also review any estimates derived from his or her own HRA cases. Such HRQ-technique-derived data is generally termed 'synthetic data', and has less 'face validity' than more empirically founded data. However, some of the case studies in Appendix VI *do* contain two or even three sets of estimates for required HEPs, and in some cases these independent estimates show a considerable degree of agreement or convergence. In such cases, the face validity of the data in question increases markedly, as a result of which such data may be used as a benchmark, or guideline, until better data are obtained.

---

* Data in the tables are drawn from Kirwan (1982), Kirwan *et al* (1990), and the database used in the Kirwan (1988) validation experiment.

Table II.1 Generic guideline data

Description	Human-error probability
1. General rate for errors involving very high stress levels	0.3
2. Complicated non-routine task, with stress	0.3
3. Supervisor does not recognise the operator's error	0.1
4. Non-routine operation, with other duties at the same time	0.1
5. Operator fails to act correctly in the first 30 minutes of a stressful emergency situation	0.1
6. Errors in simple arithmetic with self-checking	0.03
7. General error rate for oral communication	0.03
8. Failure to return the manually operated test valve to the correct configuration after maintenance	0.01
9. Operator fails to act correctly after the first few hours in a high-stress scenario	0.01
10. General error of omission	0.01
11. Error in a routine operation where care is required	0.01
12. Error of omission of an act embedded in a procedure	0.003
13. General error rate for an act performed incorrectly	0.003
14. Error in simple routine operation	0.001
15. Selection of the wrong switch (dissimilar in shape)	0.001
16. Selection of a key-operated switch rather than a non-key-operated switch (EOC)	0.0001
17. Human-performance limit: single operator	0.0001
18. Human-performance limit: team of operators performing a well-designed task, very good PSFs, etc.	0.00001

Table II.2 Data from operational plants

Description	Human-error probability
1. Invalid address keyed into process-control computer	0.007
This error occurred in a computer-controlled-batch chemical plant. When a valve sticks, or another malfunction occurs, the operator goes through a sequence on the computer which includes entering an address code for the component to be manipulated. The operator could, however, enter the wrong address, i.e. either an address for which there is no item, or the address for the wrong item; the HEP reflects the sum of these two alternative errors. There is a plant mimic available, prompt feedback is given of control actions and the task occurs in normal operations.	

(*Table II.2 Continued*)

Description	Human-error probability
2. Invalid-data error in process-control task	0.003

In the same plant as above, changing set-points is achieved by entering the new set-point on the computer. The computer shows the old and new set-points, and the operator then either accepts the new one he or she has just entered or else he or she changes it again. The error in question involves entering a set-point which is outside the allowable range for that parameter or equipment item. In this case, the computer will not accept the new set-point, and will give immediate error feedback to the operator.

3. Control error in process-control task	0.002

In the same plant as above, an operator may attempt to open a valve which is currently under automatic control and therefore not manually alterable via the computer. This is termed a control error. The operator should be aware of the state of the plant, and of associated equipment, via the plant mimic.

4. Precision error: incorrect setting of chemical interface pressure	0.03

In this event, an interface-pressure setting was incorrectly set, allowing an aqueous solution to pass into a stock tank, where it subsequently crystallised – which was a highly serious consequence. This error was largely caused by the failure on the part of the operator to be precise enough when setting the equipment.

5. Nuclear-fuel containers stacked above their limit	0.001

In this event, two containers had been temporarily double-stacked, in contravention of operating procedures.

6. Welders worked on the wrong line	0.04

Welders at a chemical plant worked on a vent line by mistake and holed a pipe.

7. Alarms disabled on large incoming equipment	0.0005

In this event, nuclear-fuel containers coming into the plant had had two of their alarms disabled (a serious rule violation).

8. Erroneous discharge of contaminants into the sea	0.0007

In this event, material was erroneously discharged into the sea, partly due to a communication failure across two shifts.

9. Fuel-handling machine moved whilst still attached to a static fuel flask	0.0005

In this event, the fuel-handling machine in question, resembling a large overhead crane but with a very limited view, from the crane-cab, of the flasks it carries, was moved by the operator while it was still in fact attached to a flask via flexible hoses, thus rupturing the hoses. This accident was in part caused by a communication failure across a shift-break,

(Table II.2 Continued)

Description	Human-error probability
i.e. a failure of communication between the first operator who was carrying out the flask operation and the second operator who came on shift and went immediately into the next operation.	
10. Critical safety system not properly restored following maintenance	0.0006
In this event, a US boiling-water-reactor (BWR) core-spray-pump system was left in an incorrect line-up configuration after testing. Testing is done by the operator in the CCR five times per year on five similar systems. This particular error occurred on the control switches on the CCR panels. The consequences are serious, since the effect is to disable a back-up safety system.	
11. Wrong accumulator drained in a US PWR task	0.0021
In this event, as part of a safety-injection task in a US pressurised water reactor (PWR), the wrong accumulator was drained. This task is controlled by the operator in the CCR via control switches, and takes place twice per year (there are only four accumulators).	
12. Emergency-core-cooling-system valve misaligned	0.0003
In this event, the ECCS was left after maintenance with a valve misaligned. The consequences of this error would be very serious if a loss-of-coolant accident were to occur.	
13. Valves mis-set during calibration task	0.001
In this task, an operator must carry out a calibration task in which valves must be set to the correct value in order to ensure the correct functioning of a monitoring system. Both feedback and procedures would benefit from any improvements made to this task.	
14. Operator works on wrong pump	0.003
In this event, an operator on a plant was instructed to work on a pump in the west part of the plant but instead erroneously worked on the identical *east* pump.	
15. Wrong fuel container moved	0.0007
In a supervised and heavily logged operation, the wrong fuel container was moved via a crane. The operator in the crane-cab did not have a direct view of the containers but could only see them via a CCTV facility. The operator was, however, in communication with local operators who *could* see the containers directly.	
16. Failure manually to close a valve at the end of a task	0.01
At the end of this 10-step task, the operator must close a valve. This final part of the task is a critical one, and yet the procedures and training programmes do not strongly emphasise this fact.	

Table II.3 Data derived from ergonomics experiments

Description	Human-error probability
1. Human-recall performance with digital displays	0.03

A six-digit sequence was presented for 2.6 seconds. The subject then had to write down the digit sequence in the intervening ten seconds before the next sequence was presented. Seventy-two slides of 6-digit sequences were shown to each subject in this experiment, taking roughly 12 minutes per subject. The error in question involved not writing down the correct sequence.

2. Inspectors' level of accuracy in spotting soldering defects in a complex system	0.2

In a study of the capabilities of quality-control inspectors, a complex unit with 1500 wires soldered to various terminals was examined by the inspectors over a three-hour period. Thirty defects had been placed in each unit, and the inspectors had to find all these defects, which were similar to the kinds of defect they would find or look for every day.

3. Typing performance	0.01

Each touch-typist in this experiment was instructed to type out a 1000-character piece of text as fast as possible without exceeding an error rate of 1 per cent.

4. Keyphone error	0.03

A standard keyphone layout (i.e. with '1–2–3' as the top line) was examined in this experiment. Subjects were asked to dial 40 seven-digit numbers on the keyphone. The error here was an incorrect sequence of digits (including wrong or missing digits, etc.)

5. Network problem-solving: a premature diagnosis	0.07

Subjects were required to find the faulty component in a network of AND units. If two units feed into another unit, and one or both of the first two units are unhealthy, then the third unit will read 'unhealthy'. However, the operator can only see which units are healthy or unhealthy at the end of a line of connected units, though he can also see how all the units are interconnected. Thus, the correct diagnosis involves determining which unit is unhealthy and is affecting other units (only one unit is unhealthy in each network), and requires the operator to perform a type of fault diagnosis known as 'backward chaining'. This task is very similar to a fault diagnosis for electrical maintenance panels. The number of units per network ranged from 16 to 24, always with four main 'lines' leading to four final output states (healthy or unhealthy). A *premature* diagnosis implied that the operator identified the faulty unit without first having carried out enough tests conclusively to determine which unit was faulty (irrespective of whether the premature diagnosis was correct or not: the task cannot afford the operator to make premature guesses).

(Table II.3 Continued)

Description	Human-error probability
6. Novel fault diagnosis in simulated process-control task	0.34
In a simulated process plant using back-projection slides of panel instrumentation, etc., each process-control trainee was faced with 33 instruments and up to 15 alarms. The process plant that was simulated included reaction and distillation functions, as well as conventional instrumentation and control features associated with pre-VDU-based control systems. Trainees were instructed in diagnostic rules and systems used for scanning the panel, etc. (e.g. how to check control loops in the affected area). Having practised a number of failures, the trainees were then given eight new failures, and allotted five minutes to diagnose each one correctly.	
7. Fault diagnosis using rules	0.16
As part of the previous experiment, the trainees were also tested on previously seen faults, with no feedback or supervision and with five minutes for each problem.	
8. Failure to carry out a 1-step calculation correctly	0.01
9. Failure to carry out a 7–13-step calculation correctly	0.27

Table II.4 Simulator-derived data

Description	Human-error probability
1. Emergency manual trip in a nuclear control room	0.2
Prior to a fault appearing, the operator would be occupied with normal operations in a simulated control room. Initially, when, in this case, a fault appeared, the operator was expected to try to control the fault, but as it quickly became apparent that this was not possible, the operator was required instead to shut down (trip) the plant. The faults in question comprised a control-rod run-out, a blower failure, a gas-temperature rise and a coolant-flow fault. Tripping the plant required a single push-button activation. The fault rate in this scenario was 10 signals per hour (normally, it would have been of the order of 1 in 10,000 hours). The operator only had between 30 and 90 seconds to respond by tripping the reactor, during which time the operator would also have had to detect and diagnose the problem and then take action almost immediately.	
2. Omission of a procedural step in a nuclear control room	0.03
This HEP is based on a number of different scenarios which were faced by shift teams in a full-scope nuclear-power-plant (NPP) simulation in the USA. The shift teams, all of whose members were being recertified as NPP operators, were required to deal with a number of emergency scenarios.	

(Table II.4 Continued)

Description	Human-error probability
3. Selection of wrong control (discrimination by label only)	0.002

This HEP, which was derived from a number of NPP simulator scenarios, was based on 20 incorrect (unrecovered) selections from out of a total of 11 490 opportunities for control selection.

4. Selection of wrong control (functionally grouped)	0.0002

As above, but this time the HEP is based on only 4 unrecovered errors out of 27 055 opportunities for error.

5. Incorrect setting	0.0002

As above, and based again on the unrecovered errors.

6. Equipment turned in wrong direction	0.0002

As above, based on the unrecovered errors again, and with equipment that does not violate a population stereotype (i.e. with normal, expected turning conventions).

7. Diagnostic response rates for various scenarios and times

The following estimates for responses within a short time-frame are all taken from US NPP simulator studies, and relate to a set of known emergency-event scenarios. The figures relate to the diagnostic failure probabilities for various different short time-frames. It is generally assumed that the more time there is available for diagnosis, the more reliable the diagnosis will be. In practice, however, this assumption appears to break down, since there appears, in many cases, to be an asymptote, after which, if a diagnosis has not been achieved correctly it probably *never* will be, irrespective of how much more time is available – unless *new* personnel arrive (as arguably happened in TMI).

7.1 Nuclear-instrument failure	0.08/5 mins
7.2 Temperature failure	0.08/5 mins
7.3 Loss-of-coolant accident	0.30/3 mins
7.4 Steam-generator-tube leak	0.08/4 mins
7.5 Control-rod accident	0.40/3 mins
7.6 Inadvertent safety injection	0.4/10 mins
7.7 Small-break LOCA	0.5/15 mins

Appendix III

Validation of HRQ techniques

Introduction

This appendix reviews evidence for the accuracy and usefulness of the various HRQ techniques described in Section 5. A qualitative review by the Human Reliability Assessment Group (HRAG) is summarised, followed by a consideration of some recent quasi-experimental studies which compared the practical use of these various techniques.

Few proper validation experiments have actually taken place (see Williams, 1983). Most of those which have occurred are described or summarised (or both) in the HRAG (1988) reference and the Swain (1989) review discussed in the previous section. Two further experiments are briefly reviewed in this section, and these add some insights into the validity of techniques, as well as the difficulties involved in the validation process itself.

III.1 Human Reliability Assessor's Guide (HRAG)

The HRAG is the result of a lengthy peer review carried out in the UK under the auspices of the Human Factors in Reliability Group, a multi-industry body aimed at promoting an awareness, in industry, of human-factors (HF) and human-reliability (HR) approaches. The HRAG reviewed eight techniques against a number of carefully defined criteria. This review produced a section on the selection of appropriate techniques for different types of situations, or, for example, for different levels of available resources. In addition, the HRAG contains a number of case studies showing how each technique has actually been used in practice, and for this reason the document is particularly useful for the HRA practitioner. The following focuses on the results of the review concerning the selection of a technique for a particular application. The HRAG assessed the techniques against a set of criteria, namely:

- Accuracy   – numerical accuracy (in comparison with known HEPs)
                       – consistency between experts and assessors

- Validity   – the use of ergonomics factors/PSFs as aids to quantification
             – the theoretical basis in ergonomics, psychology
             – the degree of empirical validity
             – the degree of validity as perceived by assessors, experts, etc.
             – the degree of comparative validity (perceived after comparing the results of one technique with the results of another, for the same scenario)
- Usefulness   – the degree of qualitative usefulness in generating error-reduction mechanisms
             – the degree to which the sensitivity analysis allows the effects of the error-reduction mechanisms to be assessed
- Effective use of resources
             – equipment and personnel requirements
             – data requirements; e.g., the SLIM and PC approaches require calibration data (containing at least two known HEPs)
             – the training requirements of assessors or experts – or both
- Acceptability – i.e. to regulators/the scientific community/assessors
- Auditability – i.e. of the quantitative assessment
- Maturity   – the current maturity and development potential

The HRAG review has been criticised as being too favourable towards the SLIM-MAUD system. Its overall results did not, however, suggest that there was any one single technique that was applicable to every HRA situation. What it instead suggested was that the practitioner, rather than looking for the best techniques, should remain flexible in his or her approach. (The TESEO and HCR methods were *not*, however, recommended by the HRAG.) The study showed that whilst there was significant room for improvement in all the techniques reviewed, and despite the lack of validation, there *are* sufficiently developed techniques available that allow the assessor to carry out professional assessments on most, if not all, types of human-error scenario. Quantitative human reliability had 'come of age'. The results of the HRAG review are encapsulated in Figures 5.5.18 and 5.5.19 in the main section on quantification techniques.

Swain has recently (1989) reviewed a large number (14) of techniques on an even larger set of (over 50) sub-criteria. This review, which is a very detailed one, is not, as it turns out, consistent with the HRAG document in a significant number of cases. The practitioner who wishes to become totally immersed in quantitative HRAs is recommended to take the trouble to read Swain's review, as it is too detailed to try and summarise here.

III.2 Validation experiments

The purpose of validation experiments is very simple: to establish the *credibility* of HRA techniques, so that they may be used with confidence. This requires, of course, that all the relevant data to be used for validating such techniques be made available. Unfortunately, not all so-called validation experiments actually follow this requirement. Sometimes, instead (e.g. Metwally *et al*, 1982), the predictions of one technique are compared with those of another technique (e.g. the THERP), a process which is called convergent validation. It is all very well if the two sets of predictions *do* converge, but of course both sets of results could be wrong in the first place! Sometimes, again, expert opinions or simulator data are used to decide if the results are adequate. What is preferable, of course, is to test the techniques against 'real' field data. However, as the third part of this section will show, the area of data collection and generation is itself fraught with difficulties.

Two studies are considered, the first of which compared six techniques (including a simple variant on the (Rasmussen, 1975) WASH-1400 human-error database, called *HED*, which was developed in BNFL and used by two expert assessors) firstly in terms of their degree of accuracy against known data and secondly in the context of scenarios which were very PSA relevant but for which data was not available (this study, therefore, took partly a true-validation and partly a convergent-validation approach. The second study, the European Benchmark Exercise, compared a number of approaches in the context of PSA-relevant scenarios for which data was not available; this was therefore a convergent-validity study.

III.2.1 *Experimental comparison of six HRQ techniques*

The experiment detailed in Kirwan's 1988 publication (from which this section draws) evaluated a total of six Human Reliability Quantification approaches, namely:

- APJ
- PC
- THERP
- HEART
- SLIM
- HED

ACCURACY

Each technique was used to calculate a maximum of 21 HEPs for which the 'true' values were known. (The PC and SLIM methods, failed to predict all of these, due to resources limitations.) This exercise was carried out to determine the degree of accuracy of the techniques, and thus to provide an empirical validation of the techniques themselves. Figures III.1 to III.6 show the results of this first validation exercise. The human-error database performed relatively well by comparison, and the APJ and THERP techniques also performed well. The highest correlations were achieved as follows, in order of significance.

Figures III.1 and III.2 Experimental results (Kirwan, 1988)

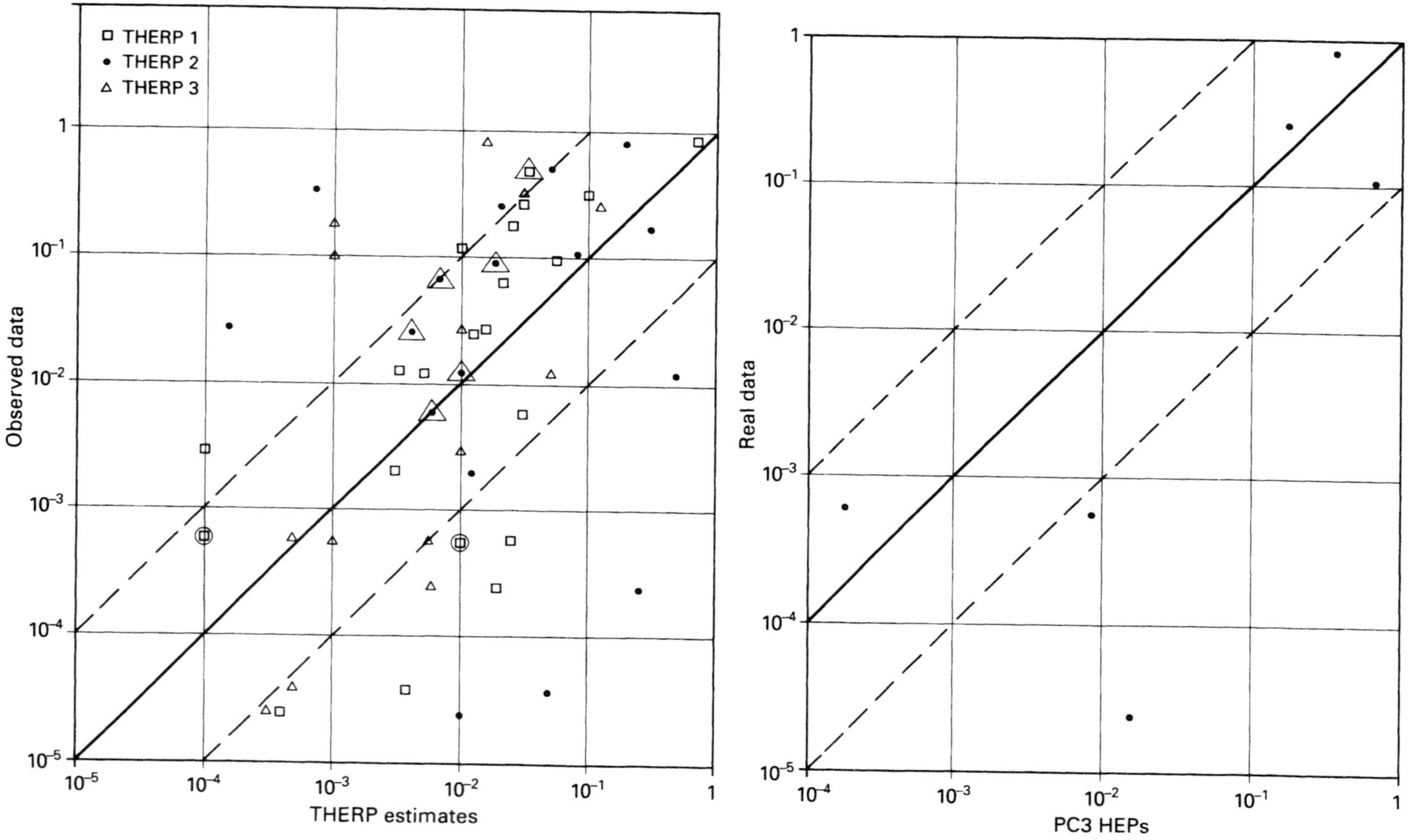

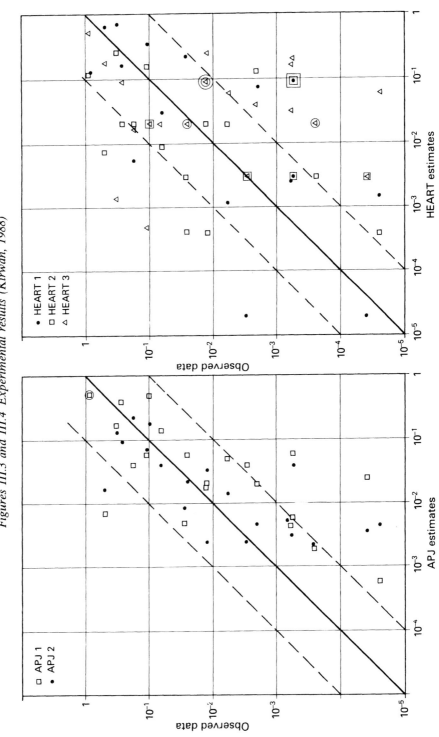

Figures III.3 and III.4 Experimental results (Kirwan, 1988)

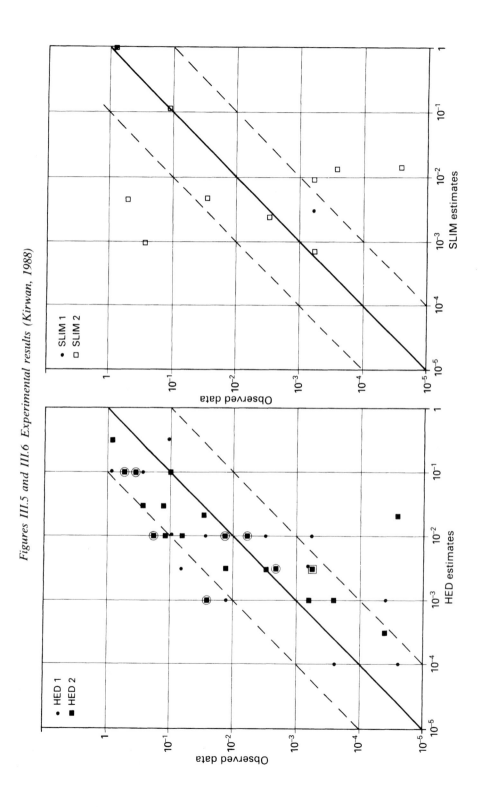

Figures III.5 and III.6 Experimental results (Kirwan, 1988)

Technique		Correlation co-efficient	Significance
HED	(Assessor 1)	0.7705	p<0.00004
APJ	(Group 2)	0.7340	
HED	(Assessor 2)	0.6958	
HEART	(Assessor 2)	0.6553	
APJ	(Group 1)	0.6446	
THERP	(Assessor 3)	0.6311	
THERP	(Assessor 1)	0.6095	p<0.003

None of the other techniques' results were significant.

These results suggested that the HED, APJ, HEART and THERP techniques are the most accurate ones. Unfortunately, the evaluation carried out for the SLIM and PC approaches proved to be less robust. In the case of the SLIM method, this was because the technique took longer to apply than anticipated, as a result of which not enough known HEPs were quantified to make a proper correlation possible. With the PC approach, as with the SLIM approach, having a greater number of data points would have served to enhance the effectiveness of the test.

The degree of precision of the techniques varied as shown below, an adequate degree of precision here being ascribed to any figure falling within one order of magnitude of the 'real' value:

Technique	Total no. HEPs assessed	Optimistic	Pessimistic	Degree of precision
HED	42 (2 assessors)	7 (16%)	3 (7%)	77%
APJ	42 (2 assessors)	2 (5%)	8 (19%)	76%
HEART	63 (3 assessors)	8 (13%)	18 (29%)	59%
THERP	63 (3 assessors)	12 (19%)	14 (22%)	59%
SLIM	11 (2 groups)	2 (18%)	3 (27%)	55%*
PC	6 (1 'group')	0 (0%)	2 (33%)	66%*

* The precision estimates of the SLIM and PC approaches are not robust due to the low number of HEPs derived.

Thus, overall, the techniques enjoy a 55–77 per cent degree of precision, with the HED and APJ approaches being the most precise methods.

The second part of the experiment compared the techniques in more realistic PSA settings, using five safety-case human-error fault and event trees. Here, the highest level of inter-technique/assessor agreement was exhibited by the THERP, HEART and APJ methods. Figure III.7 shows the global results (i.e. the summated fault and event trees) for all the various techniques, compared against the BNFL (HED-based) estimates (a maximum of seven scenarios – 1 to 4, and 5a, 5b and 5c – were assessed). It is noticeable that these summed estimates are generally in agreement, with three exceptions: two pessimistic assessments by the HED approach (global estimates of 1.0 and 10^{-5}) and one severely optimistic assessment of 10^{-6} (in contrast to the other techniques' estimates, which centred around the 10^{-2} range. Whilst the pessimism of the HED estimate on the first two scenarios (pessimistic by just over one order of magnitude) perhaps reflects an acceptable measure of conservatism, the optimism displayed in one of the scenarios clearly, however, presents a cause for concern. This issue is currently dealt with via HPLVs (human-performance limiting values).

Table III.1 summarises the results of the experiment in terms of its findings on the accuracy dimension. The three components of accuracy in this table are the degree of precision, as defined above, the degree of consistency, i.e. how well each assessor/group correlated with the others when using the same technique, and the degree of 'convergence', i.e. the degree of correlation between two different techniques' results for the same scenario(s).

USEFULNESS

In the introduction, it was noted that the HEART and SLIM techniques are potentially highly useful techniques in aiding the assessor in determining how cost-effectively to reduce the human-error factor, if such a reduction is required. In the above experiment, neither the PC nor the HED methods produced any particularly useful insights in this respect – as might have been predicted. Interestingly, the THERP and APJ techniques, which are usually seen as not especially 'useful', *did* produce this type of information. One THERP assessor produced a set of useful procedural information on how to improve the situation, while the APJ sessions involving expert groups generated, indeed, a surprising amount of useful and practicable information on risk reduction and safety practices (this also occurred in the SLIM sessions – as expected, however). Whilst this information was not linked quantitatively to the HEPs (which meant that a sensitivity analysis could not be easily carried out as a means to determining the degree of cost-effectiveness of the recommendations produced), the THERP and APJ methods were nonetheless worthy of note as being of potentially significant *qualitative* usefulness. The techniques are therefore evaluated on this criterion as follows:

Most useful – HEART*
              – SLIM*
              – APJ
              – THERP
Least useful – PC/HED

* Based on HRAG.

RESOURCES USE

The techniques are rated below on this criterion simply as a function of the number of person-hours the technique utilised in carrying out the analysis of

Figure III.7 Global prediction results (Kirvan, 1988)

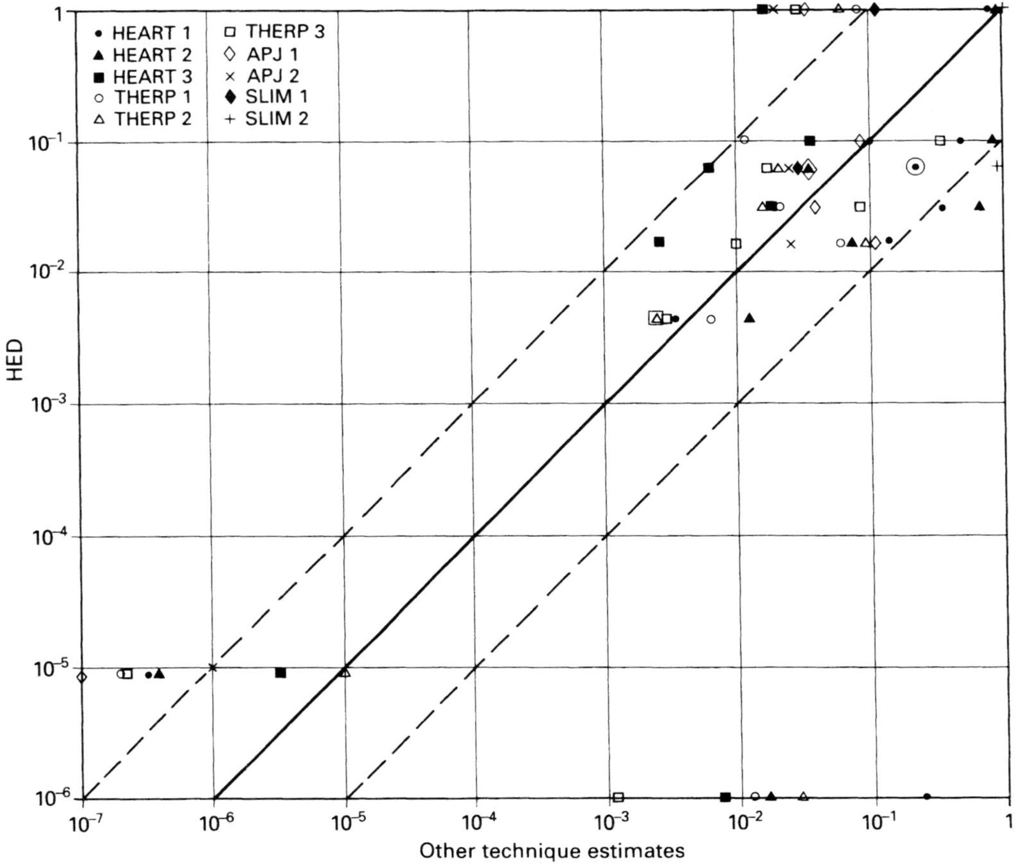

Table III.1 Summary of accuracy findings

Technique	Precision	Consistency	Convergence
APJ	High	Moderate	High/Moderate
HED	High	Undetermined	Low
THERP	Moderate	High	High
HEART	Moderate	Moderate	High
PC	Moderate*	High	Moderate
SLIM	Moderate*	Low	Moderate

* Not robustly determined.

the different scenarios (this was calculated *pro rata* for those techniques which did not complete all scenarios). Other resource costs, such as the cost of software (e.g. for the SLIM), are not included since, for any company, these would represent only a one-off payment, and would thus be less significant than the personnel time expended in each of the sessions.

	Technique	Total/projected time for all scenarios (one assessor or group)
Least resources:	HED	2 days
	HEART	3–5 days
	PC	10 days
	THERP	8–30 days
	APJ	15 days (five personnel, 3 days each)
Most resources:	SLIM	20 days (five personnel, 4 days each)

It is difficult to determine the PC approach's resource requirements, since the PC participants in this study benefited each from a detailed discussion about the scenarios. The actual time it takes to fill out a questionnaire for all scenarios would have been approximately one day, and with preferably at least 10 subjects, this yields a total of 10 days. There was also wide variance in the use of the THERP method, in that some very detailed analyses were, for example, produced by one assessor (who, incidentally, also gave relatively accurate results). The SLIM technique was marginally slower than the APJ approach. Additional weight may be given to the fact that the APJ and SLIM approaches require the identification and co-location of a panel of experts, as well as the fact that the SLIM and PC approaches also, in theory, require calibration data, with which to generate absolute HEPs, to be available.

Overall, resources requirements span a whole order of magnitude, demonstrating that techniques do differ significantly in terms of resource-use levels. Individuals should consider this fact alongside other criteria, particularly that of the degree of usefulness, when determining the level of resource-effectiveness of different techniques.

CONCLUSIONS ARISING FROM EXPERIMENTAL EVALUATION

- The THERP and APJ techniques were found to be the most accurate techniques.
- The HEART method showed promise on the accuracy dimension, and it was suggested that, for the purposes of further development, it should focus on achieving a higher degree of inter-assessor consistency.
- The HED method was shown to be reasonably accurate.
- The SLIM and PC approaches require a more robust calibration procedure that would make the HEPs generated at least as valid as the *scale values* would appear to be.
- The HEART and THERP systems both appear to achieve their best results when they make use of trained assessors.
- The APJ approach, and to a lesser extent the THERP approach, were both found to produce useful *qualitative* forms of information. Such information was not, however, quantitatively linked to the HEPs as it was with the SLIM or HEART techniques (the most useful techniques).
- The HEART and HED techniques required the least amount of resources, while the APJ and SLIM techniques, on the other hand, required the most. The THERP and PC techniques, on their part, required a moderate degree of resources, with the amount of resources required for the THERP method being fairly variable because they are dependent on the individual assessor.
- The effectiveness of the quantification process is critically dependent on the degree of accuracy achieved in the modelling of scenarios, and in the light of this it has been recommended that the focus of attention be switched, to a greater extent, away from pure quantification and onto error-identification and error-representation issues. In particular, the results of this experiment raised the important question of whether a more detailed modelling process is necessary if accurate results are to be gained.

III.2.2 European Benchmark Exercise (Waters, 1989)

The human-factors-reliability benchmark exercise (HF-RBE), reported by Waters (1989), was one of four benchmark exercises organised by the Joint Research Centre (ISPRA), Italy. This exercise involved 15 teams, from both Europe and the USA, who were asked to model and quantify various error scenarios. Six techniques were used in the HF-RBE: APJ, HCR, HEART, SLIM, TESEO and THERP (the THERP technique was used more than any other technique). The system under investigation was the Kraftwerk Union (KWU) pressurised water reactor at Grohnde. The assessed scenario of interest here concerned the Emergency Feed System, and involved both the routine testing of components and of fault-management procedures.

Unlike the previous study, the HF-RBE allowed the teams themselves to model the situation, and this process of qualitative analysis – which included a task analysis, an FMEA-type approach (i.e. one similar to a human-error-analysis table), and an error-identification/-representation process – exhibited a great deal of variety among the different teams. Unfortunately, there was a distinct lack of convergence between some of the team results generated, and this appeared to be largely because of the existence of different modelling assumptions. In a sense, the failure to ensure consistent modelling between different teams meant that the experimental validation was confounded. And what the study as a whole showed, regrettably, was that a real HRA, carried out from start to finish (i.e. from task analysis through to quantification), would not prove reliable, in particular due to the lack of a consistent methodology for the qualitative analysis. This result echoed the results of Brune *et al*'s (1983) study in which over 20 participants carried out a THERP analysis. At the worst, the estimates arrived at varied by several orders of magnitude for the same scenario. This was attributed by Swain to the different assumptions that assessors had made during the qualitative-analysis stage.

The HF-RBE also found, in one case, that the SHERPA and CADA techniques could be put to good use. However, this happened to be the case for a *real plant*, where operators could be interviewed, video-taped, etc.; it may, on the other hand, be less true for a design project. Another result was that no single HRA method of quantification appeared to serve *all* the assessment's requirements, i.e. a method might be good at fault-management error-quantification but not be good at normal operations/maintenance assessments, or vice versa. In addition, event-tree modelling was felt to be more useful than fault-tree representations.

Overall, the results of the study did not really help to validate any of the approaches. Rather, they pointed unequivocally to the current weak point in the HRA process, namely the qualitative analysis.

Appendix IV

Human factors checklist

Introduction

The following pages constitute a basic human-factors checklist for process control industries. The checklist is mainly aimed at control-room applications in the process-control field. There are also a number of exemplary figures:

- Figure IV.1: an example of the recommended dimensions for a console (for both UK male and female operators) that is used for sit/stand operations.
- Figure IV.2: recommended dimensions for a vertical console.

There are also a number of tables. Table IV.1 is a table of recommendations for control selection; Table IV.2 shows the required character height as a function of the viewing distance; Table IV.3 (organised in three parts) gives some recommendations on colour coding – as well as both a sample colour-coding format from a plant and some recommended foreground–background colour combinations; Table IV.4 gives some recommendations on VDU-screen-format design; and Table IV.5 outlines the pros and cons of typical VDU input devices.

In addition, there is a section in the bibliography listing guides and handbooks on human-factors engineering. In particular, the books by Ball (1991), Kincade & Anderson (1984) and Sanders & McCormick (1992) are recommended.

Control-room layout

1. Are controls and displays placed in functional groups, such as systems or subsystems?
2. In cases where instruments are always used in a standard sequence, are they laid out in that sequence?
3. Are displays and controls which are used together located next to one another?

Figure IV.1: Seated Console Design

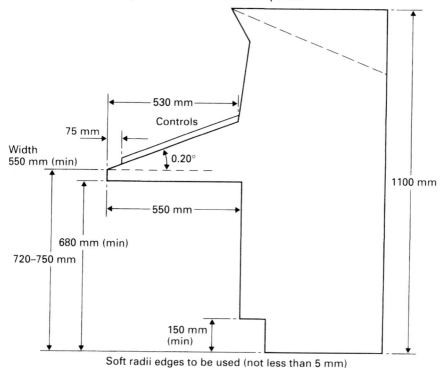

The top of console should slope downwards
if the operator is required to view over the
console – so as to accommodate the sight-line
requirements of the shortest operator

Soft radii edges to be used (not less than 5 mm)

4. Have those displays that present important information been given a prominent location, so that any urgent information will almost immediately be detected by the engineers?

5. Will the range of body dimensions accommodated in the control-room design encompass at least the fifth-percentile female and the 95th-percentile male?

6. Have seated-console profiles been implemented in cases where the engineers must interact with a number of instruments that are within their stationary reach, and that may also require precise adjustment?

7. When the engineer is at the seated console, are any instruments that he or she needs to see obscured, or too small to see?

8. Has sufficient space (e.g. 50 inches) been left for any future maintenance work on instrument panel racks?

9. Are the desks oriented so that the engineers are not sitting with their backs to instruments?

10. Are the supervisors' offices within direct voice-contact range of the

Figure IV.2: Standing (panel) console design

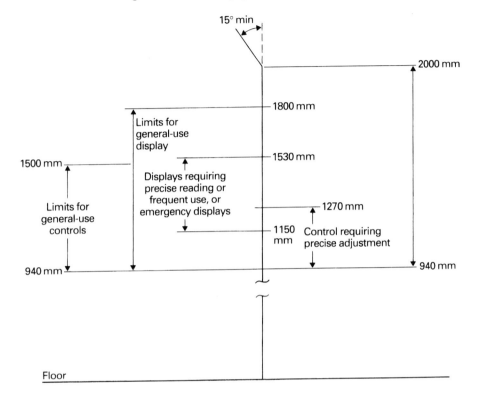

engineers' workplaces, and do they allow the supervisors to interact with other plant personnel or visitors without distracting engineers in the primary area?

11. Are all displays on vertical consoles placed in a band between 41 and 70 inches from the floor, and are those displays requiring frequent or precise readings placed between 50 and 65 inches above the floor level?

12. Are furnishings designed so that there are no sharp corners that could injure personnel?

13. Is there a rest room (with communication facilities) within the control-room enclosure, or close by, that is situated in such a way that it does not interfere with operations?

14. Is it easy to block visitors' access to the consoles situated within the control room?

15. Are the passageways between major items in the primary area at least 30 inches wide?

16. Are emergency doors accessible, and easily opened?

17. Is the supervisor able to see all the system-state displays?

18. Are documents and procedures easily retrieved?

Table IV.1 Recommendations for control selection

Required control function	Control type
To access important or central functions at a high level in the system (not emergency safety functions)	Key switches
Standard on–off function	Push-button Selector switches
For immediate communication, or action, that is frequently required	Push-button
Choice between three or more discrete positions	Push-buttons Selector switches
For continuous variability	Control knob Thumb wheel Push-buttons
Movement direction	Hand levers Joysticks Push-buttons
Accurate control Precise movement	Hand wheels Rotary selector switch Joystick Continuous adjustor knob
Alterations of speed	Direct-position joystick Relative-position joystick Piezoelectric/strain-gauge joystick Hand lever Foot pedal Push-button (the type that takes note of force)

Table IV.2 Guidelines for character height

Viewing distance (mm)	Character height (mm min)
900	4.0
1200	5.0
1500	6.5
1800	7.5
3000	12.5
6000	24.5
7500	30.5

Table IV.3.1 Colour of indicator lights and their meaning

Colour	Meaning	Examples
Red	Danger/Alarm	Control-system fault Radiation level high Excess load on crane
Yellow	Caution	Crane in collision zone Shield door open Movement of machine Approaching travel limit
White	Information only	Locking bolt engaged Control offered Machine parked
Blue	Information only Power on/Panel energised Control accepted	As for white, but use sparingly for grouping/linking together items of information

Note: green should *not* be used as a colour for indicator lamps

Table IV.3.2 Example of VDU colour-coding system

Colour	Process material identification	Current state	Alarm information/abnormality
White	Utilities	–	–
Yellow	Gaseous feed	–	2nd-level alarm: 'Caution'
Magenta	Reagent feed	–	Under maintenance calibration
Orange	Solvent	–	–
Cyan (light blue)	–	Symbol for equipment moving (if not used for low-priority alarm)	Low-priority alarm, abnormality notification
Red	–	Equipment stopped (Symbol with red in-fill)	Emergency High-priority alarm
Green	Aqueous	Equipment running (Symbol with green outline)	–
Blue	Ventilation/Air	–	–

Table IV.3.3 Colour-combination recommendations

Background	Foreground	Avoid using
White	Black	Red/orange
Dark blue	Yellow or white	Blue/green
Green	White	Red/blue
Light grey	Black	Red/black
Dark grey	White	Green/yellow
Red	White	Blue/orange
		Yellow/white

Table IV.4 Information and recommendations on design of VDU screens

1. **Screen layout**

   - In general, a complete picture should be presented on each screen, so as to avoid a situation where an operator carrying out a task has to move from one screen to another.
   - Displays should be consistent, such that menus and titles are on the same place on each page.
   - Size, centrality and labelling should be used to show the relative importance of different items. For example, the most important item on a screen should not be small and stuck in a corner of the screen, with, at the same time, a large 'unimportant' item in the middle of the screen.
   - The general layout of the display should be representative of the actual plant and its equipment, particularly if it is visible from the normal operating position (e.g. via CCTV monitors). The relative sizes and shapes of the actual equipment, and the location of different pieces of equipment in relation to each other, should be represented on the screen.
   - Successive screen displays should maintain spatial continuity, so that where certain vessels are repeatedly displayed on a number of screen pages, they maintain their relative positions on the screen.

2. **Screen structure**

   - The screen displays should be hierarchical in structure, so as to facilitate an understanding of the process in question, should be broad rather than deep and should not involve more than five levels.
   - A particular title format should always be used, so that the operator always knows where he or she resides in the process.
   - A simple means for following information from page to page should be provided.

3. **Messages**

   - If the VDU response time is longer than three seconds, a delay message should be presented.
   - Equipment under maintenance or inhibited should be 'flagged' in some way.
   - Messages should be concise and unambiguous.
   - The information displayed on the screen should be sufficient so as to allow the operator to carry out his or her task effectively, and should not include any unnecessary detail.

(Table IV.4 Continued)

- All messages should conform to the standard language used by the operators. This includes all terms used for items of equipment, sequences and processes.
- Short sentences should be used, and for example, two short sentences should be used rather than one long sentence – but beware of sentences becoming cryptic.
- The active voice should be used rather than the passive voice – for example, 'Isolate the valve' rather than 'The valve is to be isolated'.
- Affirmative sentences should be used; and do not use double negatives. For example, it is preferable to say 'Do . . . only if . . .' rather than 'Do not . . . unless . . .'. Avoid the use of 'unless' and 'except'.
- The use of negatives should be restricted to the emphasising of a forbidden activity. Warnings and cautions should be prohibitive in nature, and the syntax should be standardised. For example, use 'DO NOT . . . [Action] . . . [Condition]'. Overuse of these forms of warning will reduce their impact. They should therefore only be used in situations where there is the possibility of either a threat to operator safety, damage to the plant damage or a loss of production.
- If an alphanumeric code consists of more than five characters, it should be split into groups of 3–4 characters (e.g. 578–516–390). If there is a mixture of letters and digits, then some sort of regular pattern should be used (e.g. 123A 456B).
- Characters should be at least 4 mm high.
- Status messages should be provided, so as to indicate the current functions of multi-purpose, special-function keys.

4. **Symbols and labels**

- Irrelevant digital detail should be avoided. For example, a value of '79' may suffice instead of a value of '79.006'.
- A process flow should be represented by arrows. Arrows should be provided for a line that leaves a symbol – or that joins or leaves the screen – and should be placed at least every 80 mm so that the direction of the flow is apparent even from a glance.
- Symbols should be meaningful to the operators who will run the plant.
- Symbols should be standardised, and used consistently to represent different pieces of plant. They should also be easily distinguishable from each other.
- The use of abbreviations should be avoided. However, if they cannot be avoided because of limited space on the screen, then they should be used consistently, and should be restricted to abbreviations which are common to the operators.
- Each label should be closer to the item it refers to than to any other item on the screen, and each label should be 'tagged' to the item by means of a straight line of neutral colour.
- Labels should be in upper case, and on a single horizontal line. Vertical labelling should be avoided, except for the identification of low-importance items, and except for cases where there is only very limited space available.
- Redundant labelling should be avoided; e.g.:
    MSF REACTOR PXA 778
    MSF REACTOR PXA 788
  If such labelling cannot be avoided, highlight the non-redundant part of the label by, for example, underlining it, or by increasing its brightness.
- Labels for levels and temperatures should, where possible, be positioned next

(Table IV.4 Continued)

to an item in the same relative location as that in which they are to be found on the real item itself. The information given by such labels may prove to be of critical importance during a fault diagnosis.

- The relative sizes of the symbols used should be representative either of the item's own relative size or of the relative importance of the different items shown on the particular screen display. A maximum of *three* discriminable levels of size should be used.
- Where major-process pipelines cross from one screen page to another, text 'target' areas should be provided, allowing access to the next screen display on the same layer. Where practicable, these target areas should appear at the same vertical height on both respective screen pages.
- The alphanumeric visual symbols used should be clearly distinct from each other. For example, do not use both '*' and 'x' as symbols.
- If an inverse video is used, ensure that the level of contrast is adequate, so that legibility is maintained. The characters should all fit properly inside the inverse video – i.e. they should not touch the surrounding black area.

5. **Alarm systems**

- The flashing of messages should be used only to attract the operators' attention to important information. The excessive 'flashing' of VDU information is distracting, and will serve to reduce its attention-gaining properties.
- It should be possible to call up an alarm listing that gives chronological and prioritised alarm information on a single VDU screen.
- The presentation of alarm messages on the VDU screen displays should be standard in terms of positioning and colour coding.
- Alarm messages should be easily distinguishable both from ordinary feedback messages and from labelling.
- Alarm messages should flash at a rate of 2–3 hz, with equal intervals for 'off' and 'on'. To ensure that the messages can still be read, a 'semi-blink' facility may be utilised in which the message oscillates between a high- and low-brightness level, or, alternatively, an alarm 'flag' or symbol can be made to flash next to the alarm message.
- A display should be provided which depicts alarms in terms of their plant area or process function, as well as in the form of a chronological listing.
- Alarm messages should flash synchronously.
- Upon the occurrence of an alarm, every screen display should indicate that an alarm has occurred.
- It should only be possible to acknowledge/accept an alarm when the VDU is displaying a screen page that gives details of the alarm.

6. **Colour**

- Due to the possibility that an operator or operators may have some form of colour-vision deficiency, colour should not be the sole method of coding, nor the only means of identification.
- The meanings associated with different colours should be unambiguous and non-contradictory, and should be applied consistently across all interfaces.
- Colour should be applied to the screen displays in such a way as to enable operators to obtain machine-status information rapidly, to be less prone to reading or interpreting information incorrectly, and to establish compatibility

(Table IV.4 Continued)

---

between information conveyed via the VDU system and that coming from any hardwired controls or displays located in the plant.
- Coloured, or monochrome, foreground symbols, or characters, should not be used against coloured backgrounds if they are *small* in size since one's visual sensitivity to colour decreases with objects with a diameter of less than 3 mm.
- A 'help' screen display for colour usage can usefully be provided.
- Less than nine, and preferably a maximum of five, readily discriminable colours should be used.

---

19. Is there sufficient protective gear available for all conditions and for all employees, and is it checked regularly?
20. Is there space at the seated consoles for writing?
21. Are checksheets used in the procedures?

Controls

1. Are controls designed to follow personnel population stereotypes (pressing a switch *down* for ON in the UK)?
2. Is the degree of force required to operate controls high enough to avoid inadvertent activation but low enough, on the other hand, to avoid muscular fatigue?
3. Are controls that are critical to emergency operations clearly distinguishable?
4. Can important and frequently used controls be reached and operated without strain from the normal working position?
5. Are controls located so that they cannot be *accidentally* activated?
6. Is adequate control-response feedback provided (e.g. in the form of clicks)?
7. Does the control movement match the corresponding display movement?
8. Does the precision setting on the control correspond to the precision setting on the display?
9. Is the control/response rate adequate?
10. Are the *pointing* ends of switches clearly marked?
11. Does the movement of the control, either forward, to the right, upward or clockwise, result in increasing values or in a starting-up process?
12. Is the text on the control fully legible at normal operating distances?
13. Have the most frequently used and important controls been given an appropriately central location, and are they at a comfortable distance from the engineer?
14. Are the controls perpendicular to the engineer's line of sight?
15. Could the engineer, when seated, see over the console, if he or she were required to do so?
16. Are all controls easily identified (by colour, labels, demarcation)?

Table IV.5 Advantages and disadvantages of input devices

Device	Advantages	Disadvantages
Keyboard	Useful for a large number of commands, or where there is a high command rate. Is highly versatile, and usable for most input functions, such as state change or item selection. Function keys ideal for rapid selection.	Is not always as quick as other devices at selecting and changing variables on mimic screens. Errors made in the inputting of text need to be recoverable.
Trackerball (with a 'select' key)	Can be used for function selections if 'soft keys' exist on the screen. Quick and accurate for item selections on mimic screens. Can be used to select a single value from a continuous variable.	Not usable for inputting of text, nor is it useful for precise-value selections. Speed of movement must meet a certain level for any trained operation.
Mouse	Used in similar way to trackerball, though possibly a little quicker.	The mouse needs a clean, flat surface or else it does not work properly and thus becomes a source of frustration to the user. The wire connecting the mouse to the system may get caught or damaged relatively easily.
Light pen	Can perform similar functions to mouse and trackerball, and is fairly quick.	Not appropriate either for a small number of commands or for the inputting of text. While holding the light pen against the screen, the operator's hand momentarily *obscures* part of the screen. May *lose* the light pen. Tends also to be less accurate than a touch screen, and if insufficient pressure is applied, the screen may not register the input.
Touch screen	Can be very quick. Used in similar way to trackerball, etc., but cannot be used to select a single value from a continous variable. Possibly the most rapid and accurate method for item selection, provided that the item areas in question are large and clearly segregated.	Inappropriate for a high number of commands. Cannot be used either for inputting of text or for precise-value selections. Screen should be tilted.

17. Are the actual labels used consistent with those specified in the relevant documents?
18. Are abbreviations, symbols and labels used consistently?
19. If activation by a key is required for any of the controls, are the keys easily retrievable?

Displays

1. Are illuminated buttons used to present qualitative information, and are they used only to provide information about the state of the system?
2. Is the degree of luminance coming from the illuminated buttons sufficiently higher than that found in the surroundings (i.e. 10–300 per cent higher)?
3. Is a flashing light only used when an action is required of the engineer?
4. Can the text on an annunciator still be read when the light is out?
5. Are pointers on scales placed close to the scale so as to avoid parallax problems?
6. Are linear scales used in preference to non-linear scales wherever possible?
7. Have indicators with normal stable values within a limited part of the scale been colour-coded so as to show the limits for normal values and improper values?
8. Are counters used to present quantitative information in cases where precise and quick information is needed but where information about trends, etc., is not needed?
9. If plotters are being used, can the information on the plotters be used directly, without the need for calculation or interpolation?
10. Is all relevant information displayed?
11. Is there an alarm VDU?
12. Is the engineers' central display restricted to showing information relevant only to engineers?

Alarms and annunciators

1. Are the alarms prioritised, so that there are *different levels* of alarm?
2. Are both audial and visual signals used for important alarms?
3. If an alarm signalling a particular event is cancelled but the event recurs, does the alarm also recur?
4. Can the engineer request an alarm-temporal-sequence print-out?
5. Can the engineer respond to a given alarm on the basis simply of which alarms are occurring, rather than having to diagnose the event itself before he or she can respond?
6. Has the number of high-priority alarm sources been adequately restricted?
7. Is each major console provided with its own distinctive sound-generating source, co-located with its annunciator panels?
8. Have no more than three sound-frequency codes (i.e. centre frequencies

widely spaced apart within a range of frequencies running from 500 to
3000 hz) been selected?
9. Have provisions been made for immediately alerting control-room personnel if, for any reason, the alarm-and-annunciator system is deliber-
ately disabled or not operating as intended (or both situations)?
10. Will the failure of the flasher to function result in constant illumination
on the annunciator window?

Communication systems

1. Are there communication links between personnel in the control room
and personnel on the plant?
2. Are the communication links located near to those instruments and pieces
of equipment (e.g. display panels) that are connected with communications activities?
3. Has the need for back-up communications equipment – to be brought
into action if a piece of communications equipment breaks down – been
considered?
4. Is there a method for immediately paging personnel in emergencies?
5. Do plant-wide alarms differ for different hazards, and are these tested
regularly so that everyone knows what each alarm means, as well as what
to do in the event of an alarm?
6. Is there a hot line from the control room to the local fire station?
7. Are there communication procedures detailing who to communicate with,
and *what* to communicate, in the event of an emergency?
8. Do plant-roving engineers carry radio VHF sets, are several channels
available for the use by the plant (including an emergency channel) and
are all these channels constantly monitored in the control room, even
during an emergency?
9. Is there a general public-address system available for giving information to
plant operators?

Environment

1. Is the level of background noise in the control room low enough to allow
intelligible speech, without having to raise one's voice?
2. Is the temperature relatively uniform in the control room?
3. Is the temperature permitted to range from a minimum of 65°F to a
maximum of 85°F, with a corresponding permissible humidity level of 20
per cent (usually, 72–78°F and 20–60 per cent humidity is acceptable)?
4. Does the lighting system in the control room incorporate a 'luminous
ceiling' effect, in order to provide diffused ambient light?
5. Is the height of the luminous ceiling at least 10 degrees up from the line
of sight required for viewing displays?

6. Are the temperature, humidity, vibration, noise, light and cleanliness levels acceptable, and consistent with the legal requirements?
7. Are non-reflective surfaces used on the consoles and panels, so as to avoid glare?
8. Is indirect lighting used?
9. Is the air-conditioning system located so that hot and cold air is not directly discharged on to personnel?

Task organisation

1. Are the procedures involved in switching from automatic control to manual control clear and unambiguous?
2. Are there clear and unambiguous guidelines concerning both who takes control at each stage of an escalating emergency situation and who remains part of the team during a crisis?
3. Are there no conflicts in an engineer's task, e.g. between safety and productivity?
4. Does an attitude of non-penalisation prevail when an engineer says that a failure has occurred?
5. Are the required periods of sustained concentration shorter than one hour, and are the specified periods of continuous mental or physical *in*activity shorter than half an hour?
6. Are high-speed, high-accuracy or highly repetitive tasks done automatically?
7. Are there clear procedures for the transfer both of information and of various responsibilities from one shift to another, as well as between people with *different* responsibilities?
8. Are the engineers able to point out the errors of others, including those of the supervisor, without embarrassment?
9. Do the tasks carried out by the engineer give him or her a feeling of personal value, and a feeling of being responsible for the safety of the plant?
10. Do the tasks that are given to engineers form a meaningful combination?

Colours

1. Are the colours used unambiguously and consistently applied across all consoles?
2. Is colour not the sole method of coding or identification (in particular when different system states are being referred to)?
3. Are colours used to indicate the different extremes involved in the various operating ranges set out on the analogue displays?

Appendix V

Error mechanisms and error reduction mechanisms

EEM/PEM listing*

This appendix documents a full set of external error modes (EEMs) and psychological error mechanisms (PEMs) that can be utilised in human-error analyses. First, the EEMs are listed, in order of processing stage. Then, the various PEMs are listed, after which they are cross-referenced back to the individual EEMs to which they are related. A table is then presented which contains ideas on error-reduction measure, that could be associated with each PEM. The analyst can use these ideas as prompts that help to identify error-reduction measures if such ideas are produced via an error-identification phase (see Section 5.7). These error-reduction ideas need not be adhered to rigidly. The assessor should indeed exercise his or her own judgement in deciding what actually to recommend, and he or she may have to obtain advice from a human-factors professional if the error reduction process required is either not clear-cut or if it is significant. The assessor who is not a human-factors professional is in any case advised to seek specialist advice whenever carrying out error-reduction measures: what sometimes seems 'clear-cut' may *not* be). Also, in this appendix (see end of Appendix), an algorithm for the classification of tasks is presented i.e. determining whether they have cognitive error potential or not.

EEM Listing

1. Activation/detection

1.1 Fails to detect signal/cue
1.2 Incomplete/partial detection
1.3 Ignore signal

---

* The EEMs and PEMs in this section are based on those contained in HRMS, Kirwan, (1990).

1.4 Signal absent
1.5 Fails to detect deterioration of situation

2. Observation/data collection

2.1 Insufficient information gathered
2.2 Confusing information gathered
2.3 Monitoring/observation omitted

3. Identification of system state

3.1 Plant-state-identification failure
3.2 Incomplete-state identification
3.3 Incorrect-state identification

4. Interpretation

4.1 Incorrect interpretation
4.2 Incomplete interpretation
4.3 Problem solving (other)

5. Evaluation

5.1 Judgement error
5.2 Problem-solving error (evaluation)
5.3 Fails to define criteria
5.4 Fails to carry out evaluation

6. Goal selection and task definition

6.1 Fails to define goal/task
6.2 Defines incomplete goal/task
6.3 Defines incorrect or inappropriate goal/task

7. Procedure selection

7.1 Selects wrong procedure
7.2 Procedure inadequately formulated/short-cut invoked
7.3 Procedure contains rule violation
7.4 Fails to select or identify procedure

8. Procedure execution

8.1 Too early/late
8.2 Too much/little
8.3 Wrong sequence
8.4 Repeated action
8.5 Substitution/intrusion error
8.6 Orientation/misalignment error
8.7 Right action on wrong object
8.8 Wrong action on right object
8.9 Check omitted

8.10 Check fails/wrong check
8.11 Check mistimed
8.12 Communication error
8.13 Act performed wrongly
8.14 Part of act performed
8.15 Forgets isolated act at end of task
8.16 Accidental timing with other event/circumstance
8.17 Latent error prevents execution
8.18 Action omitted
8.19 Information not obtained/transmitted
8.20 Wrong information obtained/transmitted
8.21 Other

PEM listing, definitions and ERMs

Activation/detection

1. Vigilance failure: lapse of attention.

   Ergonomic design of interface to allow provision of effective attention-gaining measures; supervision and checking; task-organisation optimisation, so that the operators are not inactive for long periods, and are not isolated.

2. Cognitive/stimulus overload: too many signals present for the operator to cope with.

   Prioritisation of signals (e.g. high-, medium- and low-level alarms); overview displays; decision-support systems; simplification of signals; flowchart procedures; simulator training; automation.

3. Stereotype fixation: operator fails to realise that situation has deviated from norm.

   Training and procedural emphasis on range of possible symptoms/causes; fault-symptom matrix as a job-aid; decision support system; shift technical advisor/supervision.

4. Signal unreliable: operator treats signal as false due to its unreliability.

   Improved signal reliability; diversity of signals; increased level of tolerance on the part of the system, or delay in effects of error, which allows error detection and correction (decreases 'coupling'); training in consequences associated with incorrect false-alarm diagnosis.

5. Signal absent: signal absent due to a maintenance/calibration failure or a hardware/software error.

   Provide signal; redundancy/diversity in signalling-design approach; procedures/training to allow operator to recognise when signal is absent.

6. Signal-discrimination failure: operator fails to realise the signal is different.

   Improved ergonomics in the interface design; enhanced training and procedural support in the area of signal differentiation; supervision and checking.

Observation/data collection

7. Attention failure: lapse of attention.

   Multiple signal coding; enhanced alarm salience; improved task organisation with respect to back-up crew and rest pauses.

8. Inaccurate recall: operator remembers data incorrectly (usually quantitative data).

   Non-reliance on memorised data, which would necessitate better interface design – as data are received, they can either be acted on whilst still present on a display (controls and displays are co-located) or at least be logged onto a 'scratch pad'; sufficient displays for presenting all information necessary for a decision/action simultaneously; printer usage; training in non-reliance on memorised data.

9. Confirmation bias: operator only selects data that confirms given hypothesis, and ignores other disconfirming data sources.

   Problem-solving training; team training (including training in the need to question decisions, and in the ability of the team leader(s) to take constructive criticism); shift technical advisor (diverse, highly qualified operator who can 'stand back' and consider alternative diagnoses); functional procedures; high-level information displays; simulator training; high-level alarms for system-integrity degradation; automatic protection.

10. Thematic vagabonding: operator flits from datum to datum, never actually collating it meaningfully.

    Problem-solving training; team training; simulator training; functional-procedure specification for decision-timing requirements; high-level alarms for system-integrity degradation.

11. Encystment: operator focuses exclusively on only one data source.

    Problem-solving training; team training (including training in the need to question decisions, and in the ability of the team leader(s) to take constructive criticism); shift technical advisor; functional procedures; high-level information displays; simulator training; high-level alarms for system-integrity degradation.

12. Stereotype fixation revisited: need for information is not prompted (by either memory or procedures).

Emergency procedure enhancements, and emphasis of key symptoms and indicators to be checked; team training; problem-solving training; alarm re-prioritisation; simulator training.

13. Crew-functioning problem: allocation of responsibility or priorities is unclear, with the result that data collection/observation fails.

    Improved crew coordination, and allocation of responsibilities; team training; emergency training; accident-management procedures; remote-incident-monitoring/back-up centre; high-level displays; 'crash-shutdown' facilities.

14. Cognitive/stimulus overload: operator too busy, or being bombarded by signals, with the result that effective data collection/observation fails.

    See 2 above.

System state identification

15. Recognition/recall failure: operator fails to recognise plant state due to memory lapse.

    Emergency procedures; emergency training; team training in diagnoses; shift technical advisor.

16. Incomplete mental model: operator's level of knowledge inadequate for state recognition.

    Training; procedural support; decision-support system; shift technical advisor; automation; high-level displays.

17. Familiar-association short-cut: operator responds only to familiar signals, even though these form only *part* of the picture (and hence prove misleading).

    High-level displays; task-based displays; alarm prioritisation; procedural support using flowcharts/fault-symptom matrices; team training; simulator training; problem-solving training; decision-support system or procedure-tracking system.

18. Integration failure: operator fails to aggregate the data meaningfully.

    Overview displays; trend/predictor displays; training; team training; shift technical advisor; hard-copy recording (e.g. via printer, chart recorder).

19. Similarity matching: operator matches scenario to one with similar symptoms.

    Problem-solving training; decision-support system; symptom-based procedures; high-level displays; simulator training; shift technical advisor.

20. Frequency gambling: operator matches scenario to a frequently occurring one.

    Problem-solving training; decision-support system; symptom-based procedures; high-level displays; simulator training; shift technical advisor.

Interpretation

21. Bounded rationality: scenario is too complex for operator to comprehend within the available time-frame.

    Function-based procedures; problem-solving training; team training; provide 'crash-shutdown' facilities; automatic protection; simulator training; decision-support system; shift technical advisor.

22. Imperfect rationality: operator jumps to unfounded or illogical conclusion.

    Problem-solving training; decision-support system; symptom-based procedures; high-level displays; simulator training; shift technical advisor; team training.

23. Confirmation bias: operator interprets signals as concurring with his or her diagnosis, as well as re-interpreting any disconfirming evidence.

    See 9 above.

24. Incomplete mental model: operator does not have enough knowledge to solve the problem, but, not realising this, he or she may go on to derive a false solution.

    See 21 and 22 above.

25. Out-of-sight bias: operator ignores non-centrally located signals.

    Provide centralised signal; improve ergonomics of interface; use diverse back-up alarms; improve task and crew organisation; enhance signal salience; training; procedural emphasis on signals/symptoms that need to be checked.

26. Availability bias: operator jumps to most readily available hypothesis without either working out diagnosis or considering less obvious but potentially more accurate solutions.

    See 19 and 20 above.

27. Similarity matching: see 19 above.

28. Frequency gambling: see 20 above.

29. Recognition/recall failure: operator fails to remember possible interpretative states, or how to interpret the situation.

Emergency procedures (symptom-based, with flowchart); emergency training; team training in diagnoses; shift technical advisor; decision-support system.

30. Thematic vagabonding: see 10 above.

31. Encystment: see 11 above.

32. Information assumed/stereotype fixation revisited: information is assumed without being checked.

    See 9, 12 and 25.

33. Cognitive overload revisited: operator cannot cope with the large number of signals.

    See 2, 18 and 21.

Evaluation

34. Oversimplification: operator oversimplifies the situation, and this leads to inadequate goal/plan formulation.

    Problem-solving training; team training; structured recovery procedures, with function-based and symptom-based components; shift technical advisor; decision-support system; remote-incident monitoring and support facilities; high-level displays; hierarchical displays linking critical safety parameters to more detailed status displays.

35. Overconfidence: operators are overconfident, and underestimate the situation in terms of its seriousness, the imminence of danger, etc.

    Team training (with emphasis on giving and accepting feedback within the team); training on consequences of scenarios; high-level displays; high-level alarms; predictor displays; remote monitoring; automatic protection; single-sheet-flowchart EOPs; simulator training.

36. Risk-recognition failure: operators, through lack of training or just through disbelief, fail to realise the potential level of risk inherent in the situation.

    As for 35.

37. Cognitive overload: operators overwhelmed by information and thus unable to evaluate the situation.

    See 9, 12 and 25 above.

38. Bounded rationality: complexity of the problem exceeds operator's mental capacity within the available time-frame.

    See 21 above.

39. Inadequate mental model: operator's model is inadequate due to lack of knowledge of either special circumstances or side-effects of actions.

    See 10, 16, 21 and 22.

40. Reluctant rationality: operators reluctant to face the reality of the situation, or else find it hard to go through with the evaluation process.

    See 35, but make the important addition of stress training.

41. Responsibility failure: failure to allocate decision-making responsibility effectively.

    See 13 above.

42. Risk-taking: operator takes a larger risk than advisable.

    See 35 above.

Goal selection /task definition

43. Attentional resources: operator, due to workload, cannot devote enough time to the problem.

    Better task organisation and crew re-organisation; re-allocation of duties; automation; team training; more staff/flexible crewing strategies.

44. Memory failure: operator forgets one or more aspects of the solution when defining the goal/procedure.

    Flowchart or columnar procedures; team training; supervision; remote monitoring; shift technical advisor; decision-support system.

45. Reluctant rationality: operator fails to define fully the tasks required due to excessive mental effort required.

    Flowchart or columnar procedures; team training (emphasising crew-supporting functions); simulator training; problem-solving training; decision-support system; automation; high-level displays; supervision; remote monitoring; stress/emergency training.

46. Bounded rationality/side effects not considered: operator fails to consider side effects of task.

    See 21 above.

47. Inadequate mental model: operator understands goals but has insufficient knowledge to derive practicable solution.

    See 10, 16, 21 and 22.

48. Oversimplification: operator derives simplistic solution that fails to deal either with all aspects of the situation, or with the true causes of the situation.

    See 21, 34, 35 and 45.

49. Overconfidence: amount of resources or time (or both) required to resolve the situation is underestimated.

    See 13 and 35.

50. Risk-taking: a rule violation or risk is built into the solution.

    Training; team training; procedural emphasis on consequences of risks; non-risk-taking safety culture; supervision; remote monitoring; hard-copy recording; shift technical advisor.

Procedure selection

51. Inadequate recall: memory fails to offer up entire procedure, or offers up wrong one.

    See 8 and 15. Emphasis on training and procedures, as well as on supervision/crew functioning.

52. Stereotype takeover: operator selects familiar procedure with similar characteristics.

    See 17, 19 and 20. Additionally, add discriminable signals in the interface which clearly indicate the relative inappropriateness of the procedure.

53. Risk-taking: procedure selected incorporates rule violation or short-cut.

    See 50 above.

54. Attentional resources: other demands prevent procedure from being selected.

    See 2, 13 and 18. Re-allocate duties, and use flexible crewing strategies.

55. Intrusion: procedure from previous/concurrent operation selected.

    Use of task-based displays or mimics; flowchart/checksheet procedures; training (but not *over*training, as this can be too intrusive); supervision; hard-copy recording.

56. Lapse: inattention or momentary lapse leads to procedure not being selected, or else to part of the procedure specification being omitted.

    See 1 and 7. Additionally: checksheet procedures; supervision and checking (2-person task utilisation); team training; interface prompts during the procedure; shift technical advisor; high-level displays.

Procedure execution

57. Inadequate time perception: operator fails to perceive duration of time during or prior to task.

    Time feedback in workplace (dual-clock system: one clock 'normal time', the other showing amount of time that has elapsed since onset of event); procedural reference to event timings; team training; flexible crewing; supervision; shift technical advisor; simulator training; automation.

58. Memory failure: operator forgets item, or loses place in a sequence.

    Flowchart procedures; checksheet procedures; team training; supervision; shift technical advisor; training; remote monitoring; decision-support system; procedure tracking system.

59. Motor variability: lack of manual precision or skill.

    Improved training (perhaps via use of part-task trainer); improved ergonomics in layout/labelling/interface design; error-tolerant system design; protection from accidental activation.

60. Inadequate mental model: operator lacks knowledge about task's necessary sequence, precision, timing, or about any other such task constraints.

    Training (classroom, part-task, simulator); procedures (columnar or flowchart); supervision and checking.

61. Random fluctuation: momentary aberration (slip) or unintended action during execution of task.

    Error-tolerant system design; protection from accidental activation; ergonomic interface design; supervision and checking; recovery procedures.

62. Habit intrusion/stereotype takeover: familiar task 'takes over' during execution.

    See 17, 19, 20 and 55.

63. Misperception: operator misreads, mishears or otherwise misperceives.

    Improved interface that incorporates redundant and diverse signal sources; non-reliance on purely oral communication, particularly in face-to-face situations; improved feedback and displays at the interface; communication protocols; training.

64. Mis-cueing: operator misses a cue and instead responds to another one out of sequence.

    Training; accurate and timely feedback; ergonomic interface design; error-recovery procedures; columnar or flowchart procedures; supervision.

65. Mistakes alternatives: aberration of memory; mistakes left for right, east for west, etc.

    Good labelling practices; maps of layout and location of equipment; ergonomic design of interface; discriminable controls.

66. Topographic misorientation: operator mistakes or confuses similar patterns, rooms, controls, etc.

    See 65. In addition: training on layout; enhanced labelling (emphasis on non-redundant information concerning locations, e.g. 'V1034<u>W</u>', denoting the *west* valve); general use of non-redundant-labelling scheme (e.g. all west valves numbered 1–3000 and all east valves numbered 4000–6000).

67. Recognition failure: operator fails to realise that signals/feedback indicate error.

    Training; improved error feedback; high-level displays; supervision and checking; shift technical advisor; system alarms and prompts; procedural reference to explicit feedback; see 6.

68. Cue absent: cue giving operator the required feedback is absent from displays or procedures.

    See 5.

69. Signal-discrimination failure: operator fails to realise signal has changed or is different from that expected.

    See 6 and 67.

70. Reluctant rationality: operator assumes feedback/information rather than checking it.

    See 35 and 45.

71. Other.

Expansion on key ERMs

This section briefly expands on some of the key ERMs found in the above subsection, since some or all of these may be relatively unknown to the reader.

Shift technical advisor

The STA concept, which was developed post-TMI, involves having a 'diverse' mind present, in the event of an emergency. The STA is an independent member of the shift team, and one who *acts* independently so that he or she does not get drawn into any misdiagnoses or 'cognitive tunnel vision' (encystment) biases, as the rest of the shift team may do. The STA needs to have a sufficient degree of authority, so that his or her advice is taken seriously during a real event. The implementation of the STA concept in the nuclear-power world has met with mixed results, and mixed feelings from the normal shift staff.

High-level displays

These displays are aimed at showing only the key safety parameters for the system. Two types of display in evidence in the nuclear-power world are the safety-parameter-display system and the critical-function-monitoring system. The intention with such displays is to give unequivocal feedback as to whether the event is being recovered or is in fact worsening.

Team training

Team training involves developing a healthily functioning crew, i.e. one in which members enjoy appropriate authority while at the same time complying with the necessary authority structure, but also one in which the members can rely on each other, and can give, and take, both advice and criticism in a positive fashion.

Problem-solving training

Problem-solving training aims to teach operators how to handle novel faults, via enhanced theory, as well as rules of thumb, and guidance on typical strategies for problem-solving in process-control situations, maintenance-trouble-shooting scenarios, etc.

Emergency/stress training

This type of training aims to enable operators to experience high-stress situations, so that if the real thing occurs, they do not panic or become otherwise ineffective. Achieving this effectively is something that needs more research, particularly in terms of deciding how stressful such exercises/simulations should be (too much stress may have negative consequences or may lead to a loss of credibility). Well-trained operators may therefore be placed (via simulators or plant exercises) in situations which represent advanced accident scenarios (core melt, fast depressurisation in a gas-cooled reactor, etc.).

Flowchart, columnar and checksheet procedures

Flowchart procedures are particularly useful if they involve decision points with branching outcomes, i.e. if the operators have several options at various points in the procedure. Such flowcharts, if not too complex, can act as a decision aid for the operating crew.

Columnar procedures usually have two columns, the first containing the operational step and the second noting any expected feedback, warnings or error feedback that may be witnessed.

Checksheet procedures allow the operators to tick off the steps as they are each achieved. The effectiveness of such a system depends very much on how seriously the operators adopt the checking procedure, i.e. it depends on the training and safety-culture aspects of the system.

Accident-management, function-based and symptom-based procedures

Accident-management procedures are procedures which assume that some damage has *already* occurred. The aim of human intervention is then to try to contain the event, minimise the consequences and deal with evacuation/fatalities, etc.

Function-based procedures are aimed at dealing with advanced events, i.e. events which are usually close to having accidental consequences. Such procedures, which usually imply that, to some extent, the control of the situation has been lost, or is in danger of being lost, involve an attempt to *regain* control by dealing with the critical safety functions of the plant at a 'macro' level. The implementation of such procedures may have drastic consequences for the plant in terms of its future operability, but if implemented, they do, nonetheless, allow the greatest chance of avoiding losses of life and a severe impact on the environment.

Symptom-based procedures enable the operator to act in a developing event according to what symptoms are present, without necessarily having to wait to act until the event occurring has actually been identified.

Decision-support system, and procedure tracking

A decision-support system (DSS) is usually a rule-based expert system that can assist the operator in diagnoses and decision-making, provided that the operator has sufficient time and the inclination (i.e. trust in the system) to use it. Ideally, such systems are also capable of justifying the rationales they use in connection with any recommended decision.

Procedure-tracking systems take the DSS concept one step further: they actually monitor what the operator does, compare his or her actions against the procedures and then warn the operator if a step has been omitted, or is out of sequence, etc.

Remote monitoring centres (RMC)

Remote monitoring centres, which are utilised in offshore and other industries, represent a way of creating a more diverse level of input into the decision-making process: the remote monitoring centre staff are able to review the situation more objectively since they are not in immediate physical danger.

Crash-shutdown facilities

These allow the operators to shut down the plant or system by means of a single act, or via a very small combination of acts, within a matter of seconds.

Task classification algorithm

The following question structure shows the mechanics of the task-classification system in the HEA module. For each question listed below, it is noted whether the response leads to a CEP (cognitive-error potential) classification, an SRB (skill- or rule-based, and therefore non-cognitive, action) classification or the requirement to answer a further question.

Q1   Does it [the task] involve any problem-solving, diagnoses or judgement-based decision-making? (If 'yes', then CEP classification; Q2.)

Q2   Will the operator have to resort to basic theoretical principles or abstract knowledge to work out a solution? (If 'yes', then CEP classification; otherwise, Q3.)

Q3   Does it involve highly unusual or novel characteristics not covered by procedures/training? (If 'yes' then CEP classification; else Q4.)

Q4   Is the operation routine? (If 'yes', then Q5; otherwise, Q10.)

Q5   Do the procedures cover this case? (If 'yes', then Q6; otherwise, Q7.)

Q6   Are the procedures well understood by the operators? (If 'yes', then SRB classification; otherwise, CEP.)

Q7   Are the operators so familiar with the situation that their responses could be considered virtually automatic? (If 'yes', then SRB classification; otherwise, Q8.)

Q8   Will the situation involve creating new procedures? (If 'yes', then Q9; otherwise, SRB classification.)

Q9   Will such a development of new procedures involve returning to theoretical concepts and making use of detailed knowledge of system? (If 'yes', then CEP; otherwise SRB classification.)

Q10   Is the operation or situation unambiguously understood by the operators? (If 'yes', then Q11; otherwise, CEP.)

Q11   Do the procedures cover this case? (If 'yes', then Q12; otherwise, CEP classification.)

Q12   Are the operators trained so that they understand these procedures? (If 'yes', then SRB classification; otherwise, CEP.)

Appendix VI

Case studies

Introduction

This appendix describes a number of HRA case studies in which the author has been principally involved. Each case study is based on a real HRA application, though in some cases details have been altered. The aim of this appendix is to show the various HRA tools, methods and considerations in action in a range of industries (chemical, offshore, nuclear-power and chemical, plus one case study in the field of marine transport). A number of tools are demonstrated via the case studies, and each case study focuses on certain aspects of the HRA process. To facilitate the reader selecting case studies for perusal, Table VI.1 gives information on the different tools and particular HRA-process aspects focused on in each case study. Unfortunately, no examples are given of screening: this would require reporting on a large HRA/PSA, and the only such large-scale screening assessments that the author himself has been involved with are currently confidential.

The case studies are each described in the succeeding sections, in varying amounts of detail, and have been largely ordered in terms of their industry sectors. The most generalised studies are 7 and 15, the rest having been selected to demonstrate *particular* aspects of the HRA process. Each case study generally proceeds in the same order of activities as that found in the HRA process itself. Quality assurance and documentation are not dealt with in the case studies, since each industrial organisation in which the studies occurred has its own separate QA and documentation requirements. Besides the more generalised case studies, the others are each intended to highlight a particular aspect of HRA, whether misdiagnosis analysis or EOCA, or even problematical aspects of problem definition (the last case study in particular, is qualitatively different from the others). This means that some of the case studies are very fragmentary in nature, but hopefully this does not detract from the intended illustration of the technical aspect of HRA being emphasised.

Table VI.1 Table of case studies and their HRA-process stage

Case study	Industry sector	Problem definition	TA	HEA	Representation	Screening	Quantification	Impact assessment	ERA	HRA tools utilised in study	Type of errors assessed
1. Chlorine loading	Chemical	x	x	x					x	HTA, HEA	Slips and lapses
2. Emergency shutdown	Nuclear Chemical				x		x			ETA, APJ	Error in an emergency
3. Chemical-process control	Nuclear Chemical	x		x	x		x	x		ETA, APJ	Errors and misdiagnosis
4. Chemical sampling	Nuclear Chemical					x	x	x		HEART, THERP, APJ, HPLVs	Slips and lapses
5. Spurious alarm diagnosis	Nuclear Chemical					x				HRMS	Misdiagnosis
6. Hazardous-substances control	Nuclear Chemical	x					x	x		HEART, THERP, APJ	Rule-violation assessment
7. Distributed emergency operation	Nuclear	x	x		x		x	x	x	HTA, TTA, TLA, HEART, THERP	Slips and lapses during emergency
8. Control-room misdiagnosis	Nuclear		x	x						FSMA, CMA	Misdiagnosis
9. Nuclear-power scenarios	Nuclear	x		x				x		EOCA	Error of commission

No.	Application area						Methods			Dependencies
10. Nuclear-power scenarios	Nuclear	x					HEART, FTA, dependence analysis	x	x	Dependencies – various
11. Research-reactor scenarios	Nuclear						SLIM	x	x	Emergency-response failures
12. Ship–platform collision	Marine transport	x		x	x		APJ, FTA, ETA		x	Vigilance failures
13. Offshore drilling	Offshore			x	x	x	FTA		x	Maintenance and emergency errors
14. Offshore lifeboat evacuation	Offshore	x			x		Human HAZOP	x	x	Maintenance and emergency errors
15. Offshore platform depressurisation I	Offshore	x					ETA, data usage	x	x	Maintenance and emergency errors
16. Offshore platform depressurisation II	Offshore	x	x				TSA	x	x	Emergency response
17. Offshore platform blowout	Offshore	x					–		x	Emergency response

Case Study 1: assessment of system for loading chlorine onto road tanker

Background

This assessment was concerned with a road-tanker-loading system. The chemical being transported by the tanker was in this case liquid chlorine. The aim of the assessment was a *qualitative* one, namely to consider what errors could occur during the loading operation, which took place several times a day at the chlorine plant itself. Since the plant already existed and was operational, it was possible to visit it and observe the operation, as well as to discuss aspects of the operation, and possible errors, with the operators themselves. The subset of the study documented here has to do with the prechecking task(s) carried out prior to the filling of the tanker itself. Although error reduction was not a primary requirement in the study, error-reduction recommendations *are* also illustrated in the case study – though these are not necessarily those that were made in the real analysis.

1. Problem definition

The primary concern of the assessment was to identify errors which could lead to an overfilling of the tanker, and to the subsequent rupture of filling hoses and release of liquefied chlorine. The concern was for the safety both of the public and of the operators (i.e. the plant personnel) involved. Because the plant was already in existence, both observation and access to the relevant personnel were possible. As already mentioned, the analysis was qualitative in nature, with error-reduction recommendations being of secondary interest to the study. The objective was also to analyse in detail all the task steps involved, so as to determine *all* potential errors – not merely the most important ones.

2. Task analysis

2.1 Data collection

Data collection occurred via the following methods:

- Observation: two visits to view the loading operation.
- A documentation review: procedures and related incident reports were reviewed.
- Interviews: two operators and a supervisor were individually interviewed, about the procedure, about critical incidents and about potential errors.

2.2 System description

The following describes the chlorine-loading system and its operation.

Figure VI.1.1 Chlorine loading system

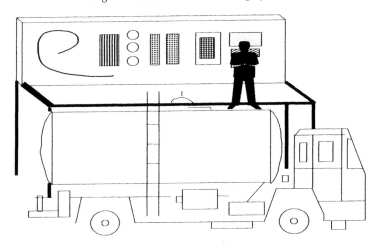

Four times a day, a road tanker is loaded with chlorine and dispatched to various chemical sites around the country. Each 20-tonne tanker, once it has arrived, parks in a special lorry park. The driver than backs the tanker into the loading bay, applies the handbrake and switches off the engine. He then gives the relevant documentation to the loading operator and takes a copy to the dispatch office on his way to lunch/dinner.

The loading operator, who carries out his or her required operations alone, has a minimum of three years' experience on this plant, as well as basic safety-training and fire-fighting skills. After working for three years on various maintenance and operational tasks in the plant, the operator spends two months working on chlorine loading with an experienced operator and is then given a one-week safety-training course on chlorine handling.

There is a gantry-type structure above the tanker on which the operator can walk. This structure houses the various flexible hoses used for the operation (see Figure VI.1.1).

The controls, instruments and manual/automatic valves are situated at a slightly higher level than the gantry, but are also to one side of the tanker rather than directly above it. Thus, the operator, when operating any of the controls or valves, can always see the connecting hoses, the valves and the top of the tanker. This loading bay is primarily used for loading onto tankers, and only occasionally for unloading chlorine.

The almost totally manual operations take place in five stages:

1. Preparation for filling
2. Coupling and purging of lines
3. Tanker-filling
4. Decoupling and purging
5. Final actions

The longest part of this operation is the tanker-filling stage (3), which takes approximately two hours. During this period, the operator has other routine tasks to deal with which may take him away from the covered tanker-loading bay and over to other plant units. Procedures are available but are not used since the operators have in fact *memorised* them – such procedures include safety actions such as checking for residual chlorine in the tanker valves before opening them fully. There are no chlorine-gas detectors (the release of liquefied chlorine would lead very rapidly to formation of chlorine gas). However, chlorine has a very strong and characteristic smell, and there are often people walking from one plant unit to another (e.g. within any hour-long period, four people, on average, might walk within a 20-metre radius of the loading bay). Loading operations occur largely during the day, and all are finished by midnight. The operators involved work in eight-hour shifts.

Operators, when questioned, seemed to be quite happy with the job, because of its variety, and because there is little time pressure on them. Once the operation is finished, the loaded tankers are then parked, and are now ready to go out onto the road at 6 am the following morning. There is only one loading bay, and so tankers are filled serially.

The main local instrumentation consists of a tanker-weight gauge, a pressure indicator on the chlorine line, a local, audible alarm – which is triggered 10 minutes before the estimated 'fully laden' weight is reached during filling – and a second alarm which emits the same sound as the first when the final desired weight is reached (the first alarm 'times out' after two minutes). Both alarms can be heard even when the operator is working in the adjacent plant. The final desired weight is inputted by the operator.

There are four emergency-stop buttons: on the gantry near to the flexible hoses; at the controls; on the outside of the loading bay; and on the next plant-unit building 50 metres away. The operation of any one of these buttons shuts off the chlorine supply both in the loading bay and at the source, and signals a general plant alarm. The loading bay itself is situated 200 metres from the nearest houses. Chlorine is toxic and could relatively quickly overpower a man. It has an unmistakeable odour that is pungent and will cause irritation to the eyes, throat and lungs. In a large enough dose, it is fatal.

The tasks of particular interest in this study are the two precheck tasks: 'Check the test valve to see if any chlorine is present in the tanker; and 'Check the weight of the tanker'. The test valve is a small valve (not credibly confusible with the much larger chlorine-fill valve) which is opened by hand, and which lets out of the tanker, very slowly, any residual gases contained therein. Once this valve has been opened, the operator should hold a cloth soaked in a (NaOH-based) testing agent close to the valve. If there is any chlorine present, the testing agent will turn white (i.e. salt crystals forming). (The operator may also be able to smell chlorine, but this cannot be guaranteed since the platform is in the open, and if there is any breeze, it will disperse the chlorine before it reaches the operator's nose.) The operator

Figure VI.1.2 HTA for chlorine-loading system

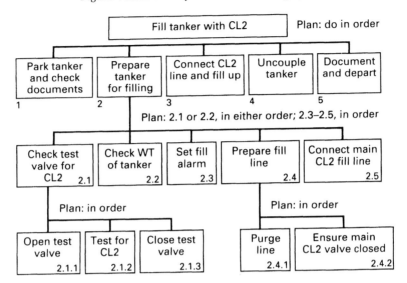

then closes the test valve. The second check (though the order of the checks is in fact at the discretion of the operator) involves reading a weight gauge which shows the total weight of the tanker. This should be 13 tonnes *un*laden and 20 tonnes *laden* (all tankers are of the same design). Each of the tests is required to be carried out as a *back-up* to the other, i.e. one or the other of the two tests will, if chlorine is present in a significant amount, detect that presence.

2.3 Hierarchical task analysis

A HTA was developed for the operation, and is shown in part in Figure VI.1.2. This HTA diagram focuses on the prechecks carried out prior to filling taking place. The full HTA would run to four pages. The HTA was developed and verified with the help of the operators themselves, as well as via observation of the task itself.

3. Human error analysis

The SHERPA system was utilised to analyse the error potential in this system. A single assessor carried out the analysis, and a second assessor then performed a quality-assurance (QA) function to ensure that no error modes had been omitted.

The resultant error analysis is depicted in Table VI.1.1. This table considers only errors made during the precheck tasks and operations.

Table VI.1.1 Part of HEA table for chlorine-loading system

Task	EEM/PEM	Consequences	Recovery	Error reduction
2.1 Check test valve for chlorine	Action omitted – assumption (risk-recognition failure)	Fail to detect CL2 – overfill if CL2 present	2.2 Check tanker weight	1. Safety-training emphasis on dual checks 2. Checking and supervision 3. Interlock, linking opening of main valve to prior, test-valve opening
2.1.1 Open test valve	Too little – motor variability	Fail to detect CL2 – overfill if CL2 present	2.2 Check tanker weight 2.1.3 Close test valve	1. Clear feedback on open and closed valve positions 2. Audio-tactile 'click' feedback on valve stem 3. Training
2.1.2 Test for CL2	Information unobtained – memory failure	Out-of-spec reagent – CL2 not detected	2.2 Check tanker weight	1. Scheduling of replacement of test agent 2. Provision of 'sell-by' date 3. Supervision and QA procedures
2.1.3 Close test valve	Action omitted – forget isolated act	Local release of CL2 – may build up, and then overwhelm operator	None unless operator survives encounter	1. Local CL2 gas detectors 2. Spring-closing test valve 3. Improved feedback for 'Closed' status

The focus of the case study shown here is, as already mentioned, the prechecks. It should be noted that these are important in ensuring that the tanker *is* in fact empty (the tanker has no contents gauge on it, and so it will not be apparent by any other means that it is empty). The tanker should not only be empty but should also in fact have been purged with nitrogen, so as to ensure that no traces of chlorine are left in the tanker. If there is a small amount of residual nitrogen left in the tanker, this will not lead to the top event of interest (i.e. the tanker overfilling). If, however, there is a substantial amount of chlorine left in the tanker (e.g. more than 10 per cent), then, before the alarm sounds to prompt the operator to stop filling the tanker, or at least to monitor the tanker's weight, the tanker will overfill and the hoses will rupture, as a result of which there will be a pressurised release of liquefied chlorine. This substance could rapidly incapacitate the operator (if he does not reach the ESD button first), and then proceed to spread within the plant and beyond its boundary, until a remote ESD is activated.

The tanker should, of course, be totally empty before reaching the site, and the event of its *not* being empty (due to a human error committed at another site) will be rare (e.g. less than once in 10 000 tanker operations). However, the consequences of the accident sequence are sufficiently serious to warrant the existence of, as well as any attempts to improve the *effectiveness* of, these prechecks. Unfortunately, the operators will always be carrying out these prechecks with the knowledge that the likelihood of any chlorine content being left in the tanker is extremely remote. The analysis identified the operator's potential error of failing to carry out one of the checks (essentially, a routine rule violation).

In general, the errors identified are of the slip/lapse variety, involving omissions or errors of quality during skill-based tasks. In one case, however, there is a knowledge-based or knowledge-related error, one which involves a failure to realise that the testing agent is out of date. In total, over 30 errors were identified for the whole procedure.

4. Error reduction analysis

The HEA table (VI.1.1) contains the error-reduction mechanisms determined for each individual error mode. Within the study, there was no requirement to carry out a cost–benefit analysis of these ERMs, though it *is* recognised that they do vary tremendously in cost terms – from the procedural/training recommendations through to the design solutions.

5. Impact of study

The study, in general, raised many potential errors of which the plant's personnel were aware. However, one or two were a little more subtle than those that would have been identified via a more superficial analysis – e.g. the testing agent being out of date, and, on the other hand, the potential routine-violating error of simply omitting the use of the test valve in the knowledge that the presence of chlorine is, after all, highly unlikely and would in any case be detected by the weight-check method (as long as the weight gauge is working correctly). Also, the leaving open of the test valve, which could lead to consequences irrespective of the state of the tanker prior to filling, was highlighted as a concern for operator safety.

Similarly, the approach drew attention to some useful potential ERMs, many of which were seen as being helpful and practicable. The analysis took place over a period of four weeks, primarily using one analyst. The utilisation of the SHERPA system was seen as a success in this study, because the system created an auditable pathway which not only facilitated the QA process but also enhanced the justifiability and transparency of any ERMs raised.

Case Study 2: emergency shutdown (ESD) scenario

1. Background

This assessment was for a nuclear chemical plant still at the detailed design stage. The plant had a comprehensive crash-shutdown system which was manually initiated from the central control room (CCR), or from a diverse back-up location (in the event of a fire, etc., in the central control room itself). However, a rare-event scenario was identified in which the shutdown system could only be partially effective. In this unlikely situation, the operators would have to detect the partial failure and then manually locate and rectify the problem by determining which automatic ESD valve had failed to close and then sending an operator to close a manual, back-up ESD valve on the same pipeline.

This scenario was evaluated in two ways, firstly by the use of the technique THERP and secondly by using two independent APJ groups, operating in a group-discussion mode, whose individual estimates were aggregated into a geometric mean value (i.e. the root of the product of the HEPs). This case study therefore usefully demonstrates two independent methods for assessing the same system via two independent APJ teams and an independent THERP assessor (see Kirwan, 1988).

2. Problem definition

The question being posed by the study was a purely quantitative one: how reliable would the operators be in a partial-ESD-failure situation? The operators would have two hours to resolve the situation. No error-reduction analysis (ERA) was required, since if the results were found to be unacceptable, a design solution would be implemented to increase the reliability of the ESD system itself (which, it should be noted, was already highly reliable).

Since the plant was at the detailed design stage, no operators were available, and there was no similar plant in existence. However, operators from the intended site *were* available to participate in the study.

3. Task analysis

3.1 Task analysis approach

The task analysis firstly proceeded by researching the relevant documentation: the ESD-system description, the review of the ESD panel-instrumentation drawings, CCR layout and staffing plans, local panel-instrumentation plans (not at as an advanced design level as the CCR) and general plant-layout plans. It should be noted that the plant in question was in fact a very large plant with an integrated modular design, and as a result was therefore divided into, and controlled by means of, individual plant units. This factor is important since it makes the ESD system more complex than a typical single functionary plant.

Once these details had been reviewed, discussions were held with ESD design personnel, safety-assessment personnel and future operations personnel as to what the correct sequence of actions was to be. There was a good deal of consensus on this issue, and a simple linear (sequential) task analysis was developed (the tasks are discussed in detail below, after more has been said about the actual details of the system).

3.2 System description

3.2.1 Control room

All activities are based in a Central Control Room (CCR). Information, presentation, alarms and control functions for the plant's operations are all coordinated via a group of VDUs accessed by a console. There will be one console, one operator and 4–6 VDUs for each of the eight operating areas. There will be 1–2 supervisors, always present. Important operating parameters and control functions will be displayed continuously on dedicated VDUs. Other less important functions will need to be accessed by means of a numerical code obtainable from a manual. Equipment status will be accessed in the same way (e.g. 'Feed Pump A running').

Alarms and trip functions will be indicated on a dedicated VDU via a flashing line of print accompanied by an audible alarm. An 'Accept' button will be operated to silence the audible alarm until the next event, but the line of print is retained, and is steadily illuminated. It is possible to obtain print-outs of alarm sequences. The alarm/trip hierarchy (i.e. in order of importance) uses different colours for three different priority levels of alarm (cyan, yellow, red, the latter higher priority). When a VDU alarm occurs, the operator targets 'Fetch alarm' at the bottom-right-hand side of the screen and presses 'Enter'; this takes the operator directly to the VDU mimic 'page' giving details of the alarm and of the relevant area. The operator may also, in multi-alarm situations, dedicate one VDU screen to the function of an alarm VDU. Pictures of the various types of console, etc., were given to all assessors, as were a schematic representation of the control room, and a schematic representation of a typical ESD-panel layout.

3.2.2 Plant manning and work allocation

For each operating area there exists the following:-

- A shift supervisor, with a graduate level of qualification. Although not necessarily experienced in reprocessing operations and standard procedures, (s)he is fully trained in the radiological, or other, consequences involved in a plant's operating outside its normal limits and the design format for major items of equipment and processes (e.g. pulsed columns), and understands the mode of operation of plant instruments used in the interpretation of readings. Finally, the SS also has a knowledge of calibration and maintenance requirements for equipment.
- A control-room operator (one per area), highly trained in all normal and plant-operating procedures (including some simulator training); basic science education (minimum 'O' level); previous site (not plant) experience, and a knowledge of standard procedures, technical phraseology and concepts at a cause–effect (knowledge-based and rule-based) level.
- A local operator (usually more than one per area), well trained for all activities he or she is required to undertake, with an especial awareness of correct operating procedures for specific tasks carried out at local (i.e. non-control-room) control and equipment stations.

3.2.3 Procedures

A full complement of operating manuals and plant-engineering flow diagrams are available in the CCR. There are also detailed procedures for start-ups, shutdowns and plant-wash operations, and these outline, in detail, the different duties carried out by each operator. And a number of important procedures are dealt with, including Permit to Work (PTW) procedures subject to written check-off lists and duties undertaken by two operators: one carrying out the operations in question, the other checking them off on a written list. In cases where it is important to perform the procedures in the correct sequence, a signed check-off list is used as a 'tag' for each next step of the process.

Emergency operating procedures are practised at regular intervals, by following the written procedures. (Certain critical, more complex operations are practised on a simulator.)

3.3 Scenario description, and associated tasks

The emergency-shutdown (ESD) system should be used in the event of either a power failure in the DCU (distributed control unit) console or a failure on the part of the DCU itself. Both of these events will be rare – e.g. one such event may be expected in ten years. When such a failure occurs, there are, in the relevant area of the plant, five liquor feeds which need to be shut off so as to prevent a dangerous build-up of material. If the ESD is *not* activated, this build-up will occur within approximately 20 minutes. The ESD system is designed to achieve this action, i.e. to shut off the five feeds, in a matter of seconds.

The operator will be made aware of the DCU failure or DCU power failure by alarms on the console, by the VDU system (if still semi-operational) and by the *loss* of information signals, as well as by the possible switch-over to emergency lighting. The supervisor will also be made aware of the situation via certain alarms on his console.

The situation requires the operator to go to his ESD panel and activate an ESD (see the typical ESD-panel arrangement) by simultaneously pushing in a button and turning a small handle – a one-handed operation. When the ESD has been initiated, the shutdown is then confirmed by a set of five lamps on the ESD panel. If one of these lights is *un*lit, then a total ESD has not occurred and the operator must rectify the situation.

However, due to the *loss* of the DCU, the rectification of the fault will require a diagnosis from the control room, as well as further actions to be carried out by another (outside) operator in a local equipment room, as described below.

The control-room operator, having realised that only a *partial* ESD has occurred, will either act on the basis of his training, utilise the existing procedures in the control room or work with the supervisor (though the supervisor may have other problems of his or her own if the power failure has also affected other areas) – or some combination of these.

The operator should read the indication on the ESD panel denoting which equipment room contains the panel corresponding to the failed ESD actuators, and then transmit this information to an outside operator by telephone. There are telephones on the console, but telephones can also be *plugged into* the ESD panel. The outside operator will either be called directly or 'bleeped', or else called over the tannoy system. Once the line of communication has been established, the outside operator will be asked to go immediately to the particular equipment room.

The outside operator must now find the correct panel in the equipment room, scan the ESD lights on the panel and identify which one (or more) of these are *not* illuminated. He or she should then communicate his or her findings to the control-room operator, either by radiophone or by the telephone in the equipment room itself (which may be a fairly large room, e.g.

40 feet long). He or she may, however, omit this communication if the time available is too short. The control-room operator must now check the equipment listing to identify the engineering flow diagram or emergency operating procedure that contains the relevant details of the failed actuator(s). Once he or she has identified the relevant EFD or EOP (available in the control room), he or she must diagnose which manual valves are to be shut if the continued feed to the process is to be isolated. The operator must then contact, again, the outside operator (unless the contact has remained intact during this process) and instruct the outside operator to proceed to the relevant actuator station. Once there, the control-room operator must tell the outside operator which valve to close, citing the relevant valve number (e.g. 'V70986'). Once the relevant valve has been closed, the outside operator should report this to the control-room operator (it may be assumed that valves only 'close' in one direction, and that this is consistent across the plant, although there may be *many* valves present in the actuator station). When the task has been completed, the control-room operator should check the flow indications (on the VDU) to ensure that the valve has indeed shut.

This is a rare event, and it is assumed that training for this scenario will amount to a walk-through of the procedures on the part of the operators, but with no actual simulation of the event. The operators will be aware of the serious consequences of a failure to achieve a total ESD, namely a risk both to personnel and plant.

4. Human-error analysis

The following human errors were identified directly from the task analysis as requiring quantification:

- HEP 1.1: operator fails to initiate ESD within 20 mins.
- HEP 1.2: given 1.1, supervisor fails to initiate ESD within 20 mins.
- HEP 1.3: operator fails to detect only a partial ESD success within 2 hours.
- HEP 1.4: given 1.3, supervisor fails to detect only a partial ESD (from the operator's panel) within 2 hours.
- HEP 1.5: operator fails both to identify appropriate equipment room and to communicate location of this room to an outside operator.
- HEP 1.6: outside operator fails to get to correct ESD actuator panel, to identify failed actuator(s) and to communicate this information to the control-room operator.
- HEP 1.7: control-room operator fails both to determine which valves must be closed, in order to achieve a total ESD, and to communicate this information to outside operator.
- HEP 1.8: outside operator fails to close correct valves within 2 hours of DCU failure/DCU-power failure.

5. Representation

The event tree used to represent the scenario is shown in Figure VI.2.1.

6. Quantification

The HEPs were quantified using the THERP and APJ approaches, as discussed below.

6.1 THERP analysis

The following are the assessor's notes concerning the quantification of each HEP involved in the scenario where the operator responds to an ESD failure. (Please note that all the tables referred to are contained in the THERP manual, *not* in this book. The THERP quantification was carried out by A. Smith.)

(a) HEP 1.1

From (THERP) Table 12.4:

Probability of failure to diagnose single abnormal event in 20 mins = 1×10^{-2}

(The nominal level of error, when correlated with the level of training assumptions, would indicate this to be appropriate.)

Given that a correct diagnosis has been made, the likelihood of a failure to activate the ESD handle correctly would therefore be as follows:

$= 1 \times 10^{-4}$ (item (8), Table 13.3) $\times 5$ (PSF: extremely high stress; item (6), Table 17.1)
$= 5 \times 10^{-4}$

(NB: Selection Error 1×10^{-3} (item (3), Table 13.3) $\times 5 = 5 \times 10^{-3}$)

Thus the following probability for operator's failing to initiate ESD within 20 mins:

$$= (1 \times 10^{-2}) + (5 \times 10^{-4})$$
$$= 1.05 \times 10^{-2}$$

(b) HEP 1.2

If it is assumed there is a strong relation of dependence between HEP 1.1 and HEP 1.2, then the probability for the supervisor's failing to initiate ESD within 20 mins is:

Figure VI.2.1 Event tree for ESD scenario

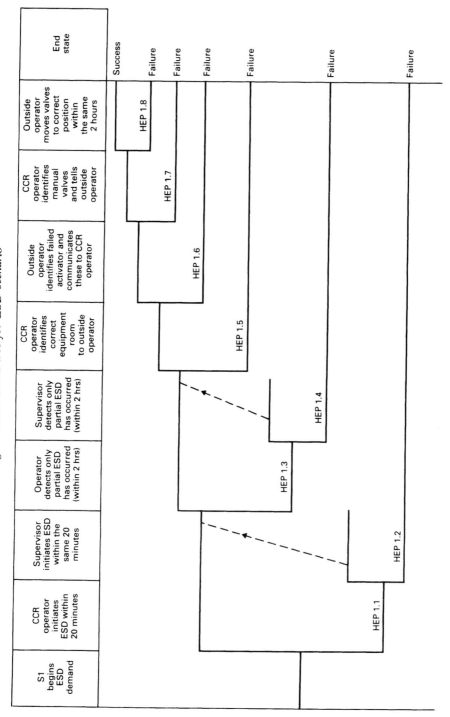

$$= (1.05 \times 10^{-2}) + 1 \text{ (Table 10–2)}$$
$$= 5.11 \times 10^{-1}$$

(c) HEP 1.3

From Table 11.4, assuming a high level of stress:

$$\text{Failure to detect a partial ESD} = 1 \times 10^{-3}$$

Assume that the failure to detect the first line will also mean that the probability of detection on further opportunities within 2 hours is zero. Thus:

Probability of operator failing to detect only a partial ESD within 2 hours
$= 11 \times 10^{-3}$

(d) HEP 1.4

Assume a strong relation of dependence between HEP 1.4 and HEP 1.3. Thus:

Probability of supervisor's failing to detect only a partial ESD within 2 hours $= (1 + 0.001)/2$

(e) HEP 1.5

From Table 11.2:

Probability of operator's selecting wrong indicator lamp $= 3 \times 10^{-3}$ (item (4)) $\times 2$ (high stress, Table 17.1, item (4))
$= 0.006$

From Table 15.1:

Probability of miscommunication error $= 0.001$ (item (1))

Thus:

Probability of operator's failing to identify equipment room and communicate this to outside operator $0.006 + 0.001 = 0.007$

(f) HEP 1.6

The error of selecting the wrong equipment room is included in the miscommunication error in (e) above.

Probability of operator's selecting wrong indicator lamp = 0.006 (see (e) above)

Probability of miscommunication error = 0.001 (see (e) above)

Thus,

Probability of outside operator's failing both to identify failed actuator(s) and to communicate this to CR operator = 0.007

(g) HEP 1.7

Assume that the probability valve for selecting the wrong EFD is negligible, since the operator will recover the error when trying to diagnose which valves require closing.

From Table 11.2:

Probability of CR operator's selecting wrong valves = 0.003 (item (4)) × 2 (high stress, Table 17.1, item (4))
= 0.006

Assume that the probability of a recovery being made thanks to the outside operator's realising that these are incorrect valves comes to 0.5. To arrive at this value, the (judgemental) assumption is made that the outside operator will recover the error *if* the valves he has been directed to close *are* already closed, and that there is a *50 per cent* chance that the valves he has been directed to close are already closed.

From Table 15.1:

Probability of miscommunication error = 1×10^{-3} (item (1))

Thus,

Probability that the CR operator fails both to determine which valves must be closed and to communicate his or her findings to outside operator
= $6 \times 10^{-3} \times 0.5 + 10^{-3}$
= 0.004

(h) HEP 1.8

From Table 14.1:

Probability of outside operator's selecting wrong valve = 3×10^{-3} (item (2)) × 2 (high stress)

Figure VI.2.2 Quantified THERP tree

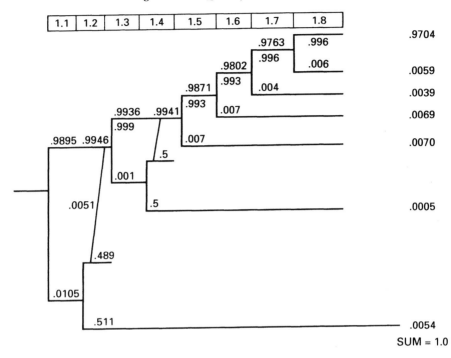

This assumes that the operator has to close only one valve.

Both reversal errors and errors made in connection with stuck valves are assumed to be negligible compared with the selection error. Assuming that no feedback on the valve status is available in the CR:

Probability of outside operator's failing to close correct valve = 0.006

The combination of these HEPs in the event tree is shown in Figure VI.2.2, with the likelihood of failure being approximately 0.03.

6.2 APJ analysis

The two independent APJ sessions were run on separate days. Each group assessed the scenario in the space of about 2 hours. The groups had an identical personnel structure: a facilitator, who did not give any quantitative input into the exercise but encouraged and controlled the discussion on PSFs and differences in experts' initial judgements; a safety analyst familiar with the plant design; two operators with 10–35 years of experience on the intended plant site (i.e. on other plants); and a human-factors/reliability assessor. The group were given information as described in this case study plus

Table VI.2.1 APJ results for ESD scenario

HEP	Group 1	Group 2	Difference factor (HEP1/HEP2)
1.1	0.01	0.0005	20.0
1.2	0.005	0.00053	9.0
1.3	0.001	0.00067	1.5
1.4	0.001	0.00084	1.2
1.5	0.001	0.00038	2.6
1.6	0.01	0.0015	6.7
1.7	0.01	0.01	—
1.8	0.01	0.008	1.3
Combined HEP	**0.03**	**0.02**	**1.5**

additional documentation (concerning ESD-panel drawings, CCR layouts, etc.).

After an initial discussion of the scenario, each member of the group made an individual assessment of each HEP (all members did this simultaneously, and without knowing what the other members' estimates were). These were then tabled and discussed. After approximately an hour's further discussion, each member then re-assessed the HEPs (again privately and simultaneously). Additionally, a consensus value for each HEP, which the group all agreed upon, or at least could 'live with', was determined. The results for two of the APJ groups are shown, in Table VI.2.1.

7. Impact

The results from all the assessments made (in fact, the HEART and SLIM methods were also used to calculate HEPs for this scenario) suggested that the overall probability of failure was between 0.03 and 0.3. A further reliability analysis was undertaken to evaluate in more detail the likelihood of a partial failure and its potential consequences. From this further work, it was concluded that, given the low likelihood of the initiating event, on the one hand, and the high reliability of the system on the other, and in the light of a consideration of any potential consequences, the system hardware did *not* need to be changed. However, due to the relatively high values of the HEPs, the ESD panels *were*, however, rendered more ergonomic, particularly with respect to the signals denoting whether the ESD had been effective or not. This redesign was extensive, affected every ESD panel in the CCR and back-up ESD room and involved a hybrid team of human-factors personnel, designers and operators. It is possible that the panels might have been redesigned according to human-factors principles even *without* the ESD HRA, i.e. due to other reasons; but the HRA certainly added weight to the arguments for such a redesign. New training recommendations (including those for refresher training) were also accepted by the future plant personnel.

Case Study 3: chemical process-control study

1. Background

This study has the same plant assumptions and plant description as those found in Case Study 2, but in this case it concerns an operator in a *local* equipment room, not the central control room. This operator is not under supervision, although he or she can make direct contact with the CCR. The operator uses a local VDU and console to obtain information on the state of the plant, but all necessary response actions are manual, and are carried out in the equipment room – or an adjacent equipment room. The scenario has a very rare anticipated frequency – e.g. 0.02 per year, or once in 50 years. Essentially, this case study focuses on knowledge-based failures occurring in a emergency situation involving considerable time pressures. The HEPs derived were quantified via the APJ method, as again with Case Study 2. No ERMs were required to be developed for this scenario, nor was any task analysis requirement – beyond a definition of the different elements of the scenario.

2. Problem definition, task description and human errors

In order to understand the knowledge-based errors involved in the scenario, the scenario in question is presented in some depth. The system has been slightly simplified for the purposes of the case-study write-up. No formal task analysis was produced since this analysis was primarily PSA-driven, the PSA requiring an estimate of the probability of non-response within a certain time-frame for the scenario. The HRA analyst then expanded this single EEM requirement into a number of potential error modes, as will be seen below. No formal technique was used for the error analysis – merely the analyst's own experience.

2.1 The system and the scenario

The scenario concerns three tanks (see Figure VI.3.1), which are fed by steam to keep the chemical reaction at boiling point. Two fans extract the by-product gas from the reaction, and maintain the tanks under a slight vacuum. Only one of the fans is required to extract the gas and to maintain this partial vacuum, but both fans are nonetheless kept running. The scenario therefore concerns the failure of both these fans. When this happens, within 8–13 minutes the tanks will reach atmospheric pressure and the 'off-gas' will enter a potentially manned area. This is a highly undesirable event.

Figure VI.3.1 Tank-system diagram

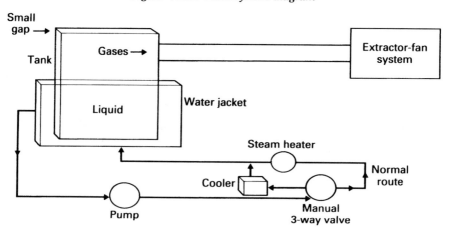

In order to prevent this event, the operator will receive an audible and flashing alarm on his or her VDU. He or she will target the 'Fetch alarm' area at the bottom right-hand corner of the VDU screen, and will then be taken to another 'page' showing the message that the two fans have stopped. A fan failure causes a trip of the following functions also, which will each set off alarms on the VDU:

• Inlet pipe
• Sparge air to tank
• Steam to heater

However, these are not true clear indicators of a fan failure, since a *spurious* alarm set off in relation to the fans could *spuriously* trip this function (i.e. whilst the fans are actually still running).

The operator must therefore check to see (via the VDU mimic information system) whether there is:

• A low-pressure alarm on the fan discharge manifold.
• Differential-pressure low alarms on three treatment columns in the off-gas system.

In a true fan-failure scenario, these alarms should, in fact, occur on the VDU system *without* the operator's having to search for them. This represents the operator's 'diagnosis', and if it is carried out correctly, the operator will have decided that the fan failure is real and not spurious.

The operator may, at this point, either revert to the procedures themselves or take action on the basis of his or her *memory* of the procedures.

2.2 Scenario's task requirements and potential errors

The three tanks will be in three different operational states, namely 'Input', 'Bleach' and 'Transfer'. The operator must first determine which tank is in which state, via the VDU mimic system. The operator must next proceed to check that the circulating pump (see Figure VI.3.1) is still in operation, again via the relevant VDU display. The operator then goes to an adjacent equipment room where the 3-way circulatory valves for the three tanks are housed. Each 3-way valve (see above figure again) allows the operator to cut off the heat source from the tank and replace it with a *cooling* source. This prevents the build-up of pressure in the tank and hence stops any off-gas from entering the potentially manned area. The operator, having determined which tank is in which stage, should first move the 'Input' 3-way valve and finally the 'transfer' valve. This is the order of priority in safety terms (although a failure to move the transfer valve will *not* have any significant effects).

It is possible for the circulating pump to trip. If this has happened, the tanks will not be heated by the steam, and 'decay' heat alone will, instead, cause the off-gas to reach atmospheric pressure after about 40 minutes, instead of 8–13 minutes. Thus, if the operator fails to check that the pump is still running and it is *not*, then the turning of the 3-way valves will do little to stop the event, unless the operator restarts the circulatory pump within 40 minutes of the start of the scenario. There is, thus, the possibility that the operator may *believe* he or she has managed to curtail the event when in fact the event is still unfolding. Following his or her actions on the 3-way valves, the operator should return to the VDU console and check that the system is moving towards a safe state.

It is also possible for the operator to misdiagnose the problem and actually stop the pump, failing to realise that the decay heat alone will still cause the off-gas to enter the potentially manned area. This misdiagnosis would imply that the operator has not referred to the procedures, or recovered, within 40 minutes.

Lastly, it is possible for the operator to stop the pump *intentionally* to give himself more time to identify and move the valves, but then to fail to restart the pump within 40 minutes simply out of forgetfulness. The stopping of the pump would indeed be a knowledge-based method of gaining more time since this action is not proceduralised. At the same time, however, it is *not* an 'illegal' action.

For those cases where the circulatory pump does not stop, i.e. where recovery must occur within 8–13 minutes, no supervisory intervention is assumed. In the case where the pump is stopped, it is assumed both that the operator *would* be likely to contact the supervisor to advise him or her that although the fans had stopped the situation was under control, and that the supervisor would *not* check that it was under control within the 40-minute period. Hence, in all scenarios, no allowance for effective supervisory intervention is made. This is a conservative (pessimistic) assumption.

3. Human errors identified

Based on the above description of both the scenario and the required human involvements, the following human errors were put forward as requiring quantification:

(a) HEP 1.1: operator fails to diagnose fan failure within:

- 1.5 minutes
- 6.5 minutes

i.e. either the operator fails to respond in time to the alarm, or he or she assumes the alarm is spurious.

(b) HEP 1.2: given HEP 1.1, the operator fails to arrive at the secondary solution and stop the circulating pump.

(c) HEP 1.3: given correct diagnosis, the operator fails to move 'Input' and 'Bleach' valves, to their correct (cooling) positions, within:

- 8 minutes
- 13 minutes

(d) HEP 1.4: given that the pump has tripped, the operator fails both to check that it is on and to restart it.

(e) HEP 1.5: given that the pump is off, the operator fails to move valves within 40 minutes.

(f) HEP 1.6: operator adopts strategy of turning off the pump to give him or her more time, but fails to reactivate the pump before 40 minutes has elapsed (this assumes that the operator has already successfully moved the valves).

4. Representation

The scenario was represented in an event-tree format, as shown in Figure VI.3.2.

5. Quantification

Quantification occurred as for Case Study 2: via an expert panel. The results are shown in Table VI.3.1.

6. Impact

The results show a significant difference between the two time-frames. This is primarily because the first time-frame is less than 10 minutes, while the

Figure VI.3.2 Event-tree analysis for fan-failure scenario

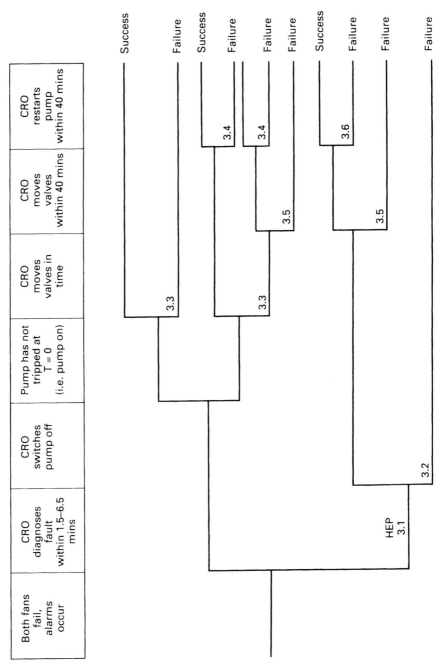

Table VI.3.1 APJ results for DOG (dissolver off-gas) scenario

HEP	Group 1
1a	0.18
1b	0.04
2	0.1
3a	0.046
3b	0.025
4	0.1
5	0.1
6	0.1
Combined HEP:	
(t + 8)	0.09
(t + 13)	0.042

second is more. This discrimination on the part of the expert panel is consistent with work that has been carried out on time-response curves for emergency responses. Essentially, knowledge-based behaviour is not held to be very reliable, in process-control contexts, at the best of times, and this is especially the case in very short time-frames.

It is interesting to note that all four judges gave the operator the same value for the likelihood of his or her adopting the knowledge-based behaviour route of extending the time-frame via the unorthodox process of switching off the pump; and that this probability value, moreover, had a success rate of 90 per cent. This implies that the judges believed that the operator knew the system reasonably well at a functional level.

The figures themselves were not high in system-reliability terms, and this suggests a need for training in emergency-recovery procedures, or, more probably, for some form of *automatic*-protection system, with the operator acting as a back-up only.

Case Study 4: chemical-sample analysis

1. Background

This scenario concerns a chemical-sample analysis carried out as part of a wider operation. The analysis would be carried out by a separate laboratory at the plant site, one which deals with many such sample analyses on a day-to-day basis. The operators of the plant request the analysis, take the sample, transmit the sample to the lab and then receive, by computer, the results of the analysis a few hours later. Many plants of various kinds will involve this type of laboratory arrangement. While not usually thought of as being part of operations themselves, the results of such analyses *can* have a significant impact on certain operations, especially if they are wrong, or have been misinterpreted.

2. Problem definition

In this scenario, a grossly excessive amount of a chemical reagent is inadvertently supplied to a chemical plant. Since this mishap would lead to the possibility of *contamination* later on in the process, which in turn would cause the very rapid corrosion of the process tanks, as well as possible leaks in pipes, etc., it is regarded as a highly serious event. The reagent level supplied would first, however, need to be far greater than is normally required, say a factor of 5 or 10 higher. This error could occur during the delivery of the out-of-specification substance/reagent to the plant from another plant on site. The frequency for such an event is very low, but it *has* happened before. The crucial question in this scenario is: given that a grossly out-of-spec batch *has* been inadvertently delivered to the chemical plant, will it be detected as such and then rejected or will it 'slip through' and be processed? The scenario, as described below, involves the taking of a sample of the batch to be processed, its analysis by a separate laboratory and a decision on the part of the operator, based on the results of that analysis, as to whether or not the batch is acceptable for immediate processing.

Error-reduction mechanism (ERMs) were not required in the analysis.

3. Task analysis

A limited task analysis took place, via telephone interviews. Observation was not possible in this study. No formal HTA was developed, but instead a textual description was produced for the task. This was essentially a *rapid* assessment rather than a detailed analysis.

The scenario begins with the outside operator going to the reagent-batch tank to take a sample. Having located the sample point on the tank, he or she opens the sample drain to the waste container, drains off about half a litre and then shuts off the sample valve. He or she now collects about half a litre from the sample valve and transfers it into the sample container.

He or she now rinses out the sample container with the sample, reconnects the sample container to the sample drain and collects the actual sample (approximately 1 litre's worth). He or she then labels the sample container with local details, date, time, etc., and takes the sample to the laboratory.

He or she hands over the sample to the laboratory, and asks for the appropriate test. The laboratory analyst then tests for the concentrations of various chemicals, using a variety of standard chemical-analysis techniques (e.g. filtration or pH testing). He or she records the results and then communicates these back to the control-room operator via a computerised spreadsheet, with a final statement at the end of the brief report, in bold and boxed, as to whether the sample is okay or not.

This test may be carried out daily, or even more than once a day. The sample will be taken from a room containing 10 similar sample points, each of which will be labelled. In this scenario, the analyst must be able to detect a gross excess of reagent.

4. Human error analysis

A small group of analysts, using their own judgement, identified the following major errors:

- The analysis sample is not taken.
- The analysis is not performed correctly (i.e. failure to detect excess reagent).
- The sample is taken from the wrong point.
- The operator responds incorrectly to the results of analysis (i.e. he or she fails to realise that the analysis reports an *excess* of the product).

Any of the above could lead to the failure involved in the scenario, and because of this fact, a fault- or event-tree representation was not required; instead, the *summed value* of the HEPs for all four errors was all that was required.

5. Quantification

Quantification was carried out using the HEART technique (assessor: S. Whalley). The results of the HEART assessments are shown in Figure VI.4.1, and the HEPs are summarised below:

HEPs:
1. 0.02
2. 0.00064
3. 0.044
4. 0.00064
Total: 0.07

6. Impact

The overall failure rate was considered credible. Since the initiating event-frequency rate was very small, an overall HEP of 0.1 could technically be tolerated, but the value of 0.08, on the other hand, was undesirably high.

Since HEART can sometimes be overconservative, THERP and APJ analyses were also carried out. The total resulting HEP ranged from 0.01 to 0.1, and this supported the original HEART figure, i.e. it was, at least, in the right ballpark. Consideration was then given both to methods of improving the

Figure VI.4.1 HEART technique's results for sampling scenario

HEP 1	Sample not taken
Type of task	E – 0.02
EPCs selected	None
HEP	0.02
HEP 2	Analysis performed incorrectly
Type of task	G – 0.0004
EPCs selected	No diversity of information (× 2.5)
Assessed proportion	0.4
HEP	((2.5 – 1) × (0.4 + 1)) × 0.0004 = 0.00064
HEP 3	Sample taken from wrong point
Type of task	E – 0.02
EPCs selected	Inadequate feedback (× 4)
Assessed proportion	0.4
HEP	((4 – 1) × (0.4 + 1)) × 0.02 = 0.044
HEP 4	Operator responds incorrectly
Type of task	G – 0.0004
EPCs selected	Misperception of risk (× 4)
Assessed proportion	0.2
HEP	((4 – 1) × (0.2 + 1)) × 0.0004 = 0.00064

Author's note
At the time of these assessments, the rate of incorrect analyses had been established to be very low on account of both well-designed laboratory procedures and a multiple-checking procedure. These considerations support, to a certain extent, the low HEP for HEP 2 above. Similarly, the operator receives very strong and clear confirmation via the computer spreadsheet; and HEP 4 also incorporates the fact that a supervisor makes a further check of the operator's performance, the results of which are then included in his or her own computer-transmitted report.

sampling-system's effectiveness (supervisory checking of sample requests, increased labelling of sample valves, etc.) and to ways of preventing the out-of-spec batch from being presented in the first place. A recalculation was *not* carried out, since this work was not part of a formal PSA but merely a semi-quantitative analysis addressing a particular potential problem that had been identified.

Case Study 5: operator fails to respond to alarms

1. Background

This study took place during the detailed design phase of a plant, and concerned alarm responses in the Central Control Room (CCR). In the early phase of analysis (i.e. the conceptual phase/preliminary engineering phase), the human-error potential had been identified, and a screening estimate of 1E–1 had been given for the HEP, based simply on the failure to respond to a single alarm, given various time constraints, etc. This meant that the HEP warranted further investigation via a more detailed approach, since the error

Figure VI.5.1 Task analysis

Assessor: BIK
Task: Alarm response
Safety case: Example
File name: Spurious

1	Operator detects fault condition
1.1	Operator detects SO2 alarm
1.2	Operator detects FQN alarm
2	Operator diagnoses fault condition
2.1	Operator checks instruments
2.2	Operator infers failure from both instrument readings
3	Operator responds to fault condition
3.1	Operator stops pump
3.2	Operator verifies effectiveness of action

in question contributed significantly to the level of risk shown in the fault tree. Furthermore, as more details became available, it was realised that there were *two* alarms involved, and that one of these might occur spuriously with respect to this alarm scenario. Furthermore, the operator would have to link these two alarms together, on the basis of his or her functional knowledge of the system, and this meant that the scenario was primarily knowledge-based. It was therefore necessary to evaluate this scenario in more detail. This case study shows some aspects of the HRMS approach, as outlined in section 5.5.

2. Problem definition

The problem was therefore to analyse the scenario in more depth than had been possible at the early design stages, and to calculate the associated error probabilities. The HRMS approach was utilised, which required a highly detailed analysis of both the task and its Performance Shaping Factors. This analysis necessitated a large number of assumptions, since the plant was not yet built, and there was no similar one to observe, or from which to interview operators. These assumptions were logged in the analysis and then confirmed as the plant-design and operational details developed. No error-reduction measures were required at this stage of the analysis.

3. Task analysis

Although a detailed analysis was called for, only a limited amount of task-analysis detail could be derived, and the task did, in fact, involve only a relatively small number of steps. The task analysis shown in Figure VI.5.1 was therefore developed, which is a simple linear, nested-loop HTA.

4. Human error analysis and qualitative screening analysis

The HRMS human-error analysis module was utilised to identify the various error modes involved in the different task steps, and the results of the analysis

Figure VI.5.2 Human-error analysis

Assessor:	BIK
Task:	Alarm response
Safety case:	Example
File name:	Spurious

1　　　OPERATOR DETECTS FAULT CONDITION
No errors considered possible

1.1　　OPERATOR DETECTS SO2 ALARM

02160　OPERATOR FAILS TO DETECT ALARM
Ambiguous/contradictory information collected/Confirmation bias

1.2　　OPERATOR DETECTS FQN ALARM

03160　OPERATOR FAILS TO DETECT 2ND ALARM (BUSY WITH 1ST)
Ambiguous/contradictory information collected/Confirmation bias

2　　　OPERATOR DIAGNOSES FAULT CONDITION

04010　OPERATOR MISDIAGNOSES CAUSE
Failure to interpret plant state
Recognition/recall failure

04260　OPERATOR DISTRUSTS FQN ALARM
Risk-recognition failure
Memory failure/bounded rationality

2.1　　OPERATOR CHECKS INSTRUMENTS

05020　OPERATOR FAILS TO CHECK BOTH INSTRUMENTS
Incorrect/incomplete interpretation of state of plant/Imperfect rationality

05280　MAINTENANCE ERRORS COULD MAKE READINGS INCORRECT
Motor variability

2.2　　OPERATOR INFERS FAILURE FROM BOTH INSTRUMENT READINGS

06010　OPERATOR FAILS TO MAKE CONNECTION BETWEEN TWO ALARMS
Failure to interpret state of plant
Bounded rationality

06020　OPERATOR ASSUMES FQN IS FAULTY
Incorrect/incomplete interpretation of state of plant
Bounded rationality

06090　FQN MAY BE UNRELIABLE, OR HAVE OTHER MEANINGS
Incomplete diagnosis

are shown in Figure VI.5.2. For each step, the figure logs the errors considered possible/credible by the assessor, in terms of both the external error mode (shown in capitals) and the potential underlying psychological error mechanisms (PEMs). The table does not show *all* the errors identified (e.g. a failure to stop the pump), but rather is intended to show, in particular, the results of the HEA for the more cognitive-related errors related to either distrusting the FQN alarm or to failing to realise the *connection* between the two alarms. The reader may also note the absence of errors in Task Step 1. This is because it was considered incredible that there would be no response whatsoever, since the CCR is permanently manned (a back-up CCR is also available) and there are two back-up operators who would also respond, if

Figure VI.5.3 Human-error-screening report

Assessor:	BIK
Task:	Alarm response
Safety case:	Example
File name:	Spurious

Step	No 2.2: OPERATOR INFERS FAILURE FROM BOTH INSTRUMENT READINGS
Error	No 06010: OPERATOR FAILS TO LINK TWO ALARMS
Error mechanism	Bounded rationality
Recovery	—
Dependence/exclusivity	—
Screening	No screening appropriate
Comments	—

Step	No 2.2: OPERATOR INFERS FAILURE FROM BOTH INSTRUMENT READINGS
Error	No 06020: OPERATOR ASSUMES FQN IS FAULTY
Error mechanism	Bounded rationality
Recovery	—
Dependence/exclusivity	—
Screening	SUBSUMED
Comments	HEP contribution accounted for by 0401

Step	No 2.2: OPERATOR INFERS FAILURE FROM BOTH INSTRUMENT READINGS
Error	No 06090: EPM MAY BE UNRELIABLE OR HAVE OTHER MEANINGS
Error mechanism	Risk-recognition failure
Recovery	POSSIBLY BY SUPERVISOR
Dependence/exclusivity	MAINTENANCE ERRORS (0528)
Screening	SUBSUMED
Comments	HEP contribution accounted for by 0401

required, to the high-level SO2 alarm. Therefore, there would *always* be a response. The main question, of course, was whether it would be the *correct* one or not. In the fault tree, there would, in any case, be a HPLV to cover those common-mode human failures which could lead to total non-response.

The numbers in Figure VI.5.2 are unique identifiers produced automatically within the HRMS system to discriminate between any two errors.

Screening took place next, during which some errors were merged with others to prevent double-counting; e.g. 06020 and 06090 were subsumed by 0401, the more general diagnostic-failure error. The screening analysis results for three errors are shown in Figure VI.5.3.

5. Quantification

The HRMS quantification results for a subset of the errors are shown in Figure VI.5.4. Figure VI.5.5 shows the computer logging of the PSFs involved in one of the errors, with the assessor's responses for each question being either 'YES' or 'NO', or else being already implicit in the answer. These

Figure VI.5.4 Human Error Quantification results

Assessor:	BIK
Task:	Alarm response
Safety case:	Example
File name:	Spurious

1.1	OPERATOR DETECTS SO2 ALARM
02160	OPERATOR FAILS TO DETECT ALARM
HEP	0.010
1.2	OPERATOR DETECTS FQN ALARM
03160	OPERATOR FAILS TO DETECT 2ND ALARM (BUSY WITH 1ST)
HEP	0.10
2	OPERATOR DIAGNOSES FAULT CONDITION
04010	OPERATOR MISDIAGNOSES CAUSE
HEP	0.30

answers are then used by the HRMS algorithms to generate a HEP. The number in brackets after each PSF heading refers to the value (between 0 and 10, the latter being extremely unfavourable) ascribed to the error scenario by the algorithms, as a function of the PSF responses entered into the system by the analyst.

6. Impact

The results for this particular scenario were high in human-error terms, the major contribution to this high level of unreliability being the diagnosis problem. The most important ways of improving the level of performance in this scenario therefore involved both providing diagnostic aids and training so as to enable the operator to make the *link* between the two alarms – a computer-displayed message, appearing whenever one of the two alarms occurred would also enable the operator to check for the presence of the other – and the rendering or re-engineering of the FQN alarm so as to make it more reliable – or, if this is not possible, the provision of a diverse verification alarm for the FQN function.

Case Study 6: hazardous-substance control

1. Background

This case study is related to Case Study 4 in that it involves a failure to realise that a particular batch is inadequate and should not be further processed. However, in this study it was decided to evaluate not merely the fact of failing to recognise the results of the sample analysis (see Case Study 4) but

Figure VI.5.5 Performance Shaping Factor answers

Assessor: BIK
Task: Alarm response
Safety case: Example
File name: Spurious

Error: 04260   OPERATOR DISTRUSTS FQN ALARM AND REACTS TOO LATE

TIME (8)

The time available is about 10 minutes.
More than twice the estimated required time is available.

QUALITY OF INFORMATION (4)

The information is utilised in the Central Control Room.
A DCS VDU is utilised.

Is information presented on a hardwired panel?	NO
Is information presented as a written/typed document?	NO
Are the information sources highly reliable?	YES
Are indicators well-laid-out and well labelled?	YES
Are indications or confirmations achieved via oral communication only?	NO

TRAINING/EXPERIENCE/FAMILIARITY (4)

The event is infrequent (approximately yearly).

Have operators been trained in this scenario?	YES
Have operators received simulator training in this scenario?	NO
Is refresher training given?	NO
Do operators have an appropriate knowledge of the system (including a knowledge of chemistry/physics) that would enable them to understand what is occurring?	YES

PROCEDURES/INSTRUCTIONS (6)

Written procedures are available.
Procedures are rated as average.

Is there enough work space for the procedures to be utilised?	YES
Are checksheets/ticksheets used in the procedures?	YES
Are the conditions that require the use of mandatory procedures clearly specified?	YES
Do procedures contain warnings about possible hazards arising during the scenario (and in particular about the event being considered here)?	NO
Do procedures give warnings not to take certain actions or not to take short-cuts, explaining their negative consequences?	NO
Do procedures highlight the checks to procedural sequences that need to be made, particularly when these checks occur at the end of task sequences?	YES
Are procedures misleading or incorrect?	NO

TASK ORGANISATION (6)

The personnel involved is just a single operator.
There are likely to be some minor distractions.

Is a computerised checking of results utilised?	NO
Does the operator understand the task's significance, i.e. the importance of a reliable performance and the cost or consequences of any errors?	YES

Figure VI.5.5 (Continued)

Is the environment hazardous (i.e. not that found in a normal plant or control room)?	NO
Is protective clothing required?	NO
Is there a high degree of noise or heat?	NO
Is there any physical danger to the operator?	NO
There is possibly some production/safety conflict.	
Is the allocation of responsibility clear?	YES
Is self-checking built into the task?	YES

TASK COMPLEXITY (4)

Does the task involve following a set of pre-ordained rules which are few in number (i.e. less than 10), and which do not require many interpretations, or much decision-making?	YES

also a *rule violation* associated with the task. The rule violation here involves deciding not to carry out the test at all, but instead to assume that the batch is okay. This behaviour represents what Reason (1990) would call an *extreme rule violation*. It would only be likely to occur, if at all, at times when the shift team is under significant production-pressure to deliver the product that amounts to the output of the process.

2. Problem definition

The problem is therefore one of quantifying an extreme rule violation, one that could have serious consequences, for that case where it happens to co-incide with the delivery of an out-of-spec batch, which latter event is itself, also, a rare, but not unheard of, event.

3. Task and human error analysis

There was no formal task analysis, since the task had already been analysed for Case Study 4, and the error here involved simply a conscious decision not to carry out the task, so as to save time and to keep up with production-quota targets.

4. Quantification

It is rare to see such extreme violations being quantified, whether in a PSA context or not (this study, as with Case Study 4, occurred *outside* a PSA framework), and it was not clear as to which, if any, of the HRQ techniques would be valid for such a quantification exercise. As a result, a number of different approaches were enlisted, with experienced assessors and expert groups working independently. The results (but not the workings) are shown below:

Technique		HEP
HEART		0.0004
THERP	(Assessor 1)	<1E−6
THERP	(Assessor 2)	0.002
APJ	(Group 1)	0.0003
APJ	(Group 2)	0.00003
Geometric mean		<u>0.00009</u>

(The THERP assessor's estimate is taken as 1E−6.)

Although the geometric-mean value is approximately 1E−4, a higher value would be used if this were a PSA estimate, since there is obviously a significant degree of uncertainty and divergence in the estimates (e.g. a value of 0.0005 might even be utilised).

5. Impact

This brief case study has been presented firstly to show that a rule-violation quantification is possible, and secondly to illustrate the level of human reliability which assessors and techniques with cases of extreme rule violations. In the author's point of view, the estimates produced by the APJ groups should be given the most credibility in connection with this error. This is because the two groups involved within the approach (see Case Study 2) were made up of some very experienced operators, and because, to some extent, neither the THERP nor (to a lesser extent) the HEART techniques were equipped to deal with rule violations of this nature. The SLIM method could also have been used for this quantification, and it would have been interesting, indeed, to see what PSFs would have been elicited from the experts carrying out this method.

In terms of its degree of impact on the system, this level of rule violation, given the frequency of the initiating event, would not prove problematical, unless, of course, it became a routine thing to omit the batch-testing procedure.

It is interesting to note that if rule violations of this type are quantified, there is a danger of implicitly assigning them a probability value which makes them appear acceptable in risk terms – as, in fact, this case study would have done if it had been carried out as part of a PSA. The danger here is therefore that of deciding that this rule violation is okay if it only occurs with a probability of less than 0.001. Clearly, any rule violation, however improbable, is, by definition, unacceptable. Since such violations are only recently being addressed within HRAs, it remains to be seen whether more violations of this type will begin to appear in PSAs, or whether they will in some way be dealt with outside of the PSA framework, via safety culture audits.

Case Study 7: distributed emergency operation

1. Background

This scenario concerns a loss-of-cooling-medium accident (in this case the coolant is a gas) in a nuclear-power plant (NPP). This is a serious but highly unlikely event, one which requires many associated operations for protecting the plant against the consequences of such an event. This study aims to look in detail at emergency operations which are distributed around the plant, i.e. which do not just occur in the CCR, or locally, but in several locations, being directed from the CCR.

2. Problem definition

2.1 The system

The nuclear-power plant (NPP) in question houses a gas-cooled reactor. A nuclear chain-reaction occurs and is controlled by the reactor staff via control rods which maintain the reaction at an efficient power-producing level, which is also a safe level for operations. Gas is circulated around the core containing both the control rods and the nuclear-fuel rods, and moves heat away from the core. This heated gas then goes through a heat-exchange system, principally via boilers which contain water. As the gas passes through the boilers (through piping, so that the gas does not actually mix with the water in the boilers), the water is rapidly heated to steam. The steam then leaves the boilers and *drives* the turbines. The turbines thus produce electric power, which, via a set of transformers, etc., is eventually fed into the power-supply grid system. The steam that drives the turbines then passes through a condenser system which condenses the steam back to water; this water then returns to the boilers. Similarly, the coolant gas, having passed through the boilers, and having cooled down as a result of passing its heat to the water in the boilers, now returns to circulate around the core. These are the essential principles of the gas-cooled reactor. The system is illustrated in Figure VI.7.1.

The system is therefore a gas-cooled nuclear-power reactor. There are two of these on the site and they are controlled from the same CCR by two different control-room operators (CROs), supervised by a control-room supervisor (CRS), who in turn is supervised by a shift supervisor (SS). Only the latter is not normally to be found in the CCR; the others, however, *are* permanently in that location. There are various local operators (LOs)/engineers allocated to various areas of the plant, areas such as the cooler house (which houses the gas coolers or circulators), the reactor buildings and the diesel house (DH), which latter houses the diesel generators present in case of a loss of power supply from the grid. Reactors are staffed on a 24-hour

Figure VI.7.1 Diagram of gas-cooled reactor (diagram courtesy of Hickson et al, 1993)

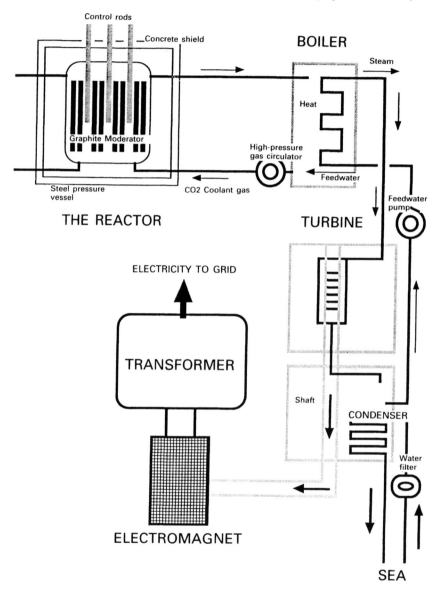

basis – usually with extra administrative-support staff, working a normal day-shift (e.g. 8 hours). The reactors' operational staff will work on rotating 8-hour shifts.

2.2 The scenario

This scenario assumes a fast depressurisation process. What this means is that a main gas pipe running to and from the reactor core itself is ruptured, for whatever reason (such a rupture is highly unlikely to occur, as such pipework is very robust, and is checked frequently for cracks, etc.). If this happens, the coolant gas will escape. This has two major effects. Firstly, the gas will no longer be cooling the core, and the core will thus rapidly heat up, causing the temperature to rise rapidly. This rise will be automatically detected (as will the loss of gas pressure and a number of other symptoms), and the NPP will automatically shut down the reactor by dropping all the control rods into the core. If the rods do *not* drop, the operator can take a number of *other* actions to shut down the reaction.

The second major effect is that radioactive gas from the core will escape into the reactor building or the cooling house or both places (depending on where the break occurs). This gas can cause asphyxiation if it is found in a large enough concentration in a confined area. Some gas is also likely to escape to the biosphere, and in this event, because of the gas's radioactive properties, all staff found in the main plant areas will have to be evacuated to safer areas (e.g. the CCR block and the administration blocks). In extreme conditions, an off-site evacuation from neighbouring villages may also have to be considered. Clearly, this type of event amounts to a 'worst credible case' scenario for any such NPPs. Such an accident has never yet happened for this type of reactor.

There would be other problems, such as likely fatalities for personnel who are in the affected area at the time of the depressurisation break, a break which would involve an initially explosive energy, due to the release of the gas under pressure. There would, as a result, be local debris, and possible injuries or even fatalities as a result of people being hit by flying objects (this would only occur in the case where personnel happen to be working in the affected area at the time: in most cases, however, there would *not* be personnel situated by the gas ducting or the pipeworks).

The CCR personnel must ensure that the reactor is shut down as rapidly as possible, and must try to minimise the extent of the accident. The main problem they face, as with any loss-of-coolant accident occurring in almost any type of NPP, is that of cooling the reactor *after* shutdown. This is necessary since, although the nuclear chain-reaction in a reactor can be stopped in a matter of seconds, the core will still contain what are called *fission products*: by-products of the nuclear-fission (energy-releasing) process. These fission products will continue to generate heat within the core (albeit at a much slower rate than the reactor itself), and this 'decay heat' can lead to severe core damage within hours (this is a conservative estimate). Thus, although the reactor *is* shut down, the cooling process must still be applied to the reactor core within a certain time-frame, or else the situation will worsen. Once the reactor has been safely 'tripped', the CCR staff will then, amongst other duties and functions, set about establishing post-trip cooling.

Post-trip cooling can be achieved by two main methods. Firstly, back-up circulators (BUCs) will automatically start pumping gas into the core. The operators must determine where the break has occurred and close off this coolant leg, so that gas being supplied to the core does not have the chance to reach the core and transfer heat away from it. For long-term cooling, however, at least one of the main circulators (MCs) – which will have tripped once the reactor has tripped since in effect, their power supply comes from the reactor-turbine system itself – must be started, via a back-up power link (BUPL).

The problem definition in this study was therefore concerned with whether the post-trip cooling process would be at all effective in this worst-credible-case scenario.

3. Task analysis

The following first describes, in more detail, the actions required in this scenario and then discusses the task-analysis methods utilised.

3.1 Details of tasks

Initially, the control-room supervisor (CRS) and the control-room operator (CRO) (as well as anyone else in the CCR) need to detect the fault indications and diagnose the problem. There are many indications that would suggest a fast depressurisation, e.g. noise and debris at the plant, an alarm on the boiler panels indicating that a duct is down, a reduction in the indicated pressure and gas-flow levels, control-rod signals and alarms and other various urgent alarms.

Once a fast depressurisation process has been diagnosed, the CRO needs to check that the reactor has tripped automatically, and if it has not, he or she must trip it *manually*. He or she will then need to check that the power level is reducing, that all the control rods have gone in and that temperatures are dropping. If the reactor has still not tripped, then it will be necessary to implement the auxiliary shutdown device (ASD) on the CRO's desk, which is used as an alternative shutdown system if the control rods are unable to enter the core.

The CRO now needs to establish the post-trip cooling of the reactor. Post-trip cooling should be initiated automatically (via the BUCs), but if this has not occurred, it can instead be initiated manually, either on the plant or in the CCR. The manual method takes longer to implement.

Establishing the post-trip cooling of the reactors involves three initial main goals: ensuring that the diesels are operating, ensuring that all of the BUCs have started and are running properly, and establishing that there is water going into the boilers.

The CRS can check from the CCR whether the diesels have started, and if they have not, he or she will need to start them manually from the CCR. If this is not possible, then they must instead be started locally, from the diesel house.

The CRO needs to communicate to the turbine-house engineer the order to check that the emergency (water) feed pumps have been initiated automatically and are operating correctly – and if they have not, to start them manually. Once the feed is under way, the CRS will instruct someone to set up the BUPL.

Throughout the scenario, the CRO, the CRS and the Diesel Engineer (DE) will be monitoring and maintaining fuel and core temperatures, and adjusting BUC motor speeds and the boiler-feed level, if necessary.

Whoever is available and knows how to do the task (there will be a number of operators/engineers converging on the CCR following the plant-wide announcement, by the CRS, that a site emergency, in particular a depressurisation, has occurred) can then go and close the BUPL emergency-feeder circuit-breaker in the BUPL local equipment room (LER). This involves pulling a switch to the correct position, manually closing the circuit breaker, putting the switch back and ensuring that the control selector is on 'Remote'. This activates the BUPL link. A similar operation then takes place to connect one of the main circulators to this link.

The CRS in the CCR can then run up the BUP motor to its maximum speed and start up the main circulator. This is done by pressing two buttons in sequence in the CCR.

During all this, there needs to be an operator in the cooler house, monitoring circulator speeds, the bearing temperatures and the flow of oil to the bearings.

The analysis described in this case study took place in the operational phase of the system; i.e. it was based on an existing plant.

3.2 Task analysis methods

The HRA approach utilised a number of task-analysis approaches, as follows:

- Data-collection methods: documentation review; observation; interviews; Critical Incident Technique.
- Representation techniques: hierarchical task analysis; tabular task analysis; horizontal timeline analysis; vertical timeline analysis; sequence-dependence tables.
- Task-simulation methods: walk-through; talk-through.

The task-analysis phase involved two levels of task analysis, a detailed task analysis for a select number of tasks and a less-detailed analysis for other PSA-identified human-error scenarios (involving the use of an ITA format, as described on p. 120–121).

Figures VI.7.2–VI.7.4 show, respectively, extracts from a tabular task analysis (TTA), a horizontal timeline analysis (TLA) and a vertical timeline analysis.

4. Human error analysis

In this study, a detailed identification of human errors was not carried out. Instead, the HRA was PSA-driven; and the actions that had to be carried out in post-trip and pre-trip conditions were, for example, specified by means of detailed PSA analyses. The emphasis was on determining whether the performance would be effectively carried out within the required time-scales for the various different scenarios. However, for each separate scenario, the potential human errors were all considered from a *qualitative* angle, so that, if required, later ERMs could be generated. Qualitative CRA methods were also utilised (see Case Study 8), as were HAZOP approaches, the latter for a consideration of errors of commission (see Case Study 9). In this particular scenario, errors were considered by reference to error taxonomies (e.g. the HRMS system), as well as via detailed discussions of the steps carried out when interviewing the operators (a form of informal HAZOP, supplemented by a walk-through and a talk-through). A number of possible key errors were then identified. These are not reported in this case study, but to give an example, during the securing of the BUPL, there are two opportunities to change the circuit-breaker's status but then *fail* to switch the mechanism back from local-control (LER) status to remote-control (CCR-controlled) status. This simple error (of omission: the operator forgets an isolated act at the end of the task) would lead to delays if the CCR operators tried to use the link later in the scenario.

5. Representation

Generally, the PSA utilised fault trees in which the errors or task failures were inserted. Aspects of dependence related to the fault trees are discussed in Case Study 10.

6. Quantification

Two quantification methods were utilised, namely the HEART method, as the main method of quantification (this author), and the THERP method (assessor B. Kennedy) – used on a small number of scenarios to check whether

Figure VI.7.2 Tabular Task analysis

Task goal	Task duration (min:secs)	Source of task times	Decision/action/ communication	Personnel involved, and location	Information available	Feedback	Comments
0. Commission back-up power link, start main circulator and close gas-duct valves.							
1. Detect and identify problem.	5:00		Detect indications and identify problem.	CRS DE 1/CCR annexe, anyone else in CCR.	Safety lines will indicate trip via rate of change of pressure and via fuel-element protection. Noise and debris. Alarm on boiler panels to indicate duct down. Reduction in pressure indications and in mass flow. Temperature increases, followed by temperature decreases. Various other alarms. Printer will be running (temperature alarms). Hardwired group alarm. Bellow sniffing high CO_2.	Boiler panel alarm indicating duct down. Get indications of top or bottom duct from reports, plus knowledge of where debris is, where cladding has come off, etc.	The 1st-duct-down alarm on the boiler panel locks all the others on the panel. This information can later be used to help identify which circuit is affected.

EJ – Engineer's judgement
WT – Walk-through
EE – Emergency exercise

Figure VI.7.2 (Continued)

Task goal	Task duration (mins:secs)	Source of task times	Decision/action/ communication	Personnel involved, and location	Information available	Feedback	Comments
2. Ensure reactor is shut down.	1:00	EJ	Ensure power reduced. Check all control rods are in. Press manual trip button, if necessary. Check temperatures are dropping. Trip reactor (only if reactor still not tripped).	CRO 1/CCR, anyone else available. SS, CRSup/ CCR, reactor desk.	Power indicator. Nuclear power and temperatures. Temperature falling.	Red light on CRO console indicates if reactor tripped. All control-rod heights at zero. Reactor trip light in CCR – indicates.	CRO would expect all control rods to drop in.
3. Implement emergency plan.	Ongoing throughout, from 2:00 mins onwards.	EJ/WT	Once SS decides it is a site incident and nuclear emergency, he then contacts emergency controller and station manager. Puts nuclear emergency siren on. Personnel will muster and carry out pre-designated tasks under emergency plan.	SS/CCR emergency controller. SS acts as emergency controller until relieved by senior staff.			Emergency plan is put into action within 5 minutes of realisation that nuclear incident is occurring, and continues until emergency is over.
4. Establish cooling of reactors.							

EJ – Engineer's judgement
WT – Walk-through
EE – Emergency exercise

Task goal	Task duration (mins:secs)	Source of task times	Decision/action/ communication	Personnel involved, and location	Information available	Feedback	Comments
4.1 Ensure diesels have started.		EJ/EE	Ensure diesels have started - from mimic, check diesels have started (CCR.) - if not started, can start from CCR or diesel house. - press button. - load diesels. - supply fuel to diesels.	CRS/CCR communicates with turbine engineer or diesel operator. Turbine engineer in turbine hall; diesel operator/ diesel house.			Once diesels have started, may need to go into annexe, since governor controls are located here.
4.2 Ensure all back-up motors have started and drive to circulators taken up.		EJ/WT	Check back-up motors have started. If not, then: - check DC side of panel. - start manually, if necessary. Run down steam-supply system. - transfer from main motor to pony motor	CRO 1/CCR CRS/CCR communicates to CRO 1/CCR	Lamp indicator at red on boiler panel.		

EJ – Engineer's judgement
WT – Walk-through
EE – Emergency exercise

Figure VI.7.2 (Continued)

Task goal	Task duration (mins:secs)	Source of task times	Decision/action/ communication	Personnel involved, and location	Information available	Feedback	Comments
			Ensure drive is taken up – monitor circulator speed. – raise speed of back-up motor until drive is taken up.	CRS/CCR communicates with CRO 1/CCR. CRS/CCR communicates with CRO 1/CCR.	Amps on generator. Blower speed indicator (digital). Amps indicator.	If less than 400 rpm and amps indicate 0, then drive not taken up.	Back-up motor has a centrifugal clutch and so should automatically take up drive, but in practice, clutch may slip.
			Close off circuit – key exchange. – close bottom valve. – close top valve. – or use emergency close procedures (special key). Send out damage-assessment team and then damage-repair team to close off affected circuit.	CRS/CCR communicates with CRO 1/CCR. CCR. CRS/CCR communicates with CRO 1/CCR.			Close off circuits of any pony motors not running or that have not taken up drive. If blower not going, only need to close bottom valves.

EJ — Engineer's judgement
WT — Walk-through
EE — Emergency exercise

Task goal	Task duration (mins:secs)	Source of task times	Decision/action/ communication	Personnel involved, and location	Information available	Feedback	Comments
			Ensure air-compressor temperatures are okay in both blower houses – monitor pump. – re-establish cooling water for air compressor. Send in damage-repair team to plug leak and ensure sealing of valves.	Assistant Shift Supervisor in CCR communicates with the local plant operator.			Operator in communication with CCR. Feasible to run back-up motors at 500 rpm without air cooling.
4.3 Ensure feed is on, and available for boilers.	5:00	EJ/WT					
4.3.1 Ensure emergency feed pumps are operating as necessary.			– inspect pump to check it is running. – start pump manually, if required. – open pump discharge valves. – check second pump, and start manually, if required.	CRO 1/CCR communicates with TH operators/TH. CRO 1/CCR communicates with TH operators/TH. CRO 1/CCR communicates with TH operators/TH. CRO 1/CCR communicates with TH operators/TH.	Running indications on pump. Suction and discharge indicators.	Tank-level feed flow. Feed water pressure (boiler panels).	

EJ – Engineer's judgement TH = Turbine hall
WT – Walk-through
FF – Emergency exercise

Figure VI.7.2 (Continued)

Task goal	Task duration (mins:secs)	Source of task time	Decision/action/ communication	Personnel involved, and location	Information available	Feedback	Comments
4.4 Monitor and maintain fuel and core temperatures.	Ongoing throughout.		Monitor fuel and core temperatures. If necessary, adjust pony-motor speed.	CRO 1/CCR	All reactor instrumentation.		Ongoing throughout whole scenario.
4.6 Make back-up power link available, and use for driving main circulators.	30:00	EJ/WT					
4.6.1 Shut down unaffected reactor.	20:00	EJ/WT	– prepare operator for shut down. – ensure diesels and pony motors are running. – other minor preparatory actions. – trip reactor. – contact 2 operators. – shut down 1 main fan and damper. – pull in damper on other fan to reduce flow.	CRS/CCR and CRO 2. CRS communicates with plant operator in 3rd-floor reactor building.	Electrical supplies, mimic panels.		Controlled shut down to avoid overcooling of unaffected reactor.

EJ – Engineer's judgement
WT – Walk-through
EE – Emergency exercise

Task goal	Task duration (mins:secs)	Source of task times	Decision/action/ communication	Personnel involved, and location	Information available	Feedback	Comments
			Shut down cooling water pumps.	Normally on plant, but can be done in electrical annexe.			To reduce electrical load.
4.6.3 Prepare and close circuit-breaker to re-energise back-up power link.	10:00	EJ/WT	Go to BUPL feeder CB: – get keys. – follow instructions on CB to change to required status. – return to remote status. – return to CCR to close 2 CBs on BUPL panel.	Whoever is available. Plant BUPL.			Instructions on this procedure are on the CB frame.
4.6.4 Run up back-up motor to maximum possible speed.	10:00	EJ		CRO 1/CCR.	Boiler panel.		
4.6.5 Snatch to main motor.	10:00	EJ	Close main motor, BUPL switch. Open back-up motor, ACB switch.				

EJ – Engineer's judgement
WT – Walk-through
EE – Emergency exercise

CB = Circuit breaker

Figure VI.7.2 (Continued)

Task goal	Task duration (mins:secs)	Source of task times	Decision/action/ communication	Personnel involved, and location	Information available	Feedback	Comments
4.6.6 Monitor circulator.	Ongoing throughout, from 105:00 mins onward.	EJ Do for each blower house.	Someone in blower house to monitor speed, bearing temperatures and flow of oil to bearings. If no good, shut down, and close off circuit.	Local operator /blower house.			May not always be safe to have operator in blower house.
4.6.7 Ensure back-up motor clutch is disengaged.	2:00		Blower operator communicates with CCR if clutch disengaged. If not, trip clutch motor and close off circuit (see earlier). – inspect motor on plant.	Blower operator/ blower house.	CCR speed and current indications.		Closing off circuit will avoid preferential flow through circuit, instead of through core.
4.7 Monitor reactor, and ensure sufficient cooling.	Ongoing throughout.		Ongoing monitoring.	CRO/CCR.	Indicators in CCR. Printers.		To avoid bypassing core.

EJ – Engineer's judgement
WT – Walk-through
EE – Emergency exercise

Figure VI.7.3 Horizontal timeline analysis

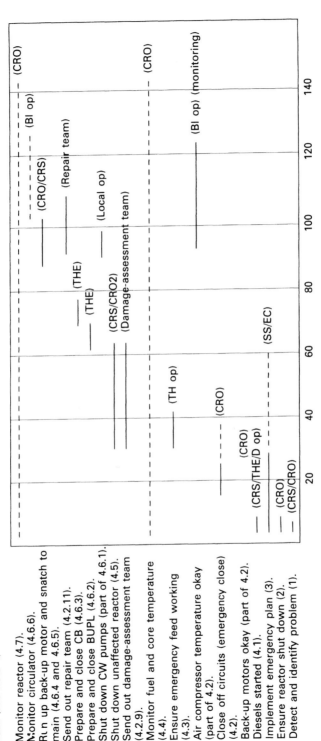

Task (see note 1)

Monitor reactor (4.7).
Monitor circulator (4.6.6).
Run up back-up motor and snatch to main (4.6.4 and 4.6.5).
Send out repair team (4.2.11).
Prepare and close CB (4.6.3).
Prepare and close BUPL (4.6.2).
Shut down CW pumps (part of 4.6.1).
Shut down unaffected reactor (4.5).
Send out damage-assessment team (4.2.9).
Monitor fuel and core temperature (4.4).
Ensure emergency feed working (4.3).
Air compressor temperature okay (part of 4.2).
Close off circuits (emergency close) (4.2).
Back-up motors okay (part of 4.2).
Diesels started (4.1).
Implement emergency plan (3).
Ensure reactor shut down (2).
Detect and identify problem (1).

Note: task number refers to task goals in HTA/TTA in Appendices 1 and 2.

THE = Turbine-hall engineer
D op = Diesel-house operator
EC = Emergency controller

Figure VI.7.4 Vertical timeline analysis

Commissioning of back-up power link, starting of main (gas circulator) motor, and closing of gas-duct valves following a depressurisation fault

Scenarios: Vertical TLA (see note 2)

Task number (see note 1)	Sub-task	Time begin task	Time finish task	MCR-based SS	MCR-based CRS	MCR-based CRO 1	MCR-based CRO 2	MCR-based Other	Local Op 1	Local Op 2
1.	Detect problem.	0	5:00	If in MCR ×	×	× (any one else in MCR)				
2.	Ensure reactor shut down.	1:00			×	×				
3.	Implement emergency plan.	2:00	Continuous throughout emergency							
4.	Establish cooling of reactors.									
4.1.1	Diesels started.	4:00								
4.1.2	Start manually.	4:10			×	×				
4.1.2.2	Locate buttons on panel.	4:20			×(1)					
4.1.3	Load diesels.	8:50	9:20							
4.2	Check back-up motor started.	10:00								
4.2.1	Check lights.	12:00				×				
4.2.2	Check DC side of panel.	15:00				×		THE (1)		
4.2.3	Start back-up motor manually.	16:00			×(1)	×(1)		×		
4.2.5	Transfer from main to back-up motor.	21:00				×		×		
4.2.6	Ensure drive taken up.	21:00				×				
4.2.6.2	Monitor circulator speed.	22:00				×				
4.2.7	Trip all CBs.	26:00				×				
4.2.6.3	Raise speed of back-up motors.	26:00	28:00			×				

Notes:

1. Task number refers to task goals in HTA/TTA in Appendices 1 and 2.
2. Items with the same number in parentheses indicate communication between these individuals.

Commissioning of back-up power link, starting of main (gas circulator) motor, and closing of gas-duct valves following a depressurisation fault

Scenarios 3: Vertical TLA (see note 2)

Task number (see note 1)	Sub-task	Time begin task	Time finish task	MCR-based					Local	
				SS	CRS	CRO 1	CRO 2	Other	Local Op 1	Local Op 2
4.2.8	Close off 2 circuits using emergency-close procedures.	28:00	34:00			×				
4.6	Make BUPL available.	32:00			Whoever available			THE (probably)		
4.6.1.1	Shut down unaffected reactor.	32:00	62:00		×					
4.2.9	Send out damage assessment team.	32:00	60:00	×			×			
4.3.1	Ensure emergency feed pump operating.	35:00			×(1)	×(1)(2)			T. hall op (1) (possibly only 1)	T. hall op (2)
4.3.1.2	Start pump manually.	35:30							×	×
	Check other pump.	37:30							×	×
	Start 2nd manually.	38:00	40:00						×	×
4.1.4	Supply field to diesels.	60:00	Ongoing					Probably, THE	Diesel op	
4.6.2	Prepare and close BUPL.	62:00	69:00		Anyone available			Probably, THE		
4.6.3	Prepare and close CB to re-energise BUPL blower board.	69:00	79:00		Anyone available	×				
4.6.4	Run up pony motor.	85:00			×					
4.6.1.6	Shut down cooling water pumps.	90:00	100:00		×(1)				×(1)	

Notes:
1. Task number refers to task goals in HTA/TTA in Appendices 1 and 2.
2. Items with the same number in parentheses indicate communication between these individuals.

Figure VI.7.4 (Continued)

Task number (see note 1)	Sub-task	Time begin task	Time finish task	Commissioning of back-up power link, starting of main (gas circulator) motor, and closing of gas-duct valves following a depressurisation fault						
				Scenarios 3: Vertical TLA (see note 2)						
				MCR-based					Local	
				SS	CRS	CRO 1	CRO 2	Other	Local Op 1	Local Op 2
4.2.1.1	Send out repair team.	90:00	105:00	×						
4.2.10	Ensure compressor temperature okay.	90:00	114:00		Ass. SS				Local op × (1)	
4.6.5	Snatch to main motor.	95:00				×				
4.6.6	Monitor circulator speed.	105:00							Local op × (1) ×	
4.6.7	Ensure back-up or clutch disengaged.	On-going				×				
4.4	Monitor reactor.	On-going								

Notes:
1. Task number refers to task goals in HTA/TTA in Appendices 1 and 2.
2. Items with the same number in parentheses indicate communication between these individuals.

Figure VI.7.5 THERP analysis

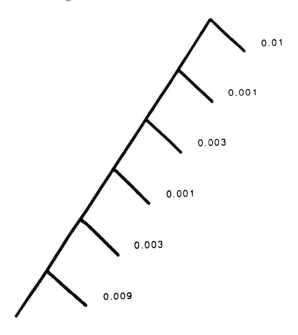

Probability of failure
= 0.027

or not the HEART-derived HEPs were acceptable. Note that this verification of the HEART method by means of the THERP method does not imply that the latter method is more accurate than the former. HEART was *more resources-efficient* than the THERP method, and was utilised for the majority of the scenarios. It was desirable to check some of the HEART methods results against the estimates of another technique. The THERP method was selected on the grounds of its credibility in the NPP HRA arena, and because the technique did not require the use of expert groups.

Figure VI.7.5 is a simplified THERP HRAET for the scenario, the data utilised in the HRAET being defined in Table VI.7.1. Figure VI.7.6 shows a corresponding HEART analysis (the numbers used in both of these figures are not necessarily those used in the real study). In the actual study, the numbers derived for this scenario by two independent assessors fell within a factor of 2 of each other – a rather high level of agreement.

Table VI.7.1 THERP data

Task step	THERP table	BHEP	EF	HEP used
1. Fails to appraise situation.	20–3(4)	0.001	10	0.01
2. Fails to ensure system has shut down.	20–11(7)	0.001	–	0.001
3. Fails to ensure that diesel electric supply is operating correctly.	20–11(2)	0.001	3	0.003
4. Fails to ensure that all back-up motors have started.	20–11(7)	0.001	–	0.001
5. Fails to ensure that emergency coolant pumps are operating as necessary.	20–11(2)	0.001	3	0.003
6. Fails to ensure that coolant route is established.	20–12(2)	0.003	3	0.009

Figure VI.7.6 HEART analysis

Fault/event tree	EMERGENCY FEED
Operator-action identifier	FEED1
Operator-action description	Operator fails to commission emergency feed, following loss of main feed.

Generic task	F: 0.003

Error-producing conditions	Maximum effect	Assessed proportion	Assessed factor
Unfamiliarity	× 17	0.1	((17–1) × 0.1) + 1 = 2.6
Time pressure	× 11	0.3	((11–1) × 0.3) + 1 = 4

Assessed probability of failure

$0.003 \times 2.6 \times 4 = 0.03$

Assessment assumptions and notes

The HEP implicitly includes possibilities of EOCs – e.g. the inadvertent closure of the valves.

Dependence relationships

N/a

Assessor's comments

Unfamiliarity and time pressure can, to an extent, be mitigated against by emergency training, and, in particular, by a clear flowchart procedure and discussion tree aiding the operators. This procedure should also takes note of the required status and position of various valves.

7. Impact and error reduction

The task analyses (L. Robinson had substantial input on the TA side) carried out enabled detailed analysis of the scenarios involved, and although they took some time (many months, in fact), the end result was a robust understanding, and a detailed modelling, of human errors within the scenarios. Furthermore, detailed ERMs were provided for the plant, and a number of these were adopted by the plant.

Case Study 8: misdiagnosis analysis

1. Background

This case study formed part of a qualitative analysis examining the potential for a misdiagnosis in a number of nuclear-power-plant scenarios. (This particular analysis did not feed any *quantitative* information into the PSA.) The study was therefore determining the overall level of vulnerability to a misdiagnosis, and evaluating whether there was a need for a greater level of support from the operators involved. There are a number of methods for evaluating misdiagnosis vulnerability, such as IMAS, CES, etc. (see the section on Human Error Identification), but these have been estimated to be too resource-intensive for this study, and in any case, at the time of the study, they were only prototypical in nature (i.e. they had not yet been used in a PSA). It was therefore decided to utilise more practicable approaches for this study, namely *confusion matrices* and *fault-symptom matrices*.

2. Problem definition

It was judged useful and resource-effective to treat the PSA-bounding scenarios as being representative of the major anticipated accident sequences. Trip and post-trip faults were therefore amalgamated, producing 28 scenarios in total.

The analysis itself was aimed at answering several questions:

- Are any of the main scenarios confusible?
- Is there enough instrumentation to allow discrimination between scenarios?
- Is there enough procedural and training support to facilitate an accurate diagnosis?
- Are the operators likely to fail to make a diagnosis?

The study took place on an existing nuclear-power plant (NPP).

3. Task analysis

Prior to the human-error analyses, a good deal of detailed task analyses had been carried out for all human aspects of the scenarios, including a review of procedural support and the identification of relevant interfaces (the former via an ITA, the latter via a TTA). Additionally, timelines (vertical and horizontal) were available for *four* of the major scenarios.

In order to gain an appreciation of the level of adequacy of training, the central training facility was visited. The Critical Incident Technique (CIT) was also utilised, which raised a number of issues, and one misdiagnosis (involving a partial diagnosis) was examined, with the shift team, in detail. The observation of a full-scale simulated emergency scenario (fast depressurisation) also proved useful for assessing the diagnostic 'style' of the operators.

4. Human error analysis

The human-error analysis breaks down into two parts, namely a fault-symptom-matrix analysis (FSMA) and a confusion-matrix anlysis (CMA). Since the CMA makes use of the results of the FSMA, the FSMA is described first.

4.1 Fault symptom matrix analysis (FSMA)

The FSM is designed to show what indications (symptoms) would be available for each scenario (fault). The FSMA took about a day and a half, and involved two analysts (carrying out a PSA and a HRA respectively) and two experienced control-room operators. The analysis was driven initially by a PSA analysis of the scenarios, which had identified what alarms should occur for each scenario. The operators then identified back-up alarms or indications, and commented on their degree of reliability and informativeness when the need arises to diagnose each particular scenario.

An extract from the FSMA is shown in Figure VI.8.1. This FSMA was extensive, with over 50 indications being considered in relation to the different scenarios. The local indications involved were listed in three special 'local' columns. The letters P, S and T refer to whether the indications are primary, secondary or tertiary in nature as far as the operators are concerned; 'primary' denoting a very high-priority alarm, one which is a key indicator for a particular scenario; 'secondary' denoting that the indication is of high importance but relates to a range of scenarios; and 'tertiary' referring to back-up corroborative indications of lesser importance. In practice, only the primary and secondary categories were generally used by the experts, and on the example shown, if there is no 'P', then the indication is to be taken as primary – i.e. only if there is an 'S' is the indication to be taken as secondary and not primary. The letter 'U' refers to whether or not the indication is unreliable, either because it suffers from a false-alarm tendency or because it proves ambiguous in a particular scenario.

The FSM showed the diverse nature of the indications available, as well as the range of CCR symptoms available. In some cases (e.g. the case of fast

Figure VI.8.1 FSMA example

SCENARIOS (fault)	Environment	Alarms (Elec)	RT	CR Slack wire	Back-up and DGs start	Safety line	SF6	DHP	Gas duct down	Main flow	Main power	Temperature
1. Loss of grid.	√ Lights (U)	√ 275kV unless tie-line	√	√	√	√ Temperature	√			√ S	√	√
2. Flooding in annexes.		√ various	√	√	√	√ Temperature	√			√ S	√ S	√ S
3. Loss of all diesels (fire).	√ Smoke Fire (1)				√U Diesels un-available indicator							
4. Reactor-cable-tunnel fire.	√ Smoke (1)		√	√	√	√ Main motors				√	√	√
5. Spurious trip.			√	√	√	√ Various						
6. Loss of circulator lube oil in one compressor house.			√	√	√	√ Main motors				√	√	√
7. Loss of air in one compressor house.			√	√	√	√ Main motors				√	√	√
8. Loss of air in both compressor house.			√	√	√	√ Main motors				√	√	√
9. Bearing-cooling-system failure.			√	√	√	√ Loss of feed indicator and power loss indicator				√ S	√ S	√ S

depressurisation), the scenario was clearly non-confusible with others, due to its severity and the high number and range of diverse symptoms (including a large, audible, explosive bang). A small number, however, were *less* discriminable in the CCR itself, relying largely, for their identification, on either one non-primary CCR-indication or else local (i.e. on-plant) instrumentation. These less discriminable symptoms must be reported back to the CCR via oral communication.

The FSM was only carried out for one time-frame, since the scenarios do not develop significantly over time, in the sense of involving lots of changes of symptom, and the FSM participants felt that with time the scenario would get clearer not more confusing, unless the scenarios compounded (i.e. multiple events occurred which were outside the scope of this analysis). Therefore, the FSMA and the CMA were aimed largely at the initial misdiagnosis, a premature diagnosis, and a diagnostic failure.

When key indications were not available at the outset of the scenario (i.e. within the first 10 minutes of the scenario), this was noted by the word 'Later' appearing in the respective FSM cell.

4.2 Confusion matrix analysis (CMA)

The FSM was first used by the analyst to cluster the scenarios into two major groupings for CMA purposes. This was done for two reasons: firstly the two groups of scenarios were clearly not confusible, as one group in particular had significant environmental effects, most of which would be immediately perceived in the CCR or very quickly reported to the CCR – or both; and secondly, if the scenarios had not been broken down, the resulting CM would have required a minimum of $(28 \times 27)/2 = 378$ comparisons, which puts excessive demands on participants and runs the risk of causing non-robust comparisons to be made. The major distinction between the two groups of scenarios was thus that one group had significant environmental effects whereas the other did not. After the CMA exercise – which used two expert participants and two analysts, as before, over approximately two days – had been completed, another CRO then reviewed the results of this CMA – in particular, the confusible scenarios identified.

One example of a confusion matrix is shown in Figure VI.8.2. The nomenclature is as follows:

X = No comparison required: identical scenario or subset of scenario.
L = Low or negligible chance of misdiagnosis.
M = Moderate chance of misdiagnosis.
H = High chance of misdiagnosis.
HI = Initially (i.e. within first 5–10 mins) high chance of misdiagnosis.
S = Significant impact on scenario if misdiagnosis occurs.
N = Negligible impact on scenario if misdiagnosis occurs.

Figure VI.8.2 Confusion-matrix analysis

Scenarios	4	6	7	8	9	10	11	12	13	14	15	18	Comments
4 Reactor-cable-tunnel fire.	X												
6 Loss of circulator lube oil in one compressor house.	L	X	HI M/N	L									
7 Loss of air in one compressor house.	L	HI M/S	X	L									
8 Loss of air in both compressor houses.	L	HI M/S	HI M/N	X	X								
9 Bearing-cooling-system failure.	L	L	L	L	X			H/S					
10 Loss of main feed to 3 boilers.	L	L	L	L	L	L	X						
11 Loss of main feed to all boilers.	L	L	L	L	HI M/S	HI M/S	L	X					

Each entry in a cell denotes a confusion of the real scenario in the respective row with the confusible column scenario – i.e. the row scenario occurs, but the operators make a misdiagnosis and believe instead that it is the *column* scenario that is occurring. If the confusability is rated as low (L), then the subsequent impact of the misdiagnosis is not considered. If such a misdiagnosis is moderately or highly likely, however, then the impact must be assessed. Initially, only one diagonal half of the matrix was utilised, although misdiagnoses in both directions are considered. Respective mirror-image cells are only recorded in the case of an M or H misdiagnosis potential. Thus, if a cell is blank in one diagonal half of the matrix, then the confusibility represented by that cell is of a low order. These mirror-image cells must be assessed particularly where the misdiagnosis of confusing one scenario with a second one may not be reversible to the same extent, i.e. a confusion of A with B might be highly likely, whilst a confusion of B with A, on the other hand, might have only a moderate or negligible likelihood. Also, the consequences may be significant in one confusion 'direction' but not in the other (such a situation actually occurred during the analysis).

The results of interest were only those scenarios rated M/S or H/S. Only 3 of these were identified.

5. Impact

The first result concerned the diagnostic 'style' of the operating teams, and was recorded via the various task analyses (observation, interviews, CIT, incident records). The conclusion reached was that a diagnostic failure was seen to be highly unlikely, and that a prompt diagnosis was, in fact, much more likely to be the case, whether that diagnosis was initially right or wrong.

This appeared to be for two main reasons: firstly, the procedures and training involved were, at the time of the study, largely event-based, and so until the event was actually diagnosed, little use could be made of these support systems; and secondly, the operators' wish was to act *promptly* in order to tackle the accident and reduce its effects.

However, whilst the operators (CCR staff) did not afford themselves the 'luxury' of taking 30 minutes to diagnose the event (as was potentially implied in NII SAP 124), at frequent intervals they did discuss, albeit briefly, their diagnosis, as well as reviewing both the relevant indications and the progress/chronology of the event. This particular diagnostic-style aspect is raised because, as the CCR engineers were aware, it leaves the way open for a premature diagnosis (be this a partial diagnosis or a misdiagnosis). What is less likely, however, with this diagnostic style is an actual failure to reach *any* diagnosis.

The FSM showed that a number of the scenarios involve environmental symptoms (e.g. a loud bang, steam, flooding, debris or a seismic disturbance), and it was clear that these would be taken as significant indicators by the

CCR team, since few of these indicators are confusible. Similarly, many of the scenarios required investigation on the local plant itself, so as to corroborate the MCR's diagnosis. Whilst this may be more likely to lead to either a premature diagnosis or a misdiagnosis – since there is time-pressure to identify the scenarios involved – the cognitive problem-solving style of all those interviewed was clearly one which placed heavy emphasis on making local investigations before 'concretising' the diagnoses made in the CCR. Lastly, although some of the scenarios could yield confusing or spurious signals (e.g. 'Reactor-cable-tunnel fire'), it was in the very nature of the diverse pattern of signals that were likely to arise that they did not fall into any recognisable, specific reactor-system-accident pattern, a fact which itself amounted to a large diagnostic cue for the operators.

The FSM was therefore a useful exercise in explicating the discriminability of these scenarios. It may be that scenarios occurring *within* the wider generic scenarios specified (i.e. subsets of related scenarios) may not be so 'clear-cut', but the HRA was limited in its resources, and to have extended the FSM so as to include any such sub-scenarios would have been very resource-intensive.

The CMA, as already noted, only raised three significant potential misdiagnoses. Their probability of occurrence was not, however, quantified, since this was not the objective of the study. However, various means of enhancing their discriminability *were* determined, largely via a process of relaying certain local indications into the CCR, and then ascribing to them appropriate degrees of salience. Largely, therefore, the instrumentation did give adequate support to the process of making an accurate diagnosis.

Training and procedural recommendations were made, in an effort to improve the level of diagnostic performance. In particular, training in the carrying-out of distributed actions, achieved by incorporating extra tasks requiring *local* operations and communications with *local* CCRs into the realistic exercises already held at the plant, was suggested. Also, with respect to procedures, it was recommended that the top-level, symptom-based procedure already being developed at the plant be finalised and implemented, so as to facilitate a more rapid and accurate diagnosis and response.

Case Study 9: fault tree and dependence example

1. Background

This case study discusses how dependence was handled in a large fault-tree assessment, one that was carried out as a part of a wider assessment for a nuclear-power plant (NPP) assessment (S. Scannali was the other developer of the approach in this study). The trees are not presented, since they would take up too much space. Having said that, the whole approach is, however, quite straightforward. The Swain (THERP) dependence model was utilised, but the quantification technique used in the actual case study, on the other hand, was HEART.

2. Problem definition

No detail of the tree's basic events is given, since such detail is not required for the purposes of the case study. Whilst dependence can be modelled easily in event trees, its modelling between different human-error events is less common, however, in fault trees (refer to the section on fault vs event trees in the main text). One of the main reasons for this is that fault trees are *static* (i.e. time-independent) models of system failures, whereas relations of dependence usually involve a *sequential* element – i.e. Event A may be dependent on Event B only if B *occurs first*, whilst Event B, on its part, may not be dependent on A at all (i.e. whatever the order of events). In a large fault tree with many human-error events, it may be only after the cut-set analysis has been completed that the various combinations of human errors in a single cut-set may become apparent. At present, therefore, human-dependence analyses in the context of fault trees, unless they are very simple, may be best carried out via an examination of all the various cut-sets, even though this may prove to be a laborious process (e.g. there may be thousands of cut-sets for even a *moderately* complex fault tree).

It should also be noted that HPLVs were not utilised in the study. CMF values *were*, however, utilised, and it could be argued that these include a human-error component.

3. Task analysis

In order to know which event was dependent on which other event(s), it was imperative, in the study, that task analyses be first carried out, and those that were carried out involved a lot of detail. By means of these analyses, the following minimum information was ascertained:

- What were the basic steps involved in the task, and what was its goal?
- Who performed the operation?
- What equipment was used?
- When was the operation performed?
- Was its cue to begin/end/continue contingent upon any other task?

This information fed forward to the dependence analysis described below.

4. Representation

The trees, which were developed over a period of some months, contained a number of HEPs, hardware events, CMF values, etc. When the trees were mathematically evaluated by computerised analytic methods, thousands of cut-sets were developed.

In many cases, these cut-sets contained more than one HEP. In these cases, a set of simple criteria were then proposed for determining whether any relations of dependence existed between any two HEPs within a cut-set. The criteria for judging whether or not there was a relation of dependence between two HEPs were essentially as follows:

(i) The same usage of personnel.
(ii) The same location.
(iii) The same, or a similar, time-frame for the operation in question.
(iv) A logical connection between the tasks involved.

Varying relations of dependence were then approximately allocated as follows:

(i) or (ii) only (i.e. not both)   = MD (moderate dependence)
(i) and (ii)   = HD (high dependence)
(i) and (ii) and (iii)   = CD (complete dependence)
(iv) only   = HD or CD, depending on the strength of the connection

Low dependence was not utilised in the study, on the grounds that since the HRQ technique being utilised was HEART, and since this technique yields more *conservative* estimates than those of the THERP technique, the effects of low dependence in quantitative terms were already effectively accounted for, to a greater or lesser extent. It should be noted that although these guidelines/criteria were developed with Swain's dependency model in mind, this present interpretation may differ from that given in the dependence model developed by Swain himself.

Following the development of these guidelines, a matrix was created showing the different levels of dependence (CD, HD, MD, zero) between all HEPs. This matrix also noted the *direction* of the dependence relation – i.e. it noted whether A and B were mutually independent or whether one event would be dependent only if it came after the other.

The next stage in the dependence analysis was the analysis of the cut-sets themselves, of which there were literally thousands. To make this process more manageable, it was decided that if two HEPs occurred together in a cut-set that was more than 2 orders of magnitude below the probability of the top cut-set, then the dependence relations involved in such cut-sets would generally be regarded as having a negligible impact, and would therefore not be modelled. Similarly, when three or more HEPs occurred more than 4 orders of magnitude below the top cut-set's probability, then all the dependence relations involved in these cases would similarly be regarded as having a negligible impact. Thus, all cut-sets were screened, but these two criteria were, at the same time, applied, so as to minimise re-calculation requirements. The assessor did check representative cut-sets that were being screened in this way to verify that the impact of the dependence relations in question would indeed be negligible, and in all cases this was confirmed.

5. Impact

Once the requantification requirements had been assessed, the base-event probabilities were altered so as to take into account any relations of dependence (e.g. a HEP might have been modified so as to include high-dependence relation). The re-evaluation of the tree resulted in some significant re-ordering of important cut-sets (e.g. some cut-sets 'jumped up', by 2 orders of magnitude, in their probability, becoming far more dominant in their impact on risk), and certain HEPs thus became far more important, individually, than they had been prior to the dependence analysis. This had significant implications for the error-reduction analysis: in particular, not only were certain tasks now seen as key tasks whose failure probability must be minimised, but in addition, the ERA was now targeting the dependent relationships themselves (between two or more errors) as well as the HEPs themselves.

Overall, the results demonstrated a measurable increase in the human-error contribution to the system's overall likelihood of failure, and the value and meaningfulness of dependence analyses, in the context, at least, of this case study, was certainly verified.

Case Study 10 – error of commission analysis

1. Background

This study formed part of a qualitative HRA carried out for a nuclear-power plant (NPP). It was decided to attempt a short analysis of the potential for errors of commission (EOCs), or extraneous acts. However, instead of using one of the prototype techniques being developed at the time, a HAZOP approach, as defined in the chapter on human-error analysis, was adopted.

2. Problem definition

The problem was to determine the plant's level of vulnerability to errors of commission. However, two additional and important criteria were identified: the analysis had to identify credible EOCs; and the method had to be resource-efficient. It was also important that the EOCs did not overlap with problems already being addressed within the PSA going on at the time.

The problem with an EOC analysis is that virtually anything can happen, and the identification of problems is inevitably limited by the amount of resources available and by the assessors' imagination and level of knowledge of the system. In reality, however, EOCs are rare, and it is unwise to devote too many resources to their analysis, when there are other more likely errors, in the areas of procedural and diagnostic performance, that require both assessments and error-reduction measures. Therefore, the error-of-commission analysis (EOCA) involved in the present case study was based only on the main scenarios dealt with in the PSA, so as to give the analysis some 'direction' or focus. Otherwise, the analyst might have ended up considering every single component on the plant, and asking if each one could be inadvertently opened, closed, substituted for another, etc.

The problem was therefore to identify credible and safety-scenario-relevant EOCs in a resource-efficient manner.

The study took place on an existing installation.

3. Task analysis

The HRA had carried out detailed task analyses (TAs) for all those PSA scenarios which included a human-error contribution (i.e. most of them), and since the plant was already in existence, procedures and operational staff were also available. However, a TA, by definition, encodes what should be done, not what should *not* be done. Therefore, whilst the TAs were useful starting points, a more generative approach was also required, one which would involve a HAZOP-style team approach.

4. Human error analysis: EOCA

The EOCA took place over a period of two days, and considered approximately 15 main PSA scenarios. The team consisted of a HRA analyst, one (and sometimes two) experienced CRO(s) and a HAZOP/safety assessor involved in the main PSA. The sessions lasted approximately 3 hours each, and there were two sessions a day.

The GASET was utilised to allow EOCs at each stage of the scenario to be considered, i.e. to allow the identification of EOCs of all kinds – be they latent errors, initiators, misdiagnoses or exacerbators (errors which worsen scenarios). A sheet listing external error modes (EEMs) and relevant PSFs was available and 'on the table', as were all relevant procedures and task-analysis information. Some scenarios were dealt with rapidly, while others required lengthy discussions. Information about the system was also available – for example, P&IDs, electrical diagrams and general layouts – and the CCR was within walking distance. In some cases, to resolve a discussion point, the EOCA team, based on the plant, went on a 'walk-about' to inspect the item of equipment or control interface being discussed.

Once an EOC had been identified, along with its possible causes and consequences, it would be decided whether, in theory (since the EOCA was not part of the quantitative PSA), the EOC would significantly influence the likelihood of occurrence of the main scenario's top event, or whether it would instead lead to a new top event or a new branch in the event tree (or even to a new event tree altogether).

Two examples are given in Tables VI.10.1 and VI.10.2. In these cases, the

Table VI.10.1 EOCA for loss-of-grid scenario

Date:	HAZOP team: (Page 1 of 10)
Bounding fault	Loss of all off-site grid supplies.
Task/action required	Set diesels to duty.
Equipment item Involved	Diesel generators.
Pre-trip Trip/SD Post-trip	√
Error of commission	DGs not set to auto-start following test/maintenance – must be set on the deadband on the flywheel.
Guideword	Setting – Latent
PSFs	Procedures, interface, task organisation.
Causes (other)	
Consequence	DGs do not start automatically, LOG, pony motors will quickly drain the batteries.
Affects probability	Y
New event/sequence	Affects many scenarios. Worst scenario is loss of grid.
FTA/ETA?	FTA
Recoverability	Should have 'diesel unavailable' alarm in CCR. Start diesels manually in CCR.
Error-reduction measure(s)	Deadband marking/alternative solution.
Comments	Only likely if maintenance or testing has occurred recently. Would really need two diesels mis-set to have a significant impact. This error mode is unlikely to be accounted for within the testing-failure-rate data, since the desk engineer knows when testing is to be carried out, and will properly prepare the DGs for a test.
Actions	Investigate qualitatively

main scenario or bounding faults are the loss of grid (power) supplies and a boiler-tube leak. The tasks/actions required are shown, and the relevant EOC is identified. The EOC in the loss-of-grid scenario would lead to a new event in the fault tree, while the EOC for the boiler-tube leak would lead to an increase in the probability of failure of the feedwater system, best accounted for by increasing the CMF value for this system (but only if required, and this would depend on the probability of the EOC, since the CMF value may already implicitly account for it).

Table VI.10.2 EOCA for boiler-tube leak

Date:	HAZOP team: (Page 6 of 6)
Bounding fault	Boiler-tube leak.
Task/action required	Identify leaking boiler.
Equipment item involved	Feedwater.
Pre-trip trip/SD Post-trip	√
Error of commission	Shutting off feed to all boilers (by inadvertently closing the wrong feed system valve).
Guideword	Misinterpretation – cognitive tunnel vision, perseverance.
PSFs	Lack of clear feedback; workload of simultaneously shutting down the reactor.
Causes (other)	Operators trying to identify leaking boiler. If cannot decide via instrumentation, may therefore try to identify one at a time with feed off.
Consequences	Affects decay heat removal.
Affects probability	Y
New event/sequence	CMF – of FW system.
FTA/ETA?	FTA
Recoverability	Would have to be within 2 hours.
Error-reduction measure(s)	Data-log moisture indicators in CCR to indicate first-up. Would record time at which first boiler failed.
Comments	Recovery-potential high.
Actions	Investigate qualitatively for ERMs – recommend control measures to prevent inadvertent valve closure, e.g. valve-locking mechanism, permit-to-work.

5. Quantification

No quantification took place, nor would it have been clear how to quantify these EOCs. They were not included quantitatively in the PSA.

6. Impact

Eighteen EOCs were identified which were considered to have a credible (though very small) chance of occurrence. Of these, 10 would occur as latent errors (in maintenance – e.g. a miscalibration of the alarm settings), seven would occur during task performance and one would occur during a diagnosis in an emergency. Whilst these were all seen to be credible, their probabilities were not, however, quantifiable. Therefore, they were considered for an error-reduction analysis.

The method proved useful in identifying some previously unforeseen failure paths, and at the same time was also relatively resource-efficient (8 person days for the actual analysis – assuming that the task analyses had already been carried out). Moreover, the analysis was seen as a credible approach to a very difficult problem, even given the obvious difficulties it experienced when it came to ensuring comprehensiveness and consistency.

Case Study 11: research-reactor sensitivity analysis

1. Background

This study concerned a human-reliability analysis for a research reactor. In this analysis, which used the SLIM-MAUD technique, one particular error, related to a valve closure, was found to be contributing significantly to the estimated level of risk inherent in the system. It was therefore desirable to calculate how to reduce the HEP via error-reduction measures (ERMs), by using the SLIM system. This case study only details this error-reduction-analysis/sensitivity-analysis part of the study, so as to exemplify how such an error-reduction analysis (ERA) may be carried out, whether by means of the SLIM, HEART or HRMS systems or by means of other such systems with ERA capabilities T. Warers and D. Embrey were the co-assessors.

2. Problem definition

The problem facing the assessor was the quantification of a failure to close a valve in a specified period of time, occurring within a loss-of-coolant type of accident scenario. If the resultant HEPs were found to be too high, then an ERA would be required for a reduction of the HEP to an acceptable value.

3. Task, human error, representation and quantification analyses

These are not detailed here. A detailed task analysis was, however, undertaken – via interviews, walk-throughs, observations and documentation reviews – by the client. A fault tree had also been developed by the client, in which the particular error of concern was required to be calculated. A small panel of expert judges was then convened, and, using the SLIM-MAUD system, the HEP was then calculated.

Table VI.11.1 SLIM-based error-reduction analysis

	Original	1	2	3	4	5	6	7	8	9
Training	6	6	3	3	1	6	3	3	3	6
Procedures	8	2	8	6	8	8	2	2	8	2
Interface	6	6	6	6	6	3	3	6	3	3
Time	5	5	5	5	5	5	5	5	5	5
Distraction	8	8	8	8	8	8	8	8	8	8
SLI	.22	.54	.46	.50	.40	.34	.96	.80	.60	.70
HEP	.10	.03	.07	.04	.02	.09	.001	.005	.01	.007
Reduction factors	*	3	1.4	2.5	5	.1	100	20	10	14

* = Indeterminate reduction factor

4. Impact assessment

The resultant HEP was found to be 0.10. Although this is not an uncommon value for a time-pressured emergency response made with little procedural support, there was a desire, nonetheless, to reduce this figure to a value of 0.001 – a two-orders-of-magnitude change, and a significantly large error reduction. If this could be achieved, then the results of the analysis would indicate a tolerable level of safety.

5. Error reduction analysis

The ERA is shown in Table VI.11.1. The original HEP ratings are shown in the first column, against the PSFs identified as important by the expert group. The SLI value originally estimated was 0.22, and when the experts' assessments were calibrated against two known calibration points (incidentally, with a fairly good regression 'fit'), the HEP of 0.10 was derived.

The ERA proceeded with two members of the original expert group. Using a part of the SLIM-MAUD software system called SLIM-SARAH, a series of alternative PSF profiles were generated, profiles which could help to reduce the HEP. It is important to stress that this was not a purely numerical exercise – in fact, far from it. The exercise proceeded by first determining which PSF-rating/-weighting combination was contributing most to the HEP value. In this case, it was the *procedures* PSF, with an undesirably high PSF rating value of 8 – which in this case signified that procedures were virtually non-existent for this particular scenario. Ways of improving procedures were then discussed. Such procedural improvements would amount to the provision of a symptom-based flowchart procedure with check-points, etc. The specific improvements were noted down as assumptions underpinning the to-be-revised figure. The two experts agreed what new rating value would be gained if such improvements were implemented (in this case, a rating of 2).

This new value was substituted into the SARAH system, and the HEP was recalculated, giving a new figure of 0.03, and an error-reduction factor of 3.

Since this was not enough of a reduction, an alternative PSF was tried, namely that of *training*. Various retraining improvements were considered which together could reduce the rating to a value of 3. However, when the original HEP was now recalculated, the results yielded a reduction factor of only 1.4 and a new HEP of only 0.07.

Other combinations of improvements were considered, including the provision of a new alarm in the interface. This ERM, together with the greatest degree of improvement in training and procedures that could be realistically achieved within the client's budget, yielded a value of 0.001 in Iteration No. 6, shown in the table. Several more iterations were tried, but only Iteration 6 produced the desired reduction factor, and this was the case therefore accepted for the error-reduction analysis. Its assumptions, in terms of changes to be made to training, procedures and the interface, were also accepted for implementation by the client.

Several points are noteworthy about the analysis. Firstly, whilst it had a HEP 'target value', each iteration was driven not by numbers but by a consideration of what could be *feasibly* improved. Secondly, at each iteration, the improvement assumptions were documented, so that the commitments required, if the new value was accepted, could always be made known. Thirdly, there were two PSFs that were not modifiable in this scenario, namely those of *time* and *distractions*, since the former was a function of the reactor physics in the scenario – and therefore no predictor displays were feasible – and the latter, if it were to be reduced, would require an enormous financial outlay, one which would allow a total redesign of the control interface (this was, after all, a research reactor). Lastly, the client, who was also involved in the process, undertook the ERM-implementation process, ensuring the proper implementation of the quality-assurance aspects of the various ERMs.

Case Study 12: human factors in ship platform collision (SPC)

1. Background

This project was concerned with the risk of an offshore platform being struck by a vessel in the UK Continental Shelf (UKCS). Several such collisions have already occurred in the Gulf of Mexico, but so far, in the UK Continental Shelf, such collisions have been restricted to collisions involving small supply boats and fishing vessels. However, the discovery of oil in more congested waters could now intensify the risk of a collision with larger vessels. To gain an insight into how such collisions could be avoided, the study concentrated on analysing the human-error aspects of watch-keeping and navigation. These aspects appeared to be the critical ones, judging by discussions with marine-safety experts and by an analysis of statistics for the most similar kinds of accident, namely ship groundings (involving running aground or hitting underwater objects).

2. Problem definition

It has been estimated that there are approximately 100 fixed or floating offshore structures within the UK sector of the North Sea. A considerable number of these installations are in close proximity to heavily used shipping lanes, and there is thus some degree of concern regarding the possibility of a merchant vessel's colliding with an offshore structure.

As with many accidents, or unacceptable performance levels, connected with aircraft, ships or power plants, human error is frequently cited as the cause. In many of these cases, it has been estimated that 70–90 per cent of all accidents can be traced to human error. Drager, in a (1981) statistical analysis of marine accidents, suggested that 80 per cent of these accidents were attributable to human error.

The ship–platform collision 'scenario' is quite unacceptable, in terms of all the potential consequences – fatalities, economic losses and pollution – that such a collision could cause. This study therefore attempted to define what exactly could cause such an accident, and how it could be prevented.

3. The system

The ship can be a demanding workplace: operators often have to work, within a restricted space, in an unforgiving and even hostile environment, with added problems of ship movement and excessive noise. Although avoiding a collision is obviously a primary safety function, the navigator or watch-keeper will have other tasks to fulfil at the same time, to do with port work routines, etc. Additionally, since manning levels are generally decreasing on ships, and since ships are under significant pressure to meet arrival deadlines, there can be significant pressure, in turn, on those responsible for watch-keeping duties. The sailor is also faced with the additional problems associated with the social isolation of small groups.

The above problems become even more critical when the rate of technological change is considered. The navigator on the bridge is now confronted with a mass of information coming from highly complex electronic equipment. The engineer now operates in a control room that bears a close resemblance to the control room of a nuclear or process-control room, though at night the engine-control room may operate in an unmanned condition, with the control of the information systems being transferred to the navigator on the bridge. This latter development has significantly increased the work-load of the navigator, especially during transit operations in heavily congested waters. Not only must the navigator perform anti-collision manoeuvres, he or she must also monitor the status of the engine-room equipment. Furthermore, the manoeuvres themselves must be well-planned on large ships, which may take a mile to slow down, and significant time actually to turn to avoid another ship or obstacle.

A ship today is therefore a highly complex and sophisticated man–machine system within which technological, social and environmental variables intermingle to produce situations which may give rise to the occurrence of a human error. Since humans are not particularly good at vigilance tasks for extended periods, particularly those with a low 'signal' rate, clearly, watch-keeping errors in particular must be considered a significant potential problem worth further investigation.

Usually, a study of this kind would start by analysing a database of previous such accidents. In the case of ship–platform collisions, however, there was virtually no relevant database at all – at least, not for the UK Continental Shelf. A small group of researchers were therefore assembled whose expertise was considered germane to the problem area. Their experience lay in the fields of:

- Human factors
- Human reliability
- Ship-collision modelling and analysis
- Marine-traffic routing

The group developed a set of seven ship–platform-collision scenarios (see Technica, 1985) which were considered to be realistically possible. The technique of *paired comparisons* was utilised to achieve a rank ordering of the likelihood of occurrence of these various scenarios. Results indicated that the three most frequent collision events would tend to occur with vessels answering to the following descriptions:

- The *errant* vessel: a ship which is approaching a platform but which, as a consequence of an inadequate level of watch-keeping, is *unaware* of its collision course.
- The *blind* vessel: a ship whose radar is operating at an inadequate level (because of a hardware malfunction, or an error on the part of a human operator, or the weather affecting it adversely), and which is also experiencing bad-visibility conditions, as a result of, for example, fog.
- The *drifting* vessel: a ship which has lost (power and steering) control in the vicinity of a platform, and is thus 'drifting' towards the platform.

It was concluded, on the basis of ship–ship and ship–light vessel-collision data, that the errant-vessel-collision scenarios presented the highest degree of risk to an offshore installation.

In order to quantify the degree of risk posed by an errant-vessel collision, and in order to suggest measures to reduce this degree of risk, a more detailed analysis of errant-vessel behaviour was carried out, as described below.

4. Task and error analysis

No formal task analysis was produced for this study. Instead, information was elicited as to the nature of the watch-keeping duties and use of equipment such as radars, while at the same time, the relevant error potential was identified. This information was then entered directly into the event and fault trees for the study. By its very nature, the watch-keeping duty is mainly a very *passive* one, and in the light of this fact, a task representation was not deemed appropriate. Most tasks relating to the study were in fact *recovery* tasks, in which the operator suddenly realises (too late) the danger and attempts anti-collision manoeuvres. Such tasks, also, appear in the fault- and event-tree representations discussed later.

An expert group was convened consisting of four experts: a human-factors consultant with significant marine expertise, an expert both in shipping casualties and in the statistical analysis of worldwide incidents, a professor, from the marine-safety and marine-training areas, specialising in ship–platform collisions (SPCs) and an expert on coastal-traffic analysis and pilotage systems. All four had significant marine experience.

The expert group produced various insights into watch-keeping problems associated with work overloads, on the one hand, and work *under*loads on the other. Work overloads occur as a result of a combination of factors, such as minimum manning levels, short transit times, the documentation required for each port or a level of visibility requiring a high degree of concentration over extended periods. Work underloads, on the other hand, occur because watch-keeping can be a highly unstimulating task, due to the relative infrequency of events – especially in uncongested open waters.

The analysis of the errant-vessel scenario proceeded in two stages. Firstly, watch-keeping was investigated as a subtask of the overall operation of the ship. Watch-keeping, although a relatively easy task to perform, is carried out in a highly demanding environment, one which may, on occasions, compromise the efficiency of the watch-keeper, whose primary role is always the safe navigation of the vessel.

The second stage involved the analysis of ship-grounding casualties, which were considered similar, in their aetiology, to ship–platform collisions. In contrast, however, this particular area of analysis had a *well-developed* database from which to draw information for investigative purposes.

The group's expert opinions and the grounding-casualty analysis were both used to derive a set of 'failure modes' for inadequate levels of watch-keeping (see Figure VI.12.1), and below. Each one of these, taken alone, can lead to an inadequate level of watch-keeping, and therefore to the occurrence of an errant vessel.

- Watch-keeper absorbed in secondary task: the crew member that should be watch-keeping may become absorbed in another task, such as updating charts, and as a result fail to realise the passing of time, as well as the corresponding increase in the level of risk.

Figure VI.12.1

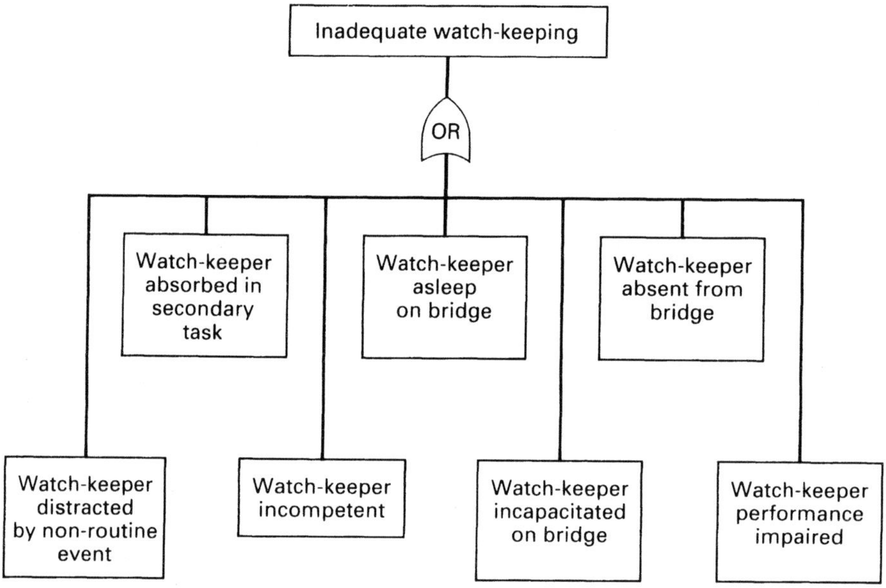

- Watch-keeper's performance impaired: due to excessive and prolonged fatigue, the watch-keeper, while not actually asleep, may still be effectively 'inert' as a watch-keeper.
- Watch-keeper asleep: several of the Gulf of Mexico incidents (including groundings) involved watch-keepers who had fallen asleep on the bridge, again due both to excessive fatigue and to prolonged watch-keeping duties.
- Watch-keeper absent from bridge: at least one collision and several groundings have occurred because the watch-keeper had to leave the bridge for a short period of time, either to eat, or to call the next watch, or to go to the toilet, etc.
- Watch-keeper distracted by non-routine event: the watch-keeper may be distracted on the bridge when an unusual event occurs, such as a minor emergency below, or problems in the engine room.
- Watch-keeper incapacitated on bridge: the watch-keeper may become ill, suffer a cardiac arrest, etc., while on duty, and his or her incapacitation may not be discovered until the next watch-keeper comes to take over.
- Inadequate watch-keeping performance: due to manning problems, personnel who have little or no experience in watch-keeping or the use of radars may have to perform watch-keeping duties.

The derivation of these failure modes permitted quantification of the errant-vessel-collision risk.

5. Representation

The two fault trees shown in Figures VI.12.2 and VI.12.3 list the causes of errancy, as well as the events contributing to a failure to prevent an errant-vessel collision course, while Figure VI.12.4 shows an error-recovery event tree. The focus on error recovery was warranted because of a small but significant number of near-miss incidents reported by platforms in the UK sector, during which, for whatever reason, the ship averted disaster at the last minute, missing the platform but still coming well within the platform's 500-metre exclusion zone (in one unofficial report, the ship actually sailed right underneath the heli-pad).

The fault trees were not quantified directly but instead used by the experts to depict the different contributions to the risk of errancy, as well as the probability of a failure to recover (and to this extent, they *did* aid the process of quantification). The event trees *were*, however, directly quantified.

6. Quantification

The target probability which needed to be quantified was the proportion of vessels that may be errant for a given period of time (usually 20 minutes to 4 hours: less than this is unlikely to be a problem, and more than 4 hours is very unlikely, due to shift systems). It was recognised that different failure modes would have different errancy times attached to them. For example, the distraction involved in performing a secondary, non-routine task may last 20 minutes, whereas someone who is not competent at using a radar in fog conditions will *remain* incompetent throughout the 4-hour watch.

The expert panel considered a good deal of evidence, i.e. grounding data and statistics (see Reference 3 at the end of this case study). The causes of the grounding, in terms of watch-keeping failure modes, was then statistically estimated (see Table VI.12.1). A problem with such data is that, by definition, most groundings occur in coastal waters, whereas most platforms are situated in the open seas. The expert panel, after a day spent discussing the various factors, agreed, in consensus fashion, to the figure of 1.0 per cent as the proportion of errant vessels for the UK Continental Shelf.

As a means of calibration or corroboration, collision data for UK light vessels were examined, and again collisions which had involved the same failure modes as those found in the SPC errant-vessel fault tree were pin-pointed. This research yielded a figure of 1.6 per cent (see Technica, 1985). Being very close to that agreed by the panel, and being the more conservative of the two estimates, this figure was accepted. The panel also quantified, directly, the event tree, again using a direct APJ (in the consensus mode) approach, and taking about half a day to quantify the tree.

The specified failure modes were then incorporated into a predictive fault-tree model of a ship–platform collision. This model (which the HRA fed into) made use of a previously developed computer simulation of vessel-traffic

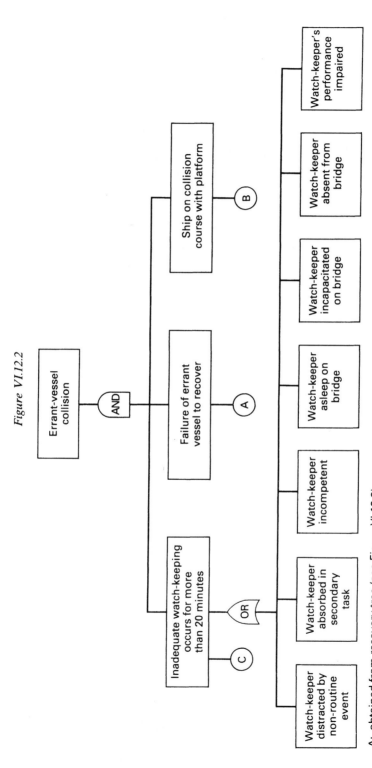

Figure VI.12.2

A: obtained from recovery tree (see Figure VI.12.3).
B: obtained from crash and traffic-survey data.

Figure VI.12.3

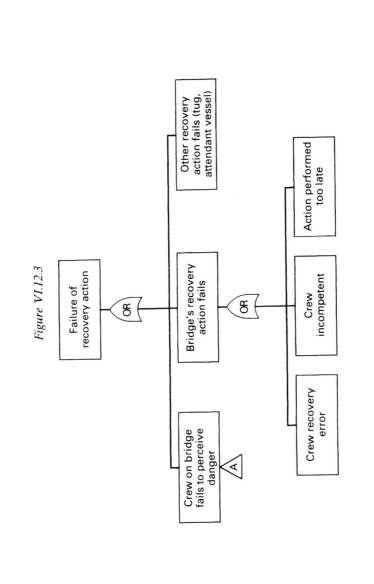

Figure VI.12.3 (Continued)

Figure VI.12.4 Recovery event tree for errant vessels

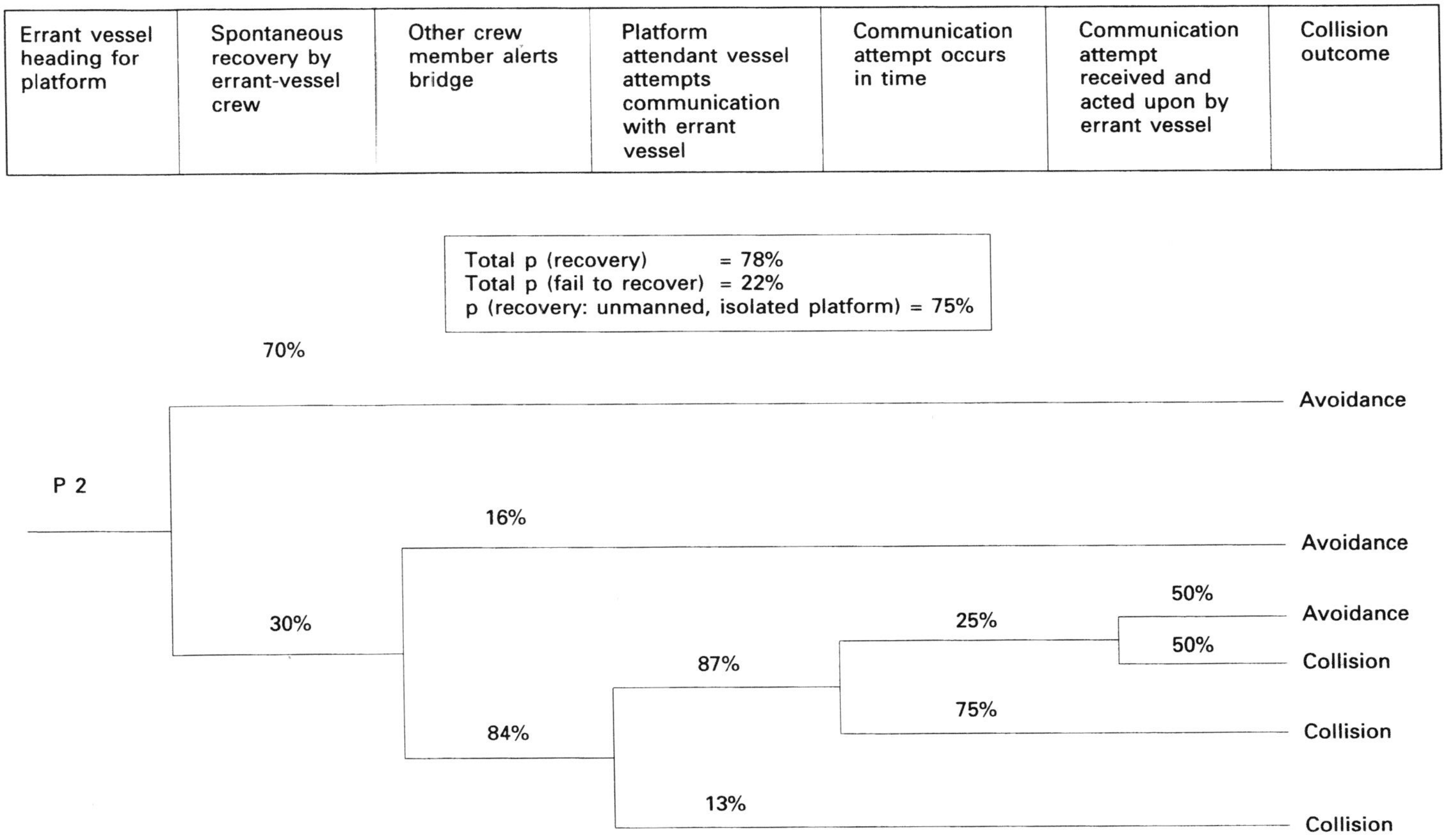

Errant vessel heading for platform	Spontaneous recovery by errant-vessel crew	Other crew member alerts bridge	Platform attendant vessel attempts communication with errant vessel	Communication attempt occurs in time	Communication attempt received and acted upon by errant vessel	Collision outcome

Total p (recovery)   = 78%
Total p (fail to recover) = 22%
p (recovery: unmanned, isolated platform) = 75%

P 2

70% ── Avoidance

30%

16% ────────────────────────────── Avoidance

84%

87%

25%   50% ──── Avoidance
       50% ──── Collision

75% ──── Collision

13% ──── Collision

variability on predefined shipping routes to determine the likelihood of a ship's being on a collision course with a platform. This approach generated a risk-based model which can estimate the level of errant-vessel-collision risk for various different platforms on the UK Continental Shelf – depending on their degree of proximity to UK shipping lanes (Pyman *et al*, 1983).

7. Error reduction analysis

The fault-tree- and event-tree-modelling approach, breaking down the SPC scenario into meaningful failure modes, led easily to the specification of SPC error-reduction methods. These ranged from automatic anti-collision warnings (via radar) on the ships to vessel-traffic monitoring and control systems in those areas in which there is a significant build up of traffic near to a platform. Manning and training levels on ships was also raised as an obvious ERM, but it was recognised that such measures as these may not be implemented in the increasingly competitive market, and that in fact such levels may well *decrease* with the passing of time, without necessarily being compensated by advances in computerised navigational aids.

8. Impact

The analysis took an ill-defined problem – but one that was causing concern – and rationally explained how it was that a large ship with nav-aids could collide with a platform. It is likely that this modelling was the most important part of the study, rather than the quantification aspect. In particular, this modelling process meant that practical and effective ERMs could now be identified.

However, while human-factors analyses have undoubtedly contributed to the identification and quantification of the degree of risk inherent in errant-vessel-collision scenarios, the human-factors aspects of the situation are believed, unfortunately, to be *endemic* to the marine world, and amenable to no easy and immediate practical solution. This does not imply that human-factors solutions could *not* help to reduce the level of risk; it only implies that their effective implementation in the marine world would require a large degree of commitment on the part of international legislative and commercial organisations.

Case Study 13: offshore drilling

1. Background

This study involved a fairly large, human-error-based fault-tree analysis of offshore drilling operations. The top event of consideration was that of a *blowout*, which involves an uncontrolled release of hydrocarbons via the drilled

Figure VI.13.1 Schematic diagram of drilling system

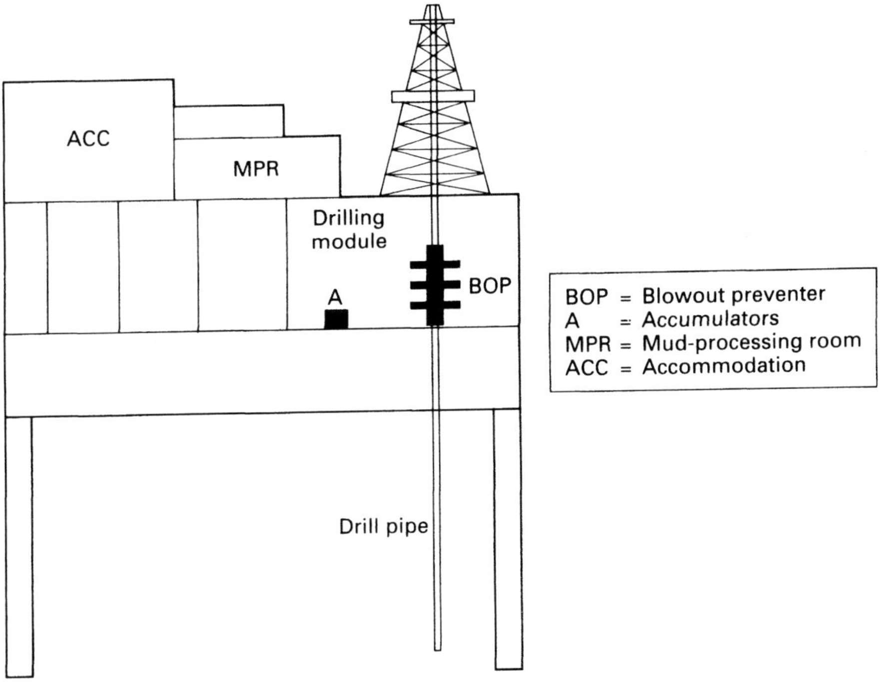

well. Blowouts occur rarely, but when they do occur, they can have dramatic and catastrophic consequences, leading, for example, to the loss of the platform. Essentially, a blowout occurs as a result of a failure to maintain the balance of pressure associated with the underground hydrocarbons fields. An imbalance of pressure results in hydrocarbons rushing to the surface, either underneath the platform, or right onto the platform itself. Once this happens, it becomes rapidly uncontrollable, and there is a concomitant high risk of fire and explosion.

2. Problem definition

A blowout can occur at several stages in drilling. Prior to drilling taking place, survey work will be carried out, to determine the approximate depth and size of the hydrocarbon field. If this work is inaccurate, it may lead to the field's being hit at too early a point in time. When this happens, there may be a sudden and unanticipated pressure surge, or 'kick'. If this kick is not quickly compensated for by pumping more mud down the drilled hole, it may then develop into a blowout (see Figure VI.13.1). Alternatively, during the early drilling stages, the drilling may encounter a gas 'pocket' which is under pressure, and fairly close to the surface compared with the field at which the drilling is being aimed. This 'shallow gas' kick can very rapidly become a shallow-gas *blowout*, and as such it is a serious cause for concern during the early stages of drilling. Even if there were originally no gas present at the commencement of drilling, under certain circumstances, because of a phenomenon known as *gas migration*, gas pockets will develop around the drilling sites.

Assuming that the field is detected as planned, a blowout may still, nonetheless, occur, as a result of pressure variations, or of drilling operations themselves. The 'tripping' operation, in particular, can lead to, or exacerbate, a kick. Tripping occurs when, for example, it is necessary to change the drilling bit. In such a situation, the drill bit must be raised from the bottom of the hole up to the level of the platform itself – perhaps a distance of some 1000 feet or more. There is a natural danger that pulling the drill bit and drill pipe out of the hole in this way will cause a siphoning effect. It is this potential siphoning that can lead to a kick developing which, if unchecked, can then develop into a blowout.

On the platform itself, there will be a protection device known as a *blow-out preventer* (BOP). This can take various shapes, but it usually consists of three sets of vertically stacked, high-pressure hydraulic rams. The lower set has a set of pipe rams which close around the drill pipe itself, thus preventing hydrocarbons from rushing up through the drilled hole around the drill pipe. The top set of rams are also of the pipe-ram variety. The middle set of rams are different, however. These are *shear* rams, and they are strong enough to cut or squash the drill pipe itself, thus firmly blocking the flow of hydrocarbons. In practice, if the blowout occurs and initiation of the rams is delayed for any reason, these will no longer be effective since their edges will become worn very rapidly. They should therefore be put into action only when the drill crew realise that they are losing control of the kick, and that a blowout is therefore imminent. The choice of whether to utilise the pipe rams only, or the shear rams as well, will be up to the driller or the tool-pusher on the drill floor, whoever is in charge at the time.

The objective was therefore to identify all errors which could contribute to the likelihood of a blowout, whether latent failures or active (emergency-response) failures. The resulting fault tree was to be quantified, and error-reduction measures were to be carried out in order to reduce the level of risk where appropriate.

3. Task analysis

There was no formal task analysis (such as, for example, an HTA). Some limited documentation review did take place, and this was followed by substantive interviews with a highly practised driller with over 30 years' offshore drilling experience. A fault tree was constructed directly, on the basis both of these interviews and of the judgement of the author, which made use only of external error modes (EEMs).

4. Human error analysis and representation

A drilling-blowout fault tree was developed in order to evaluate human factors in drilling, as well as to allow recommendations to be made for the reduction of the level of risk involved. The tree itself was fairly large, and it featured five major subsidiary top events:

- A drill-pipe blowout
- A blowout through the BOP
- A blowout through the choke system
- A blowout at the mud-processing level
- A shallow gas blowout

There are other types of blowout which are not included in the human-factors analysis, such as a well-head rupture or a sub-surface blowout. Completion blowouts and production/work-over blowouts (which occur at other stages of the drilling process) have not been considered. It was not possible to look at the human factors involved in all the blowout scenarios occurring within this study. In the light of this fact, and given the objectives of the study, it was decided that drilling blowouts occurring *above* the surface would be the most useful kind of blowout to analyse.

The fault tree is shown in part in Figure VI.13.2 (25 pages).

The structure of the tree can best be understood by considering the chain of events that lead to a blowout. These are:

Event	Important considerations
Kick occurs	Induced or spontaneous; drilling or tripping.
Kick is detected	Speed of onset of kick; detection-system failure.
Kick is controlled	Drilling or tripping; degree of kick.
Blowout prevented	Ram/diverter operation.

These are the main stages in a blowout occurrence for all the five blowout scenarios labelled above, although in the case of a shallow gas blowout, there is no control method practicable since the event will occur so rapidly that only prevention methods can be carried out. As the overall fault tree is very large, the structure is briefly described below, relating the different stages involved in the occurrence of a blowout to each of the different blowout scenarios.

Kick occurs

A kick can occur during drilling, as a result of hitting a high-pressure formation without increasing the mud weight, or as an after-effect of the loss of circulation during drilling. During tripping operations, the kick can occur as

Sheet 1

Figure VI.13.2 FTA Drilling

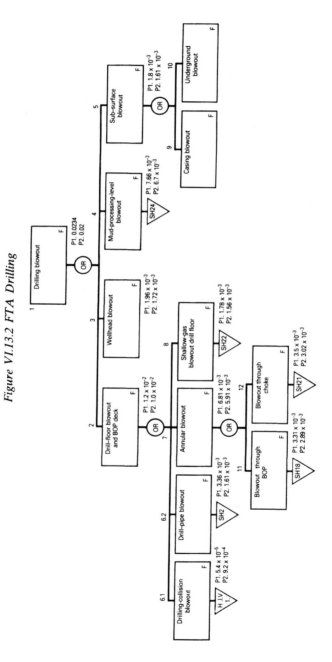

Sheet 2

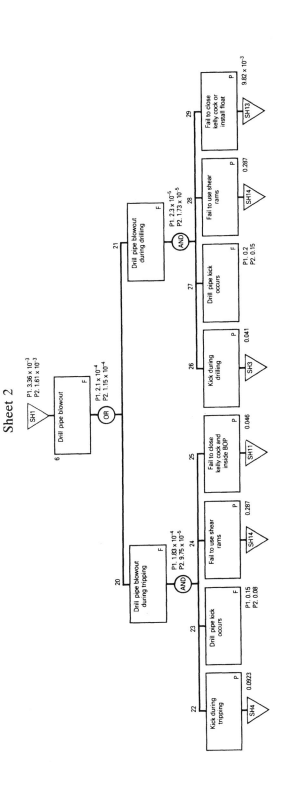

Sheet 3

Figure VI.13.2 (Continued)

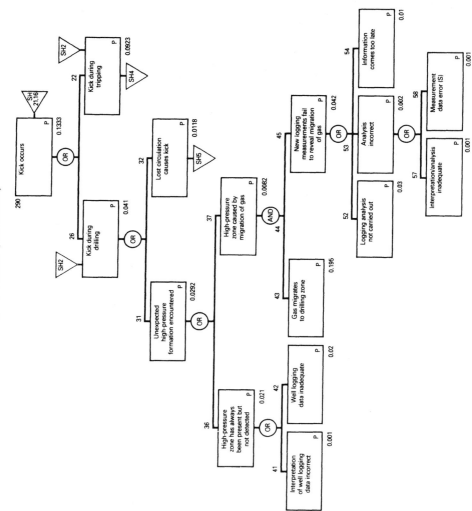

Sheet 4

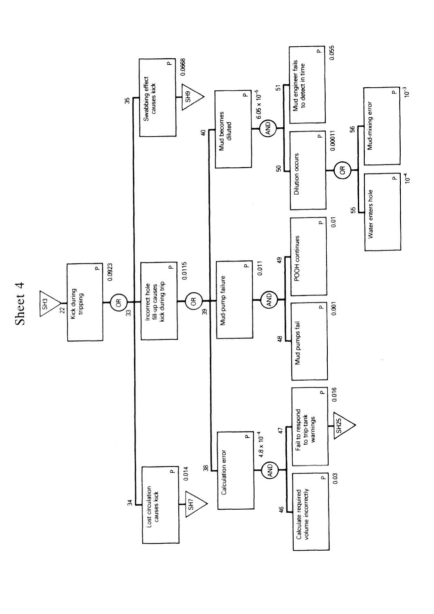

Sheet 5

Figure VI.13.2 (Continued)

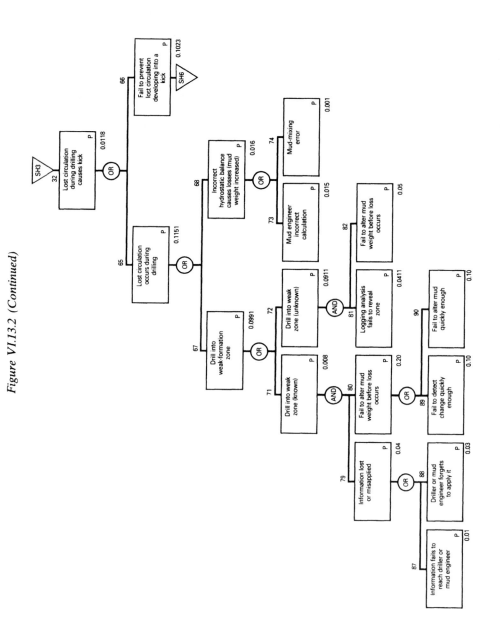

Sheet 6

Fail to prevent loss circulation developing into a kick — SH5 — 66 — P — 0.1023

AND — 70

Fail to kill well — 70 — P — 0.341

Fail to plug lost-circulation zone — 69 — P — 0.3

OR — 78

Fail to keep hole full — 78 — P — 0.301

Incorrect mud weight used — 77 — P — 0.04

OR — 76

LCM does not plug hole — 76 — P — 0.2

Fail to use lost-circulation material (LCM) — 75 — P — 0.10

OR — 85

Complete losses — 85 — P — 0.10

Pump failure — 86 — P — 0.001

Fail to supply mud quickly enough — 148 — P — 0.2

OR — 84

Use incorrect mud weight — 84 — P — 0.01

Unable to determine correct kill weight — 83 — P — 0.03

Sheet 7

Figure VI.13.2 (Continued)

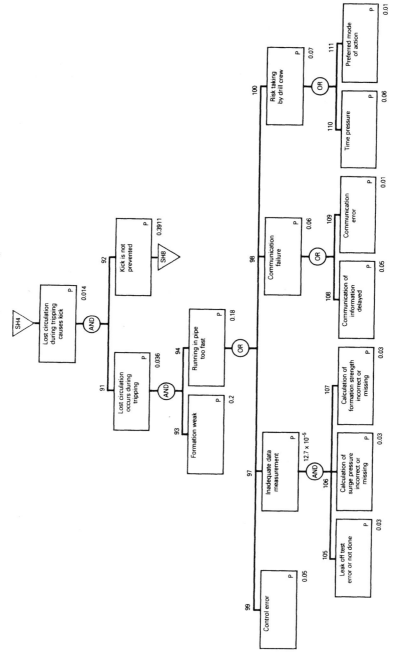

Sheet 8

Sheet 9

Figure VI.13.2 (Continued)

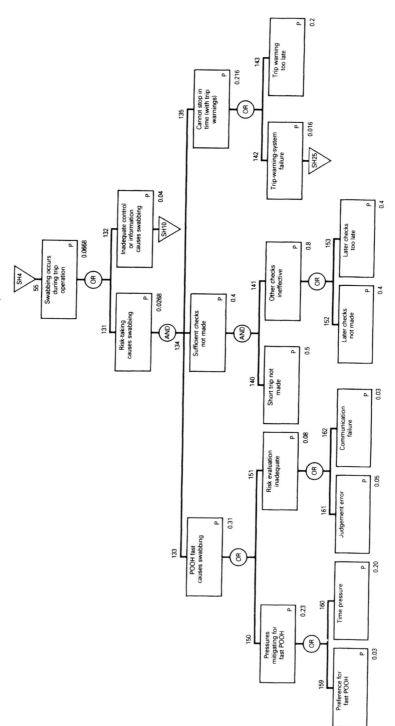

Sheet 10

Sheet 11

Figure VI.13.2 (Continued)

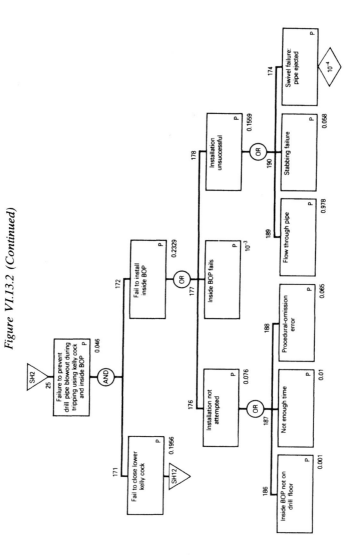

Sheet 12

Fault tree (Sheet 12). Nodes and labels:

- SH11 — 171 — Fail to close lower kelly cock — P — 0.1956 (OR)
  - 173 / SH13 — Fail to raise kelly — P — 0.0355 (OR)
    - 179 — Ram's closed too early, kelly cannot be raised — P — 0.0153 (AND 192)
      - 192 — Rams closed before kelly raised — P — 0.05
      - 193 — Rams cannot be opened again — P — 0.306 (OR 198)
        - 197 — Blind rams closed in error — P — 0.001
        - 198 — Pipe rams closed, and not enough power to open them again — P — 0.2
        - 199 — Pipe rams closed, and power low so driller does not want to open them again — P — 0.1
        - 200 — Rams fail closed — P — 0.005
    - 180 — Power failure — P — 10⁻⁴
    - 181 — Draw-works failure — P — 10⁻⁴
    - 182 — Procedural-omission error — P — 0.01
  - 174 — Swivel failure: pipe ejected — P — 10⁻⁴ (diamond)
    - 183 — Pipe cannot be stripped through preventer (closed on tool joint) — P — 0.01
  - 175 / SH13 — Fail to close valve — P — 0.026 (OR)
    - 184 — Drill fails to close valve — P — 0.021 (OR 196)
      - 194 — Not enough time — P — 0.01
      - 195 — Turn the wrong way — P — 0.001
      - 196 — Procedural-omission error — P — 0.01
    - 185 — Valve stuck — P — 0.005
  - 123 — Fail to install lower kelly cock — P — 0.134 (OR)
    - 124 — Installation not attempted — P — 0.016 (OR 128)
      - 126 — Not on drill floor — P — 0.001
      - 127 — Not enough time — P — 0.01
      - 128 — Procedural-omission — P — 0.005
    - 125 — Installation fails — P — 0.118 (OR 260)
      - 129 — Valve left closed and flow in pipe — P — 0.06 (AND 262)
        - 261 — Valve left closed — P — 0.12
        - 262 — Flow in pipe — P — 0.5
      - 260 — Installation failure — P — 0.058

Sheet 13

Figure VI.13.2 (Continued)

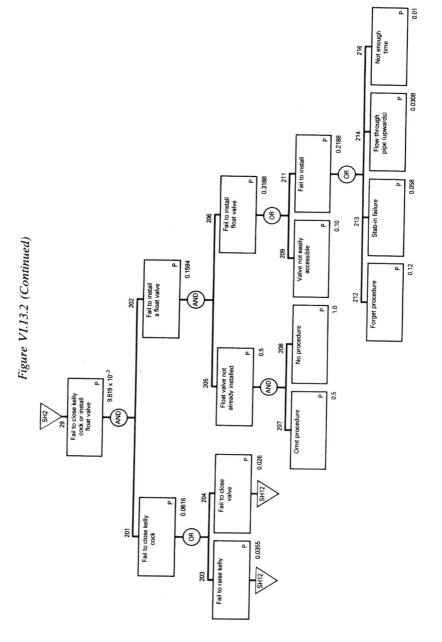

Sheet 14

Sheet 15

Figure VI.13.2 (Continued)

P1. 3.31×10^{-3}
P2. 2.89×10^{-3}

SH1
7

Annular blowout through BOP (BOP deck/drill-floor)
F
291
AND
2.07×10^{-4}

290
Annular kick
F
SH3
0.133

292
Hydrostatic control fails
P
SH17
0.1958

Failure of BOP to prevent blowout
P
294
OR
7.92×10^{-3}

293
BOP system not present
P
10^{-5}

295
BOP system leaks/ruptures
P
10^{-4}

Failure of BOP operations
P
297
AND
7.81×10^{-3}

296
Failure of upper pipe rams
P
SH2
0.08

Failure of shear rams
P
SH2
0.287

298
Failure of lower pipe rams
P
SH2
0.34

Sheet 16

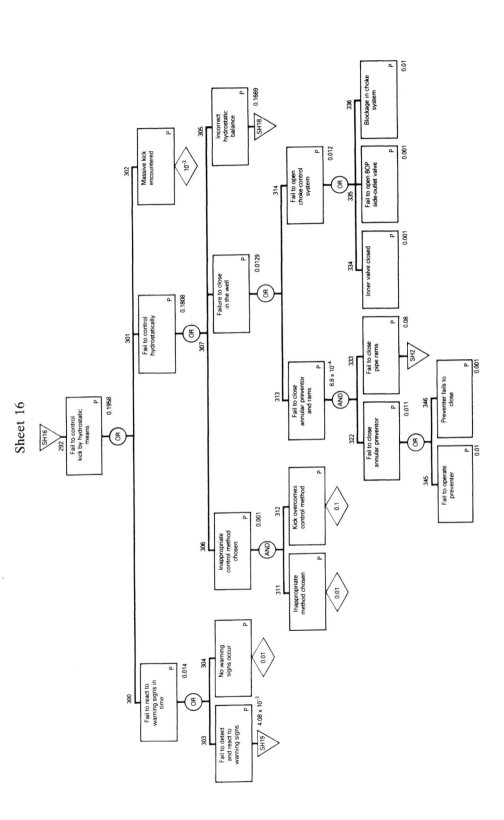

Sheet 17

Figure VI.13.2 (Continued)

Sheet 18

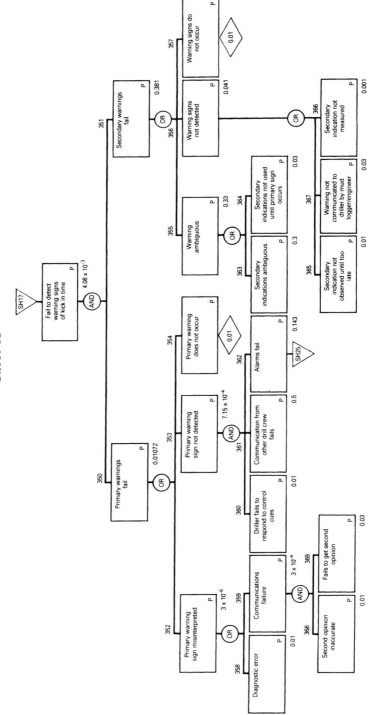

Sheet 19

Figure VI.13.2 (Continued)

Sheet 20

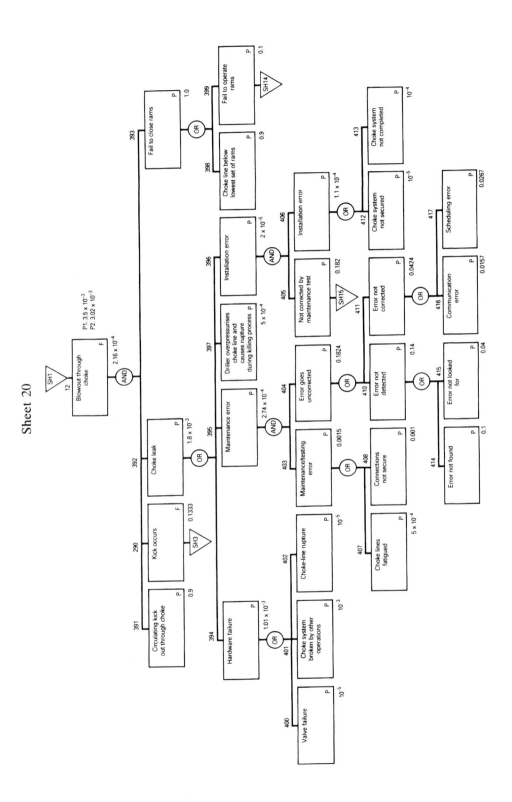

Sheet 21

Figure VI.13.2 (Continued)

Sheet 22

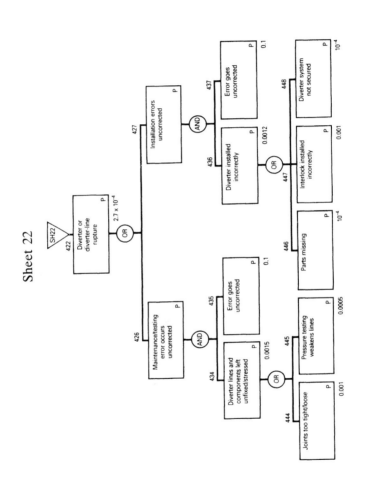

422 Diverter or diverter-line rupture — P — 2.7×10^{-4}

OR

426 Maintenance/testing error occurs uncorrected — P

427 Installation errors uncorrected — P

AND

435 Error goes uncorrected — P — 0.1

434 Diverter lines and components left unfixed/stressed — P — 0.0015

OR

445 Pressure testing weakens lines — P — 0.0005

444 Joints too tight/loose — P — 0.001

AND

437 Error goes uncorrected — P — 0.1

436 Diverter installed incorrectly — P — 0.0012

OR

446 Parts missing — P — 10^{-4}

447 Interlock installed incorrectly — P — 0.001

448 Diverter system not secured — P — 10^{-4}

SH22

Sheet 23

Figure VI.13.2 (Continued)

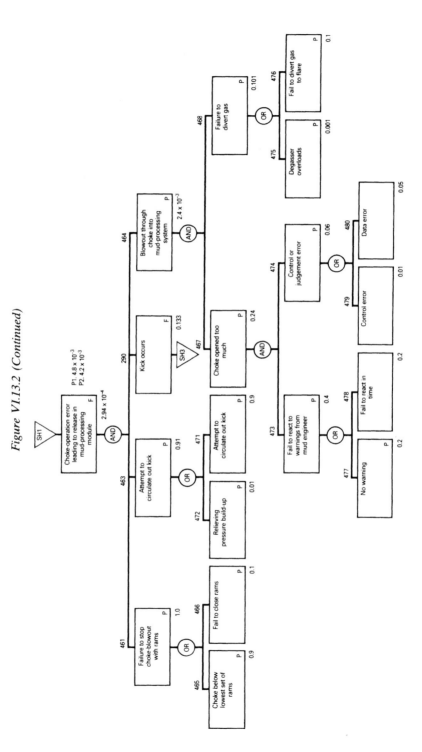

Sheet 24

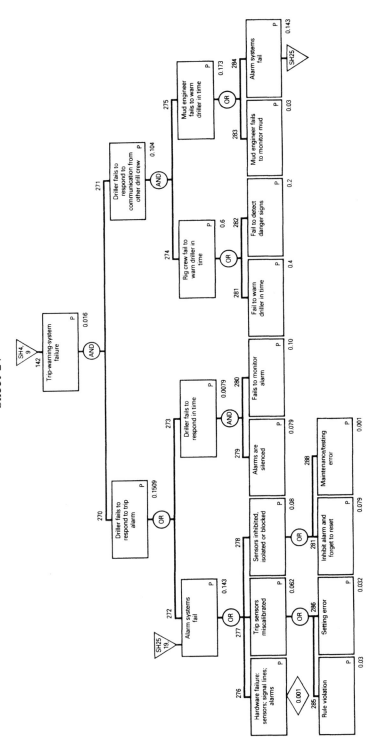

a result of swabbing (a siphoning effect), or of failing to fill up the hole as the pipe is pulled out. Alternatively, when the pipe is being run back into the hole, lost circulation can again occur, leading later to a kick. These are the main failure modes leading to the occurrence of a kick in all the scenarios except for the shallow-gas-blowout scenario.

Shallow gas kicks, which occur more spontaneously, are best prevented by adequate surveying of the area to be drilled before drilling starts. The occurrence of a shallow gas kick is therefore due more to geological than to human factors.

Kick is detected

In tripping operations, a trip-tank alarm system will be used to detect any signs of an improper fill-up in the hole, or of swabbing. In drilling, the primary signals of kick detection – such as a drilling break (i.e. suddenly, the drilling penetration rate speeds up, perhaps as a result of the drill's breaking through rock, etc., to a chamber, or to a softer medium), gas-cut mud, etc., – often occur too late to prevent the kick, whereas secondary signs (e.g. what is known as the 'd' exponent) *do* provide more of a timely warning. These latter, however, are less reliable predictors. Hardware failures and human errors constitute the main failure modes shown in the subtrees for kick detection and trip-warning-system failure. These trees, particularly the general kick-detection failure tree, are simplistic representations of a highly complex process of monitoring, interpreting and aggregating 'fuzzy' information.

In the case of shallow gas blowout, the kick-detection process must be rapid since the kick occurs quickly, and frequently, little warning is given (e.g. only a minute or so).

Kick is controlled

No method of control is given for a shallow gas kick. For the other types of kick, the main method of hydrostatic control considered is circulation, using the balanced-mud method. There are three other major methods used (the drillers method, the volumetric method and bullheading), but the development of the tree so as to incorporate the use of each of these, and the various ramifications arising, would have been very time-consuming, and was therefore not considered to be cost-effective at this stage of the study. The balanced-mud method is the most likely method to be used for hydrostatic control. The most likely failure to control the kick hydrostatically will involve an incorrect hydrostatic balance, caused by the mud weight, the pumping rate, or the choke control.

Blowout is prevented

In a shallow gas kick situation, only the diverter is available to prevent a blowout. This system, simple in design and in operation, has four failure modes. The main problem will be responding in time to the kick. Approximately, the diverter diverts the gas back over the side of the platform.

For the other blowout scenarios, pipe and shear rams represent an effective barrier system in scenarios where control is lost. If choke control is attempted, however, a loss of control over the choke system can lead to a blowout either through the choke system itself or through the mud-processing module. The choke system used will normally be situated below the lowest set of rams, so this failure does normally lead to a blowout.

5. Quantification

In this study, the frequency of blowouts was predicted in two ways, via the analysis of frequency data on blowouts and via a fault-tree analysis. The former method, as would be expected in a normal total-risk analysis, was the main input for the evaluation of risk, and for calculations of consequences. However, this normal total-risk-analysis (TRA) approach cannot identify *human* contributions to the level of risk because the frequency data rarely defines the cause in enough detail to give frequencies for human errors.

To overcome this problem, a fault-tree analysis was carried out so that the human errors could be identified *ad hoc*.

There are two important reasons why blowouts are not normally analysed in a TRA using a fault-tree analysis. Firstly, unless one wants to make detailed design recommendations, there is not necessarily any benefit involved in the accurate use of a fault-tree analysis. Secondly, much of the data required for a fault-tree analysis of blowouts is not available.

In this study, however, detailed design recommendations *were* required, but there were still areas where no information could be provided on various events such as the probability of drilling into a weak formation zone, or the degree of expectation of a shallow gas migration. Human and hardware factors can be estimated, but it is often the environmental considerations that tend most of all to undermine the accuracy of the fault-tree-derived frequencies.

The fault-tree analysis thus has both a great advantage and a disadvantage. The advantage is that the causes of failures, and therefore recommendations, can be made explicit and specific. The disadvantage is the limits to the accuracy and reliability of the ultimate answers – due to large uncertainties in the trees. The design recommendations can still be prioritised, but the fault-tree-derived frequencies for blowouts may not be as reliable as the historically derived frequencies. Therefore, the fault trees are a design tool only; it is the historical data which serves as the main input for the calculation of consequences.

The fault-tree analysis of a drilling blowout was therefore conducted using two methods, a 'bottom-up' approach – using failure data to quantify the relative degree of risk imposed by human and hardware failures – and a top-down approach – using historical information on blowout frequencies to quantify environmental factors. The net result is a fault tree which has a top-event frequency the same as the historical frequency (i.e. a tree which has been 'calibrated'). The contributions of human and hardware failures to this top event frequency, on the other hand, are specific, explicit and quantified. This approach was at the time considered optimum for this study.

The main exception to this approach was to be found in the specific case of drillings made into a live-well (a so-called *drilling collision*). No data existed for this type of failure. This event was therefore outside the scope of this analysis, and was carried out instead by another assessor (see Bellamy *et al*, 1986).

The actual quantification of the HEPs used a combination of human-error data and analyst judgement, and was, as already mentioned, to an extent calibrated by world-wide experience relevant to North Sea drilling operations.

The major failure modes, in terms of their degree of impact on risk, were identified to be the following:

Major failure modes causing a kick

A1. Incorrect hole fill-up during a trip, causing a kick. Mud problems, and a failure to stop pulling out of the hole (POOH) quickly enough.
A2. Pulling out of hole too fast, due to risk-taking without sufficient checks during, and at the start of, the trip.
A3. Pulling out of hole too fast without adequate calculation of effects of hole conditions and past wells' pulling problems, etc.
A4. Running-in of the pipe too fast during tripping, due to risk-taking and inadequate consideration of hole conditions.
A5. Drilling into a weak-formation zone without changing mud weight, resulting in lost circulation.

B Major failure modes in kick control

B1. Wrong mud weight calculated for hydrostatic control.
B2. Pumping mud too slowly.
B3. Incorrect choke control.
B4. Blockage in choke system.
B5. Failure to plug a lost-circulation zone (this is a zone in which mud being circulated is lost into the formation, rather than circulating back up to the platform).
B6. Failure to respond to the secondary warnings about a kick before the primary kick warnings then occur.
B7. Failure to calculate correctly prevailing mud weight and mud properties when trying to control a kick that occurs when the bit is off the bottom.
B8. Failure to keep hole full during a lost-circulation problem, and to calculate correct mud weight.

C Major failure modes for preventing a blowout

Drill pipe blowout (blowout through the centre of the drill pipe itself)
C1. Failure to install lower kelly cock during tripping, due to failing to install it correctly.
C2. Failure to install an inside BOP during tripping, due to flow through the pipe (making installation difficult), or an omission error.
C3. Failure to raise the kelly, as a result of closing the rams too early (this means that the kelly cock (a valve inside the drill pipe) cannot be closed, and hence, unless the shear rams have been effective, the blowout will continue its course through the trapped drill pipe).
Shear ram failure
C5. Failure to operate shear rams manually when required.
C6. Failure to switch up shear-ram pressure rating during operation.
C7. Failure to operate shear rams in time.

Blowout through choke system
C8. Failure to close chokes quickly when choke system leaks.
C9. Setting up choke system below lowest set of rams.
C10. Choke system fractured by traffic movements in the area (e.g. moving-around of heavy equipment on the drill floor).
Shallow gas blowout
C11. Failure to operate diverter in time.

Blowout at mud processing level
C12. Choke connected below lowest set of rams.
C13. Failure to divert gas to flare.

6. Impact

The fault tree, in total, ran to 25 sheets, with over 100 base events of a human-error nature, all quantified and documented via descriptions of the assessor's rationale for their quantification. A number of ERMs were suggested, based on the relative human-error contributions to the overall risk of a blowout. Some ancilliary interest was also taken in the study as a result of an incident that actually occurred at the time it was being carried out: a mobile platform sank in the North Sea sector, as a result of a sub-sea blowout.

The areas which, to the client, appeared to have the most impact were the potential maintenance failures – leading to the failure of the shear rams to operate at the required time – and the potential for the integration of certain parameters – covering the primary and secondary indicators of an impending kick/blowout – into an integrated display. The author does not know if this idea was ever developed further.

Case Study 14: offshore platform lifeboat evacuation

1. Background

This study concerned the evaluation of errors and error-recovery opportunities involved in the preparation and launching of an offshore-platform-evacuation lifeboat. The lifeboat system in question was a freefall-lifeboat system. There are two main types of freefall lifeboat. The first is a type that is lowered to a point above the sea, is then released and drops until sea contact is made, after which the boat drives away. The second type, that which was under evaluation here, is a lifeboat which runs off a sloping 'skid' or ramp, drops into the sea at an angle, briefly submerges, comes to the surface away from the platform and then drives under its own motorised control. Both systems were in part designed to overcome problems associated with launching lifeboats in cases where, for whatever reason, the platform itself has begun to list (i.e. tilt) – as happened in the Alexander Kjelland disaster (see Bignell and Fortune, 1984). In such conditions, launching a conventional lifeboat by lowering it into the sea and then releasing it may simply not work, because of the angle of the platform. The skid-launched freefall-lifeboat system has the added potential advantage that it launches the boat away from the platform. The crew are thus quickly removed from any impending danger on the platform, even if the lifeboat motor fails.

2. Problem definition

The analysis formed part of a risk assessment for the entire lifeboat-evacuation system. The objectives of the HRA were to assess what could go wrong and to quantify such error/failure modes. It also aimed to consider error-recovery measures, where these could benefit the risk reduction process.

The platform was in its preliminary design stage, and as a result no operators were available. Lifeboat trainers and personnel who had carried out evacuations *were*, however, available for discussions.

3. Task analysis

The task-analysis phase involved the use of several techniques:

- A documentation review: a review of descriptions/diagrams, etc., associated with the system.
- A review of documented offshore-platform-evacuation incidents.
- The observation of prototype freefall-lifeboat launch.
- Interviews with lifeboat-evacuation trainers.
- Interviews with offshore personnel who had carried out evacuations in emergency situations.

Practice with the lifeboat was likely to be held every week, and would involve the following basic steps:

- A hydraulic test cylinder is attached to the rear of the boat; this records the test and prevents the boat from launching.
- The crew assembles by the boat ('musters'); the names of the crew members are checked.
- A retaining pin is taken out of the side of the skid.
- The crew (except for a few who remain outside the boat to monitor the practice) then board the boat.
- The coxswain operates a lever inside the boat at the same time as another man or woman operates a hydraulic pump. This opens a pair of locks called the *skidlocks*.

The boat then lowers slightly, but not fully – i.e. enough for the correct working of the skid-lowering mechanism to be tested.

After the practice, the boat is 'stored', or put back into the 'ready' position (i.e. ready for an emergency launch). Then, the stowing procedure (which is carried out by those outside the boat): the test cylinder is used to retrieve the boat, and then all personnel disembark as the boat is secured and the test cylinder is removed.

A hose is attached to the back of the boat. This carries the hydraulic-flow line from the inside of the boat over to the skidlock-opening mechanism outside the boat. This hose must be connected after each practice, or launch, as it *disconnects* once the skid has been lowered below the horizontal.

Various maintenance and testing jobs must also be carried out on the lifeboat-launching system – involving 'bleeding'/filling the accumulators with hydraulic fluid, etc.

4. Human error analysis

4.1 Human error analysis

A human-error HAZOP was carried out for this analysis. The HAZOP team comprised a human-factors/reliability practitioner, a fault-tree analyst involved in the PSA, a person involved in training people in lifeboat launching and a HAZOP leader. The HAZOP occurred in two sessions during one day. It focused on the P&IDs (process/piping and instrument diagrams) involved in the system, alongside a consideration of the various different tasks involved in lifeboat-launching. A number of potential errors were identified, using traditional keywords and EEMs as found in the THERP system. In particular, it was realised that not only the launching phase but also both the maintenance phases and the practice sessions themselves – for which a small number of high-risk scenarios were generated – had to be considered.

4.2 Definitions of system failures

Several degrees of success and failure were identified for the lifeboat system, as defined below:

S1: successful launch first time in an emergency situation; all crew safely away.

S2: successful launch after failed first launch (i.e. recovery from first launch failure), with all crew safely away.

F1: launch achieved on second attempt, as above, but one or more crew left behind (the crew left behind would have been instrumental in recovering the initial failed launch attempt, but in so doing, they would not have been able to board the boat).

F2: failure to launch lifeboat at second attempt. All crew left on platform.

F3: disastrous launch. Possibility that crew and boat lost (i.e. boat launches, but problems occur with skid, leading to *unsafe* trajectory of boat).

S3: safe practice launch; boat retained by test cylinder.

I1: inadvertent but safe launch during practice or maintenance.

I2: inadvertent and dangerous launch. Possibility of loss of both crew and boat.

Some of the errors identified and quantified during the evaluation of the system were as follows:

1. Failure to restore system after maintenance.
2. Failure to restore system after maintenance and strong feedback.
3. Failure to carry out an isolated act in a scenario where discovery cues are minimal.
4. Failure to carry out an isolated act in a scenario where discovery cues are maximal.
5. Failure to carry out a procedure which involves disabling a potential safety system (part of the stowing procedure).
6. Failure to perform the stowing procedure correctly after wirefall recovery.
7. Forget to perform isolated act, at time of launch, which involves disabling safety system.
8. Forget to perform isolated act involving use of a checklist.
9. Losing the retaining pin and replacing it with a nut and bolt (involves failure to consider side-effects of an action).
10. Failure to connect test cylinder before practice.

Recovery actions

1. Failure to achieve easy recovery.
2. Failure to achieve difficult recovery.
3. Failure manually to lower skid when required.

5. Representation

The latent errors were represented in a fault tree, while the actual errors occurring during the launch process, together with the recovery opportunities involved, were represented by an event tree. A diagram of the fault tree, showing not its detail but the prevalence of human errors throughout the tree (and particularly near the top event of the tree (where it matters)), is shown in Figure VI.14.1. One of the event trees used in the analysis is shown in Figure VI.14.2.

6. Quantification

The HEPs were quantified using structured expert judgements (the actual values are not included in this case study). The approach used was the Influence Diagram approach (D. Embrey ran the IDA session).

7. Impact

The assessment made a significant impact on the design proposed for this system, and a number of error recommendations were put forward. In particular, the evaluation highlighted the strengths of the human-HAZOP approach both in dealing with relatively small amounts of detail – as, for example, is the case at the early, design stages of a system – and in making use of personnel who have direct experience in the types of event being assessed. The impact of maintenance errors on the level of risk inherent in the system was significant, and it is questionable whether such an impact would have been identified at the time without close collaboration between the various HAZOP parties involved. The other risk analysts involved in the study (which occurred in the early 1980s) were *also* somewhat surprised by the extent to which human error dominated the determination of the risk factor for this system.

Case Study 15: offshore platform depressurisation I

1. Background

This case study occurred in the late, detailed, design stage of a large offshore-platform-development project. The full case study is detailed in Kirwan (1987). This was a relatively early HRA, and the methodology was crude in comparison with some of the others in this appendix. The study focused on decision-making performances carried out within a group of offshore emergency scenarios. In such scenarios, the operators in the Central Control Room (CCR) would find the platform under threat (e.g. from a fire). Their chosen course of action should be to get rid of any hydrocarbons being processed or stored on the platform (in case these should catch fire or explode) by venting them

A Guide to Practical Human Reliability Assessment

Figure VI.14.1 Fault-tree-analysis diagram

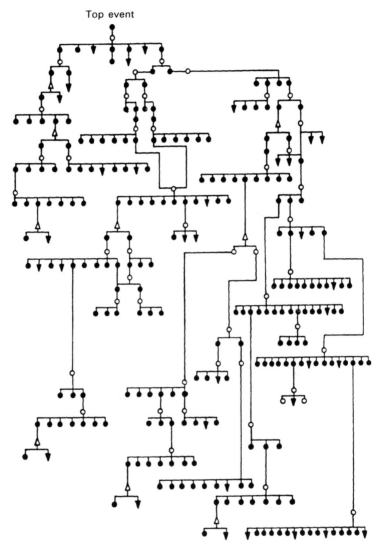

Key:
● Failure events
○ 'OR' gates
△ 'AND' gates
▼ Human errors

Figure VI.14.2 Event-tree analysis for freefall system

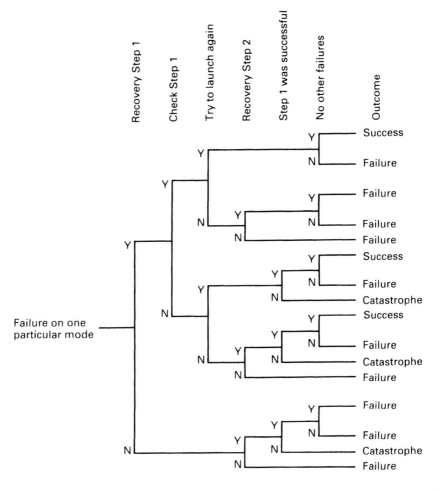

off the platform, either to the sea (*not* the most desirable option, because of the risk of a sea fire) or through the flare stack. This process is known as depressurisation or 'blowdown'. This case study concerned the estimation of the degree of human reliability involved in blowdown performances (blowdowns would be initiated and controlled *manually*).

2. Problem definition

2.1 The problem

The particular platform in question was a large platform, and had two sectors of blowdown, A and B. The flare stack was, however, *not* designed to accommodate the simultaneous venting of both sectors (this would have required a very large and robust flare stack). In an emergency, therefore, it would be necessary to be able to pinpoint the nature and location of the emergency, and then depressurise the platform accordingly. For example, if there were a fire in Sector A, this sector would be depressurised *first*, followed by the other sector four minutes later (it takes four minutes to blow down each sector). If there were a fire which affected both areas simultaneously, then the operators would have to make a decision as to which to blow down first. In all scenarios, however, the main error of interest was a simultaneous blowdown, which could lead to a rupture in the flare mechanism.

To complicate matters slightly, if the *compressors* were under the threat of a fire, the local operator at that point could initiate their depressurisation locally – basically, in an effort to preserve his own life. A problem would therefore arise if he or she initiated a blowdown and, in the same four-minute time-frame, the CCR operators also initiated the *other* sector blowdown.

The analysis was to consider ERMs only if these were seen as necessary. It was also initially intended that the study would only focus on blowdown performances, i.e. responses to the initiating event, rather than on the contribution of human error to failures which could then lead to the initiating events themselves. As will be discussed, however, a maintenance error *was* discovered which could have led to a particular initiating event.

2.2 The system

The system comprised the following main components relevant to the assessment (see Figure VI.15.1):

- The compressors and separators containing flammable fluids.
- The computers controlling the processing of these fluids.
- The emergency-shutdown system which depressurises these fluids and then vents them to the flare stack.
- The control-room operators monitoring both process conditions and the general status of the platform.
- The local control-room operator monitoring the compressors.

The compressors and separators, which are essential to the oil-production process, contain flammable hydrocarbons. In the event of an emergency, the hydrocarbons in both the compressors (located in Sector 1) and the separators (located in Sector 2) must be depressurised and vented (blown down) so that the risk of an explosion or fire is reduced.

The blowdown of both systems may be manually initiated from the Central Control Room (CCR). A Local Control Room (LCR) operator can blow down the compressors, since these are under his or her manual control during normal operations. If a blowdown is initiated from the CCR, then the relevant signals feed through a computerised control system (CCS). If, for any reason, the power to the CCS fails, or is cut off, the system could lose its ability safely to shut down pressurised systems on the platform. To overcome

Figure VI.15.1 (Kirwan, 1987)

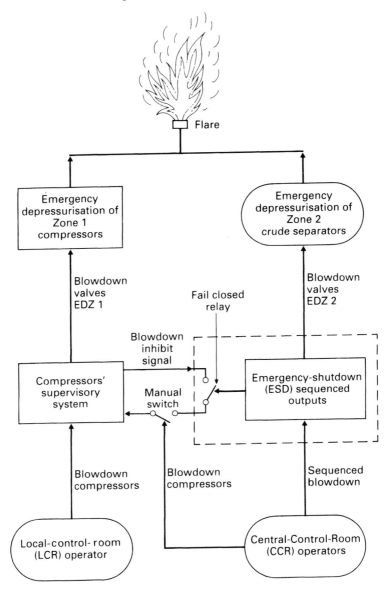

this weakness, the CCS, which controls the emergency shutdown (ESD) process, has largely been designed to be 'fail-safe'. This means that, generally, in the event of a CCS (or ESD) failure, all systems will automatically depressurise, switch off, etc. In this study, the separators (located in Emergency Depressurisation Zone 2) will depressurise *instantly* in the event of a CCS/ESD failure.

However, with this platform, not all the systems could be designed to be fail-safe. In particular, the flare stack was not (for other reasons, unrelated to a blowdown scenario) made large enough to handle the simultaneous blowdown of both the compressors and the separators. Thus, if both these systems were designed fail-safe, then, as soon as the CCS lost power, the flare stack and associated piping would be overloaded, and this would possibly lead to explosions, pipe ruptures, fires and radiation hazards. Therefore, the compressors are not fail-safe in their design. In the event of a CCS/ESD failure, therefore, a relay in the CCS fails closed, leaving a blowdown inhibit signal, which prevents compressor blowdown. After four minutes, it would be safe to depressurise the compressors manually (the separators would have largely finished depressurising) via a hardwired switch that does not depend on the CCS. Alternatively, the *LCR operator* could manually initiate compressor blowdown, if, for some reason, the CCR operators could not do so themselves.

Therefore, this system could only be called 'fail-safe' if the level of reliability shown by the operator were sufficiently high enough that, in the event of an emergency, the flare stack would not be overloaded; and, on the other hand, if the compressors would be depressurised soon after the separators had vented. The most important concern became not the risk of the compressors remaining pressurised but the risk of a flare overload. A flare overload was seen as a credible event because of the general assumption that human operators are prone to make errors during emergencies. If the risk of a flare overload was found to be sufficiently high, it would be necessary either to reduce the level of risk, design an alternative blowdown-control system or redesign the flare stack. The latter two options, at this late stage in the design and construction of the platform, would prove very costly. It was, therefore, essential to quantify the risk of a flare overload – and, hence, the probability of human error in several emergency scenarios.

The section of the overall project which is described in this paper involves the calculation of the frequency, per year, of flare overloads. Other aspects, namely the *consequences* of such overloads, were calculated separately, and are not detailed in this report (see Reference 2 at the end). Because controlled blowdowns require human-operator actions, consideration of human error was recognised as an essential part of the process of estimating accurately the frequency of a flare overload.

The scenarios – fire, explosion, gas leak, etc. – which make it necessary to depressurise or 'blow down' the platform occur infrequently – indeed, typically less than once a year. In such situations, the two operators in the Central Control Room (CCR) would be required to blow down the platform sequentially in two stages: either compressors first – followed, at least 4 minutes later, by separators – or vice versa. The decision as to which 'zone' to depressurise first (i.e. compressors or separators) would depend, in some circumstances, on where the incident was occurring; for instance, if there were a fire in the compressor area, it would be sensible to blow down this area first. In certain cases (for example, if there were a general fire or a large gas leak), it would not matter which zone is initiated first. The operators have a fairly simple choice to make, but the circumstances in which this choice is required are themselves likely to be stressful and confusing, with many alarms occurring and many automatic safeguards actuating. Personnel on the platform, and in some cases the control-room operators themselves, would be in some degree of danger.

A depressurisation panel exists in the control room, but at the time when this study took place, it had not yet been decided exactly where this panel would be located within the Central Control Room. Two buttons labelled Zone 1 and Zone 2 are used to depressurise the platform sequentially, starting with either Zone 1 (compressors) or Zone 2 (separators). Once one of these buttons is pressed, the operator does not need to do anything else with respect to depressurisation; the computer's logic system carries out the time-delayed sequence involved in the process of depressurising both zones.

There are two process-control computers, either one of which alone can keep the system going. If both computers fail, then the compressors and separators trip and the separators vent to the flare system. The compressors, however, remain pressurised, so as to avoid overloading the flare. Two additional buttons, which act independently of the computer's logic, are therefore added to the panel for this contingency, so that the operators can manually depressurise compressor trains A and B (Zone 1), since otherwise, these will remain pressurised (because the computer is down). Therefore, the operators should only activate these 'hardwired' buttons at least 4 minutes after the separators, since otherwise a flare overload will occur. Furthermore, if, for any reason, these hardwired buttons are pressed while the computers are working, and at the same time the sequential button initiating the separator depressurisation first (Zone 2) is pressed, a flare overload will, again, occur. Thus, the above gives an example of a safety solution (the 'hardwired' buttons) that have been designed to overcome a potential problem (a computer failure) which, in turn, creates a new possible problem (the initiation of a flare overload).

A further complication to the system's functioning is the existence of a Local Control Room (LCR) operator located next to the compressors. This LCR operator can initiate a compressor blowdown at any stage, independently of either CCR control or the computers.

The CCR operators will know when a blowdown is required since various emergency-shutdown (ESD) alarms will be sounding. Training and procedures will also ensure that the operator knows when a blowdown is required.

2.3 Scenarios

It was considered that the 'spontaneous' and simultaneous depressurisation of both zones by the operators during normal operations was a non-credible event (especially as the depressurisation control panel was stationed away from other instrumentation), but that the occurrence of an operator's error during an emergency *was*, on the other hand, credible. The hardware-reliability analysis identified several scenarios which would require, or which could be perceived as requiring, a blowdown:

Scenario 1: ESD system failure

This scenario involves the failure of the computer-controlled ESD (emergency shutdown) system, resulting in the simultaneous blowdown of the whole platform – except for the compressors. If an operator in either the CCR (Central Control Room) or the LCR (Local Control Room) were to vent the compressors within 4 minutes of the ESD-system failure, then a flare overload would occur. An ESD failure has some similarities to a power failure since a great deal of instrumentation will switch off and equipment all over the platform will go into 'fail-safe' mode.

Scenario 2: compressor leak or fire

A major leak from the compressors, as could be caused by a seal failure, would cause an automatic blowdown of the compressors. If the operators in the CCR subsequently depressurised Zone 2 (containing the separators) within 4 minutes, a flare overload would result.

In this scenario, an automatic blowdown of the compressors would occur 2 minutes after the seal failure has been automatically detected. If, at the same time, however, the *gas detectors* signalled the presence of a serious gas leak to the control room, the operator could certainly then initiate a blowdown within 1 minute of receiving such a signal from the gas detector (as well as information about the platform ESD). This would lead to a flare overload.

Scenario 3: platform emergency – resulting in ESD II*

The operator's response to a major emergency situation is very likely to include the depressurisation of the platform. This would normally be carried out sequentially, such that the area most at risk would be blown down first. However, an error on the part of the operator could result in the simultaneous blowdown of *both* the separators and the compressors, which in turn could result in a flare overload.

If the operator decides to blow down the platform by pressing the Zone 2 (separator) button, but then also presses the compressor blowdown button, then a flare overload would, again, occur.

*ESD II requires total platform shutdown and mustering of personnel, though not actual evacuation unless the situation worsens.

Scenario 4: common cause failure

There are events which, on their own, would conceivably cause the simultaneous blowdown of both Emergency Depressurisation Zones. A number of

---

* ESD II requires total platform shutdown and mustering of personnel, though not actual evacuation unless the situation worsens.

such possible events were considered, of which only one was found to be a possible mechanism for a common-cause failure, namely the loss of power to both the ESD system and the compressor-supervising system, as a result of an explosion in the battery room. Human errors in this Scenario 4 failure were not considered, and as a result the rest of the case study deals only with the first three scenarios.

3. Task analysis

The requirements imposed by the system on the operators were investigated for each of the three scenarios. Very little information was available concerning the workload that would be imposed on them, since the platform had not been built at the time of this study, and procedures were not yet available for review. There was no access available, at the time, either to operators from similar kinds of platform elsewhere or to future potential operators. The task descriptions developed therefore simply comprised lists of tasks which the operators had to perform; and no further detailed task analysis was possible, given the scope of the study. These task descriptions provided little information that would have helped to identify errors, but they are included nonetheless, to show the overall kinds of task that the operators have to perform. One of the simple task analyses developed for Scenario 2 is shown below.

Scenario 2: compressor leak or fire
In this scenario, alarms will occur on the CCR's fire and gas panel, indicating a release of hydrocarbons. The CCR operators' tasks will be:

1. To identify the location of the fire or leak.
2. To verify the release of hydrocarbons (by means of a number of coincident alarms).
3. To activate fire-fighting systems, if necessary.
4. To give a general alarm.
5. To send a field operator to the area.
6. To inform the OIM (Offshore Installation Manager).
7. To shut down other systems that could be affected by the fire or explosion.

4. Human error analysis and quantification

The overall methodology adopted in this study was as follows:

- The identification of human errors by means of: operator-action event trees; task-description analyses; and the consideration of Performance Shaping Factors (PSFs).
- The quantification of human-error probabilities (HEPs) by means of both relevant data and the absolute-probability-judgement (APJ) approach.

- The specification of error-reduction mechanisms (ERMs) based on the causes and mechanisms of errors.

Since this case study has been described in full elsewhere (Kirwan, 1987), the identification and quantification aspects will not be described. Only the results and the error-reduction mechanisms will be discussed; these showing the degree of impact that the study made on the design of the system.

5. Results

The results showed an overall flare-overload frequency of approximately once per 1000 years (the probable production life of the platform would, on the other hand, be only 15–20 years). The consequences were calculated separately, and were found not to be excessively severe – leading to injuries rather than fatalities. The system's design was therefore, on a risk basis, found to be a tolerable one.

The highest error probabilities had to do with errors of coordination/communication between the LCR/CCR operators during an ESD II (platform-emergency) situation occurring in Zone 2A (near the compressors, but in the separator zone). Overall, errors committed in the ESD II situation have a higher probability than those committed as part of the ESD-failure or compressor-failure scenarios, and this is to be expected, due to the generally higher stress levels involved in the ESD II scenario. In fact, the human errors committed in the ESD II scenario were found to make the greatest impact on the overall risk of occurrence of a flare overload. The human-error issue is therefore the most important issue to address in the business of reducing the likelihood of a flare overload.

In addition, a particular human-maintenance error involved with the ESD computer system was found, not surprisingly, to increase significantly the likelihood of an ESD-computer-system breakdown. This simple error, which involves opening the wrong cabinet during maintenance, could lead to an instantaneous ESD-system failure, and as a result, this particular error was considered for a design change (not dealt with further below).

6. Error reduction measures

Although the overall flare-overload frequency predicted in the total study was tolerably low, recommendations were formulated according to which further improvements could be achieved, and at a low cost. These are detailed below. The recommendations arose directly from earlier on in the modelling stage, largely as a result of a study of the major PSFs identified for this scenario.

Emergency depressurisation panel
The first problem identified with the panel had to do with the colour of the compressor LEDs that operated during the ESD-failure scenario. These

appeared *red*, thus signalling danger. As these buttons are intended to be used rarely, it was recommended that the LEDs be coloured *yellow* instead during this situation and other situations. In general, an offshore-operating alarm-signal convention is to use yellow to denote *caution*. The rationale behind this recommendation was that if all the LEDs appeared red, this might signal to the operator that they must be pressed if they are to be brought to a safer state (as indicated by *green*). Such action might therefore lead to all the zones being activated by the operator *simultaneously*, 'in the heat of the moment'.

Caps over the compressor-depressurisation buttons were also recommended, so as both to reduce the chances of an accidental activation and lessen the chances of the operator, in a state of panic, pressing all the buttons at once.

Written warnings on the panel itself were recommended, reminding the operators how a flare overload is caused, as well as reminding them not to depressurise the compressors manually until 5 minutes after the start of the separator's depressurisation.

A simple clock was also recommended which would start when one zone was depressurised. On the face of the clock, the first 5 minutes would be shown in a red 'danger' zone, together with explanatory warnings underneath.

Control-feedback or system-pressure indicators on the panel could be used to tell the CCR operators if the LCR operator had initiated any action. Indeed, a simple reminder, on the panel, of the existence of an LCR-control capability would also be helpful.

An interlocking device was suggested that would prevent any of the dangerous sequences from occurring. Although, in most cases, this would only involve a simple timing device, making such an interlock work reliably in *all* situations would prove a difficult task. Besides, in certain scenarios, the occurrence of a flare overload would be preferable to the threat posed by a gas explosion a heated, pressurised compressor.

Allocation of responsibility
It was an oversimplification to suggest that the LCR operator should in no case depressurise the compressor without explicit authorisation from either the CCR or the OIM. There would always be the possibility that a situation where compressors come under threat from fire happens to coincide with the loss of CCR control. However, it was felt that in emergency situations, the LCR operator would not necessarily be present anyway, and as a result, the main responsibility should lie with the CCR operators rather than with the LCR operators.

It was recommended that the CCR operator instruct the LCR operator about what to do if the LCR were to come under threat. More importantly, the LCR operator should get the go-ahead first from the CCR personnel if he or she intends to carry out depressurisation, or any other such action.

Emergency procedures
As all these events have been predicted to be infrequent, reliance cannot be placed upon *familiarity* and *practice* factors alone. Simple concise procedures

should also be made available in the CCR and LCR. In fact, the CCR panel is used so infrequently that the procedures could actually be put on the panel itself. These procedures should be practised during emergency-response training.

Training

The use of high-fidelity training simulations was recommended: most of the scenarios studied involve high-stress situations, and these are probably best simulated by means of high-fidelity simulation technology. A knowledge of all different possible scenarios, and of correct and incorrect operations, should also be promoted by means of the training scheme, and there should be special simulator exercises to deal with the scenarios themselves. As these scenarios involve relatively infrequent events, refresher training would be necessary – at, for example, yearly intervals.

7. Impact

This study demonstrated how the human-error factor makes a considerable contribution to the probability value for the flare-overload-occurrence top event, as well as to that for a previously overlooked failure mode which leads to an increased initiating event frequency. A number of the ERMs identified were accepted by the design and operations groups, though the flare-overload risk was, however, considered to be a tolerable one. This is in line with the principle of ensuring that a system's inherent risk level is as low as reasonably practicable (known as the ALARP principle). The study took 11 days to carry out, from start to finish.

Case Study 16: emergency blowdown task analysis

1. Background

This study concerned the evaluation of human-performance levels in connection with the initiation of depressurisation in the event of an emergency requiring a platform blowdown (K. Rea was the other assessor in this study).

2. Problem definition

The platform in question had been operational for some time, and the company wished to evaluate its emergency-response capability with respect to emergency depressurisation. The technical requirements were that, in the event of a major emergency event – whether platform-wide or localised (e.g. a fire) – a blowdown be initiated within *five minutes*.

The intention of the assessment was to carry out a qualitative analysis, involving a task and human-error analysis, and then quantify the likelihood of non-response (i.e. of a failure to initiate a blowdown) by utilising both the SLIM and the human-cognitive-reliability (HCR) approaches. ERMs *could* also have been made, but these were not specifically required by the study.

The system involved was a typical offshore platform in the North Sea sector, and unlike the situation studied in the previous case study, a blowdown *could* be effected for the whole platform simultaneously, if required.

It should be noted that, with respect to this kind of scenario, there are frequent false alarms on offshore platforms concerning gas detection and even fire detection. This means that operators have a natural tendency to try and verify that they really are facing a true emergency before initiating functions such as a blowdown.

3. Task analysis

Since the platform was already in existence, its relevant details were already well-documented, and it could, of course, be *visited* for the purposes of this study. The data-collection approaches utilised therefore comprised a documentation review (there was a significant amount of documentation to be reviewed, including documentation on related incidents and near-misses), observations (via walk-throughs), interviews and the use of the Critical Incident Technique. A number of personnel on two of the shift teams were also interviewed. They were:

- The Offshore Installation Manager (OIM)
- The operations supervisor
- The shift supervisor
- The control-room technicians
- The area technician (AT)

Walk-throughs and talk-throughs were conducted with all these personnel. In addition, the instrumentation found both in the CCR and in local areas was reviewed in terms of its level of adequacy as regards human factors.

The type of task analysis used was a (sequential/event-driven) tabular scenario analysis (TSA), since the emergency-blowdown operation is itself both sequential and somewhat unproceduralised in nature (it only involves a few operations). Since the blowdown operation also depends a good deal on the operators' expectations – as well as, potentially, on the resolution of production/safety conflicts (i.e. on decision-making) – the *expectations* of the operator in question were also recorded (a similar format to that used is shown in Table VI.16.1).

The task analysis concentrated on a significant unignited gas release in a particular location where gas would be likely to occur accidentally but where the operators can neither locate nor hence isolate the leak, as a result of which they must eventually resort to a blowdown.

Table VI.16.1 Blowdown tabular scenario analysis

Time	Operators	System's status	Information available	Operator's expectations	Procedure (written, memorised)	Decision/communication action	Equipment/location	Feedback	2ary duties; distractions; penalties	Comments
0:00	CR	Yellow alert; gas leak in HEM.	Audible alarm, yellow flashing light on F&G pane.	Uncertain: could be real.	1) Accept alarm. 2) Call area technician. 3) Make PA announcement. 4) Determine location of gas detector in alarm state.	Stop whatever operator is doing. Scan panels for flashing yellow or red light. Turn around to HEM F&G panel. Press 'Accept' button.	CCR layout. HEM F&G panel.	Visual and audial.	Whatever operator is doing when alarm occurs.	Initially disorientating because HEM does not have its own sound source. Alarm could be missed if second simultaneous alarm occurs on main bank of F&G panels.
						Go to radio. Radio AT to go to HEM and verify gas leak.	Radio-channel availability.		Noise, interference. AT may be busy.	Channels may be unavailable.
						Make PA announcement for personnel to stop hot work and stay away from HEM.	PA system.			This message takes up time, and distracts operator from watching F&G panel. Consider taped messages.
			Gas-detector-location manual.			Check location of gas detector in book.	GD handbook.			
						Radio AT. Tell AT location of gas detector in alarm.	Radio.		Noise interference.	

HEM = Hydrogen extraction module
F&G = Fire and gas
PA = Public address
AT = Auxiliary technician
GD = Gas detector

Time	Operators	System's status	Information available	Operator's expectations	Procedure (written, memorised)	Decision/ communication action	Equipment/ location	Feedback	2ary duties; distractions; penalties	Comments
1+15_secs	CR	Gas leak growing, unignited. Platform on muster status.	Red flashing and yellow flashing alarms (1 each), audible on HEM F&G panel.	Not a false alarm.	1) Accept alarms. 2) Make PA announcement. 3) Collect hot work permits. 4) Answer incoming communications.	Accept alarms on panel. Prepare for shutdown. Use PA to recall HWPs, and call to muster. Note HWPs as they appear. Respond to OS, OIM, etc. Get rid of non-urgent callers.	F&G panel PA radio. Meet operators at entrance to CR. Respond on communication desk.	Audial and visual. HWPs.	Too many coincident actions required. Level in CR will become very hectic. Nuisance calls will cause confusion.	Very hectic in the CR, and initially the CR is undermanned. Operators returning with HWPs will clog up the entrance to the CR. No calls should come through at this stage.
					5) Advise drillers of leak.	Advise drillers to stop work a.s.a.p.	Radio.			
					6) Check HEM pressure levels.	Check pressure levels on indicator panel.	HEM control panel.			CR technician must now keep his eyes on at least three panels: HEM F&G;
					7) Await verification.	Await AT's call. Monitor other F&G panels. Accept other alarms.				HEM control, main F&G. Must also accept any other alarms as they occur.
1+15_secs	AT	Gas leak growing; unignited.	Red alarm and klaxon.	Gas leak real.	1) Investigate leak. 2) Maintain personal safety.	Makes haste towards.	HEM.	Noise of gas leak.	Avoid collision en route with evacuating personnel.	

HWP = Hot work permit
OS = Operations supervisor
OIM = Offshore Installation Manager
CR = Control Room (central)

(Table VI.16.1 Continued)

Time	Operators	System's status	Information available	Operator's expectations	Procedure (written, memorised)	Decision/ communication action	Equipment/ location	Feedback	2ary duties; distractions; penalties	Comments
1+15secs	CO	As above.	As above.	As above.	Make crane safe.	Stop crane-lifting as soon as possible.	Crane above HEM.		Lift ongoing.	No check is made on CO.
1+15secs	DR	::	::	::	Make drilling safe.	Stop drilling a.s.a.p.	Drilling module.		Drilling.	May be difficult to stop if, e.g. in middle of tripping operation.
1+30secs	SS, OS, ATs, CR2, OS2	Red alert; gas leak growing; shutdown imminent; leak unignited.	Red alarm, klaxon; CR tech's PA announcement.	Real gas leak.	1) Go to CR. 2) OS assist CR technician with HWPs. 3) SS and OS go to HEM; pick up portable gas detector en route, as well as BA. 4) Remaining OS and SS help out in CR. 5) Designated AT musters men. 6) OS call in any remaining HWPs.	Various personnel arrive in CR. First SS to arrive goes with an OS to the HEM, picking up on gas detectors en route. CR takes over collection of HWPs. SS and OSs may assist with communications and alarm acceptance. Radio.	Portable gas detector.	Number of outstanding HWPs.	Allocation of duties will be confusing at this stage, but nevertheless probably helpful. Secondary alarms will be occurring.	Still hectic in CR, a state enhanced by the sheer number of people who will be in the CR. Alarms will be occurring. Operators will be mustering also at the CR entrance/exit. Personnel will also probably be entering the CR from behind the CR panel racks, and crossing the CR, making it rather full. May block channel for HEM AT.

CO = Crane operator
DR = Drilling operator (drill. or tool-pusher)
BA = Breathing apparatus

Time	Operators	System's status	Information available	Operator's expectations	Procedure (written, memorised)	Decision/ communication action	Equipment/ location	Feedback	2ary duties; distractions; penalties	Comments
1-45secs	CR, OS, SS	Coincident alarms; leak unignited.	Two red alarms; ESD alarms.	SPS; awaiting verification.	Ensure safe SPS.	OS and CR and SS ensure safe ESD actions. Check power-shedding. Check HEM pressure levels and gas-detector readings.	Various panels.	Various alarms and indications.	Communications; returning HWPs; ancillary duties; mustering process.	Allocation of function still fuzzy; an ESD checklist might prove useful.
3:00	AT	Gas leak unignited.	Smell and noise of gas; H_2S smell present also; location of first-up gas detector.	Significant gas leak; flange or seal, etc.	Verify leak and execute evacuation in response to presence of H_2S.	Operator investigates briefly to see if source of leak is apparent, but it is not, due to H_2S level.	HEM SPS activator. Radio.	Noise; smell; no visual cue for presence of gas, since gas is warm. Acknowledgement from operator.	Channel could be busy.	Operator must be thoroughly aware of the dangers of H_2S so that he is not overpowered while he investigates too long. If message unclear, use telephone.

(Table VI.16.1 Continued)

Time	Operators	System's status	Information available	Operator's expectations	Procedure (written, memorised)	Decision/ communication action	Equipment/ location	Feedback	2ary duties; distractions; penalties	Comments
4:00	CR1, OS1, SS	Unignited gas leak; presence of H_2S (localised).	Leak verified by AT; pressure indications.	May be simple to isolate.	Blowdown.	Operators consider blowdown versus delay				Could be an easily isolatable leak, but this does not necessarily justify any delay. If source can be found, at least
					Check wind.	problems of starting up HEM, and finding leak once depressurised. Risk of ignition considered. Decide to delay blowdown until				maintenance will be quicker. Both of these points are production-oriented. Since the risk of ignition is
					Consider other reasons for not carrying out blowdown.	2 operators have had an attempt at finding and isolating the leak. Check ignition sources and execute evacuation of the area.			Mustering will still be continuing; personnel may be in leg, and will want to know what is going on.	always finite, and personnel are present in the HEM, it is difficult to justify either of them.
						Communicate with OIM.		Indications, communications.		
						Check wind direction; radiation risk; flare integrity; helicopters, etc.; CR considers how high pressure is in HEM vessels – still high. CR maintains watch on F&G panels.	ESD panels, HWPs, musterers. Telephone to radio room. Radio room.	Radio room specifies wind direction.		Could have wind-direction monitor in CR. Checklist might also help.

Time	Operators	System's status	Information available	Operator's expectations	Procedure (written, memorised)	Decision/ communication action	Equipment/ location	Feedback	2ary duties; distractions; penalties	Comments
5:00	SS, OS	SPS; gas leak unignited; large pressure of gas in HEM; wind dispersing some of gas.	Loud noise; location of first-up gas detector; high reading on portable gas detector.	Flange, seal instrument, etc., not a large bore rupture in a pipe.	Blowdown. Don BA and locate source of leak; isolate if possible.	Don BA. Search for leak source in HEM using audible and visual cues.	HEM, BA.	Noise; gas jet, if naked flesh exposed to it; visual cues.	Take care not to ignite leak.	
6:00		Gas leak; prevailing wind blowing gas away from platform; HEM gas pressure slowly dropping.	Gas detectors; pressure indicators; gas monitors in other areas.	Will find and isolate leak.	Blowdown.	Communicate with OIM at intervals to advise of status. Communicate with onshore emergency centre (OEC) Discussions will take place on blowdown, and how long they should wait, and indeed if blowdown need be used at all. Procedures are fuzzy and can be interpreted as leaving the decision to blowdown at the operator's discretion.	Radio; telephone; BA sets.	Occasional calls from OS and SS.	Other events and communications from OEC, etc.	Less hectic now. May refer to procedures. Will be aware that sooner or later they should probably blow down. But is the risk reduction involved in a blowdown seen to be significant?

Table VI.16.1 (Continued)

Time	Operators	System's status	Information available	Operator's expectations	Procedure (written, memorised)	Decision/ communication action	Equipment/ location	Feedback	2ary duties; distractions; penalties	Comments
15:00	OS, SS, CR, OS2, SS2, CR2	Gas leak still present; blowing gas.		Will not identify source of leak.	Blowdown.	Operating team decides that it may take some time to find the leak. Decide to execute blowdown. SS & OS evacuate HEM. CR technician operates blowdown buttons for HEM. CR technician monitors HEM, falling pressure, and flare pressure, and other F&G panels.		Noise, pressure-level indicators.		Event effectively ends.

The resultant TSA was fairly extensive, covering the major phases involved in the operator's emergency response to the situation:

- The response to the alarm, and the identification of the location of the leak.
- Verification (i.e. that it was *not* a false alarm).
- The identification of the source of the leak (reparable or not).
- The decision-making process – i.e. 'blowdown or no blowdown?'.
- The execution of the blowdown.

4. Human error analysis

The analysts' judgement and the SHERPA system were together used to define the various errors of judgement, decision-making and execution which could occur during the various different blowdown scenarios. These errors were also derived from a combination of questions asked and observations made during the audit, as well as from prior human-error-analysis experience. The following lists the errors in the approximate chronological order in which they would occur in the various different scenarios.

1. Prior to alarms

1.1 Fails to return fire or gas detectors to service after maintenance.
1.2 Leaves detectors inhibited.

2. Initial alarm situation

2.1 Cancels simultaneous gas indication on main panel, and fails to notice yellow alarm on specific-location panel.
2.2 CR operator momentarily absent from CR.
2.3 CR operator incapacitated in CR.
2.4 CR operator's workload (communications, etc.) very high. Delay therefore experienced in sending of AT to location.
2.5 CRO inhibits coincident alarm to prevent shutdown.
2.6 CRO gives wrong location of gas detector to AT.

3. Verification

3.1 AT believes it is a false alarm, when it is in fact a real one.
3.2 AT is asphyxiated by gas/H_2S.
3.3 AT cannot get through on radio to CR.
3.4 AT decides to investigate before radioing CR.
3.5 AT–CRO communication error.
3.6 AT believes problem is insignificant, when in fact it is serious.

3.7 AT can find no gas (detector not working properly).

3.8 AT delays, etc.

3.9 Operators (CR and local) fail to realise collateral damage to flare integrity has occurred (blowdown should not be executed if the flare has also been damaged).

4. Identification of leak source

4.1 CRO/SS team decide to delay blowdown until they have found source.

4.2 AT/SS have to return to get breathing apparatus or explosimeters.

4.3 AT/SS (shift supervisor) unable to find source.

4.4 Team decide to continue to delay blowdown.

4.5 No one sure who should make decision to blow down – all assume *someone else* will.

5. Blowdown decision

5.1 Decides blowdown unnecessary.

5.2 Procrastinates (reluctance effect/decision-making avoidance).

5.3 Allocation-of-function problem – uncertainty as to who should make the decision.

5.4 Forgets about blowdown.

5.5 Carries out other priority tasks (e.g. power-shedding).

5.6 Decides to let system depressurise on its own (erroneous assumption).

5.7 Fails to correctly ascertain the risks involved in gas migration.

5.8 Delays blowdown as a last-resort action.

5.9 Misreads pressure instruments, etc.

6. Blowdown execution

6.1 Keys missing from blowdown panel (key access).

6.2 Fails correctly to operate blowdown by means of dual-action system (separators).

6.3 Fails correctly to verify the blowdown's degree of effectiveness.

6.4 Believes someone else in CR, or locally, has operated it.

6.5 Forgets to execute blowdown (distracted).

6.6 Does not blow down for long enough (i.e. fails to *hold down* button).

6.7 Stress reaction – operator fails to activate.

6.8 Awaits high-level order from OIM.

5. Results

Once the TSA was completed, it was realised that it was highly unlikely that a blowdown would be initiated within 5 or even 10 minutes. This was because of the general problem of *false* gas alarms on platforms, and hence the subsequent perceived need to verify that the release of gas *did* in fact occur. The verification phase, unfortunately, would be found to take at least 10 minutes, not to mention subsequent communication and decision-making requirements. The TSA, therefore, in a similar way to a timeline analysis, showed that the execution of a blowdown simply would *not* occur during the required time-frame. The quantification was therefore redundant, and was not carried out.

6. Impact

The results indicated, in particular, that the need for a blowdown was not perceived as high a priority offshore as it was *on*shore, and hence that there was a need for a clearer philosophy on blowdowns. This philosophy would then need to be translated into practice via enhanced training which should be refreshed yearly rather than on a once-only basis during the operator's initial pre-operational training period. It was also noted that the operators concerned were a little 'rusty' in their appreciation of how the blowdown system worked in practice – which left open a potential for errors of execution.

Several improved systems were recommended for consideration: an enhanced manual system, a semi-automatic system and a fully automated blowdown system, all of which could, in theory, achieve the required level of performance reliability in emergency scenarios. The assessment took six person months to complete and involved two analysts.

Case Study 17: offshore platform production blowout

1. Background

This project concerned the risk of a blowout occurring during the production phase, at which stage the well has been drilled and the hydrocarbons are being extracted from the well and preprocessed on the platform, before being pumped ashore for further treatment. During production, there is the risk of a catastrophic leak or pipe rupture in the wellhead area on the platform, which could lead to a fire or an explosion. Since there are usually many such wells drilled and on a platform, if one well leaks and catches fire, there is a possibility that such a fire would cause the catastrophic failure of all other wellheads, with a significant level of danger for the platform as a whole. This assessment concerned the human-reliability aspects associated with a new protection system that was being designed to prevent such an escalation in the event of a leak.

2. Problem definition

The protection system being developed was a shear-ram system for operating wells which, like its drilling counterpart, would close off the well pipeline by squeezing/crushing it at a level beneath the level at which leakage would be a problem (e.g. beneath the wellhead deck). The intention was that such a device would be manually initiated locally, once it had been detected by the CCR personnel. There would be gas and fire detectors in the wellhead area, and there would also possibly be remote CCTV surveillance equipment, relaying information to the CCR. There would also be personnel in the well-head area, but not *continuously*. The shear-ram operating system itself involved a set of labelled levers in the module adjacent to the wellhead area, with a viewing window (protective glass) allowing visual access to the wells themselves at the place of shear-ram activation. There were 15 wells in the wellhead area, and 15 associated shear rams also.

As regards the shear-ram operation, the requirement would be a *rapid* response in the event of a leak, since escalation could occur within minutes. However, having said that, the cost of an *inadvertent* activation, or of an activation which was not required, would be huge, since essentially, once set into motion, the rams cannot be stopped, nor can their effects be reversed, and the result is that the well is lost, in terms of future production. This places a significant reluctance burden on the operators. It is also the case that, if a leak does ignite and then cause an explosion, the local operator, even if he or she is located behind protective glass, may still not survive such an explosion.

The question therefore posed in the assessment was how reliable the shear system would be in terms of its human operation.

3. Task analysis

No formal task analysis was carried out, since the platform and the shear-ram system themselves were still at the design stage, and since no similar, production-based shear-ram system existed. As a result, observations, interviews and the Critical Incident Technique (CIT) could not be carried out, and only a limited documentation review was, moreover, possible. The actual sequence of operations involved was instead therefore elicited from systems analysts and design personnel, and went approximately as follows:

1. Detect gas-warning alarms from wellhead area (in CCR).
2. Verify that a well is leaking.
3. Determine which well is leaking.
4. Send local operator (LO) to wellhead location.
5. LO to identify appropriate shear ram lever.
6. LO to operate shear ram.
7. CCR to verify whether or not actuation is effective (if not, re-try; or abort, and then evacuate area).

4. Human error analysis

There were a number of errors that could occur in this scenario, including the following:

- A failure to respond in time to the alarms.
- A failure to realise the severity of the situation.
- The CCR overestimates the degree of severity – rams closed when unnecessary.
- A failure to identify the correct well.
- A communication error between the CCR and the LO – the wrong well is communicated.
- The local operator (LO) fails to reach the module in time.
- The LO is incapacitated.
- The LO fails to operate the correct level.
- The LO operates more than one level.
- A maintenance error on the shear-ram's accumulator system leaves the system inoperational.

These errors were identified at the external-error-mode (EEM) level, due to the short time-scale involved in the study, and by using only the THERP EEM taxonomy along with the 'right action on wrong object' EEM (prior to the development of the SHERPA system). The Performance Shaping Factors (PSFs) made evident after a consideration of the scenario in question would be stress, reluctance, a production–safety conflict and training and procedures. The operator's interface would also be a strong PSF, though little was known at the time about how the system would look and be instrumented.

5. Results

An overall HEP figure was determined by comparing different sources of data on emergency scenarios. The results showed the human-error probability to be fairly high, but since the likelihood of a wellhead failure in the first place would be very low, the overall risk was determined to be tolerable, according to offshore-safety-risk regulations prevalent at the time (see the Impact section below).

6. Error reduction measures

A number of the error-reduction measures carried out focused on improving the ways of confirming the existence of a leak, of determining which well was leaking (via heat- and vibration-sensing equipment) and of allowing the *remote* activation – i.e. activation from the CCR – of the shear rams.

7. Impact

An interesting problem arose at the end of this study. The system was calculated to be safe, and the overall risk factor to be more than one order of magnitude lower than it was required to be, according to current regulatory standards. It was suggested by the client that, given this situation, perhaps some safety equipment found on board the platform could be dispensed with, given that the platform was apparently 'oversafe'. This was an unusual suggestion, though perhaps a logical one. It was countered for two reasons: firstly, there is always a degree of uncertainty in risk analyses, particularly at the design stage before any operational experience has been gained, and because of this fact the order-of-magnitude safety margin can *never* be guaranteed; and secondly, the ALARP principle dictates that the removal of such already-designed-and-accepted safety measures would be quite inappropriate. These counter arguments were accepted.

References

AAIB (1990) Report on the accident to Boeing 737-400 G-OBME near Kegworth, Leicestershire, on 8 January 1989. Air Accidents Investigation Branch, Aircraft Accident Report 4/90, London: HMSO.

ACSNI (1991) Human Reliability Assessment – a critical overview. Advisory Committee on the Safety of Nuclear Installations, Health and Safety Commission, London: HMSO.

Ainsworth, L.A. (1985) A study of control room evaluation techniques in a nuclear power plant. Applied Psychology report no. 115, University of Aston.

Ainsworth, L.R., and Whitfield, D. (1984) Are verbal protocols the answer or the question? Proceedings of the Ergonomics Conference, pp. 68–73. London: Taylor and Francis.

Apostolakis, G. (Ed.) (1990) Special issue on human reliability analysis. Reliability Engineering and System Safety, Vol. 29, No. 3.

Askren, W.B., and Regulinski, T.L. (1969) Quantifying Human Performance for Reliability Analysis of Systems. Human Factors, 11, 4, 393–396.

Ball, P. (Ed.) (1991) The guide to reducing human error in process operations. Report No SRD R484. Warrington: Safety & Reliability Directorate.

Bell, B.J. (1984) Human reliability analysis for probabilistic risk assessment. Proceedings of the 1984 International Conference on Occupational Ergonomics, pp. 35–40.

Bellamy, L.J. (1986) The safety management factor: An analysis of human aspects of the Bhopal disaster. Paper presented at the Safety & Reliability Symposium, Altrincham, September 26.

Bellamy, L.J., Kirwan, B., and Cox, R.A. (1986) Incorporating human reliability into probabilistic risk assessment, in the 5th International Symposium on Loss Prevention and Safety in the Process Industries, Societe de Chimie, 28 Rue St. Dominique, Paris, pp. 6.1–6.20.

Bello, G.C., and Columbari, V. (1980) The human factors in risk analysis of process plants: the control room operator model TESEO. Reliability Engineering, 1, pp. 3–14.

Bignell, V., Peters, G., and Pym, C. (1978) Catastrophic Failures. Milton Keynes: Open University Press.

Bignell, V., and Fortune, J. (1984) Understanding System Failures. Milton Keynes: Open University Press.

Blanchard, R.E., Mitchell, M.B., and Smith, R.L. (1966) Likelihood of accomplishment scale for a sample of man-machine activities, Dunlap and Associates Inc, Santa Monica, CA, June.

Bley, D., and Stetkar, J.W. (1988) Significance of sequence timing to human action modelling. Paper presented at the 4th IEEE Conference on Human Factors and Power Plants, Monterey, California, June 5–9.

British Standards Institute (1991) Reliability of systems, equipment and components: BS 5760: BSI, London.

Berliner, D.C., Angelo, D., and Shearer, J. (1964) Behaviours, measures and instruments for performance and evaluation in simulated environments. Presented at the symposium on the quantification of human performance, Albuquerque, New Mexico August 17–19.

Brune, R.L., Weinstein, M., and Fitzwater, M.E. (1983) – Peer review study of the draft handbook for human reliability analysis. SAND-82-7056, Sandia National Laboratories, Albuquerque, New Mexico.

Cacciabue, C., Decortis, F., Mancini, G., Mason, N., and Nordvik, J.P. (1989) COSIMO: A cognitive simulation model in blackboard architecture: synergism of AI and Psychology. In the Proceedings of the Cognitive Science Approaches to Process Control (CSAPC) Conference, Siena, Italy, October 24–27.

Carnino, A. (1988) The actions taken by EDF after the Three Mile Island Accident in the domain of human reliability. IEEE Conference on Human Factors and Power Plants, Monterey, California, June 5–9, pp. 65–68.

Chatfield, C. (1978) Statistics for technology, 2nd Edition. London: Chapman and Hall.

Coburn, R. (1971) A human performance databank for command control. Proceedings of the US Navy Human Reliability Workshop. February.

Collier, S.G., and Graves, R.J. (1986) Improving human reliability: practical ergonomics for design engineers. In proceedings of the 9th Advances in Reliability Technology Symposium, University of Bradford, 2–4 April.

Comer, M.K., Seaver, D.A., Stillwell, W.G., and Gaddy, C.D. (1984) 'Generating Human Reliability Estimates Using Expert Judgement'. Vols 1 and 2. NUREG/CR-3688 (SAND 84-7115). Sandia National Laboratory, Albuquerque, New Mexico, 87185 for Office of Nuclear Regulatory Research, US Nuclear Regulatory Commission, Washington, DC 20555.

Comer, P.J., Fitt, F.S., and Ostebo, R. (1986) – A driller's HAZOP method, Society of Petroleum Engineers, SPE 15867.

Comer, P.J., and Kirwan, B. (1985) A reliability study of a platform blowdown system, in the proceedings of Automation and Safety in Shipping and Offshore Petroleum Operations, Royal Garden Hotel, Trondheim, Norway.

Cox, S.J., and Tait, N.R.S. (1991) Reliability, Safety and Risk Management. London: Butterworth Heinemann.

Dalkey, N. (1969) An experimental study of group opinion: the Delphi method. Futures, 1, 408–426.

Dolby, A.J. (1990) A comparison of operator response times predicted by the HCR model with those obtained from simulators. International Journal of Quality and Reliability Management, 7, 5, pp. 19–26.

Dorner, D. (1987) On the difficulties people have in dealing with complexity, in Rasumssen, J., Duncan, K., and Leplat, J. (Eds) New Technology and Human Error. Chichester: John Wiley and Sons.

Dougherty, E.M. (1987) Comparison of Time Reliability Correlations, Report SAIC-NY-OR-07-368-14-1, New York.

Dougherty, E.M., and Fragola, J.R. (1987) Human reliability analysis: a systems engineering approach with Nuclear Power Plant applications. Chichester: John Wiley & Sons.

Drager, H. (1981) Cause Relationships of Collisions and Groundings Final Report. Det Norske Veritas, Report 81-0097.

Drury, C.G. (1983) Task analysis methods in industry. Applied Ergonomics, 14.1, 19–28.

Drury, C.G. (1990) Methods for direct observation of performance, in Wilson, J.R., and Corlett, N.E. (Eds) Evaluation of human work. London: Taylor and Francis. pp. 35–57.

Eils, L.C., and John, R.S. (1978) A criterion validation analysis of multi-attribute utility analysis and a group communication strategy. SSRI-Report 78-4.

Eisner, H.S., and Leger, J.P. (1988) The ISRS in South African mining. Journal of Occupational Accidents, 10, pp. 141–160.

Embrey, D.E., and Kirwan, B. (1983) A comparative evaluation of three subjective human reliability quantification techniques, in Annual Ergonomics Society Conference Proceedings, K. Coombes (Ed.), London: Taylor and Francis.

Embrey, D.E. (1986) – SHERPA: A systematic human error reduction and prediction approach. Paper presented at the International Topical Meeting on Advances in Human Factors in Nuclear Power Systems, Knoxville, Tennessee.

Embrey, D.E. (1986) – SHERPA: A systematic human error reduction and prediction approach. Paper presented at the International Topical Meeting on Advances in Human Factors in Nuclear Power Systems, Knoxville, Tennessee.

Embrey, D.E., Humphreys, P.C., Rosa, E.A., Kirwan, B., and Rea, K. (1984) 'SLIM-MAUD: An Approach to Assessing Human Error Probabilities Using Structured Expert Judgement'. NUREG/CR-3518, (BNL-NUREG-51716). Department of Nuclear Energy, Brookhaven National Laboratory, Upton, New York 1173, for Office of Nuclear Regulatory Research, US Nuclear Regulatory Commission, Washington, DC 20555.

Embrey, D.E. (1990) Outline of System for Critical Human Error Management (SCHEMA). Personal communication.

Fleishman, E.A., and Quaintance, M.K. (1987) Taxonomies of human performance: the description of human tasks. London: Academic Press.

Gall, W. (1990) – An analysis of nuclear incidents resulting from cognitive error. Paper presented at the 11th Advances in Reliability Technology Symposium, University of Liverpool, April.

Gertman, D.I., Gilmore, W.E., Groh, M., Galyean, W.J., Gentillon, C., and Gilbert, B.G. (1988) Nuclear Computerised Library for Assessing Reactor Reliability (NUCLARR), Volume 1 – Summary Description. EGG-2458, NUREG/CR-4639, USNRC, Washington DC, February.

Gertman, D.I., Gilmore, W.E., and Ryan, T.G. (1988) NUCLARR and human reliability: data sources and data profile. Paper presented at the Fourth IEEE Conference on Human Factors and Power Plants, Monterey, California, June 5–9, pp. 311–314.

Gertman, D.I. (1991) INTENT: A method for calculating HEP estimates for decision-based errors. Paper in the Proceedings of the Human Factors Society 35th Annual Meeting, pp. 1090–1094.

Goossens, L., Cooke, R.M., and van Steen, J.F.J. (1989) Expert Opinions in Safety Studies: Vol 1, Final Report. Safety Science Group, Technical University of Delft, Netherlands.

Grandjean, E. (1988) Fitting the task to the man. 4th edition. London: Taylor and Francis.

Green, A.E. (1983) Safety systems reliability. Chichester: John Wiley.

Haddon, W. (1973) Energy damage and the Ten Counter-measure Strategies. Human Factors, 15/4, pp. 355–366.

Hale, A.R., and Glendon, I. (1987) Individual behaviour in the control of danger. Oxford: Elsevier.

Hall, R.E., Fragola, J.R., and Wreathall, J. (1982). Post-Event Human Decision Errors: Operator Action Tree/Time Reliability Correlation, NUREG/CR-2010, US Nuclear Regulatory Commission, BNL-NUREG-51601, Brookhaven National Laboratory.

Hannaman, G.W., and Spurgin, A.J. (1984) Systematic Human Action Reliability Procedure (SHARP), EPRI NP-3583, Electrical Power Research Institute, Palo Alto, California.

Hannaman, G.W., Spurgin, A.J., and Lukic, Y.D. (1984) Human Cognitive Reliability Model for PRA Analysis, Draft Report NUS-4531, EPRI Project RP 2170-3, Electric Power Research Institute, Palo Alto, Calif.

Henley, E.J., and Kumamoto, H. (1981) Reliability Engineering and Risk Assessment. New Jersey: Prentice-Hall.

Hickson, J., Kennedy, R., Hardiman, T., and Milner, R. (1993: unpublished) Case study project on VDU format ergonomics for process control. School of Manufacturing Engineering, University of Birmingham. May.

Hopkins, C.O., Snyder, H.L., Price, H.E., Hornick, R.J., Mackie, R.R., Smilie, R.T., and Sugarman, R.C. (1982) Critical human factors issues in nuclear power regulation and a recommended comprehensive human factors long-range plan. USNRC, NUREG/CR-2833, Washington DC-20555.

Howland, A.H. (1980) Hazard analysis and the human element. 3rd International Loss Prevention Symposium, Basle, 15–19 September.

HSE (1992) The tolerability of risk from nuclear power stations. London: HMSO (2nd edition).

Hunns, D.M. (1982) 'The Method of Paired Comparisons'. In: Green, A.E. (Ed.). High Risk Safety Technology. Chichester: Wiley.

Hunns, D., and Daniels, B.K. (1980) The method of paired comparisons. In proceedings of the 6th Symposium on Advances in Reliability Technology, Report No. NCSR R23 and R24, AEA Technology, Warrington.

IEC (1995: in preparation) International standard on Human Reliability Assessment. International Electro-technical Committee IEC/TC/56 WG 11. London: BSI.

Kantowitz, B., and Fujita, Y. (1990) Cognitive theory, identifiability, and Human Reliability Analysis. Reliability Engineering and System Safety, 29, 3, pp. 317–328.

Kincade, R.G., and Anderson, J. (1984) Human factors guide for nuclear power control room development. EPRI-NP 3659, EPRI, Palo Alto, California, August.

Kirwan, B. (1982) A comparative evaluation of three human reliability quantification techniques. MSc dissertation, Dept of Engineering Production, University of Birmingham, September.

Kirwan, B., and Wilson, I. (1985) Human Factors in Ship-Platform Collision, in Contemporary Ergonomics, Oborne, D.J. (Ed.). London: Taylor and Francis, pp. 160–164.

Kirwan, B., and Rea, K. (1986) – Assessing the human contribution to risk in hazardous materials handling operations. Paper presented at the First International Conference in Risk Assessment of Chemicals and Nuclear Materials. Robens Institute, Surrey University.

Kirwan, B. (1987) Human reliability analysis of an offshore emergency blowdown system. Applied Ergonomics, 18.1, pp. 23–33.

Kirwan, B. (1988) Integrating human factors reliability into the plant design and assessment process, in Contemporary Ergonomics Megaw, E. (Ed.). London: Taylor and Francis.

Kirwan, B. (1988) A comparative evaluation of five human reliability assessment techniques. In Human Factors and Decision Making, Sayers, B.A. (Ed.). London: Elsevier, pp. 87–109.

Kirwan, B., Embrey, D.E., and Rea, K. (1988) Human Reliability Assessor's Guide. Report RTS 88/95Q, NCSR, UKAEA, Culcheth, Cheshire.

Kirwan, B., and James, N.J. (1989) A Human Reliability Management System, in Reliability 89, Brighton Metropole, June.

Kirwan, B., and Reed, J. (1989) A task analytical approach for the derivation and justification of ergonomics improvements in the detailed design phase. In Contemporary Ergonomics, Megaw, E. (Ed.). London: Taylor and Francis.

Kirwan, B. (1990) Human Reliability Assessment, in Wilson, J.R. and Corlett,

N.E. (Eds) Evaluation of human work. London: Taylor and Francis, pp. 706–754.

Kirwan, B. (1990) – A resources flexible approach to human reliability assessment for PRA, Safety and Reliability Symposium, Altrincham, September. London: Elsevier Applied Sciences, pp. 114–135.

Kirwan, B., Martin, B.R., Rycraft, H., and Smith, A. (1990) Human Error Data Collection and Data Generation, in International Journal of Quality and Reliability Management, 7.4., pp. 34–66.

Kirwan, B., and Ainsworth, L.K. (Eds) (1992) Guide to task analysis. London: Taylor and Francis.

Kirwan, B. (1992) Human error identification in human reliability assessment. Part 1: overview of approaches. Applied Ergonomics, 23, 5, 299–318.

Kirwan, B. (1992) Human error identification in HRA. Part 2: Detailed comparison of techniques. Applied Ergonomics, 23(6), 371–381.

Kirwan, B. (1992) A Task Analysis Programme for THORP. in Kirwan, B. and Ainsworth, L.K. (Eds), A guide to task analysis, pp. 363–388. London: Taylor and Francis.

Kirwan, B. (1992) Plant control diagnostic failure – just a matter of time? in Proceedings of the IEEE Conference on Human Factors and Power Plants, Monterey, USA, June, pp. 51–69.

Kirwan, B., Ainsworth, L., and Pendlebury, G. (1992) Task analysis: the state of the art. In Contemporary Ergonomics, Lovesey, E.J. (Ed.). London: Taylor and Francis.

Kirwan, B. (1993) A Human Error Analysis Toolkit for Complex Systems. Paper presented at the Fourth Cognitive Science Approaches to Process Control Conference, August 25–27, Copenhagen, Denmark, pp. 151–199.

Kletz, T. (1974) HAZOP and HAZAN – Notes on the Identification and Assessment of Hazards. Rugby: Institute of Chemical Engineers.

Koorneeff, F. (1990) Personal communication.

Kopstein, F.F., and Wolf, J.J. (1985) Maintenance Personnel Performance Simulation (MAPPS): User's Manual. Nureg/CR-3634, USNRC, Washington DC, September.

Kozinsky, E.J., Grey, L.H., Beare, A.N., Burks, D.B., and Gomer, F.E. (1984) Safety-related operator actions: Methodology for developing criteria, NUREG/CR-3515, USNRC, Washington DC.

Johnson, W.G. (1980) MORT Safety Assurance System. New York: Marcel Dekker Inc.

Laughery, K.R. (1984) Computer Modelling of Human Performance on Microcomputers, in the Proceedings of the Human Factors Society Meeting, Santa Monica CA, pp. 884–888.

Laughery, K.R., and Laughery, K.R. (1987) Analytic Techniques for function analysis. In Salvendy, G. (Ed.), Handbook of Human Factors. New York: Wiley and Sons.

Lucas, D.A., and Embrey, D.E. (1989) Human reliability data collection for qualitative modelling and quantitative assessment. EUREDATA Conference, Siena, Italy.

Martin, B.R., and Wright, R.I. (1987) A practical method of common cause failure modelling. Reliability Engineering, 19, pp. 185–199.

Meddis, R. (1973) Elementary analysis of variance for the behavioural sciences. London: McGraw-Hill.

Metwally, A.M., Sabri, Z.A., Adams, S.K., and Husseiny, A.A. (1982) A data bank for human related events in nuclear power plants. Proceedings of the 26th Human Factors Society Annual Meeting, New York.

Mills, R.G., and Hatfield, S.A. (1974) Sequential task performance, task module relationships, reliabilities, and times. Human Factors, 16, 117–128.

Munger, S.J., Smith, R.W., and Payne, D. (1962) An index of electronic equipment operability: Data store. AIR-C43-1/62-RP(1). Pittsburgh, PA: American Institute for Research.

Murphy, A.H., and Winkler, R.L. (1974) Credible interval temperature forecasting: some experimental results. Monthly Weather Review, 102, pp. 784–794.

Nicks, R. (1981) Probabilistic approach to problems in reactor safety risk assessment. Paper presented at Convegno Internacionale sin Foundamenti della Probabilita e della statistica, Luino, Italy.

Nobile, N. (1992) Human Factors: One of the key issues for safer nuclear plants. Paper in the IEEE Proceedings of the Human Factors in Power Plants, June 7–11, Monterey, California, pp. 564–569.

Paradies, M., and Busch, D. (1988) Root cause analysis at Savannah River Plant. Paper presented at the IEEE 4th Conference on Human Factors and Power Plants, Monterey California, June 5–9.

Pedersen, O.M. (1985) Human Risk Contributions in the Process Industry, Riso-M-2513. Riso National Laboratories, DK-4000 Roskilde, Denmark.

Perrow, C. (1984) Normal Accidents. New York: Basic Books.

Pew, R.W., Miller, D.C., and Feehrer, C.S. (1981) Evaluation of proposed control room improvements through analysis of critical operator decisions NP 1982. Palo Alto, CA: Electric Power Research Institute.

Phillips, L.D., Humphrey, P., and Embrey, D.E. (1983) A socio-technical approach to assessing human reliability. Technical Report 83–4, London School of Economics, Decision Analysis Unit.

Phillips, L.D., and Embrey, D.E. (1985) Appendix E: Quantification of operator actions by STAHR [IDA]. In Selby, D.D., Pressurised Thermal Shock Evaluation of the Calvert Cliffs Unit 1 NPP, ORNL, Oak Ridge, Tennessee, USA.

Pitblado, R., Slater, D., and Williams, J. (1990) Quantitative assessment of process safety programmes, Plant/Operations Progress, 9, 3, pp. 169–175.

Pontecorvo, L.B. (1965) A method of predicting human reliability. Annals of Reliability and Maintenance, Vol IV, pp. 337–342.

Potash, L. *et al* (1981) – Experience in integrating the operator contribution in the PRA of actual operating plants. In proceedings of the ANS/ENS Topical meeting on PRA, New York, American Nuclear Society.

Pyman, M.A.P., Austin, J.S., and Lyon, P.R. (1983) Ship/platform Collision Risk in the UK Sector. IABSE Colloquium, Copenhagen.

Rasmussen, N. (1975) Reactor Safety Study. Report WASH-1400, Washington DC: Atomic Energy Commission.

Rasmussen, J., Pedersen, O.M., Carnino, A., Griffon, M., Mancini, C., and Gagnolet, P. (1981) Classification system for reporting events involving human malfunctions. Report Riso-M-2240, DK-4000. Roskilde, Riso National Laboratories, Denmark.

Rasumssen, J., Duncan, K., and Leplat, J. (Eds) (1987) New Technology and Human Error. Chichester: John Wiley and Sons.

Rea, K. (1986) Personal communication.

Reason, J.T., and Embrey, D.E. (1985) Human Factors Principles relevant to the Modelling of Human Errors in Abnormal Conditions of Nuclear and Major Hazardous Installations. Report for the European Atomic Energy Community, Technical Report ECI-1164-B7221-84-UK, for the European Atomic Energy Community. Human Reliability Associates Ltd, Parbold, Lancs, England.

Reason, J.T. (1987) Generic Error Modelling System: A cognitive framework for locating common human error forms. In New Technology and Human Error, Rasmussen, J., Duncan, K., and Leplat, J. (Eds). Chichester: Wiley.

Reason, J.T. (1988) Human Fallibility. Evidence presented at the Hinkley C Public Enquiry on behalf of the Consortium of Opposing Local Authorities. Document COLA 22. December.

Reason, J.T. (1990) Human Error. Cambridge: Cambridge University Press.

Reed, J. (1992) A plant local panel review. In Kirwan, B., and Ainsworth, L. (Eds) Guide to task analysis. London: Taylor and Francis, pp. 267–288.

Reed, J. (1992) The contribution of human factors to the THORP project. Paper presented at SPECTRUM '92, August, Idaho, USA.

Rigby, L.V. (1976) The Sandia Human Error Rate Bank (SHERB). In Blanchard, R.E. and Harris, D.H. (Eds) Man-machine effectiveness analysis: A symposium of the Human Factors Society. June.

Rogers, W.P., *et al* (1986) Report of the Presidential Commission on the Space Shuttle Challenger Accident, June 6th. Washington DC.

Rosa, E.A., Humphreys, P.C., Spettell, C.M., and Embrey, D.E. (1985) 'Application of SLIM-MAUD: A Test of an Interactive Computer-based Method for Quantifying Expert Assessment of Human Performance and Reliability'. NUREG/CR-4016 Brookhaven National Laboratory, Upton, New York, USA, for Office of Nuclear Regulatory Research, US Nuclear Regulatory Commission, Washington, DC 20555.

Rosness, R., Hollnagel, E., Sten, T., and Taylor, J.R. (1992) Human reliability assessment methodology for the European Space Agency (STF75 F92020). Trondheim, Norway: SINTEF.

Rycraft, H., Brown, F., and Leckey, N. (1992) Operational safety review of a solid waste storage plant. In Kirwan, B. and Ainsworth, L. (Eds) Guide to task analysis, London: Taylor and Francis, pp. 355–362.

Sanders, M.S., and McCormick, E.J. (1992) Human Factors in Engineering and Design, 7th Edition. London: McGraw-Hill.

Seaver, D.A., and Stillwell, W.G. (1983) Procedures for using expert judgement to estimate human error probabilities in nuclear power plant operations. NUREG/CR-2743, USNRC, Washington DC.

Stillwell, W.G., Seaver, D.A., and Schwartz, J.P. (1981) Expert estimation of human error probabilities in nuclear power plant operations: a review of probability assessment and scaling. Sandia 81-7140, Sandia National Laboratories, Albuquerque, New Mexico, August.

Shepherd, A. (1986) Issues in the training of process operators, International Journal of Industrial Ergonomics, 1, 49–64.

Siegel, S. (1956) Non-parametric statistics for the behavioural sciences. London: McGraw-Hill Kogakusha Ltd.

Siegel, A.I., and Wolf, J.J. (1969) Man-machine simulation models. New York: Wiley and Sons.

Siegel, A.I., Wolf, J.T., and Laufman, M.R. (1974) A model for predicting integrated man-machine system reliability. Wayne, PA, Applied Psychology Services, Technical Report M24-72-C-1277, November.

Siegel, A.I., Bartter, W.D., and Kopstein, F.F. (1982). Job Analysis for Maintenance Mechanic Position for the Nuclear Power Plant Maintenance Personnel Reliability Model, Nureg/CR-2670, USNRC, Washington DC, June.

Siegel, A.I., Bartter, W.D., Wolf, J.J., Knee, H.E., and Haas, P.M. (1984) Maintenance Personnel Performance Simulation (MAPPS) Model: Summary Description, Nureg/CR-3626 Vol 1, USNRC, Washington DC, May.

Sinclair, M.A. (1990) Subjective assessment, in Wilson, J.R., and Corlett, N.E. (Eds) Evaluation of human work. London: Taylor and Francis, pp. 58–88.

Smith, A. (1990) Personal communication.

Smith, A. (1992) Personal communication.

Spettell, C.M., Rosa, E.A., Humphreys, P.C., and Embrey, D.E. (1985) 'Application of SLIM-MAUD: A Test of an Interactive Computer-based Method for Quantifying Expert Assessment of Human Performance and Reliability'. NUREG/CR-4016 Brookhaven National Laboratory, Upton, New York, USA, for Office of Nuclear Regulatory Research, US Nuclear Regulatory Commission, Washington DC 20555.

SRD (1987) SRD Dependent Failures Guide. SRD-R418, UKAEA, Culcheth, Warrington, England, March.

Stael von Holstein, C-A.S., and Matheson, J.E. (1979) A manual for encoding probability distributions. SRI International Project 7078, Arlington, Virginia, August.

Swain, A.D. (1964) Some problems in the measurement of human perform-
    ance in man-machine systems. Human Factors, 6, pp. 687–700.
Swain, A.D. (1967) Some limitations in using the simple multiplicative model
    in human behaviour quantification. In Askren, W.D. (Ed.) Symposium
    on reliability of human performance in work, pp. 17–31, AMRL Report
    AMRL-TR-67-88. May.
Swain, A.D. (1976) Shortcuts in human reliability. In Heinley, E.J. and Lynn,
    J.W. (Eds) Generic techniques in systems reliability assessment, pp. 393–
    410. Leyden: Noordhoff.
Swain, A.D. (1987) Accident Sequence Evaluation Program Human Reliabil-
    ity Analsis Procedure, Nureg/CR-4722, USNRC, Washington DC-20555.
Swain, A.D. (1989) Comparative Evaluation of Methods for Human Reliabil-
    ity Analysis, GRS-71, ISBN 3-923875-21-5, April, Gessellschaft fur
    Reaktorsicherheit (GRS)mbH, Schwertnergasse I, 5000 Koln I.
Swain, A.D., and Guttmann, H.E. (1983) A Handbook of Human Reliability
    Analysis with Emphasis on Nuclear Power Plant Applications, Nureg/
    CR-1278, USNRC, Washington DC-20555.
Spurgin, A.J., Lydell, B.D., Hannaman, G.W., and Lukic, Y. (1987) Human
    Reliability Assessment, a systematic approach, in Reliability 87, NEC
    Birmingham.
Taylor-Adams, S.E., and Kirwan, B. (1994) Development of a human error
    databank. Paper presented at PSAM II, San Diego, March 20–25.
Technica Ltd. (1985) Prediction of the proportion of errant vessels in traffic
    lanes on the United Kingdom Continental Shelf. Offshore Technology
    Report 85 214, ISBN 0-11-411911-2. London: HMSO.
Thurstone, L.L. (1927) A law of comparative judgement, Psychology Review,
    34, pp. 273–286.
Topmiller, D.A., Eskel, J.S., and Kozinsky, E.J. (1984) Human reliability
    databank for nuclear power plant operations. NUREG/CR-2744, USNRC,
    Washington DC.
Torgerson, W.S. (1958) Theory and method of scaling. Chichester: Wiley.
Trost, W.A., and Nertney, R.J. (1985) Barrier Analysis. Report No. USDOE-
    76-45/29, SSDC-29. Idaho Falls, Idaho, USA: EG&G.
Tversky, A. and Kahneman, D. (1974) Judgement under uncertainty; Heuristics
    and Biases, Science, 185, 1124–1131.
USNRC (1985) Loss of main the auxiliary feedwater at the Davis-Besse Plant
    on June 9, 1985, Nureg 1154, USNRC, Washington DC.
Vuorio, U.M., and Vuaria, J.K. (1987) Advanced Human Reliability Analysis
    Methodology and Applications, in PSA 87 Verlag, TUV Rhineland,
    GmBH, Koln.
Waters, T. (1989) – A review of recent developments in human reliability
    assessment. In Reliability 89, Brighton Metropole, June.
Watson, I., and Oakes, F. (1988) Management in High Risk Industries. Paper
    presented at the Safety and Reliability Society Symposium (SARSS '88)
    Conference, Sayers, B.A. (Ed.). London: Elsevier Applied Sciences.

Webley, K., and Ackroyd, P. (1988) Process Control Study. In Kirwan, B., Embrey, D.E. and Rea, K. Human Reliability Assessor's Guide, SRD Report RTS 88/95Q, UKAEA, Culcheth, Warrington, pp. 142–155.

Whalley, S.P. (1988) Minimising the cause of human error. In 10th Advances in Reliability Technology Symposium, Libberton, G.P. (Ed.). London: Elsevier.

Whalley, S.P., and Kirwan, B. (1989) An evaluation of five human error identification techniques, paper presented at the 5th International Loss Prevention Symposium, Oslo, June.

Whittingham, D. (1988) The design of operating procedures to meet targets for probabilistic risk criteria using HRA methodology. Paper presented at the Fourth IEEE Conference on Human factors and power plants, Monterey, California, June 5–9, pp. 303–310.

Whittingham, B. (1989) PHRASE – computerised ASEP. Personal communication.

Whittingham, B. (1990) Personal communication.

Wickens, C. (1992) Engineering psychology and human performance. New York: Harper Collins.

Williams, J.C. (1983) Validation of human reliability assessment techniques. Paper persented at the 4th National Reliability Conference, NEC, Birmingham.

Williams, J.C. (1986) – HEART – a proposed method for assessing and reducing human error. In 9th Advances in Reliability Technology Symposium, University of Bradford.

Williams (1987) Paris Air Disaster. Material from the SRD Human Reliability Assessment Course, Safety and Reliability Directorate, UKAEA, Culcheth, Warrington, Cheshire.

Williams, J.C. (1988) Human factors analysis of automation requirements, in the proceedings of the 10th conference on Advances in Reliability Technology, G.P. Libberton, (Ed.). London: Elsevier Applied Sciences.

Williams, J.C. (1992) Toward an improved evaluation analysis tool for users of HEART. Paper presented in the International Conference on Hazard Identification and Risk Analysis, Human Factors and Human Reliability in Process Safety, January 15–17, Orlando, Florida.

Williams, J.C. (1990) Personal communication.

Williams, R.J.H. (1993) An evaluation of the Paired Comparisons technique in human reliability quantification. MSc thesis, School of Manufacturing Engineering, University of Birmingham.

Wilson, I., and Kirwan, B. (1992) Training aspects of human reliability. International Journal of Quality and Reliability Management, 9, 6, pp. 39–51.

Wilson, J.R., and Corlett, N.E. (Eds) Evaluation of human work. London: Taylor and Francis, pp. 706–754.

Woods, D.D., Roth, G., and Pople, H. (1987) An artificial intelligence based cognitive model for human performance assessment. NUREG/CR-4852, USNRC, Washington DC.

Woods, D.D., Pople, H.E., and Roth, E.M. (1990) The CES as a tool for modelling human performance and reliability. Nureg/CR-5213, Volume 2, USNRC, Washington DC.

Wreathall, J. (1982) – Operator Action Trees, An Approach to Quantifying Operator Error Probability During Accident Sequences. NUS 4159, NUS Corporation, Gaithesbury, MD, July.

USNRC (1985) Loss of Main and Auxiliary Feedwater at the Davis-Besse Nuclear Power Plant on June 9, 1985 Nureg 1154, USNRC, Washington DC 20555.

Zimmerlong, B. (1992) The HEPs of HEP experts: Empirical evaluation of THERP and SLIM. Paper presented at the International Conference on Hazard Identification and Risk Analysis, Human Factors and Human Reliability in Process Safety, Orlando, Florida, January 15–17, pp. 273–280.

Acronyms

ACB	Air Circuit Breaker
ACSNI	(UK) Advisory Committee for the Safety of Nuclear Installations
AFWS	Auxiliary Feedwater System
AFWP	Auxiliary Feedwater Pump
AIR	American Institute for Research (database)
ALARA	As Low As Reasonably Achievable
ALARP	As Low As Reasonably Practicable
ANOVA	Analysis of Variance
AOA	Advanced Operator Aids
APJ	Absolute Probability Judgement
APOA	Assessed Proportion of Affect
ASAP	As Soon As Possible
ASD	Auxiliary Shutdown Device
ASEP	Accident Sequence Evaluation Procedure
ASME	American Society of Mechanical Engineers
ASS	Assistant Shift Supervisor
AT	Area Technician
BA	Breathing Apparatus
BDBA	Beyond Design Basis Accident
BHC	Basket Handling Cave
BHEP	Basic Human Error Probability
BNFL	British Nuclear Fuels plc
BOP	Blowout Preventer
BUCs	Back-Up Circulators
BUP	Back Up Panel
BUPL	Back-Up Power Link
C&I	Control & Instrumentation
CADA	Critical Action and Decision Analysis
CB	Circuit Breaker
CCF	Common Cause Failure
CCR	Central Control Room
CCS	Computerised Control System

CCTV	Closed-Circuit Television
CD	Complete Dependence
CEGB	Central Electricity Generating Board
CEP	Cognitive Error Potential
CES	Cognitive Environment Simulation
CFMS	Critical Function Monitoring System
CIT	Critical Incident Technique
CL2	Chlorine
CMA	Confusion Matrix Analysis
CMF	Common Mode Failure
CR	Control Room
CRO	Control Room Operator
CRS	Control Room Supervisor
CRT	Cathode Ray Tube
CSL	Control Station Layout
DAD	Decision Action Diagram
DCS	Distributed Control System
DCU	Distributed Control Unit
DE	Desk Engineer (CRO)
DG	Diesel Generator
DHO	Diesel House Operator
DOG	Dissolver Off-Gas
DOP	Dropped Object Protection
DP	Differential Pressure
DSS	Decision Support System
EC	Emergency Controller
EE	Emergency Exercise
EEM	External Error Mode
EF	Error Factor
EFD	Engineering Flow Diagram
EJ	Engineering Judgement
EOAT	Enhanced Operator Action Tree
EOCA	Error of Commission Analysis
EOPs	Emergency Operating Procedures
EPCs	Error Producing Conditions
ERA	Error Reduction Analysis
ERMA	Error Reduction Mode Analysis
ESD	Emergency Shutdown
ETA	Event Tree Analysis
F&G	Fire & Gas
FA	Functional Analysis
FAST	Functional Analysis of Systems Techniques

FDSC	Fully Developed Safety Case
FMEA	Failure Modes and Effects Analysis
FSMA	Fault-Symptom Matrix Analysis
FTA	Fault Tree Analysis
FW	Feedwater
GASET	Generic Accident Sequence Event Tree
GEMS	Generic Error Modelling System
GNI	Guaranteed Non-Interruptable (power supply)
GOMS	Goals, Operators, Methods, System (modelling approach)
H2S	Hydrogen Sulfide Gas
HAZOP	Hazard and Operability Study
HCI	Human Computer Interaction
HCR	Human Cognitive Reliability
HD	High Dependence
HEA	Human Error Analysis
HEART	Human Error Assessment and Reduction Technique
HED	Human Error Database
HEI	Human Error Identification
HEIST	Human Error Identification in Systems Technique
HEP	Human Error Probability
HF	Human Factors
HF	Hydro-Fluoric Acid
HFA	Human Factors Analysis
HFI	Human Factors Issues
HFRG	(UK) Human Factors in Reliability Group
HF-RBE	Human Factors Reliability Benchmark Exercise
HPES	Human Performance Evaluation System
HPLV	Human Performance Limiting Value
HRA	Human Reliability Assessment
HRAET	Human Reliability Analysis Event Tree
HRAG	Human Reliability Assessor's Guide
HRMS	Human Reliability Management System
HRQ	Human Reliability Quantification
HTA	Hierarchical Task Analysis
IMAS	Influence Modelling and Assessment System
IDA	Influence Diagrams Approach
ISRS	International Safety Rating System
ITA	Initial Task Analysis
JHEDI	Justification of Human Error Data Information
KBB	Knowledge Based Behaviour
KWU	Kraftwerk Union

LB	Lower Bound
LCM	Lost Circulation Method
LCR	Local Control Room
LD	Low Dependence
LER	Local Equipment Room
LER	Licensee Event Report
LO	Local Operator
LOCA	Loss of Coolant Accident
LOG	Loss Of Grid
LTM	Long Term Memory
MAN-AUTH	Management Authorisation
MAPPS	Maintenance Analysis of Personnel Performance Simulation
MCs	Main Circulators
MCR	Main Control Room
MD	Moderate Dependence
MF	Modification Factor
MIC	Methyl Iso-Cyanate
Micro-SAINT	System Analysis of Integrated Networks of Task (workload analysis computer package)
MMI	Man-Machine Interaction
MoD	Ministry of Defence
MPR	Mud Processing Room
MSIVs	Main Steam Isolation Valves
NAOH	Sodium Hydroxide
NII	Nuclear Installations Inspectorate
MORT	Management Oversight Risk Tree
OA	Operator Aids
OCB	Oil Circuit Breaker
OAET	Operator Action Event Tree
OEC	Onshore Emergency Centre
OER	Occupational Experience Review
OIM	Offshore Installation Manager
OO	Outside Operator
OPREDS	Operational Performance Recording Evaluation and Data System
OR	Operating Rule
OSA	Operational Safety Appraisal
OSD	Operational Sequence Diagrams
OTSG	Once-Through Steam Generator

P&IDs	Piping & Instrument Diagrams
PA	Public Address
PAQ	Position Analysis Questionnaire
PC	Paired Comparisons
PCSR	Pre-Construction Safety Report
PEM	Psychological Error Mechanism
PHECA	Potential Human Error Causes Analysis
PHOENIX	Prediction of Human Operator Error using Numerical Index eXtrapolation
POOH	Pulling Out Of Hole
PORV	Pilot-Operated Relief Valve
PRA	Probabilistic Risk Analysis
PSA	Probabilistic Safety Assessment
PSFs	Performance Shaping Factors
P&ID	Piping and Instrumentation Diagram
PI	Performance Indicator
PMI	Person-Machine Interface
POI	Plant Operating Instructions
PORV	Power Operated Relief Valves
PSA	Probabilistic Safety Assurance
PSF	Performance Shaping Factors
PTW	Permit To Work
PWR	Pressurised Water Reactor
QA	Quantified Assessments
R&D	Research & Development
RA	Risk Assessor
RBA	Rule-Based Action
RBB	Rule-Based Behaviour
RBD	Rule-Based Diagnosis
RCS	Reactor Coolant System
RHRS	Residual Heat Removal System
RMC	Remote Monitoring Centre
RPM	Revolutions Per Minute
RVA	Rule Violation Assessment
SBB	Skill-Based Behaviour
SAP	Safety Assessment Principle
SC	Shear Cave
SC/SM	Safety Culture/Safety Management
SCHEMA	System for Critical Human Error Management Analysis
SG	Steam Generator
SGTR	Steam Generator Tube Rupture

SHARP	Systematic Human Action Reliability Procedure
SHERB	Sandia Human Error Rate Bank
SHERPA	Systematic Human Error Reduction and Prediction Approach
SI	Safety Injection
SLIM-MAUD	Success Likelihood Index Method using Multi-Attribute Utility Decomposition
SLIM-SARAH	SLIM with Sensitivity Analysis and Reliability Assessment of Humans
SNUPP's	Standard Nuclear Unit Power Plant System
SOI	Station Operating Instruction
SPDS	Safety Parameter Display System
SRB	Skill or Rule Based
SRK	Skill, Rule, Knowledge Level
SRP	System Risk Perspective
SS	Shift Supervisor
STA	Sequential Task Analysis
STAHR	Socio-Technical Assessment of Human Reliability
SUB	(Error) Subsumed by Another HEP
SUFP	Start-Up Feed Pump
T&M	Test and Maintenance
TA	Task Analysis
TAKD	Task Analysis Knowledge Decomposition
TEPPS	Technique for Estimating Personnel Performance Standards
TESEO	Tecnica Empirica Stima Errori Operatori
TH	Turbine Hall
THE	Turbine Hall Engineer
THO	Turbine Hall Operator
THERP	Technique for Human Error Rate Prediction
THORP	Thermal Oxide Reprocessing Plant
TLA	Timeline Analysis
TMI	Three Mile Island
TRA	Total Risk Analysis
TRE	Turbine Route Engineer
TSA	Tabular Scenario Analysis
TT	Talk-Through
TTA	Tabular Task Analysis
UB	Upper Bound
UHF	Ultra High Frequency
UK	United Kingdom
UKAEA	UK Atomic Energy Authority
UKCS	UK Continental Shelf

VDU	Visual Display Unit
VHF	Very High Frequency
WS	Work Station
WT	Walk-Through
ZD	Zero Dependence

Index